Hydraulics of Sediment Transport

McGRAW-HILL SERIES IN WATER RESOURCES AND ENVIRONMENTAL ENGINEERING

Ven Te Chow, Rolf Eliassen, and **Ray K. Linsley** Consulting Editors

Graf Hydraulics of Sediment Transport
Hall and Dracup Water Resources Systems Engineering
James and Lee Economics of Water Resources Planning
Walton Groundwater Resource Evaluation

Hydraulics of Sediment Transport

Walter Hans Graf
Lehigh University

McGraw-Hill Book Company
New York St. Louis San Francisco
Düsseldorf Johannesburg Kuala Lumpur
London Mexico Montreal
New Delhi Panama Rio de Janeiro
Singapore Sydney Toronto

Hydraulics of Sediment Transport

Library of Congress Catalog Card Number 79-128788

07-023900-2

1 2 3 4 5 6 7 8 9 0 M A M M 0 9 8 7 6 5 4 3 2 1

This book was set in Times Roman, and was printed on permanent paper and bound by The Maple Press Company. The drawings were done by John Cordes, J. & R. Technical Services, Inc. The editors were B. J. Clark and M. E. Margolies. John Sabella supervised production.

... sanft klang der vielstimmige Gesang des Flusses ...

Ihm schien, es habe der Fluss
ihm etwas Besonderes zu sagen,
etwas, das er noch nicht wisse,
das noch auf ihn warte.

der Fluss lachte.

H. Hesse *Siddhartha*

to Christa
to my parents

Contents

PREFACE ix

Part One A SHORT HISTORY OF SEDIMENT TRANSPORT

Contents 1
Introduction 1
1 Some Prehistoric and Historic Documents, up to 1500 A.D. 3
2 Hydraulics as a Science: From L. Da Vinci to P. Forchheimer 11
References for Part One 24

Part Two HYDRODYNAMICS OF FLUID-PARTICLE SYSTEMS

Contents 27
Introduction 28
List of Symbols 29
3 General Remarks 31
4 Settling Velocity of Particles 35
5 Effect of Particles on the Viscosity 65
References for Part Two 70

Part Three SEDIMENT TRANSPORT IN OPEN CHANNELS

	Contents	75
	Introduction	79
	List of Symbols	80
6	Scour Criteria and Related Problems	83
7	The Bedload	123
8	The Suspended Load	161
9	The Total Load	203
10	The Regime Concept	243
11	Bedform Mechanics	273
12	Cohesive-material Channels	323
13	Sediment Measuring Devices	357
14	Model Laws	385
	References for Part Three	398

Part Four SEDIMENT TRANSPORT IN CLOSED PIPES

	Contents	421
	Introduction	422
	List of Symbols	422
15	Flow of Solid-Liquid Mixtures in Pipes	425
16	Measuring Devices for Solid-Liquid Mixtures in Pipes	483
	References for Part Four	496

NAME INDEX 503

SUBJECT INDEX 511

Preface

One can learn much from a river.

From the river I have learned too:
everything comes back.

H. Hesse *Siddhartha*

The understanding and formulation of movement and transportation of solid granular particles in or through liquid bodies represent an important issue within the fields of hydraulics, fluvial geomorphology, and others. The problems are complicated, as are many other two-phase and interface problems, and thus have remained often subject to semiempirical or empirical treatment. However, theoretical endeavors do exist, but they are few and are not indisputable. Immediate answers, or at least guidelines, are often necessary, and thus a careful examination of existing developments seems justified and timely. It is hoped that future research will throw more light on the sediment transport phenomenon.

The book is divided into four parts of unequal length: Part 1 is entitled "A Short History of Sediment Transport"; Part 2 deals with the "Hydrodynamics of Fluid-Particle Systems"; Part 3 is concerned with the "Sediment Transport in Open Channels"; and Part 4 describes the "Sediment Transport in Closed Pipes." It was the intent to make each of these parts as self-contained as possible. A brief introduction preceding each part should make the reader familiar with the topics to be treated within.

It may be worthwhile to point out that not all topics concerned with the hydraulics of sediment transport could be treated, because of the inadequacy of present-day knowledge in some areas and in order not to make this book too voluminous.

In conceiving and writing this book I was led by the conviction that our engineering profession is in need of having the available knowledge on the hydraulics of sediment transport organized and summarized. Only then will the practicing engineer as well as the future engineer, who now is the student, be able to make ready use of the storehouse of information in this field. It is hoped that the book will serve as a textbook for advanced undergraduate and for graduate students, and as a reference book for all concerned with this topic.

It must be clear to those who use the treatise as a textbook that it is impossible to cover the entire material in one or even two semesters. However, with a proper (personal) selection of parts and/or chapters an adequate coverage of a coherent topic may well be achieved within either a one- or two-semester course.

The inspiration of the treatise was in many respects provided by Professor H. A. Einstein who introduced me to the fascination of this field during my studies at the University of California in Berkeley. Subsequently I began to develop lecture notes for a graduate course at Cornell University. I derived stimulus and encouragement from discussions with my assistants; notable among these were Dr. E. Acaroglu, R. Weisman, and J. Reed. Year after year I improved the set of notes which were developed without immediate intention for publication in book form. Credit to encourage me to put the material into print must go to Dr. V. T. Chow. My efforts continued when I took the present position which put me in charge of the hydraulics activities at Lehigh University. Again, I had the benefit of fruitful discussions with my assistants; notable among these are Ö. Yücel, M. Robinson, and Dr. S. Ko. During this time I was invited to lecture on selected topics of this book at the University of Karlsruhe in Germany. At Lehigh University the final manuscript was put together.

For assistance in reviewing the many chapters I am greatly indebted to Dr. H. A. Einstein for Chapters 1, 2, 6 through 9, and 14; to Dr. H. W. Shen for Chapters 1, 2, 6 through 9, and 15; to Dr. V. T. Chow for Chapters 1, 2, and 6 through 9; to Dr. L. F. Mockros for Chapters 3, 4, and 5; to Dr. D. B. Simons for Chapters 10 and 11; to Dr. A. Hjelmfelt for Chapters 3 and 4; to Dr. F. D. Masch for Chapter 12; to D. W. Hubbell for Chapter 13; and to Dr. E. T. Smerdon for Chapter 12. Special thanks go to Dr. O. ElGhamry who carefully read through all the galley proofs and made most valuable suggestions. The author himself is responsible for any errors that may remain.

It is a pleasure to acknowledge the cooperation of editors of numerous publications which gave permission to use figures, quotations, and tables. It would be impossible to list all of these at this point. Much of the more recently published material is contained in the publications of the American Society of Civil Engineers (ASCE) and of the United States Department of Interior (USDI); to these two organizations I am especially thankful.

Walter Hans Graf

part one

A Short History of Sediment Transport

CONTENTS

INTRODUCTION

1 SOME PREHISTORIC AND HISTORIC DOCUMENTS, UP TO 1500 A.D.
- 1.1 Introductory Remarks
- 1.2 Mesopotamia, Land of Canals
- 1.3 Egypt, the Gift of a River
- 1.4 China, the Land of Yellow Waters
- 1.5 Persia and the Kanats; India
- 1.6 The Greeks—Their Understanding of Physics
- 1.7 The Romans and the Art of Water Supply Engineering
- 1.8 The Middle Ages

2 HYDRAULICS AS A SCIENCE: FROM L. DA VINCI TO P. FORCHHEIMER
- 2.1 Leonardo da Vinci
- 2.2 The Italian School of Hydraulics
- 2.3 DuBuat and His Contemporaries
- 2.4 Stream Hydraulics in the Nineteenth Century
- 2.5 Concluding Remarks

REFERENCES FOR PART ONE

INTRODUCTION

It may prove to be a worthwhile venture to say something about the history of sediment transport, if for no other reason but to point out the significant evidence that men have been concerned with this problem even in the remote past. Like everything else, at the beginning it was an art until it became part of the field of mechanics about 200 years ago. Neither the *art* nor the *mechanics* can as yet provide exhaustive answers to many a problem. We are in the midst of growth in this field.

In writing about the history of sediment transport in open-channel flow, an important fact must be kept in mind. Most of the (natural) watercourses have a movable bed and not as is often assumed, understandingly for good reasons, a fixed one. (With the advent of mechanics, rivers or canals were simplified in many respects and could thus be treated according to developed theories; fixed boundaries were invariably assumed.) In studying the important highlights of sediment transport throughout the history of mankind, we are compelled to assume that whenever man was concerned with rivers or canals he had a notion—at least subconsciously—of the importance of science. Indeed, this is at times expressed and recorded. It is fitting, therefore, that this chapter might well be called a history of stream hydraulics. Only in the latter part of the last century did the sediment aspects within stream hydraulics become a science in itself, if we are permitted to say so; or in other words, only recently has *real* channel hydraulics begun to replace *ideal* channel hydraulics, with the fixed boundary becoming a loose boundary.

What then is history? Let us define historic knowledge as that which has no new or further immediate influence on present or future developments. Thus we may safely assume that the last 50 years[1] are not considered "history." Since this date coincides roughly with the first appearance of FORCHHEIMER'S (1914) "Hydraulik"—this treatise is commonly considered the outstanding work of its kind of all time—we have reason to assume that 1914 is the year where we may draw a close to historic events.

[1] Discussions with H. Graf, a historian, led me to introduce this figure.

1

Some Prehistoric and Historic Documents, up to 1500 A.D.[1]

1.1 INTRODUCTORY REMARKS

At the present writing, there exists no historic evidence of scientific hydraulics prior to the Renaissance. However, there is sufficient historic and even prehistoric evidence that men did observe natural streams—at times even corrected or realined them—and that they built systems of canals for many purposes. Apparently they failed to discover or, if they did, to record those laws of nature that governed the water and sediment motion in these watercourses. We are indeed very fortunate to have many relics of hydraulic accomplishments from bygone civilizations all over the world. River channels, canals, and aqueducts are found as part of extensive irrigation systems in China, as part of a dense net of waterways in Mesopotamia, and as part of domestic water systems throughout the Roman Empire. These engineering achievements are frequently of such magnitude that neither

[1] In preparation of this section, the author freely draws on FLEMING (1957), ROUSE et al. (1957), KOLUPAILA (1961), FLACHSBART (1932), and BISWAS (1967).

nature nor the savage elements of men could destroy them entirely. Very few of these archaic structures are still or again in use. Many of them are but witnesses of the past, impressing students of archeology and hydraulics, and many more are probably still buried in the debris of subsequent cultures.

It would be truly surprising if the engineer of the remote past had not a very definite notion about the movement and forces of the water. The lack of any record does in this case prove little; the creators of such enormous, well-designed constructions were craftsmen and artisans who handed down knowledge from generation to generation and who became successful through the slow, expensive, but eventually safe, process of *cut and try*. They did not have the aid of formulas, which are based on rational deductions, but that did not hamper them in their pursuit of greatness.

1.2 MESOPOTAMIA, LAND OF CANALS

In Mesopotamia, for some the cradle of Western civilization, there exists evidence that a net of canals was in existence prior to the "deluge" (ca. 4000 B.C.). Excavations of the ruins of Nippur and others have been dated 5200 B.C., and vaulted irrigation canals have been found among these ruins. This is the oldest evidence to date that man was aware of the importance of building canals for whatever purpose he deemed necessary. The extensive system of canals between the Tigris and the Euphrates provided wealth to the dense population of the plains of Babylon. There is indication that the canals served many purposes: as waterways to exchange goods, as irrigation channels to use the favorable climate for more than one harvest, as drainage canals to control the water levels, or to reclaim new country. In fact, there is some speculation that the ancient city of Ur had been reclaimed from the sedimentary deposit of the river. Not only did the water help to keep the country fertile but also the large amounts of solid matter that moved with streams were used as natural fertilizer. Further hydraulic achievements are reported around 800 B.C., during the reign of the Assyrian Queen Semiramis. The cities of Nineveh and Babylon were beautified by the famous *hanging gardens*, which required a highly developed irrigation system. An important canal was rebuilt, originally constructed around 3000 B.C., which connected the all-important city of Babylon with the Gulf of Persia. One of the Assyrian kings, Sennacherib (707–681 B.C.), is especially well known for his keen personal interest in the development of the available water resources in the vicinity of Nineveh. But also Nebuchadnezzar II (604–562 B.C.) is accredited with having built canals, dams, and various other constructions.

The technical competence exhibited by some of these canals may be realized by the fact that the present government of Iraq is most interested in using these archaic canals for their water resources plan.

1.3 EGYPT, THE GIFT OF A RIVER

Were it not for the River Nile and the way it was made useful for mankind, the great civilization of Egypt would not have been the same.

There are records that around 3000 B.C. under the legendary reign of King Menes, first of the Pharaohs, dams were built—one at the old capital Memphis—to store the flood water provided seasonally by the Nile and use it for irrigation purposes. During the reign of King Amenemhet III, a Pharaoh of the twelfth dynasty (1841–1801 B.C.), a truly gigantic water resources project was completed. The reservoir of the neolithic Lake Moeris at Fayum was developed to store sufficient water for irrigation of a region within the Nile delta for an entire season. This Lake Moeris was filled during floods, thus controlling them from the Nile via the Canal of Joseph, presently in use, and the water was returned to the Nile during the low stages, thus ensuring the year-around irrigation waters. The oldest recorded flood stages fall into the same period (1827 B.C.). The floods of the Nile, directly or indirectly as stored water, the fertile silt carried with the water, and climatic condition ensured more than two harvests annually, and thus provided wealth to the country. Extreme stages of lows and highs caused disasters. The whole wealth of Egypt was dependent on this river of life. Little wonder that so many indications of early hydraulics and hydrometry have been found here. The remarkable water gages, called *Nilometers*, were installed throughout the irrigation territory. They served as a useful guide for the irrigation practice, and priestlike supervisors would, on the attainment of a certain Nile stage, proclaim the beginning of the annual feast. Quite justly the land tax was established by the ruler according to yearly maximum flood level. Also the old Egyptian calendar exhibits hydrometric influences.

The first attempt to provide a canal connecting the Mediterranean with the Red Sea dates back to about 1900 B.C. Another time, around 600 B.C., during the reign of King Necho II, this project was again undertaken, and some 100 years later the powerful Darius I, King of the Persians, continued these plans. With certain improvements, the canal was in use till about 770 A.D. This idea was only recently realized again when in the nineteenth century the Suez Canal was finished after the plans of A. Negrelli.

Indeed, it is more than true that this Egypt is a gift of water, of a river, of the Nile.

To the north of Egypt, the Phoenicians built aqueduct systems in Syria and Cyprus. About 1200 B.C. sinnors or water tunnels supplied Jerusalem with water. In what is now Yemen, the Marib Dam was constructed between 1000 and 700 B.C.; one of the wonders of the ancient world, it served as the key structure for an extensive irrigation project.

1.4 CHINA, THE LAND OF YELLOW WATERS

In this land, where history played such an important role, we find that a hydraulic engineer became emperor. This occurred because China's history has been full of endless and tedious struggles with disastrous floods of silt-laden waters. One of its emperors, Yau, is said to have ordered the regulation of these rivers. However, the man in charge, after building for 12 years, was unable to provide the expected protection and fell into disgrace; his son, Yu, continued this effort. The legend has it that this Yu succeeded after eight years of work to bring not only the Hwang Ho under control but also the Yangtze Kiang and its tributaries. Truly the title this man deserved was bestowed on him; he was made emperor. Emperor Yu was and still is honored by constructions of pagodas at various watercourses; one memorial plaque attached to the cliff of the Hwang Ho marks the occasion of the construction of a floodway in 2278 B.C. The great Yu speaks of himself as the one who guided nine rivers toward the sea. And Chinese historians write: "River regulation according to Yu was such that each river was treated according to its properties." These rivers remained in their newly given beds for almost 1,700 years. Disastrous floods were under reasonable control, and excess amounts of sediment were prevented from entering the main channel. The ditches, however, which filled up with sediments, mainly loess, were periodically emptied by farmers who, in turn, used them as fertilizers on agricultural soil. What a unique achievement, whether wanted or not!

An outstanding mark of hydraulic engineering is the most complicated net of an irrigation system designed by Chengkuo about 240 B.C. Some of the irrigation canals from that century and rules for their operation are said to be still in use.

1.5 PERSIA AND THE KANATS; INDIA

The oldest civilizations, those which flourished some 4,000 years ago, must be linked to the water availability; in the valley of the Nile, in Mesopotamia, in the river valleys of China, and that of the Indus. In the latter area was the civilization of the Harappans. The historic evidence of these peoples is at the present scarce, but it is to be expected that soon more archeological findings will become available. No such major river was available, however, to the Persians, and they had to tap the ground water resources in a system known as the *qanats* (Kanats or Kiariz). Underground channels carried the water from water-bearing geological basins over distances of, at times, 90 km to regions where water was needed most. About every 30 m, the channel had vertical air shafts, facilitating the construction and ensuring the quality of the water. The channel's cross section was egg-shaped and its slopes were gentle, of 1:100 to 3:100. The qanats apparently originated in Armenia around 1000 B.C., but were soon found in Persia and northern India. Saragon II (721–705 B.C.) introduced them in Assyria, and later under Darius I (521–485 B.C.) such infiltration galleries were built for irrigation purposes in Egypt; from there they spread to northern Africa and Sicily. Many of these qanats built in Persia are still in use, tapping the water of the alluvial fans of the Elburz Mountain. The estimate is that today 22,000 qanats deliver a total of 550 m^3/sec of water. Truly, these qanats must be considered as one of the most outstanding achievements of engineering in antiquity.

In the neighboring Turkestan, in the valley of the Murghab River, an extensive irrigation system was in use around the eleventh and twelfth centuries; now it is a barren desert. In the Punjab, now in Pakistan, referred to as the land of the large canals, an equally huge irrigation project was in operation. The water, originating in the mountain region as snow melt or rain, was conveyed in channels built of stone and clay, or made out of wooden trunks or bamboo when the grade required the channel to be above the terrain.

The Indian system of irrigation has been described as one where the water is applied to terraces which were built along the mountains. Should this prove to be correct, then it seems that further evidence exists of the Indian civilization spreading not only to the east, such as to southeastern parts of Africa, but also to the west, to the Philippines. In both regions, excavations have presented some evidence of the existence of such terraces. In the Philippines they are still in use. Some archeologists use the existence of terraces in the Indian empires of the Americas as a further proof of a link between the Asian and American civilization. When Cortez and Pizarro brutally conquered Mexico in 1520 A.D. and Peru in 1532 A.D., respectively, a superb and elaborate irrigation and water supply system was in existence. There is every reason to believe that much earlier, probably a few centuries, a highly developed civilization had realized the importance of water in this tropical to subtropical region.

1.6 THE GREEKS—THEIR UNDERSTANDING OF PHYSICS

The civilizations of the Near East, which preceded the Greek civilization, had an influence on its development and scientific achievements, which should not be underestimated. Especially during the Hellenistic period, after the death of Alexander of Macedonia, 323 B.C., it was the Greeks who duplicated, improved, and described many scientific accomplishments which were apparently known to the people of their colonies—such as Egypt, Persia, etc.—for some time. Thus every history of hydrodynamics and physics in general starts with the Greeks, maybe for lack of extensive written evidence of those civilizations where these sciences truly originated.

Physics of the Greeks is based on the deductive method, from the general to the specific. There is the physics of the natural philosophers (scientists), with its main figure Aristotle (384–322 B.C.), a man with many interests in all areas of man and his environment. Aristotle, the scientist, had a compelling interest in biology, medicine, astronomy, and physics, and exhibited a love of facts, not like the great predecessors, Socrates (469–399 B.C.) and Plato (429–348 B.C.), with their spiritual outlook and their love of ideas. His contributions to early hydraulics focus mainly on facts, speculations, and observations, and not on laws. There are some remarks on the weight of the various elements, on the pressure these elements produce, and on the compressibility of water; to explain suction, the *horror vacui* is introduced, and there are some remarks on the movement of one element through another, and the resistance it experiences, etc.

Three major schools of thought antedate Aristotle's appearance. These schools centered around Thales of Miletus (640–546 B.C.), Pythagoras of Croton (582–507 B.C.), and Democritus, the Thracian (460–377 B.C.); all three studied or traveled in Egypt for some time, thus the influx of methods and ideas of this civilization must not be overlooked.

Aristotle's influence on the further development in Greece was small; his influence on the Middle Ages, considerable as it was, however, was an extremely unfortunate one. His writings, not at all free of errors, were taken as dogma and seriously hampered any advancement, the very issue he himself did not want to happen.

Besides the group of scholars who used natural philosophy in order to understand physics, there was a group who employed mathematics. Its champion is Archimedes (288–212 B.C.). Two most important stimuli, Alexander's financial encouragement to the progress of research and the fusion of knowledge from Egypt and Mesopotamia with the learnings of the Greeks, produced during the Hellenistic civilization a most brilliant age in the history of sciences. Although these sciences produced nothing which directly advanced river, channel, or pipe hydraulics, the new findings were basic and, thus, of general importance. Archimedes formulated with scientific exactness the principles of hydrostatics and flotation. Although he preferred to devote his time to science itself, his engineering inventions, such as the tubular screw for pumping water, the screw propeller for ships, etc., should not be overlooked. Archimedes' work on hydrostatics was continued by Hero (ca. 150–100 B.C.), who also explained correctly the suction and the siphon action. Hero's writings also contain some reference to an expression of the principle of continuity and a respective number of inventions, such as the hydraulic organ, force pump, and various others.

Nevertheless, at the end of Greek civilization, between 146 and 30 B.C., the understanding of flow phenomena or hydrodynamics was still scarce.

A last word on engineering under the Greeks. For good reasons and lack of immediate urgency, great hydraulic engineering achievements, comparable to those in the Near East, are missing. We like to record, however, that during the Aegean civilization, the one which preceded the Hellenic one, the Palace of Knossos on the Island of Crete was designed in the seventeenth century B.C. with a most advanced knowledge of sanitary engineering. At its summit, during the "golden age," the city of Athens had 18 different water supply systems. Far more interesting to us is that a Greek by the name of Ctesias is said to have written around 250 B.C. a treatise on rivers, the oldest known written contribution to river hydraulics.

1.7 THE ROMANS AND THE ART OF WATER SUPPLY ENGINEERING

The glory of the Greek civilizations began to fade, but the importance of Rome as the future nucleus of a huge empire had started. Historians record that the intellectual, scientific, and artistic accomplishments of the Romans never equaled the ones of the Greeks, despite their having the advantage of Hellenistic science as a foundation on which to build. For good reasons the Romans were very practical-minded, which, as such, is not sufficient to carry scientific progress very far. However, certain engineering achievements, such as extensive aqueducts or sewer systems, typify their creative approach.

Around 1000 B.C. or possibly later, Indo-European and Asian-Greek invaders moved into the Italian peninsula. The aggressive, capable builder nation, the Etruscans, seems to have had the strongest impact on Roman engineering. Although the Etruscans, apparently coming from the Near East, possessed the ability of building vaulted sewers, this method of construction was known to the people of Mesopotamia considerably earlier. Also, there exists evidence that the art of irrigation was practiced by the Etruscans; in other parts of the future empire, such as in the Iberian peninsula, the traders of the Near East and of Crete introduced the idea of irrigation systems.

In the truest sense of the word, it is the art of water supply engineering in which the Romans excelled. Almost 100 significant installations are known; the city of Rome had at one time 17 systems with a water supply of 300,000,000 gallons (gal) daily. Some of these structures, certainly monumental like most Roman structures, can still be seen, and are partially still in use throughout the former Roman Empire, such as in Italy, Spain, France, Germany, Switzerland, Austria, on the Balkan, in the Near East, and in North Africa. Well known and famous are the ruins of Rome, the Port du Gard in Nimes, and the Aqueduct of Segovia. The latter has a length of 2,680 ft and its superposed 170 arches support an open channel 92 ft above ground.

Searching for the hydraulics behind these engineering facts, we might obtain some answers in two treatises written by two eminent Roman engineers. One is on "De architectura libri decem" by Vitruvius Pollio (first century

B.C.); in this ten-volume compendium, the eighth book is devoted to hydraulics and water in general. The other is a two-volume treatise on "De aquaeductibus urbis Romae commentarius" by Sextus Julius Frontinus A.D. (40–103), a water commissioner of the city of Rome. It is reported that in the water supply system, aqueducts and pressure conduits were in extensive use. The masonry canals were actually open channels with a cover to ensure water quality. Increasing water demand resulted in building one channel above the other. Being of rectangular cross section with a width of 2 to 6 ft and a height of 2 to 8 ft, flow was supposedly achieved when the slope was 1 or 2 ft/mile. Somehow the Roman engineer seemed to have had an appreciation, but not a full understanding, of hydraulic phenomena, such as flow resistance and continuity, and flow control or measurement.

1.8 THE MIDDLE AGES

The Middle Ages, the period from the fall of the Roman Empire (476 A.D.) to the Renaissance (ca. 1500 A.D.), offer very little worth mentioning in our brief account of the field of hydraulics. This period experienced the growth of Western (European) civilization, and was largely preoccupied with spiritual affairs. Furthermore, the temperate geographic-climatic location of Europe contributed little challenge for extensive hydraulic engineering feats, and essentially nothing could be added to the great engineering achievements of the ancient past. Thus the Middle Ages must be considered as a period of waiting for the hydraulic science.

Among the leading scientists of the Byzantine Empire was Joannes Philoponus, called John the Grammarian (about sixth century), who was the first to challenge the traditional theories of motion and gravity. Denying the impossibility of creating a vacuum, he also anticipated the concept of inertia and rejected the notion that the speed of a falling body is proportional to its weight. There are a few medieval scientists in Western Europe who had some influence in their respective fields. In England, Roger Bacon (1214–1294) stressed the importance of the experimental research, and W. Heytesbury (around 1360) referred to a distance-velocity and velocity-acceleration relationship. In France, J. Buridan (around 1310) made remarks on the impetus, Oresme (around 1320) was concerned with the principle of kinematics, and Albert de Saxe (around 1320) studied the free fall and resistance phenomena.

As far as hydraulic engineering is concerned, no truly monumental construction was undertaken but, nevertheless, it was at a respectable level. It is recorded that water supply systems spread widely during the Arab rule. Various hydraulic achievements were used and further developed for the ever important mining industry. Limited irrigation projects in the Alps, notable in the Canton Wallis, were developed.

2

Hydraulics as a Science: From L. da Vinci to P. Forchheimer[1]

With Leonardo da Vinci, hydraulics became a science. This does not imply that the art of hydraulic engineering ceased to exist. What it means is that empiricism of the engineer was complemented by the scientist's rationale. It was not until the twentieth century that rational engineering replaced considerably the art of engineering.

Thus is is not surprising that, at the beginning of the science of hydraulics, we find a man who not only excelled equally in the arts and in engineering but also in many other fields; truly Leonardo da Vinci was a *studiosus generalis*.

2.1 LEONARDO DA VINCI

This genius of the Italian Renaissance, Leonardo da Vinci (1452–1519), showed a keen interest in the problem of water flow, not only as a practicing

[1] In preparation of this section, the author has drawn freely on ROUSE et al. (1957), FLACHSBART (1932), FLEMING (1957), FREEMAN (1929), and FORCHHEIMER (1914); other references are quoted in the text.

engineer but also as an experimentalist. "Del moto e misura dell'acqua" was to be the monumental treatise containing nine individual books. To those of us with an interest in channel hydraulics, the first four books are of special importance, i.e., "Della sfera dell'acqua," "Del moto dell'acqua," "Dell'onda dell'acqua," and "De retrosi dell'acqua." The work was written around 1500, but was not published prior to 1649, and was only recently, in 1923, reedited. Thus it is evident that da Vinci did not share his knowledge—intentionally or not—with either contemporaries or subsequent students, and it remained hidden for many years. In fact, much had to be rediscovered, but there also exists some suspicion that later generations had derived, at least partially, some knowledge from da Vinci's scripts which for a long time remained scattered throughout libraries.

Leonardo da Vinci was a great experimentalist who uttered the wise warning: "When you try to explain the behaviour of water, remember to demonstrate the experiment first and the cause next." Sketches and comments clearly indicate the keen sense for observing an experiment—there is belief that some sort of hydraulic flume was used—and they show how many ideas he advanced and studied. Most notable among his sketches is the rather clear notion of the location and the existence of eddies. An attempt was also made to formulate the law of continuity; the problem of flow from a container was discussed quite elaborately. Observation of nature must have inspired him to study various aspects of open-channel flow. He studied the problems of waves in canals and also the hydraulic jump. The velocity distribution in canals is attributed to the resistance of the wall or the air. The continuity in a river is linked with the quantities, width, depth, slope, roughness, and tortuosity for an equal water surface width. The smaller the depth, the faster water will move; if the depth remains uniform, the wider the section, the less rapid the flow.

Much of the interests in the flow in streams and canals stemmed from da Vinci's experience as a designer of various hydraulic engineering works. As a military engineer at Milan, he was responsible for making navigable the canal of Martisana. He planned to connect the major cities of Tuscany and Florence with a system of multipurpose canals for navigation, irrigation, and water supply. Later, working in France, he developed plans to connect the Seine with the Loire. Without doubt, his experimental experience, stimulated by nature, must have helped him as a practicing engineer.

2.2 THE ITALIAN SCHOOL OF HYDRAULICS

A great interest in the motion of water was shown by Galileo Galilei (1564–1642), the creator of the science of mechanics. He and da Vinci are accredited with the inductive method of research, namely, to carry out systematic experiments.

B. Castelli (1577–1644), a pupil of Galilei, wrote what appears to be the first treatise on rivers. This book entitled "Della misura delle acque correnti," published in 1628, gave reason to consider its author as the founder of the Italian school of hydraulics. Castelli rediscovered the continuity principle and, based on experiments, he wrongly concluded that at the regulating devices and channels themselves the flow velocity is proportional to the water depth. E. Torricelli (1608–1647), Castelli's pupil, gave the more correct relation that the velocity is proportional to the square root of the head.

Systematic studies in channel hydraulics were begun with D. Guglielmini (1655–1710), who is sometimes called the father of the science of river hydraulics. This scientist was a person of many interests, one of which was hydraulics. At one time, he was superintendent of water and professor of hydrometry in Bologna. Two major contributions to our field are known. The one published around 1690 on "Aquarum fluentium mensura nova methodo inquisita" was concerned with a method of measuring flowing water with a suspended ball deflected by the current. It is the second treatise on "Della natura dei fiumi," published in 1697, which established his fame. Guglielmini's contributions to river hydraulics were obtained by field observations rather than by laboratory experiments. Correctly it was realized that the velocity is not equally distributed over the depth, but some difficulty arose when explanation of theoretical hypothesis disagreed with field observations. There is evidence that he had a notion of the importance of bed resistance, but was as yet unable to express the resistance law quantitatively. His remarks to the sediment problem, while only qualitative, are most valuable ones. The importance of the channel slope and depth was discussed in their relation to the scour and deposition problems. Furthermore, for scour to exist, it becomes necessary that the "scouring force" be greater than the "resistance of the soil." It seems that Guglielmini also expressed some ideas about the reduction of the sand grain volume along a watercourse. Thus this scientist must be accredited with having made the first basic statements about loose-boundary hydraulics.

The Italian school of hydraulics included a few more scientists, such as Frisi, Grandi, Zendrini, etc., but among all of them the contributions by Guglielmini must receive our special attention and credit.

2.3 DUBUAT AND HIS CONTEMPORARIES

P. duBuat (1734–1809) has been frequently referred to as the founder of the French school of hydraulics. He carried out many different hydraulic experiments, and wrote the fundamental (and first) treatise on hydraulics entitled "Principes d'Hydraulique." It would be unfair to his predecessors or colleagues to accredit only duBuat with the vast progress on which he reports in his writings. Certainly, the importance of contributions in general mechanics in the seventeenth century by Pascal (1623–1662), Newton (1642–1727), Leibniz (1646–1716), etc., and in the eighteenth century by D. Bernoulli (1700–1782), Euler (1707–1783), Lagrange (1736–1813), etc., were very significant and most influential. Also the contributions by engineering scientists such as Brahms, Bossut, Chezy, etc., in particular to the hydraulics of streams must be considered.

In 1753 A. Brahms published a book on "Anfangsgründe der Deich- und Wasserbaukunst." The most important contribution of this work was the uniform flow formula later established by Chezy. Brahms argued that the slope is given when the gravity component balances the resistance of the bed. As far as the initiation of sediment motion is concerned, Brahms suggested a proportionality of the bed velocity and the one-sixth power of the immersed weight of the bed material. This relation, given by Eq. (6.8), is still in use.

A. Chezy's (1718–1798) contribution to channel hydraulics is the well-known uniform flow formula, now known as Chezy's formula, which states a proportionality of flow velocity, channel slope, and hydraulic radius, or

$$\bar{u} = C\sqrt{R_h S}$$

where C is the factor of flow resistance—the Chezy C—which is a variable depending among others on the sediment transport. According to some sources, Chezy developed the previous relation in 1775, but it was either not presented or not received by the engineering community. Certainly, in the book of DuBuat (1786), where an entire chapter is devoted to the flow formula, there exists no evidence that its author had any knowledge of Chezy's formula although the project in which Chezy was involved is mentioned (¶225).

The books of duBuat (1734–1809) must be considered as a most comprehensive treatment of channel hydraulics as far as both theory and experiment are concerned. DuBuat's first book, printed in 1779, presented a uniform flow equation based on the pipe and open-channel experiments by Bossut (1730–1814). Subsequently, duBuat obtained encouragement and money to carry out his own experiments. Upon completion of the tests, duBuat wrote the famous treatise on "Principes d'Hydraulique, vérifiés par un grand Nombre d'Experiénces faites par Ordre du Gouvernement." The "Principes d'Hydraulique," as it is called, appeared in 1786 in two books and was divided into three parts. The first part is essentially a reedited version of his first book of 1779; the second part is a discussion of his own test series, mainly done with pipes and open channels; the third one deals with the fluid resistance of bodies. Most important to us are the first two (from the four) chapters of the first part.

The first chapter is an extensive, rather logical, discussion and derivation of the flow formulas for channel or pipe flow. DuBuat (1796)[1] has it well summarized in his preface (p. 19):

> It seems a general acceptable and agreeable principle: that when water moves uniformly in a bed, (that) the acceleration force, which causes the flow, is equal to the resisting forces which are due to its own viscosity, or due to friction along the bed. This law appears to me to be as old as the origin of all the water courses, and to be the key to all hydraulics. . . .

Assuming the resistance forces to be proportional to the square of the velocity and the acceleration force to be a product of gS, with S being the slope, DuBuat (1796, ¶26) wrote the relation as

$$\frac{\bar{u}^2}{m} = gS$$

where m is a factor of proportionality. Based on experimental data (¶55), the uniform flow formula is given as [¶51 and Forchheimer (1914)]

$$\bar{u} = \frac{48.85\sqrt{R_h} - 0.80}{\sqrt{1/S} - \ln\sqrt{1/S + 1.6}} - 0.05\sqrt{R_h}$$

and, most interestingly, the hydraulic radius R_h has come in use. Another remark deserves our attention; in deriving the foregoing equation, DuBuat (1796, ¶34, etc.) talks about the retarding effects of the boundaries. Can we see in this statement a forerunner of the boundary-layer theory? The German translation contains at the end of the first chapter some interesting and useful remarks by J. Eytelwein (1764–1848). This engineer scientist pointed out that, owing to the cumbersome form of duBuat's equation, R. Woltman, after having read duBuat's first treatise, suggested (in 1791) a simpler equation of the general form

$$\bar{u} = C_W\sqrt{R_h S}$$

In addition, Eytelwein himself analyzed all of the open-channel data of duBuat and presented the following comparatively simple relation:

$$\bar{u} = 50.9\sqrt{R_h S}$$

Woltman's and Eytelwein's contributions must be considered as independently (from Chezy) derived uniform-flow equations.

[1] In 1796, the "Principes d'Hydraulique" was translated into German with remarks, additions, and an introduction by J. Eytelwein. It is a copy, formerly owned by a so-called "Julius Weisbach," of this version which is presently available to the author.

The second chapter, entitled "Theory of Stream Beds; Their Establishment," contains—at least for the river hydraulician—some most interesting sections, and shows duBuat's familiarity with his subject and love of this field. DuBuat quite correctly realized that any artificial channel dug into a soil and any natural watercourse represented a solid-liquid interfacial problem. The author began this chapter with a discussion on the shape of the channel cross section. The circular cross section is recognized as the one where the least resistance occurs and, thus, it is the most economical. Then the trapezoidal cross section is discussed, because it occurs so frequently in natural and artificial open channels, and emphasis is placed on the importance of slope stability of the embankment, suggesting a slope ratio of 4 horizontal to 3 vertical. However, if the cross section is a rectangular one, the width should be twice its depth. DuBuat realized so well that none of the channel cross sections might be maintained if the channel is not in equilibrium, i.e., erosion or deposition occurs; thus, the channel bed has to be studied in more detail. DuBuat reports on what appears to be the first experiment to determine this flow velocity which at a given soil condition can be resisted. As a velocity criterion, the bottom velocity is used—a relation of bottom to average and surface velocity is also given; seven different soil materials ranging from clay to stone "of the size of an egg" were investigated. DuBuat's remarks on the transport of the sediment are so vivid that we quote them in the following (from ¶72)[1]:

> The way running water attacks the bed which cannot resist any more, and how the sand is shoved is fascinating indeed, and deserves description. At times, it is an eddy which carries with it the soil and the fine sand, like the wind might carry the smoke, and this happens when the velocity is strong enough such that the impact of the water overcomes the inertia of the solid particles. At other times, it is smooth and so-to-speak methodic working, which one may call a major work of dynamics. I shall try to describe this. When the bottom velocity is large enough to bring a body of a specific weight exceeding the one of water into motion and to transport it, then these bodies (particles) do not move uniformly, but much rather they tramp and jump. Take for example the sand. Suppose the channel bed is made up of sand of the size to be distinguished by eyes, and suppose the velocity is found to be 10 to 12 "pouces" (inches) per second, then the channel bed appears like the pattern of a tapestry, which is known under the name Hungarian, such that irregular furrows appear whose direction is perpendicular to the water course. Each of these furrows has two glacis with opposing

[1] The author has recently discovered that the same quotation is used by Hagen in "Handbuch der Wasserbaukunst" (3d ed., pt. 2, vol. 1, p. 159) to illustrate the sediment transport problem. Hagen's subsequent experiment documented duBuat's findings.

> slopes; the one which faces the water has a long slope, and has a common summit with the one which opposes the water and is much steeper and lies downstream. The profile of a furrow resembles the one of a glacis and of a covered walk-way at a fortification. At about a distance of a foot from the steep slope begins a gentle one of another furrow, and this is repeated going downstream. A sand grain, pushed by the current, moves up the gentle slope of the first talus, arrives at the summit, and rolls down due to its weight along the opposing slope; here in the shelter of the fluid's action it pauses and will be covered by other grains which arrived the same way. This work resembles the one of a ditch-digger who takes his wheelbarrow up the ramp to unload it and the soil tumbles into the depth. The sand grains thus buried, remain there loaded down and covered by other ones until the total mass of the furrow, which they have left behind, has passed over them. It is thus that the entire furrow deplaces itself and advances little by little for its own width. Thus, the particle under consideration appears at the beginning of the new slope which was just created, and since it is again exposed to the influence of the current, it climbs the slope and falls then into the depth just as it did the first time. All of the furrows move very slowly, and for moderate water velocities it may take no less than an hour that everyone progresses by 4 to 5 pouces (inches). When the water-velocity increases, this activity increases too, but decreases, however, as soon as the one diminishes. Under average work, two years are required for a sand grain to travel the distance of 2,400 toises . . . .

The previous words witness the familiarity of duBuat with the mechanics of bed forms. A subsequent paragraph (¶73) contains a brief description of the armoring of the channel bed.

> Since the smallest sand and clay becomes removed most easily, we find on the channel bed the most coarse sand, pebbles, stone, and other large objects which tumble upon it and remain there. They often give the stream bed such a strength that the resistance of it causes the stream to widen its bed or it might change it altogether, if it meets to the right and to the left less resistance . . . .

DuBuat's greatest contribution to the field of sediment transport is the *shear-resistance* concept, which is considered to be a force which a stream exhibits with respect to the motion toward its bed. This is subject of discussion in ¶74. Because of its historical importance, a photostat of these pages is shown in Fig. 2.1, where at the end of the second paragraph the

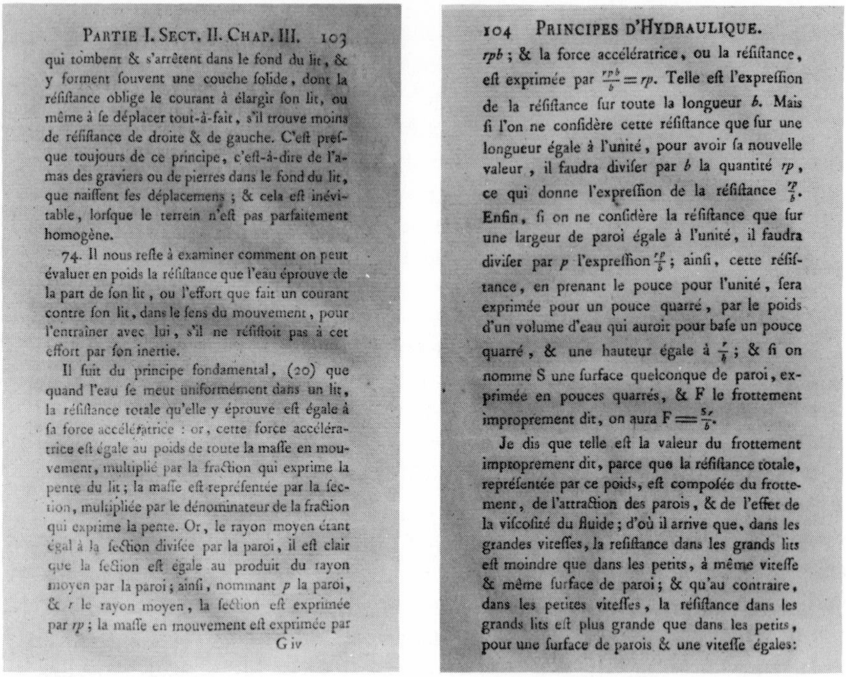

PARTIE I. SECT. II. CHAP. III. 103

qui tombent & s'arrêtent dans le fond du lit, & y forment ſouvent une couche ſolide, dont la réſiſtance oblige le courant à élargir ſon lit, ou même à ſe déplacer tout-à-fait, s'il trouve moins de réſiſtance de droite & de gauche. C'eſt preſque toujours de ce principe, c'eſt-à-dire de l'amas des graviers ou de pierres dans le fond du lit, que naiſſent ſes déplacemens ; & cela eſt inévitable, lorſque le terrein n'eſt pas parfaitement homogène.

74. Il nous reſte à examiner comment on peut évaluer en poids la réſiſtance que l'eau éprouve de la part de ſon lit, ou l'effort que fait un courant contre ſon lit, dans le ſens du mouvement, pour l'entraîner avec lui, s'il ne réſiſtoit pas à cet effort par ſon inertie.

Il ſuit du principe fondamental, (20) que quand l'eau ſe meut uniformément dans un lit, la réſiſtance totale qu'elle y éprouve eſt égale à ſa force accélératrice : or, cette force accélératrice eſt égale au poids de toute la maſſe en mouvement, multiplié par la fraction qui exprime la pente du lit ; la maſſe eſt repréſentée par la ſection, multipliée par le dénominateur de la fraction qui exprime la pente. Or, le rayon moyen étant égal à la ſection diviſée par la paroi, il eſt clair que la ſection eſt égale au produit du rayon moyen par la paroi ; ainſi, nommant p la paroi, & r le rayon moyen, la ſection eſt exprimée par rp ; la maſſe en mouvement eſt exprimée par

G iv

104 PRINCIPES D'HYDRAULIQUE.

rpb ; & la force accélératrice, ou la réſiſtance, eſt exprimée par $\frac{rpb}{b} = rp$. Telle eſt l'expreſſion de la réſiſtance ſur toute la longueur b. Mais ſi l'on ne conſidère cette réſiſtance que ſur une longueur égale à l'unité, pour avoir ſa nouvelle valeur, il faudra diviſer par b la quantité rp, ce qui donne l'expreſſion de la réſiſtance $\frac{rp}{b}$. Enfin, ſi on ne conſidère la réſiſtance que ſur une largeur de paroi égale à l'unité, il faudra diviſer par p l'expreſſion $\frac{rp}{b}$; ainſi, cette réſiſtance, en prenant le pouce pour l'unité, ſera exprimée pour un pouce quarré, par le poids d'un volume d'eau qui auroit pour baſe un pouce quarré, & une hauteur égale à $\frac{r}{b}$; & ſi on nomme S une ſurface quelconque de paroi, exprimée en pouces quarrés, & F le frottement improprement dit, on aura $F = \frac{Sr}{b}$.

Je dis que telle eſt la valeur du frottement improprement dit, parce que la réſiſtance totale, repréſentée par ce poids, eſt compoſée du frottement, de l'attraction des parois, & de l'effet de la viſcoſité du fluide ; d'où il arrive que, dans les grandes viteſſes, la reſiſtance dans les grands lits eſt moindre que dans les petits, à même viteſſe & même ſurface de paroi ; & qu'au contraire, dans les petites viteſſes, la réſiſtance dans les grands lits eſt plus grande que dans les petits, pour une ſurface de parois & une viteſſe égales:

Fig. 2-1 Photostat of paragraph 74 of duBuat' treatise "Principes d'Hydraulique." [DUBUAT (*1786*).

formula is given as

$$F = \frac{Sr}{b}$$

where F = improper so-called friction
S = area
r = average radius (hydraulic radius)
$1/b$ = slope

The remaining sections of his chapter 2 are devoted to the discussion of fluvial morphology. When one realizes the slow advance of this science, duBuat's remarks become again most profound. There are some ideas on equilibrium (*stabilité* or *Beharrungs-Zustand*) of stream beds, and a few vague remarks on nonuniform flow conditions as they affect the equilibrium of the stream bed. The sinuosity of rivers is discussed rather extensively, and it is remarked that "the water does not like straight lines" (¶84), and at another place (¶101): "Nature, which tries to achieve its objective with the simplest means, is responsible that all rivers in order to become *normal* make a great many curves." The distorted velocity distribution in bends is correctly realized, but also the importance of the local scour problem is stressed and discussed.

Chapter 3 of the first part is devoted to the application of the theory to practical problems. DuBuat said in the preface to his treatise (p. xxxiii):

> To study the laws of Nature does satisfy the mind, but it also does serve an utilitarian purpose: . . . The theory must be applied to the practice . . . .

The writings of duBuat became indispensable for the success of research in fixed- and loose-boundary hydraulics in the nineteenth century, and still remain most interesting literature.

His influence was especially strong in Germany. It was already mentioned that both Woltman, upon reading "Principes d'Hydraulique," and Eytelwein, in commentating the translation of the work, had suggested uniform flow equations. R. Woltman (1754–1837) was a most successful engineer and scientist. As superintendent of hydraulic constructions in the lower Elbe River district he was responsible for providing Hamburg with a navigable river channel. He wrote a treatise entitled "Beiträge zur hydraulischen Architektur" in 1791, but more important and original is the booklet on "Theorie und Gebrauch des hydrometrischen Flügels," published in 1790. J. Eytelwein (1764–1848) was the author of "Handbuch der Mechanik fester Körper und der Hydraulik" (1801), in which he discussed, among many other topics, the uniform flow equation.

Among the many more contributions to channel hydraulics of this time, we want to restrict ourselves to but a few. A book on "Grundlehren der Hydraulik" was written by A. Bürja. In the fifth of the eight chapters BÜRJA (1792) discussed the velocity distribution and remarked on problems of erosion. A Dutch treatise, in its German translation entitled "Abhandlungen über die Geschwindigkeit des fliessenden Wassers," was produced by C. Bruning. After a thorough historical review about the ideas on velocity distribution, BRUNING (1798) implied that no mathematics but only measurements can give a valid answer; an instrument to measure velocities was developed and is discussed in detail. Two French hydraulicians may not be overlooked, namely, P. Girard (1765–1836) and G. deProny (1755–1839). The first of the two is accredited for pointing to the existence of Chezy's research, which had apparently received no attention. A scholarly treatise on "Recherches physico-mathematiques sur la Theorie des Eaux courantes" was written by DEPRONY (1804). More than half of this book is devoted to various aspects of flow in canals, but no attention is paid to loose-boundary channels. Of historical importance are some remarks on the "motion of water in streams," which are contained in "Traité de l'Hydraulique à l'Usage des Ingénieurs" by J. d'Aubuisson de Voisins (1769–1841). On the motion of debris, D'AUBUISSON (1835) remarked:

> Almost always do these (streams in their head-water) carry rocks, stone pieces, and debris, which originate in the mountain slopes situated along the water courses. They (streams) push these masses, continuously replaced by others, at flood stages and carry them, and this they do easier and longer (distance), the more the channel bed is inclined, the smaller the specific gravity, and the smaller their volume . . . .

There are some remarks on the gradual erosion of the debris along the watercourse. These and a few more statements show the attention that sediment transport had begun to receive.

2.4 STREAM HYDRAULICS IN THE NINETEENTH CENTURY

It is appropriate to start this new section with G. Hagen (1797–1884), a civil servant, teacher, and writer. Well known to the hydrodynamicist for his contributions to the flow of liquid through pipes, this man is equally distinguished in the field of hydraulic engineering. His name is closely connected with many engineering feats in northern Germany where various harbors (Hamburg) and navigation canals (Rhine-Marne-Kanal) have been designed by him. This field experience must have been of much help to him in writing the multiple-volume treatise on "Handbuch der Wasserbaukunst," which went through several editions and English translations. The first part was entitled "Springs," the second one "Streams," and the third and last "Sea"; we shall devote our attention to the second part which contains four books and a booklet with graphs and figures.

In the first book we find an extensive discussion of the sediment transport problem, but it remains qualitative. HAGEN (1871) says (p. 159): ". . . The smaller grains remain longer in suspension, they are lifted by the inner motion of the water, while the larger grains soon settle out . . . ."

We suppose that for inner motion of the water, the word "turbulence" could be replaced. Hagen quotes duBuat's statement on the formation of the stream bed, and stresses that his own experiment gave the very same results. Furthermore, in an effort to check a relationship between critical velocity and type of soils, as suggested by DUBUAT (1796) and used since, HAGEN (1871) questions the generality of such relation.

The remaining three books of the second part are exclusively concerned with hydraulic engineering. By many standards, Hagen's "Handbuch der Wasserbaukunst" compares favorably with modern books in the field of hydraulic engineering, and it is a delight to browse through its pages.

Comprehensive experimentation and intelligently written books by

Weisbach (1806–1871) show little interest in stream hydraulics. To be sure WEISBACH'S (1845) "Lehrbuch der Ingenieur- und Maschinen-Mechanik" is an extremely well-written and complete treatise, but its overall emphasis is toward the *machine*. However, it was Weisbach who expressed for the first time a resistance equation for pipes and channels as,

$$h = \xi \frac{lp}{F} \frac{C^2}{2g}$$

where ξ expresses a number deduced from an experiment, which may be called the coefficient of the resistance of friction, and is not constant. The quotient p/F is "the perimeter of water profile and the area of the whole transverse profile." Some remarks on the value of ξ are contained in WEISBACH'S (1855) "Die Experimental-Hydraulik" (¶43).

Understandably, the question of the uniform flow formulas was of considerable importance to the hydraulician of the second part of the nineteenth century. We shall mention only the names of those scientists who contributed important experimental and/or analytical data. Notable among the researchers reporting in the German language are Lahmeyer (1845), Hagen (1871), Ganguillet and Kutter (1869), Strickler (1923), and Forchheimer (1923); in the French language are Darcy and Bazin (1865) and Bazin (1897); and in the English language Humphreys and Abbot (1861) and Manning (1890).

J. Dupuit (1804–1866) makes some interesting remarks on sediment transport in his "Etudes théoriques et practiques sur le Mouvement des Courantes" (1848). In an enlargement of this treatise, DUPUIT'S (1865) book on "Traité de la Conduite et de la Distribution des Eaux" contains the following on suspension (¶23):

> . . . It is, in fact, a rather curious phenomenon of water to carry in permanent suspension material with a density considerably exceeding the one of its own. We explain this through the differences in velocities between filaments (layers), which produce an underpressure equal to the difference in the weight of the suspended matter and an equivalent volume of water. This is not true anymore, when the water has no motion (is at rest), or even more when all filaments have the same velocity, then the diverse suspended matter will settle out . . . .

Quite clearly, these sentences may be considered as the beginning of the suspension theory.

Whereas suspension is the mode of solids transport within the liquid body, the mode of transport along the water-sand interface (on the wetted perimeter) is due to drag or traction.

The first important contribution of the drag principle was advanced by P. duBoys (1847–1924). In a paper on "Le Rhône et les Rivières à Lit affouillable," DuBoys (1879) determined the transported material by using the depth and slope of the stream rather than the stream's velocity. This, however, was not entirely new, since DuBuat (1786) had suggested the same relationship. However, what was novel in duBoys' paper was his quantitative expression for bedload transport. Unrealistic as his model was—he imagined the transport occurs in a series of layers—duBoys arrived at a useful expression, such as (p. 160):

$$q = \chi F(F - F_0)$$

where q = transport rate
F = force of entrainment
F_0 = force at the beginning of sand motion
χ = characteristic sediment coefficient

(The photostat of these so important pages in duBoys' paper is given in Fig. 2.2.) Despite violent criticism to duBoys' equation and its model, many of the presently in use bedload equations exhibit the same structure.

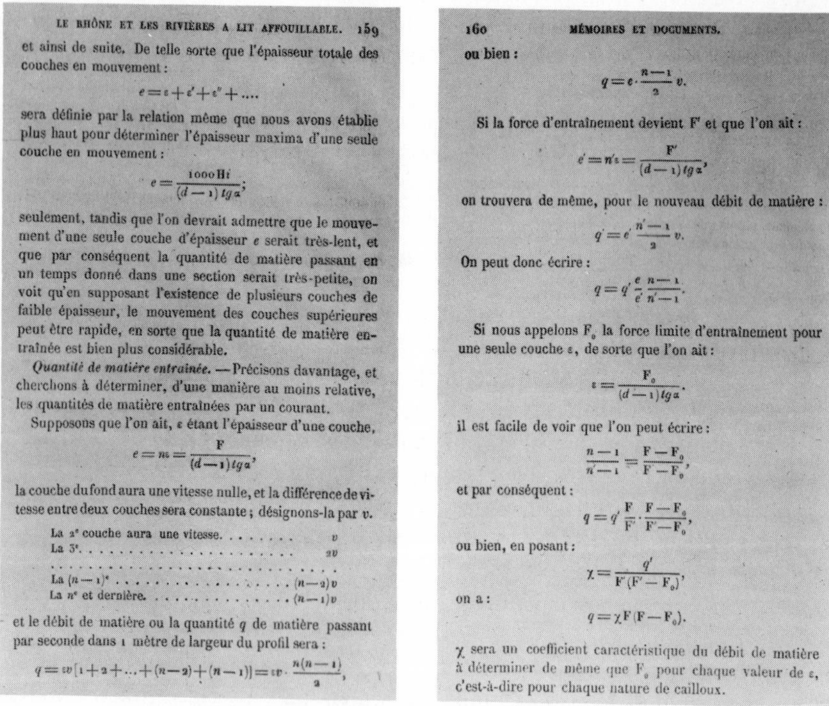
LE RHÔNE ET LES RIVIÈRES A LIT AFFOUILLABLE. 159

et ainsi de suite. De telle sorte que l'épaisseur totale des couches en mouvement :

$$e = \varepsilon + \varepsilon' + \varepsilon'' + \dots$$

sera définie par la relation même que nous avons établie plus haut pour déterminer l'épaisseur maxima d'une seule couche en mouvement :

$$e = \frac{1000 Hi}{(d-1)\,tg\,\alpha};$$

seulement, tandis que l'on devrait admettre que le mouvement d'une seule couche d'épaisseur ε serait très-lent, et que par conséquent la quantité de matière passant en un temps donné dans une section serait très-petite, on voit qu'en supposant l'existence de plusieurs couches de faible épaisseur, le mouvement des couches supérieures peut être rapide, en sorte que la quantité de matière entraînée est bien plus considérable.

Quantité de matière entraînée. —Précisons davantage, et cherchons à déterminer, d'une manière au moins relative, les quantités de matière entraînées par un courant.

Supposons que l'on ait, ε étant l'épaisseur d'une couche,

$$e = n\varepsilon = \frac{F}{(d-1)\,tg\,\alpha},$$

la couche du fond aura une vitesse nulle, et la différence de vitesse entre deux couches sera constante ; désignons-la par v.

La 2ᵉ couche aura une vitesse. v
La 3ᵉ. . . . . $2v$
. . . .
La $(n-1)$ᵉ . . . . $(n-2)v$
La nᵉ et dernière. . . . . $(n-1)v$

et le débit de matière ou la quantité q de matière passant par seconde dans 1 mètre de largeur du profil sera :

$$q = \varepsilon v[1 + 2 + \dots + (n-2) + (n-1)] = \varepsilon v \cdot \frac{n(n-1)}{2},$$

160 MÉMOIRES ET DOCUMENTS.

ou bien :

$$q = e \cdot \frac{n-1}{2} v.$$

Si la force d'entraînement devient F′ et que l'on ait :

$$e' = n'\varepsilon = \frac{F'}{(d-1)\,tg\,\alpha},$$

on trouvera de même, pour le nouveau débit de matière :

$$q' = e' \frac{n'-1}{2} v.$$

On peut donc écrire :

$$q = q' \frac{e}{e'} \frac{n-1}{n'-1}.$$

Si nous appelons F_0 la force limite d'entraînement pour une seule couche ε, de sorte que l'on ait :

$$\varepsilon = \frac{F_0}{(d-1)\,tg\,\alpha}.$$

il est facile de voir que l'on peut écrire :

$$\frac{n-1}{n'-1} = \frac{F - F_0}{F' - F_0},$$

et par conséquent :

$$q = q' \frac{F}{F'} \cdot \frac{F - F_0}{F' - F_0},$$

ou bien, en posant :

$$\chi = \frac{q'}{F'(F' - F_0)},$$

on a :

$$q = \chi F(F - F_0).$$

χ sera un coefficient caractéristique du débit de matière à déterminer de même que F_0 pour chaque valeur de ε, c'est-à-dire pour chaque nature de cailloux.

Fig. 2-2 Photostat of pages 159 and 160 of duBoys' paper. [DuBoys (*1876*).]

At the end of the nineteenth century, the first loose-boundary stream models were built. Its pioneer is L. Fargue (1827–1910), who took a section of a natural river into the laboratory by reducing, rather arbitrarily, the scales of depth, breadth, and time. O. Reynolds (1842–1912) introduced the time element, studying a tidal model, and L. Vernon-Harcourt (1839–1907) in his investigation reproduced vertical and horizontal scales.

It is out of the question to discuss every phase of the new developments in the nineteenth century. The interested reader is, however, referred to a more detailed discussion as is given by SCHLICHTING (1892) and FORCHHEIMER (1914).

Finally, attention is drawn to the achievements of two engineers whose works show great insight into the fluvial processes of rivers. J. G. Tulla (1770–1828) drew up a master plan for the regulation of the River Rhine, which was finished half a century after his death. Tulla was given the title "Bändiger des wilden Rheins." The other one, L. Franzius, lived in the second part of the last century. After having achieved fame with the regulation and correction of the River Weser, he was widely sought as a consultant.

A new awakening in river hydraulics began at the close of the last century with the construction of laboratories specially designed to study river and channel problems. From the novel kind of experimentation, the field of sediment transport profited considerably. The first of these laboratories was built by H. Engels (1854–1945). ENGELS (1929) remarked that in 1891 he commenced experimental research at Dresden, but the special *Flussbau-Laboratorium* was not available until 1898. In the latter laboratory, which was continuously updated, and in 1913 even replaced by a newer one, many experiments on loose-boundary hydraulics were investigated. These included studies on scour around bridge piers, on longitudinal cross sections of rivers, on the reproduction of river stretches, on bed configuration, on flow in bends, and others. Doubtlessly, this added a new dimension to our knowledge on the transport of sediments. This powerful tool was not overlooked, and very soon similar laboratories were opened in Karlsruhe by T. Rehbock (1864–1950) and in Berlin by H. Krey (1866–1928).[1] Outstanding theoretical work was done by P. Forchheimer (1852–1933). Besides the numerous papers on almost every aspect of hydraulics, the book entitled "Hydraulik," first published in 1914, must be taken as evidence that FORCHHEIMER (1914) ranks as one of the greatest hydraulicians of the past. The student of sediment transport who traces ideas and developments will invariably consult this book. Also, in 1914, Schoklitsch (1888–1969), at that time an assistant to Forchheimer, published a treatise on

[1] An excellent account of the development of the hydraulic laboratories is given in FREEMAN (1929).

"Über Schleppkraft und Geschiebebewegung," in which he discussed, among others, his own experiments and set forth a new bedload equation.

Experimental data on the transport of sediment material were extremely scarce till Gilbert (1843–1918) conducted controlled experiments at the University of California. These data, which GILBERT (1914) reported in "The Transportation of Debris by Running Water," remain, up to the present, a unique storehouse for researchers. Almost every theoretical or empirical sediment equation owes part of its character to the data. Gilbert's observational remarks show his deep insight, but the proposed equations were rather limited.

Important field investigations were conducted in India. KENNEDY (1895) published a paper that included the first quantitative study in the determination of the channel shape. For the desirable condition that a channel neither erode nor deposit any material, a velocity-depth relationship was suggested. The paper implied an idea which was later developed into the so-called *regime theory*.

2.5 CONCLUDING REMARKS

When and where hydraulics, in particular channel and river hydraulics, began is hidden in antiquity and probably will remain so. The relationship of man and water, indeed, must be ancient. Even the oldest discovered hydraulic achievements give evidence that man had knowledge and appreciation of what water can do to man and what man can do to water. Of course, hydraulic engineering did not become a science until the sixteenth century, but remained a creative art up to that time, yet set goals were, nevertheless, achieved by the people of antiquity. For about 200 years, science has helped in systemizing the ancient art of channel and river engineering. We are presently in the midst of this development.

REFERENCES FOR PART ONE

BISWAS, A. K. (1967): Hydrologic Engineering Prior to 600 B.C., *Proc. Am. Soc. Civil Engrs.*, vol. 93, no. HY5.

BRUNINGS, C. (1789): "Über die Geschwindigkeit des fliessenden Wassers," Translation from the Dutch, Behrens and Körner, Frankfurt a/M.

BÜRJA, A. (1792): "Grundlehren der Hydraulik," Lagarde, Berlin.

D'AUBUISSON DE VOISINS, J. F. (1835): "Handbuch der Hydraulik," translation from the French, Weidmann, Leipzig.

DEPRONY, R. (1812): "Über die Leitung des Wassers in Kanälen und Röhrenleitungen," translation from the French, Heyer, Giessen.

DUBOYS, M. P. (1879): Le Rhône et les Rivières à Lit affouillable, *Mem. et Doc., Annales des Pont et Chaussees*, ser. 5, vol. XVIII.

DuBuat, P. (1786): "Principes d'Hydraulique," 2d ed. (1st ed., 1799), 2 Books, de L'Imprimerie de Monsieur, Paris.

——— (1796): "Grundlehren der Hydraulik," Translation of the 2d ed., with introduction, additions, and remarks by J. A. Eytelwein, Belitz und Braun, Berlin.

Dupuit, H. P. (1865): "Traité de la Conduite et de la Distribution des Eaux," Paris.

Engels, H. (1929): Historical Development and Value of Experiments with Models, in J. Freeman, "Hydraulic Laboratory Practice," Am. Soc. Mech. Engrs., New York.

Flachsbart, O. (1932): Geschichte der experimentellen Hydro- und Aeromechanik, in "Handbuch der Experimentalphysik," vol. 4, pt. 2, Akademische Verlags Gesellschaft, Leipzig.

Fleming, H. W. (1957): "Wüsten, Deiche und Turbinen," Messerschmidt-Verlag, Göttingen.

Forchheimer, P. (1914): "Hydraulik," 1st ed. (2d ed., 1924, and 3d ed., 1930), Teubner, Leipzig-Berlin.

Freeman, J. R. (1929): "Hydraulic Laboratory Practice," Am. Soc. Mech. Engrs., New York.

Gilbert, G. K. (1914): Transportation of Debris by Running Water, *U.S. Geol. Survey, Prof. Paper 86.*

Hagen, G. (1871): Streams, in "Handbuch der Wasserbaukunst," pt. 2, vol. 1, Ernest & Korn, Berlin.

Kennedy, R. G. (1895): The Prevention of Silting in Irrigation Canals, *Minutes, Proc. Inst. Civil Engrs.*, vol. CXIX.

Kolupaila, S. (1961): "Bibliography of Hydrometry," Notre Dame Press, Notre Dame, Ind.

Rouse, H., and S. Ince (1963): "History of Hydraulics," Dover, New York.

Schlichting, J. (1892): Der Wasserbau, in "Handbuch der Ingenieurwissenschaften," 3d ed., Engelmann, Leipzig.

Weisbach, J. (1845): "Lehrbuch der Ingenieur- und Maschinen-Mechanik," Braunschweig.

——— (1855): "Die Experimental-Hydraulik," Engelhardt-Verlag, Freiberg.

part two

Hydrodynamics of Fluid-Particle Systems

CONTENTS

INTRODUCTION

LIST OF SYMBOLS

3 GENERAL REMARKS
- 3.1 Motion with Linear Resistance
- 3.2 Motion with Nonlinear Resistance
- 3.3 Concluding Remarks

4 SETTLING VELOCITY OF PARTICLES
- 4.1 Uniform Motion in an Ideal Fluid
- 4.2 Uniform Motion in a Real Fluid
  - 4.2.1 Theoretical Considerations for a Sphere
    - 4.2.1.1 Stokes' Solution
    - 4.2.1.2 Oseen-type Solutions
    - 4.2.1.3 Recent Developments
    - 4.2.1.4 Investigations at High Reynolds Numbers
  - 4.2.2 Empirical Equations for the Drag Coefficient of a Sphere
  - 4.2.3 Terminal Settling Velocity
  - 4.2.4 Complicating Effects on the Uniform Motion of a Sphere
    - 4.2.4.1 Particle Shapes
      - 4.2.4.1.1 Regular Shapes
      - 4.2.4.1.2 Irregular Shapes
    - 4.2.4.2 Various Boundaries
    - 4.2.4.3 More than One Sphere
    - 4.2.4.4 Particle Roughness
    - 4.2.4.5 Particle Rotation
    - 4.2.4.6 Turbulence Effects
      - 4.2.4.6.1 In the Transition Region
      - 4.2.4.6.2 Below the Transition Region
      - 4.2.4.6.3 Small-Particle Motion in Turbulent Environment
    - 4.2.4.7 Combinations of Various Effects
      - 4.2.4.7.1 Combinations of Correction Coefficients
      - 4.2.4.7.2 Investigation of the Complete Problem

5 EFFECT OF PARTICLES ON THE FLUID VISCOSITY
- 5.1 Theoretical Considerations
  - 5.1.1 Suspensions of Spherical Particles
  - 5.1.2 Suspensions of Nonspherical Particles

5.2 Experimental Considerations
5.2.1 Suspensions of Spherical Particles
5.2.1.1 Stable Suspensions
5.2.1.2 Unstable Suspensions
5.2.2 Suspensions of Nonspherical Particles
5.3 Remarks on Non-Newtonian Behavior
REFERENCES FOR PART TWO

INTRODUCTION

The interplay between the resultant of the hydrodynamic forces on a particle and of its weight represents the physical mechanism of the sediment motion. Although certain physicochemical forces may exist, it is assumed that these are negligible. This is necessary because knowledge of these forces and of their effects on the sediment-transport mechanics is still scarce. On the other hand, cohesive sediments, whose physicochemical forces do play the major role, are encountered in many practical problems. Therefore an entire chapter in another part of this book will be devoted to the discussion of the available literature pertaining to the hydraulics of cohesive sediments. However, unless specifically stated, Part 2 will be limited to the treatment of noncohesive sediment particles.

In Chap. 3, the general equation of motion of a single spherical particle moving through a fluid of infinite extent will be reviewed. In Chap. 4 the steady-state condition is discussed. At low Reynolds numbers the Stokes solution is available, whereas at high Reynolds numbers the knowledge of boundary-layer theory aids in an explanation. Empirical equations covering the whole range of Reynolds numbers and the determination of the terminal velocity are elaborated on next. Then follows a discussion of the influences of different effects on the drag coefficient, i.e., several boundary conditions, more than one particle, particle shapes, particle rotation, surface roughness of particles, turbulence effects, and combinations of these.

In Chap. 5 the effect of particle concentrations on the viscosity of a fluid is considered.

LIST OF SYMBOLS

a	radius of particle; semiaxis of ellipsoid
A_p	projected area
A_S	surface area of spherical particle
A_{SN}	surface area of nonspherical particle
b	semiaxis of ellipsoid
c	semiaxis of ellipsoid
C	concentration by volume
C_D	steady-state drag coefficient
C_{DR}	resistance coefficient
D	container diameter
d	particle diameter; $d = 2a$
d_N	normal diameter; $d_N = 2\sqrt[3]{abc}$
g	gravitational constant
K	correction factor in Stokes' equation
k	virtual mass coefficient
k_e	Einstein's viscosity constant
k_H	Heywood's volume factor
L	length, as defined
m_s	mass of a particle
m'_s	mass of liquid displaced
R	resistance, drag
R_{cyl}	radius of circular cylinder
Re	Reynolds number; $\text{Re} = (v_s - v)2a/\nu$
Re_{cr}	critical Reynolds number
s	distance, as defined
t	time
v	velocity
μ	coefficient of viscosity
ρ	density
ν	kinematic viscosity
ω	circular frequency
Ψ	shape factor, Wadell's sphericity

Subscripts

none	liquid phase
s	solid phase
0	initial stage
1, 2, . . .	index
susp	suspension, mixture

3

General Remarks

3.1 MOTION WITH LINEAR RESISTANCE

Applying Newton's second law of motion BASSET (1888, 1961), BOUSSINESQ (1903), and OSEEN (1927) derived independently an equation of motion for a particle moving under the influence of gravity in a fluid at rest. TCHEN (1947) extended this equation into an equation of motion for a small spherical particle in fluid moving with a variable velocity v. The resulting integro-differential equation, frequently referred to as B.B.O. equation, is given by:

$$\frac{4\pi a^3}{3}\rho_s\dot{v}_s = \frac{4\pi a^3}{3}\rho\dot{v} - \frac{2\pi a^3}{3}\rho(\dot{v}_s - \dot{v})$$

$$- 6\pi\mu a\left[(v_s - v) + \frac{a}{\sqrt{\pi\nu}}\int_{t_0}^{t} dt_1 \frac{\dot{v}_s(t_1) - \dot{v}(t_1)}{\sqrt{t - t_1}}\right]$$

$$- \frac{4\pi a^3}{3}(\rho_s - \rho)g \qquad (3.1)$$

where

$v(t)$ = the velocity of the liquid phase
$v_s(t)$ = the velocity of the solid particle
$\dot{v}$ and $\dot{v}_s$ = respective accelerations
ρ_s = the density of the solid particle
ρ = the density of the liquid phase
a = the particle radius
t_0 = the starting time
$\nu = \mu/\rho$ = the kinematic viscosity

In Eq. (3.1) the term on the left-hand side is the force required to accelerate the particle. The first term on the right-hand side represents the resultant of the pressures, a direct consequence of the fluid acceleration $\dot{v}$. The second term represents the surplus of inertia caused by the pressures resulting from the relative acceleration $(\dot{v}_s - \dot{v})$. This term is equivalent to the inertia of a virtual mass attached to the solid particle. The third one is a linear-resistance term, the *Stokes* term. This term as well as the next one is considered to be valid only for low Reynolds numbers. The fourth term is usually called the *Basset* term and gives the force due to the history of the particle; consequently, it is also called the *history* term. The fifth and last term is due to the gravitational force.

TCHEN (1947) in his derivation of Eq. (3.1) assumed a homogeneous fluid velocity field of infinite extent, no mutual action between particles, no particle rotation, and that the linear resistance, expressed with the relative motion, holds true. However it should be stated that Eq. (3.1) is not exact. Discussions of the approximations involved have been presented by CORRSIN et al. (1956), HINZE (1959), and SOO (1967).

3.2 MOTION WITH NONLINEAR RESISTANCE

The shortcoming of Eq. (3.1) is that it is applicable only at small Reynolds numbers. If the Reynolds numbers become larger, the resistance is proportional to the square of the relative velocity, or

$$R \propto (v_s - v)^2 \tag{3.2}$$

There is no theoretically sound approach for the derivation of the equation of motion at high Reynolds numbers, but if some basic assumptions in the derivation of Eq. (3.1) are disregarded, the slow-motion equation can be modified. The modification will be confined to the second, third, and fourth terms of Eq. (3.1).

The third term in this equation represents the drag at steady state. This steady-state drag has been the subject of many investigations in the past, and an equation,

$$R_1 = C_D a^2 \pi \frac{\rho(v_s - v)^2}{2} \tag{3.3}$$

has been established, which is at least experimentally well documented. C_D is called the steady-state drag coefficient and is in functional relationship with the particle Reynolds number, or

$$C_D = f\left[\frac{2a(v_s - v)}{\nu}\right] \tag{3.4}$$

The second term in Eq. (3.1) was labeled as the virtual-mass term. This term can be modified by multiplying the displaced volume of the fluid with a so-called *virtual-mass* coefficient. For a spherical particle, the virtual-mass coefficient k is 0.5, and we obtain the term as in Eq. (3.1). IVERSEN et al. (1951) pointed out that this coefficient depends on an acceleration parameter or

$$k = f\left[\frac{(\dot{v}_s - \dot{v})2a}{(v_s - v)^2}\right] \tag{3.5}$$

From dimensional analysis TCHEN (1947) obtained the same result. The modified resistance term becomes, therefore,

$$R_2 = k\tfrac{4}{3}\pi a^3 \rho(\dot{v}_s - \dot{v}) \tag{3.6}$$

The fourth term is the *Basset* term. This one presents the biggest problem, because at the moment we are unable to define it for high Reynolds numbers. In the literature, which is not very extensive, the term is considered by some as a low-Reynolds-number term, but more commonly it is altogether omitted. In the latter case, any importance it might have is probably carried over to the virtual-mass term. The equation of motion for a particle moving under the influence of gravity in a fluid not at rest at large-particle Reynolds numbers can thus be written as

$$\begin{aligned}\frac{4\pi a^3}{3}\rho_s \dot{v}_s = {} & \frac{4\pi a^3}{3}\rho \dot{v} - k\frac{4\pi a^3}{3}\rho(\dot{v}_s - \dot{v}) \\ & - C_D a^2 \pi \frac{\rho(v_s - v)^2}{2} - \frac{6\pi\mu a^2}{\sqrt{\pi\nu}}\int_{t_0}^{t} dt_1 \frac{\dot{v}_s(t_1) - \dot{v}(t)}{\sqrt{t - t_1}} \\ & - \frac{4\pi a^3}{3}(\rho_s - \rho)g\end{aligned} \tag{3.7}$$

A similar equation is suggested by TCHEN (1947).

3.3 CONCLUDING REMARKS

In the derivation of Eq. (3.1) the mathematical difficulties of integrating the general equation of motion forced us to exclude all terms involving squares and products of velocities. This appears to be quite legitimate, provided attention is confined to slow motion. No theoretical analysis substantiates at present the equation of motion for high Reynolds numbers, as given by Eq. (3.7). This equation was merely rationally deduced from Eq. (3.1), and thus it is an *ad hoc* equation.

Equations (3.1) and (3.7) should be good for ideal- and real-fluid flows as well as for steady- and nonsteady-state conditions. In Chap. 4 the steady-state motion of particle(s) will be discussed extensively. The non-steady-state motion of particle(s) will not be discussed in this book. The latter problem is rather involved and considered beyond the scope of the present discussion. Important knowledge to the nonsteady problem is contributed by TCHEN (1947), BASSET (1910), HUGHS et al. (1952), BRUSH et al. (1964), HJELMFELT et al. (1967), HINZE (1959), and other investigators.

4

Settling Velocity of Particles

The motion of a particle is assumed to be a steady-state one, and Eq. (3.7) reduces to

$$0 = 0 - 0 - C_D a^2 \pi \frac{\rho(v_s - v)^2}{2} - 0 - \frac{4\pi a^3}{3}(\rho_s - \rho)g \tag{4.1}$$

4.1 UNIFORM MOTION IN AN IDEAL FLUID

If the fluid is considered ideal, which implies a lack of frictional or viscous forces, the potential-flow theory is applicable. Potential-flow theory applied to a steady moving particle will predict no flow separation around this particle. This, in turn, implies that in front and after the particle the pressure will have the same magnitude and thus cancel. Therefore the resistance to a particle moving with constant velocity through an unbounded inviscid fluid is zero.

This result, which is obviously in contradiction to observations, is known as the *Euler-D'Alembert* paradox. KOTSCHIN et al. (1954) provide a mathematical proof. The paradox can be explained in such a way that flow around a particle without separation actually does not occur; vortices start to shed, changing the streamlines and the pressure distribution and accounting for the formation of boundary layers. The smallest amount of viscosity, which does always exist in a fluid, is therefore of great importance and is responsible for the observed resistance or drag force.

4.2 UNIFORM MOTION IN A REAL FLUID

For the case of a real or viscous fluid several scientists have presented us with analytical as well as more realistic theories.

As with so many problems of mechanics, it was Newton who goes on record here as having tackled the problem of drag for the first time. TRUESDELL (1960) writes:

> . . . Newton sets himself the problem of determining mathematically the nature of motion in resisting media. This, with some excursions, is the subject of Book II. To make any solid study of fluid resistance, Newton had first to learn the laws of motion.

In Book II Newton develops many a new concept, overwhelmingly ingenious but, in the end, unsatisfactory, and so gives scientists in the next centuries ideas and a vast field for research.

Let us see what Newton's drag formula really offers. Newton built himself a hypothetical model of a fluid. Whereas a fluid, according to modern concepts, is a continuous medium, Newton assumed it to be a discontinuous, rare medium, made up of a large number of small particles moving freely at equal and great distances from each other and, therefore, exerting no mutual influence. PRANDTL et al. (1934) summarize it as such when they write:

> The body moving through this medium experiences impacts from all the particles in its path and consequently imparts momentum to them. The total mass of all particles coming to impact with the body per second is $\rho A_p v$. This mass is given a velocity v_i which is proportional to the velocity v_s of the body. The amount of momentum created per second, which has to be equal to the resistance, or drag of the body, thus becomes

$$R = \rho A_p v_s v_i = f A_p v_s^2 \tag{4.2}$$

> The resultant momentum depends on the assumption of whether the impact is elastic or non-elastic. Experiments indicated a more or less non-elastic impact.

In Eq. (4.2) A_p stands for the projected area of the body or particle in the direction of flow.

It is quite generally agreed on nowadays that the drag, which is proportional to the dynamic pressure $\rho v_s^2/2$ is written as

$$R = C'_D A_p \frac{\rho v_s^2}{2} \tag{4.3}$$

The f in Eq. (4.2) and the C'_D in Eq. (4.3) are both factors of proportionality, drag coefficients as they are usually referred to, and are different for various particle shapes and other factors.

Newton also reasoned that the discrete fluid particles move in uniform rectilinear motion and, therefore, strike only the front part of the particle, while the sides and the tail end of the particle are left out of consideration.

Nevertheless, Newton's formula is correct and valid for the system he himself visualized, but such a system differs considerably from one of a true fluid. The major criticism to Newton's drag formula is a result of modern fluid mechanics as advanced by Prandtl, with the introduction of the boundary-layer concept. PRANDTL (1956) shows clearly that the entire surface of a particle is of importance and that, besides the very important pressure drag, the particle experiences also frictional drag. Whereas the former one depends largely on the form of the particle, therefore called *form drag*, the latter one is determined by the area of its surface and is referred to as *surface drag*. In other words, besides the dominating inertia forces the viscous forces have to be considered, no matter how small they might be. Investigations have shown that there does exist an intricate relationship between the inertia and viscous forces, and only under certain conditions does the problem lend itself to a mathematical theory. Therefore in such a problem one has to rely on numerous experiments, which must be performed in accordance with the principle of similarity. Geometrical and mechanical similarities exist only when the Reynolds numbers—this is the ratio of inertia to viscous forces—for two cases are the same, because only then the drag coefficients for the two cases are the same. This implies that a change in Reynolds number produces a change in the drag coefficient or, in other words, the drag coefficient is in functional relationship with the Reynolds number. This has been verified by numerous researchers. Making use of Newton's drag equation, given by Eq. (4.3), all complications of the viscosity effects are summarized in the drag coefficient, and we have

$$R = C_D A_p \frac{\rho v_s^2}{2} = f(\mathrm{Re}) A_p \frac{\rho v_s^2}{2} \tag{4.4}$$

where C_D is the steady-state drag coefficient and Re is the symbol used for the Reynolds number, defined as $\mathrm{Re} = 2a(v_s - v)/\nu$.

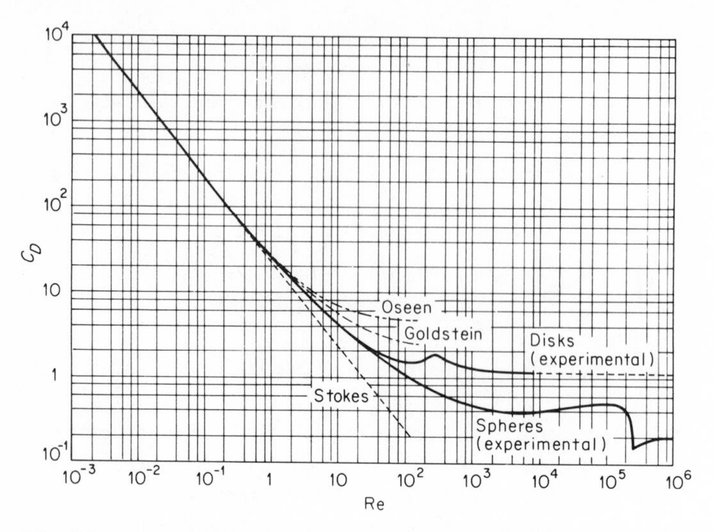

Fig. 4.1 Drag coefficients vs. Reynolds number for spheres and disks; experimental data. [*After* ROUSE (*1938*).]

From Eq. (4.4) the drag of a particle is easily calculated if the drag coefficient as a function of the Reynolds number for the particular shape of the particle is established. Figure 4.1 shows the resistance curve for a sphere; plotted on a logarithmic coordinate system is the drag coefficient C_D vs. the particle Reynolds number Re, defined by the ratio of $2av_s/\nu$. It should be stressed that this curve applies only to smooth, nonrotating spheres, moving in a fluid free of disturbances with a constant relative velocity. If any of the above-mentioned conditions are violated, deviations are bound to occur.

4.2.1 THEORETICAL CONSIDERATIONS FOR A SPHERE

4.2.1.1 Stokes' solution. At very small Reynolds number, $\text{Re} < 1$—this occurs if either the viscosity is very large (heavy oil) or the particle is very small (dust particle)—a deforming action takes place up to large distances from the particle. The particle pushes itself or creeps through a medium, and the influence of the viscosity forces on the drag becomes of greater importance than the influence of inertia. Stokes, in solving the general differential equation of Navier-Stokes, neglected the inertia terms completely. Assuming a very slow and steady moving sphere in an infinite liquid, Stokes obtained with the aid of a stream function a formula for the drag or

$$R = 6\pi\mu a v_s \tag{4.5}$$

This, in turn, gives a drag coefficient of

$$C_D = \frac{24}{\text{Re}} \tag{4.6}$$

Here we should remark that one-third of the total drag is caused by pressure differences and two-thirds of it is owing to frictional forces. That the drag in the Stokes range is proportional to the first power of the velocity is just another very interesting result. As can be seen from Fig. 4.2, Eq. (4.5) should be applied only when the Reynolds number is smaller than one-half ($\text{Re} < 0.5$), preferable only one-tenth ($\text{Re} < 0.1$).

4.2.1.2 Oseen-type solutions. OSEEN (1927) seems to have been the first who successfully included, at least partly, the inertia terms in his solution of the Navier-Stokes equation. Oseen points out that at large distances from the body the assumption does not hold that the acceleration terms are negligible with respect to the frictional forces. This approximate result for the drag coefficient is

$$C_D = \frac{24}{\text{Re}}\left(1 + \frac{3}{16}\,\text{Re}\right) \tag{4.7}$$

GOLDSTEIN (1929) provides a more complete solution for the Oseen approximation and obtains a drag coefficient of

$$C_D = \frac{24}{\text{Re}}\left(1 + \frac{3}{16}\,\text{Re} - \frac{19}{1{,}280}\,\text{Re}^2 + \frac{71}{20{,}480}\,\text{Re}^3 \cdots\right) \tag{4.8}$$

Reliable agreement with experimental data is obtained up to $\text{Re} = 2$, as can be seen in Fig. 4.2. According to Eq. (4.8) the drag formula changes gradually from a linear to a quadratic relationship with respect to the particle velocity.

After Oseen and Goldstein came many more scientists—SHANKS (1955), STEWARTSON (1955), VAN DYKE (1964), etc.—equipped with a good knowledge of mathematics and computers to improve and possibly extend their solution. To our knowledge none of these present us with a formula which predicts very accurately the drag coefficient beyond $\text{Re} = 2$.

4.2.1.3 Recent developments. The method that Prandtl applied for the formulation of the boundary-layer theory shed new light on the solution of the Navier-Stokes equation. Application of perturbation theory and, in particular, matching of asymptotic solution to the Navier-Stokes equation were suggested by PROUDMAN et al. (1957). Owing to the nature of the perturbation problem, i.e., singular perturbations, a modified perturbation

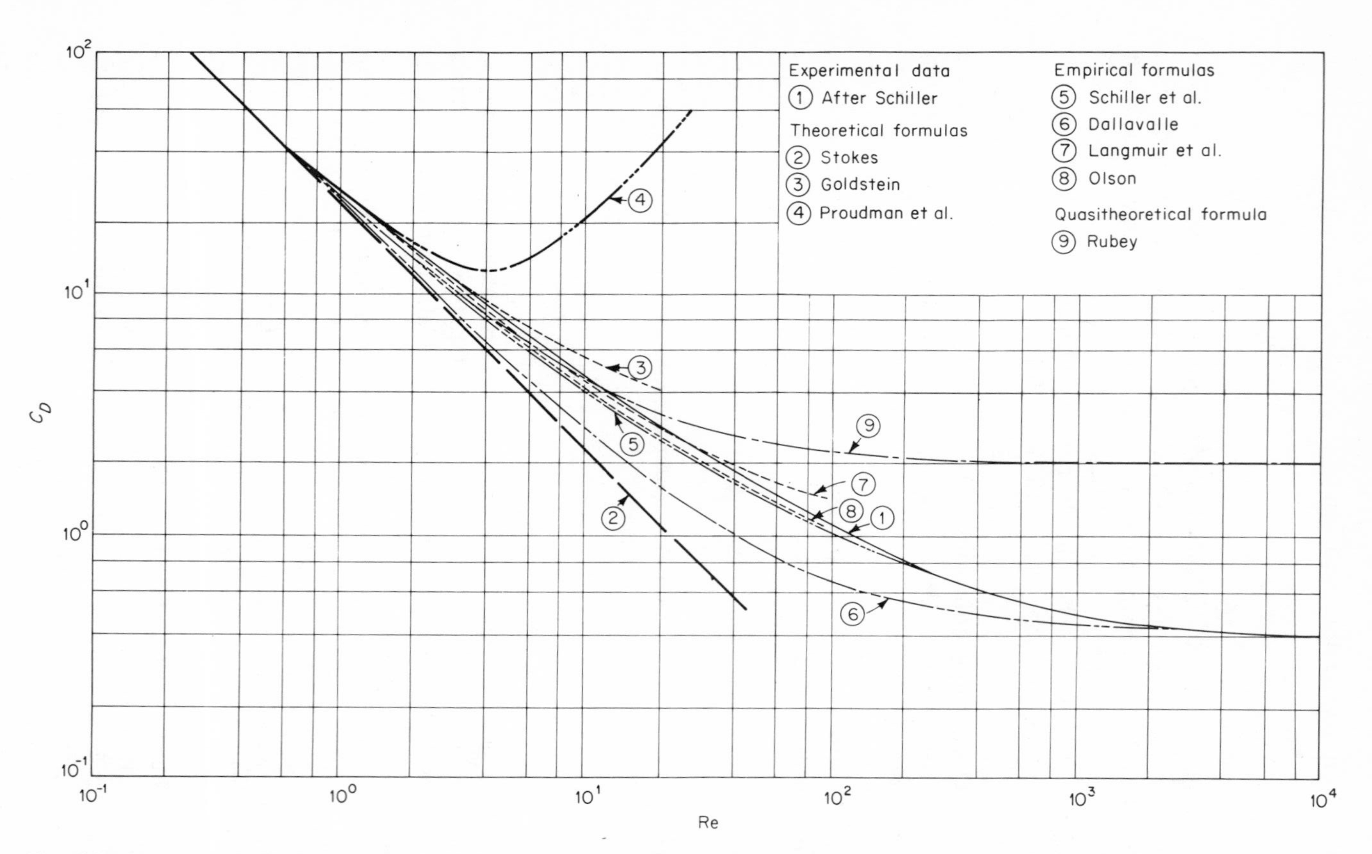

Fig. 4.2 Drag coefficient vs. Reynolds number for sphere; experimental data compared with formulas. [*After* GRAF *et al.* (*1966*).]

method must be used to solve the governing differential equation. This leads to asymptotic matching and gives a solution for small Reynolds numbers satisfying all boundary conditions. With this approach the foregoing authors used Stokes' solution as an inner and Oseen's solution as an outer one and found an overlapping region. The resulting equation for the drag coefficient is, then,

$$C_D = \frac{24}{\text{Re}}\left[1 + \frac{3}{16}\,\text{Re} + \frac{9}{160}\,\text{Re}^2 \log \text{Re} + O\left(\frac{\text{Re}^2}{4}\right)\right] \tag{4.9}$$

and is plotted in Fig. 4.2. It should be noticed that it diverges considerably from the experimental data at high Reynolds numbers.

Another approach to the problem is to solve numerically the complete Navier-Stokes equation by using relaxation techniques. JENSEN (1959) obtained in this way solutions for the drag coefficient at Reynolds numbers of Re = 5, 10, 20, and 40, which compare favorably with the experimental results. In such a way the numerical method can fill the gap between the few analytical studies and the more practical results obtained through experiments. With the advent of high-speed electronic computers, it has and will become possible to back up the presently available diverse empirical data.

It should be mentioned that FROMM (1963) gives numerical solutions—computer-simulated—for flow about obstacles in a channel for Reynolds numbers up to Re = 6,000 with emphasis given to the process of the development of the Karman vortex street. Reasonably good agreement with experimentally determined drag coefficients is reported.

4.2.1.4 Investigations at high Reynolds numbers. Whereas analytical solutions to the drag problem are only available at low Reynolds numbers ($\text{Re} < 2$), at larger ones only qualitative indications can be given. In this range the problem gets extremely complicated because separation of the flow in the aft of the particle starts to occur and vortices and wakes are shedding first periodically and then quite randomly. A boundary layer, which is created, is first laminar, while above a certain critical Reynolds number it suddenly becomes turbulent. Nevertheless, analytical reasoning is necessary to systemize, explain, and extrapolate experimental data which provide us with quantitative results. Beyond any doubt, Prandtl's boundary-layer theory presents the most powerful tool in explaining the high-Reynolds-number drag phenomenon. For an extensive discussion of the problem the interested reader should consult "Applied Hydro & Aeromechanics" by PRANDTL et al. (1934) and "Boundary Layer Theory" by SCHLICHTING (1968).

4.2.2 EMPIRICAL EQUATIONS FOR THE DRAG COEFFICIENT OF A SPHERE

In the foregoing it was shown that at $\mathrm{Re} > 2$ the drag coefficient curve, given by Fig. 4.1, has to be used. To simplify this procedure several investigators have established empirical relationships which are applicable only under certain specified conditions.

Many equations have been proposed, but only a few are mentioned here, and in Fig. 4.2 they are compared with experimental data.

SCHILLER et al. (1933) suggest a formula that gives good results for $\mathrm{Re} < 800$, or

$$C_D = \frac{24}{\mathrm{Re}}\,(1 + 0.150\ \mathrm{Re}^{0.687}) \tag{4.10}$$

DALLAVALLE (1943) in his book "Micromeritics" feels that the complete span of the experimental curve can be represented with a fair degree of accuracy by the expression

$$C_D = \frac{24.4}{\mathrm{Re}} + 0.4 \tag{4.11}$$

According to TOROBIN et al. (1959) an equation suggested by Langmuir et al., which reads

$$C_D = \frac{24}{\mathrm{Re}}\,(1 + 0.197\ \mathrm{Re}^{0.63} + 0.0026\ \mathrm{Re}^{1.38}) \tag{4.12}$$

gives accurate results between $1 < \mathrm{Re} < 100$. OLSON (1961) suggests that the drag coefficient can be well represented by an equation, for $\mathrm{Re} < 100$, such as

$$C_D = \frac{24}{\mathrm{Re}}\left(1 + \frac{3}{16}\,\mathrm{Re}\right)^{1/2} \tag{4.13}$$

The reader is invited to judge for himself from Fig. 4.2 how well Eqs. (4.10) through (4.13) check. For very accurate results the experimental data given by Fig. 4.1 will have to be used, but for many engineering problems an empirical equation chosen from the foregoing will provide sufficiently accurate answers.

A quasitheoretical equation is suggested by RUBEY (1933). Rubey feels that "Stokes' law and the impact formula may be combined very simply into a general equation . . ."; thus an equation of the form

$$C_D = \frac{24}{\mathrm{Re}} + 2 \tag{4.14}$$

is obtained. Equation (4.14) is plotted in Fig. 4.2 and shows considerable deviations from experimentally observed values, especially at high Reynolds numbers. A critical discussion of Rubey's equation is given by GRAF et al. (1967).

4.2.3 TERMINAL SETTLING VELOCITY

Under the influence of gravity a spherical particle will ultimately attain a uniform velocity. This was already expressed in Eq. (4.1) or

$$C_D a^2 \pi \frac{\rho (v_s - v)^2}{2} = \frac{4a^3 \pi}{3} (\rho_s - \rho) g \tag{4.1}$$

where the hydrodynamic forces are counterbalanced by the gravitational forces. For the drag coefficient we can introduce either the experimental finding given by Fig. 4.1 or any of the equations applicable for the range of investigation and its desired accuracy.

To find the terminal velocity $(v_s - v)$ for a given spherical particle of a density ρ_s dropped in a fluid of known properties, SCHILLER et al. (1933) multiplied Eq. (4.1) by $(2a/\nu)^2$ and obtained

$$C_D \mathrm{Re}^2 = \frac{4}{3} g \frac{\rho_s - \rho}{\rho} \frac{(a/2)^3}{\nu^2} \tag{4.15}$$

The right-hand side of Eq. (4.15) is computable and such a value for the left-hand side is obtained. With the help of a curve, plotting $(C_D \mathrm{Re}^2)$ vs. Re and given in Fig. 4.3, the Reynolds number is obtained and, in turn, the desired settling velocity is found.

Suppose that the diameter of a spherical particle is in question and this particle was dropped with a uniform velocity in a liquid of known properties. In this case it is advised to multiply Eq. (4.1) with $(v_s - v)/\nu$ or

$$\frac{C_D}{\mathrm{Re}} = \frac{4}{3} g \frac{\rho_s - \rho}{\rho} \frac{\nu}{(v_s - v)^3} \tag{4.16}$$

The right-hand side of the Eq. (4.16) is completely known, and by consulting Fig. 4.3, where C_D/Re vs. Re is plotted, the particle Reynolds number and, in turn, the particle diameter are obtained.

It should be remarked that Eqs. (4.15) and (4.16) or Fig. 4.3 represent a general explicit approach for the terminal settling velocity and the particle diameter, respectively.

For the range of Reynolds number where Stokes' resistance holds true, the solution of Eq. (4.1) simplifies, and we obtain for the terminal settling velocity

$$v_s - v = \frac{2}{9} a^2 g \frac{\rho_s - \rho}{\mu} \tag{4.17}$$

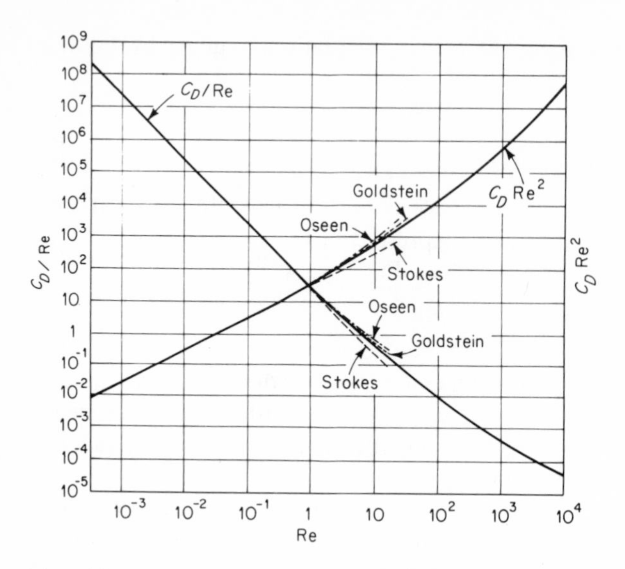

Fig. 4.3 C_D/Re and $C_D\text{Re}^2$ vs. Reynolds number for sphere; for computation of the settling velocity and 'the particle diameter. [*After* SCHILLER *et al.* (*1933*).]

In most of the problems the hydraulician is concerned with the determination of the settling velocity of quartz particles with a density $s_s = 2.65$ in water at a temperature of 20°C. This simplifies the foregoing equations considerably; a plot of relative velocity $(v_s - v)$ vs. particle diameter $2a$ is given in Fig. 4.4.

4.2.4 COMPLICATING EFFECTS ON THE UNIFORM MOTION OF A SPHERE

Up to now we have considered the uniform motion of a single spherical particle through a fluid extending to infinity and free of disturbances. In the following, the various effects which complicate the problem, such as particle shape, boundary conditions, multiparticle influences, particle rotation and roughness, turbulence, and combinations of these, will be examined.

These effects shall be expressed such that Eq. (4.5), derived and applicable for the Stokes range, will be multiplied by a correction factor K, or

$$R = K(6\pi\mu a v_s) \tag{4.18}$$

This correction factor has to be selected in such a way that its value is unity for a sphere under the assumptions of Stokes' derivation. Deviation from unity or $K \neq 1$, results therefore from the departure of the assumptions. McNOWN (1951) suggests to refer to the dimensionless quantity K as a Stokes number.

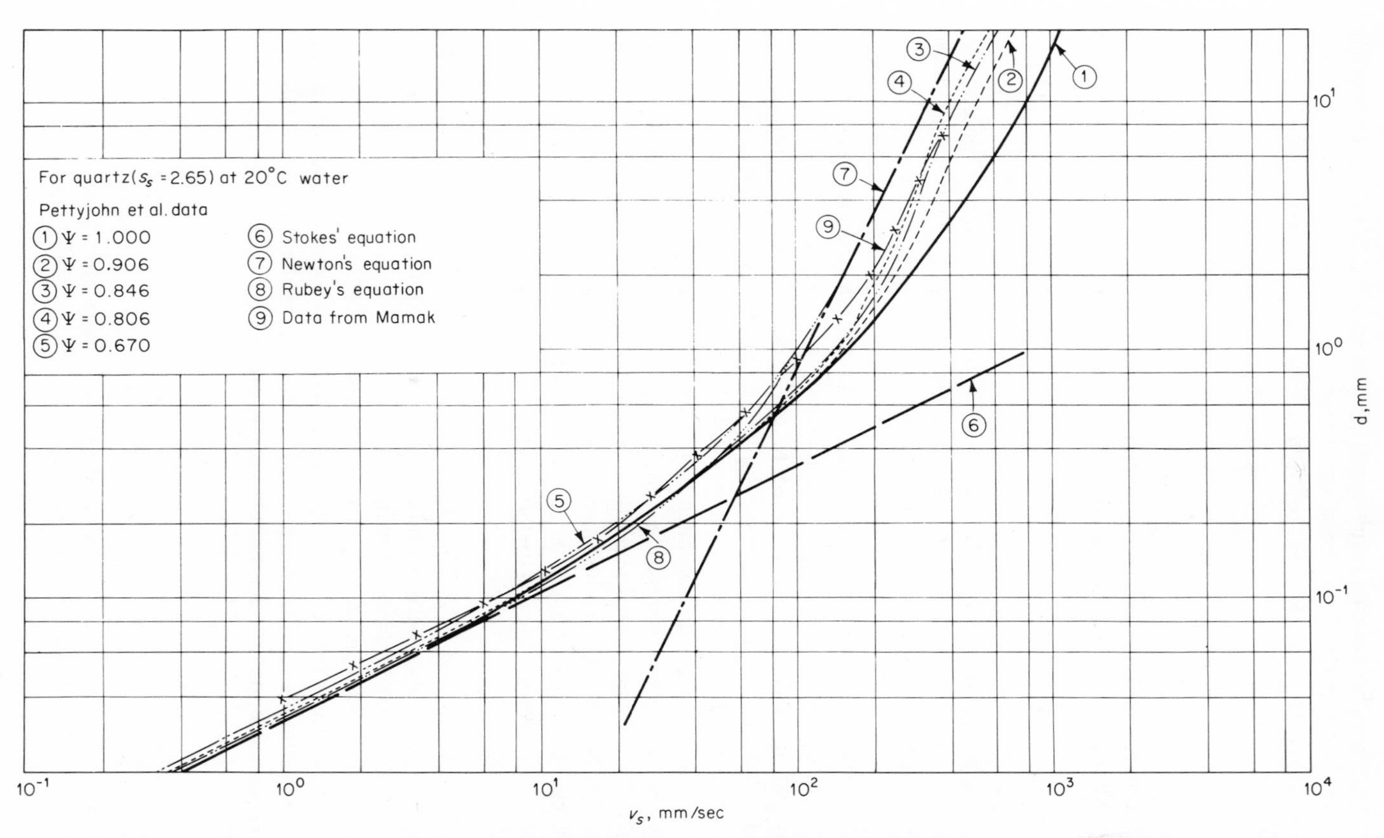

Fig. 4.4 Settling velocity vs. particle diameter; various equations and shape-factor parameters Ψ. [*After* Graf *et al.* (*1966*).]

Incidentally, Oseen's result can also be expressed with this correction factor K. OSEEN (1927) included some inertia effects and, hereby, obtained an improvement over Stokes' solution. Rewriting Eq. (4.7) with this in mind, we obtain

$$K = 1 + \frac{3}{16}\,\mathrm{Re} \tag{4.19}$$

The same is true for other inertia correction studies.

Above the Stokes' range the particle drag is expressed by Eq. (4.4), or

$$R = C_D A_p \frac{\rho v_s^2}{2} \tag{4.4}$$

with C_D being the steady-state drag coefficient. It is common practice to have this coefficient absorb the various complicating effects rather than multiplying the coefficient with another one.

4.2.4.1 Particle shapes. The shape of a solid particle can exert a profound influence on the drag. It is therefore not surprising that extensive investigations have been conducted to get a better understanding of this effect. Whereas geometrically regular shapes have been studied analytically and experimentally, irregular shapes, like those found with natural sand grains, are investigated experimentally only. The available analytical solutions, however, are limited to the low-Reynolds-number range and are excellently derived and discussed in HAPPEL's et al. (1965) book, "Low Reynolds Number Hydrodynamics." In the following we shall discuss first such regular shapes as might be considered to be approximations for natural grains. Afterwards the discussion will focus on irregular shapes and natural sand grains.

4.2.4.1.1 REGULAR SHAPES. *Cylinder* (circular). The relationship of the drag coefficient and the Reynolds number for an infinite ($L/d = \infty$) and a finite ($L/d = 5$) cylinder is given in Fig. 4.5. It can be seen that the relation is strikingly similar to the one of the sphere given by Fig. 4.1.

Again, as in the case of a sphere, a theoretical solution is available for the low-Reynolds-number ranges. This result, presented by LAMB (1945), was obtained under the assumption of Oseen's approximate condition. A correlation coefficient K for the resistance per unit length is suggested as

$$K = \frac{1}{2.0 - \log \mathrm{Re}} \tag{4.20}$$

Beyond the Stokes range the drag coefficients have to be determined experimentally. From Fig. 4.5 it can be seen that a finite cylinder causing

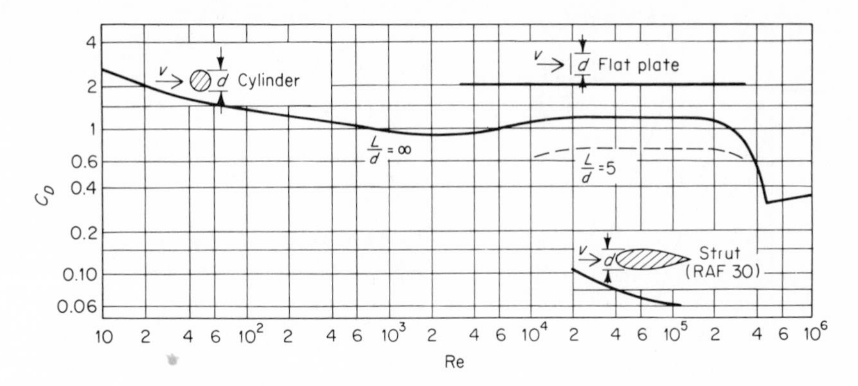

Fig. 4.5 Drag coefficient for circular cylinders, flat plate, and streamlined strut.

three-dimensional flow has a lower drag coefficient if compared with an infinite one. At a certain Reynolds number the flow around a cylinder produces vortices of first stable and then unstable nature. These vortices have been called Karman trails and are discussed in PRANDTL et al. (1934). For the sake of comparison the drag coefficients for flat plates and streamlined struts, both of finite length, have been included into Fig. 4.5.

Ellipsoids. In the Stokes range MCNOWN et al. (1950) presented[1] a solution to the equations for slow motion of an ellipsoidal particle. This investigation produced the following correction coefficient to be used with Eq. (4.18):

$$K = \frac{16}{3d_N\zeta} \tag{4.21}$$

where d_N is a nominal diameter, defined as $d_N = 2\sqrt[3]{abc}$, with a, b, and c being the semiaxes of the ellipsoid, and where ζ ought to be expressed with elliptical functions and depends on the relative magnitude of a, b, and c. The graphical evaluation of Eq. (4.21) is shown in Fig. 4.6, where the variation of the correction coefficient K with the two arguments $a/\sqrt{bc}$ and b/c is plotted. From observation of Fig. 4.6 we might conclude that a considerable variation of K does exist and that the ratio of the principal axes appears to be a significant parameter. MCNOWN et al. (1951) conducted experiments with ellipsoidal and nonellipsoidal particles, the latter being symmetrical with respect to each of three mutually perpendicular planes. The results of these studies are compared with Eq. (4.21) in Fig. 4.6. MCNOWN et al.

[1] The solution of the problem is credited to LAMB (1945).

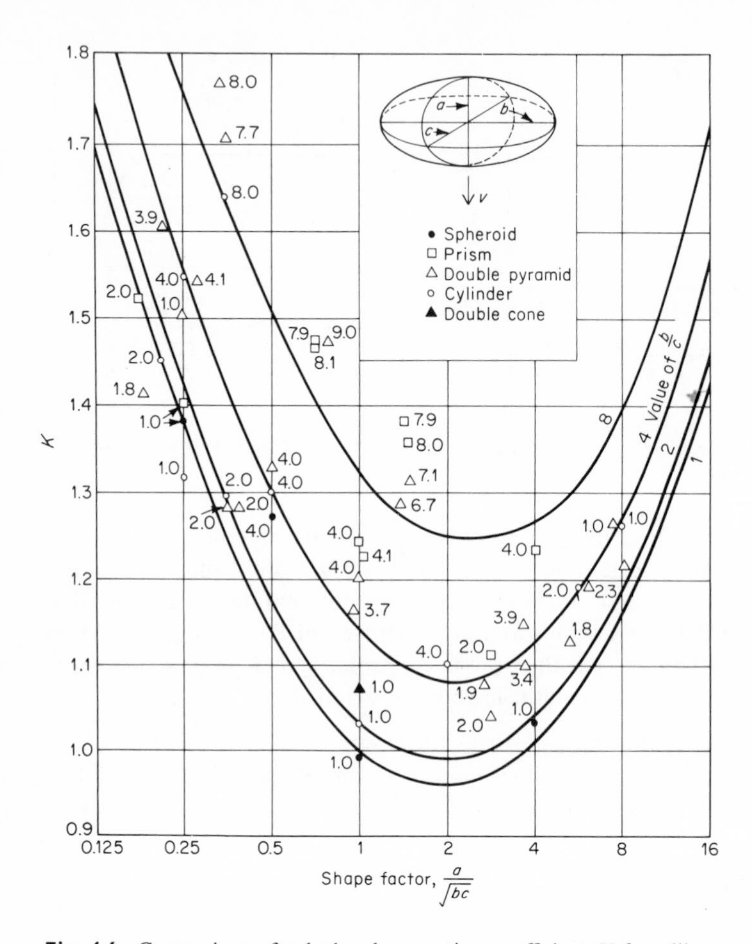

Fig. 4.6 Comparison of calculated correction coefficient K for ellipsoidal particles with experimental values at low Reynolds numbers. [*After* McNown *et al.* (*1951*).]

(1950) feel that the drag coefficient over a wide range of shapes can be estimated within 10 percent from the theoretical results for ellipsoids if the principal-axes ratio is determined.

McNown et al. (1951) extended their experimental study to Reynolds numbers as large as 10^3. The results are plotted, even for larger Reynolds numbers, in terms of the correction coefficient K rather than of the drag coefficient C_D. This was felt necessary because the K value in this intermediate range makes the gradual transition more evident. Figure 4.7 shows a summary of the experimental work. The Reynolds number, defined with the nominal diameter d_N, is plotted against the correction coefficient from

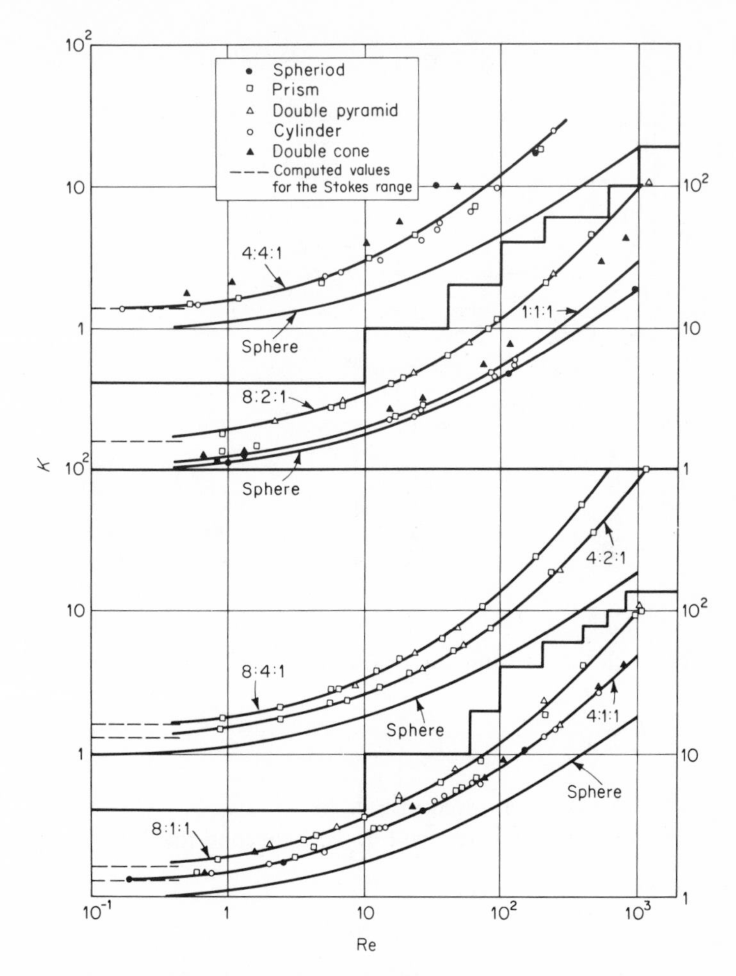

Fig. 4.7 Variation of correction coefficient K vs. Reynolds number for ellipsoidlike shapes. [*After* McNown *et al.* (*1951*).]

Eq. (4.18); the axis ratio $a\colon b\colon c$ is the parameter, and the particles are oriented such that they fall in a direction parallel to their shortest axes.

Disks. McNown (1951) presents the resistance of a disk as a limiting case of the ellipsoids. Quite generally, the correction coefficient can be summarized in a plot as shown in Fig. 4.8; depending on the orientation of the disk to the flow, a different value is obtained. The limiting value of K for a circular disk broadside-on is

$$K = \frac{8}{3\pi} \tag{4.22}$$

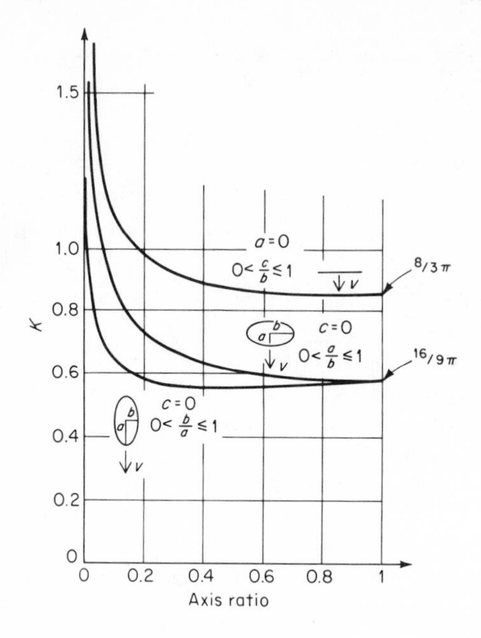

Fig. 4.8 Variation of correction coefficient K with axis ratio for a disk. [*After* McNown (*1951*).]

whereas for a disk moving edgewise,

$$K = \frac{16}{9\pi} \tag{4.22a}$$

Both values are indicated in Fig. 4.8.

Oseen (1927) extended Eq. (4.22) beyond the Stokes range and obtained

$$K = \frac{8}{3\pi}\left(1 + \frac{\text{Re}}{2\pi}\right) \tag{4.23}$$

More recently Michael (1966) obtained solutions to the disk motion from Reynolds numbers 1.5 to 50, with very good agreement with the available experimental data. By numerical iteration the difference approximation to the Navier-Stokes equation was solved. Generally speaking, for high Reynolds numbers only experimental data are available, as given by Fig. 4.1.

It is evident that every particle having unequal axes and being dropped freely gets a chance to reorient itself. Within the Stokes range any orientation is stable; whereas for larger Reynolds numbers Graf (1965) showed that an elongated body will tend to set itself broadside to the relative motion.

Furthermore, Marchillon et al. (1964) point out that at higher Reynolds numbers (Re $\approx$ 1,000) an oscillatory motion accompanies the translation of the particle. The effect of this secondary motion on the drag coefficient can be significant and was found to depend on the amplitude of oscillation of the particle and the particle density.

Isometric particles. PETTYJOHN et al. (1948) did an extensive investigation with isometric particles where all edges are of equal length. Working in a wide range of Reynolds numbers, a correlation with the drag coefficient and a shape factor as a parameter was obtained. For the definition of the shape factor a concept developed by Wadell was borrowed. WADELL (1934) expressed the ratio of the surface area of a sphere A_S of the same volume as the particle to the actual surface area of the particle A_{SN}, and referred to it as "degree of true sphericity." The sphericity of particles given with $\Psi = A_S/A_{SN}$ and investigated by PETTYJOHN et al. (1948) is given in Table 4.1.

It is suggested that in the Stokes range the experimental data can be represented by

$$K = 0.843 \log \frac{\Psi}{0.065} \tag{4.24}$$

within about 1 percent. At higher Reynolds numbers it is advised that the curves

$$C_D = f(\text{Re}, \Psi) \tag{4.25}$$

be used. For the special case of quartz particles dropping in a water environment of 20°C Eq. (4.25) is plotted in Fig. 4.4. Pettyjohn et al. in evaluating the shape effect also studied Heywood's shape factor. HEYWOOD (1938) suggests using a diameter equal to the diameter of a circle having an area equal to that of the projected area of the particle. However, this shape factor seemed to provide a poor criterion, especially at higher Reynolds numbers. More recently HEYWOOD (1962) presented a modified approach, including the flatness and elongation of the particle.

Drag coefficients for many more regularly shaped particles or objects (approximate sphere, oblate and prolate spheroid, elongated rod, etc.) may be found in HAPPEL et al. (1965), if the motion is within the Stokes ranges,

Table 4.1 Particle shape expressed by Wadell's sphericity and Heywood's shape-factor concept [*after* PETTYJOHN *et al.* (*1948*)]

Particle shape	*Wadell's sphericity,* $\Psi = A_S/A_{SN}$	*Heywood's shape factor*
Spheres	1.000	0.52
Cube-octahedron	0.906	0.58
Octahedron	0.846	0.41
Cube	0.806	0.68
Tetrahedron	0.670	0.29

and in HOERNER (1958) for the entire Reynolds-number region. Also treated in the book "Low Reynolds Number Hydrodynamics" by HAPPEL et al. (1965) is the quite general problem, entitled "The motion of a rigid particle of arbitrary shape in an unbounded fluid."

4.2.4.1.2 IRREGULAR SHAPES. One way to obtain a suitable drag coefficient for an irregular-shaped particle is to approximate it with a regular-shaped one of which the drag coefficient is obtainable. However, there have been some investigators who focused their interest especially on the determination of the resistance of irregular shapes as well as of natural sediment particles.

ALBERTSON (1953), studying the shape effect of gravel particles on the settling velocity, finds it suitable to use a shape factor suggested by MCNOWN et al. (1950) such as

$$\text{Shape factor} = \frac{a}{\sqrt{bc}} \tag{4.26}$$

with b being the maximum axis. The other shape parameter used by McNown was found to be of insignificant influence. Plots of the drag coefficient vs. Reynolds number, using the nominal diameter, for river gravel and crushed gravel were obtained as shown in Fig. 4.9. It was found that the suggested shape factor correlated rather well, but that lines of constant shape factor do not render the same results for river and crushed gravel. The Subcommittee on Sedimentation of the Inter-Agency Committee on Water Resources (1957) nevertheless suggests the use of a shape factor, as given by Eq. (4.26).

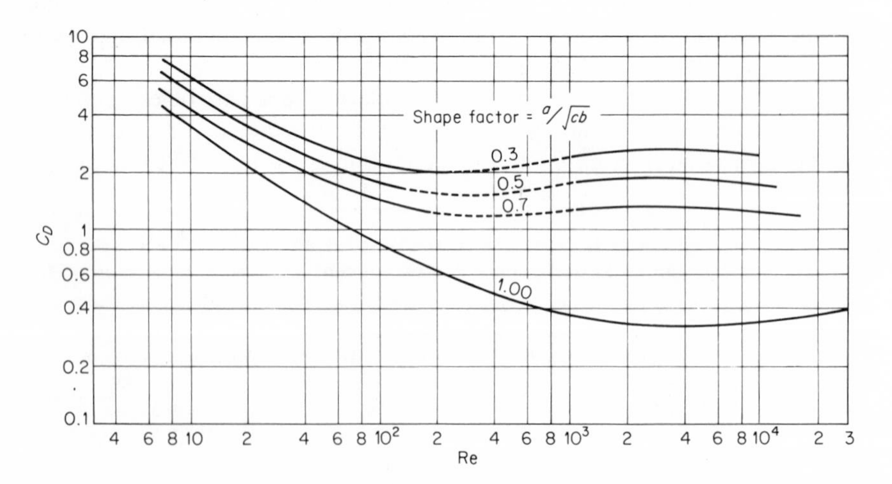

Fig. 4.9 Drag coefficient vs. Reynolds number for different shape factors. [*After* ALBERTSON (*1953*).]

HEYWOOD (1962) represents the shape effects by introducing a volume coefficient k_H expressed by

$$\text{Volume of particle} = k_H d_a^3 \tag{4.27}$$

where d_a is the diameter of the circle that has the same projected area as the particle when viewed in a direction perpendicular to the plane of greatest stability. The value of $k_H = 0.524$ for a spherical, 0.696 for a cube, and 0.328 for a tetrahedron; for most mineral particles it averages 0.25 to 0.20. Furthermore, Heywood shows an intricate relation of k_H with flatness and elongation ratio of the particle; nevertheless, in some cases, an experimental determination of the volume coefficient k_H is unavoidable.

GRAF et al. (1966) investigated the settling of natural quartz particles in water. Together with data reported by MAMAK (1964) these data are given in Fig. 4.4, where an average particle diameter is plotted against the settling velocity. From Fig. 4.4 it can be shown that the settling of natural sediment grain is neither well represented by the spherical data nor by the too popular Rubey equation given with Eq. (4.14).

4.2.4.2 Various boundaries. In most practical cases the fluid is externally bounded by either rigid walls or by a free surface, or sometimes by both. The modifications due to these influences will be discussed for several simple cases of interest.

The slow motion of a spherical particle toward or away from a single *plane surface* was studied by BRENNER (1961). Two distinct cases were investigated. If the plane surface is rigid, which is the case when a particle is approaching the bottom of a container, the correction factor in the Stokes equation [Eq. (4.18)] is:

$$K = 1 + \frac{9}{8}\frac{a}{s} \tag{4.28}$$

where s is the instantaneous distance of the center of the particle from the plane surface, and is shown in Fig. 4.10*a*. Brenner suggests the use of Eq. (4.28) only when the particle is far away from the wall, $a/s > 15$, but does provide the exact solution at the same time.

In the second case under consideration the plane toward which the sphere falls is a planar-free surface, where tangential stresses and normal velocity vanish. This case corresponds to the interface between two liquids. Brenner presents the complete solution, but the approximate one is given by

$$K = 1 + \frac{3}{4}\frac{a}{s} \tag{4.28a}$$

Experimental confirmation of the foregoing equations is reported by HAPPEL et al. (1965). It should be pointed out again that Eqs. (4.28) and

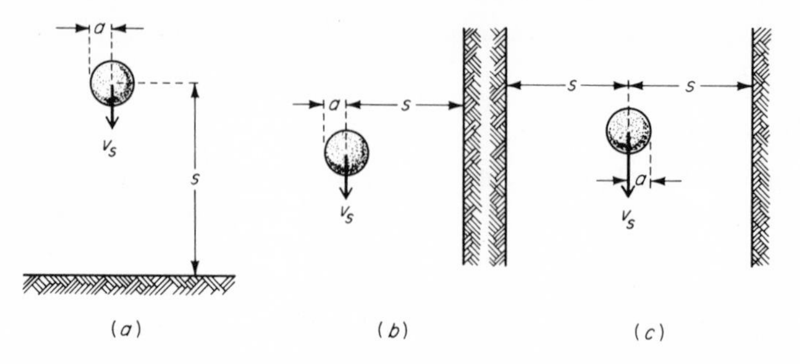

Fig. 4.10 Definition sketch for particles moving within various boundaries.

(4.28*a*) are solutions to the Stokes case. At the present there is no evidence that at higher Reynolds numbers any experiments are performed.

McNown (1951) reviews the slow-motion problem of a spherical particle settling close to a single wall as well as between two vertical walls, as shown in Fig. 4.10*b* and *c*. For a single vertical wall the expression becomes

$$K = 1 + \frac{18}{32}\frac{a}{s} \tag{4.29}$$

as a first approximation. If the particle falls midway between two parallel plane walls, a correction coefficient of

$$K = 1 + 1.006\frac{a}{s} \tag{4.30}$$

is suggested. A solution for a particle settling at quarter point, now subject to rotation, is available in Happel et al. (1965).

The case of a spherical particle moving in axial direction in a *circular cylindrical* tube is of considerable importance. In other words, it is frequently desired to determine the correction to be applied for the retarding effect of the walls on the settling particle. Although most of the literature relating to this problem is for the Stokes range, there exists also some information, mainly experimental, beyond the Stokes range.

Happel et al. (1965) solve the slow-motion problem where the sphere may occupy any preassigned position within the cylinder. For the simple case of the spherical particle settling in the center of the cylinder, their solution checks with the Faxen equation, or

$$K = 1 + 2.1\frac{a}{R_{cyl}} \tag{4.31}$$

where R_{cyl} is the radius of the circular cylinder. Experiments show that Eq. (4.31) should only be applied for the case of $a/R_{cyl} < 0.1$. However,

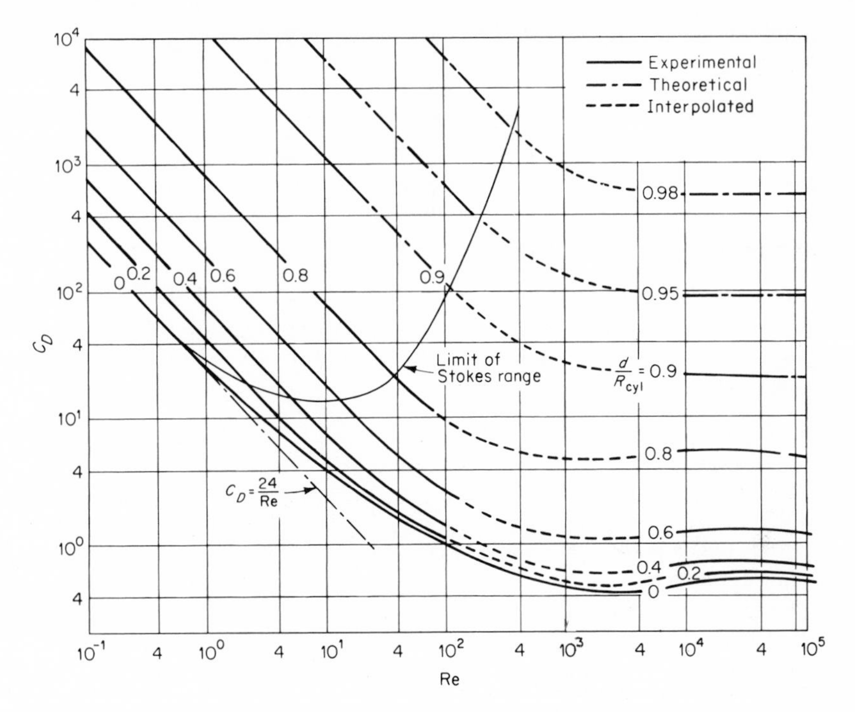

Fig. 4.11 Drag coefficient vs. Reynolds number with a/R_{cyl} as parameter; studying the influence of cylindrical tubes. [*After* McNown *et al.* (*1951a*).]

McNown et al. (1948) give an approximate theory, applicable for low Reynolds numbers, where a/R_{cyl} goes toward unity. Under these conditions the correction coefficient results in

$$K = \frac{(3\pi\sqrt{2})}{8}\left(1 - \frac{a}{R_{cyl}}\right)^{-5/2} \tag{4.32}$$

which is plotted in Fig. 4.11.

At higher Reynolds numbers the investigations are based primarily on experimental research, but McNown et al. (1951*a*) give, with simple fluid mechanics reasoning, the following equation for the drag coefficient:

$$C_D = \frac{a/R_{cyl}}{1 - (a/R_{cyl})^2} \tag{4.33}$$

Equation (4.33) was established under assumptions to be more likely fulfilled at large ratios of a/R_{cyl} and fairly large Reynolds numbers. The applicability of both Eqs. (4.32) and (4.33) can be seen in Fig. 4.11. Experimental data, provided by Fidleris et al. (1961), justify the general trend as suggested by McNown et al. (1951*a*) and by research given in Fig. 4.11.

4.2.4.3 More than one sphere. If more than a single sphere moves through an unbound fluid system, a mutual interaction will be noticed. It has been observed that the settling velocity increases if only a few closely spaced particles move, and that the fall velocity is reduced, i.e., the drag increased, if many particles are dispersed throughout the fluid.

Few closely spaced particles. In a paper by Eveson et al. (1959) the interaction of two equally sized spherical particles moving within the Stokes range is investigated. The theoretical solution by Smoluchowski gave a relationship such as

$$K = 1 - \frac{3}{8}(1 + \sin^2 \beta_s)\frac{d}{s} + \frac{9}{64}(1 + 3\sin^2 \beta_s)\left(\frac{d}{s}\right)^2 \tag{4.34}$$

where β_s is the angle between the line of centers of the sphere and the horizontal, and s is the distance between sphere centers. The experiments gave a fair agreement with Eq. (4.34) for various angles β_s. In all cases a mutual interaction was noticed, causing a decrease in the resistance, as well as the fact that there was no rotation of the spheres.

The effects on the drag coefficient caused by the interaction of unequisettling spheres, which also may be unequisized, are extensively studied in Happel et al. (1965). In the same treatise appears a survey of formulas predicting the resistance due to more than two spheres.

No information seems to be available for the determination of the resistance above the Stokes range.

Dispersed particles (concentrations). Maude et al. (1958) present a most extensive study on this topic, commonly referred to as *hindered settling*. They feel that regardless of the Reynolds number, an equation of the form of

$$K = (1 - C)^{-m} \tag{4.35}$$

can be used, where C is the concentration per volume of the solid particles and m is a coefficient, depending among others on the Reynolds number and the particle shape. Comparing Eq. (4.35) with experimental data, some of them obtained with spheres, others with other shapes, a variation of m with Reynolds numbers was expressed. For $\text{Re} < 1$, $m \approx 4.5$, while at $\text{Re} > 10^4$, $m \approx 2.2$, with a transition between these Reynolds numbers. For dilute suspensions Happel et al. (1965) discuss various models; the results are best summarized with the equation

$$K = 1 + hC^{1/3} \tag{4.36}$$

with h varying between 1.30 and 1.91, depending on the mathematical model. In a critical survey of equations and their experimental justification, Happel et al. (1965) conclude that agreement between the two was fairly poor. Even with the simplest model, the one of uniformly sized spherical particles, other variables than the concentration C, such as particle circulation or particle aggregation and segregation, have to be considered.

FIDLERIS et al. (1961*a*) study the resistance of a spherical particle moving through a stable suspension up to $\mathrm{Re} = 10^4$. They conclude that with fine suspensions ($<422\ \mu$) no deviations are noticeable on a drag coefficient vs. Reynolds-number plot. In the Stokes range, even with coarse suspension, no deviation is reported; however, at turbulent flow conditions the experimental points deviate from the pure-liquid curve. The size of the settling particle influences these deviations, reaching a maximum for spheres of the order of the size of the suspension particle. Furthermore, it is worthwhile to point to studies by DUPLESSIS et al. (1967) and VALENTIK et al. (1965), who investigate particle settling in a Bingham plastic fluid.

4.2.4.4 Particle roughness. The roughness of a spherical particle influences the resistance in two ways: it changes the value of the drag coefficient C_D and it alters the transition of the laminar-turbulent boundary layer.

This transition, a very peculiar phenomenon and extensively discussed by PRANDTL et al. (1934), takes place in the region of Reynolds numbers between $2.5 \cdot 10^5$ to $5 \cdot 10^5$ for a smooth sphere, as can be seen in Fig. 4.1, and is accompanied by a considerable reduction of the drag coefficients. These critical Reynolds numbers[1], Re_{cr}, might also be considered as a stability criterion, and the slightest change—e.g., in roughness, in rotation of the sphere, or in the turbulence level—might alter their value.

HOERNER (1958) summarizes some experimental studies in Fig. 4.12. This graph suggests that the roughness of a sphere causes an earlier transition or lower critical Reynolds numbers. The rougher the surface, the earlier will the transition take place. Below certain Reynolds numbers the influence of the roughness on the drag coefficient seems to disappear. Because of the difficulty in defining roughness, a quantitative treatment of this effect has so far been impossible.

4.2.4.5 Particle rotation. Particles might start to rotate if a velocity gradient exists in the fluid through which they settle, owing to the effect of a boundary or to a collision with a boundary. It has been noticed that such rotation does alter the drag coefficient slightly.

In the Stokes range BRENNER (1962) presents a theory for a solid particle rotating symmetrically in a bounded fluid. No experimental data appear to exist to check this theory.

[1] By convention the critical Reynolds number, Re_{cr}, is taken at the point where the resistance coefficient of $C_D = 0.3$ intersects with the data curve.

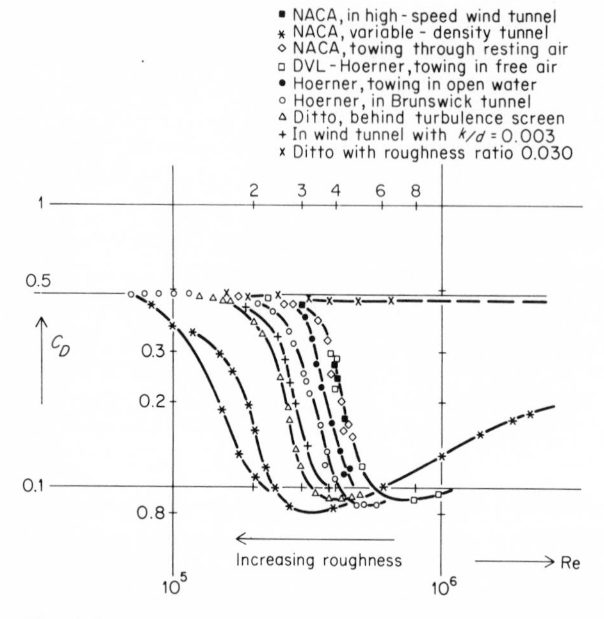

Fig. 4.12 Drag coefficient vs. Reynolds number as influenced by the spherical particle roughness. [*After* HOERNER (*1958*).]

SCHLICHTING (1968) reports on a study by Luthander et al. who investigate a rotating sphere. From Fig. 4.13 it can be seen that in the critical Reynolds-number region the rotation, expressed by the rotation parameter $a\omega/v$, caused an earlier laminar-turbulent transition.

Other pertinent literature is reviewed in TOROBIN et al. (1960*a*).

4.2.4.6 Turbulence effects. A most important aspect in evaluating the resistance experienced by a particle settling in a fluid is whether or not free-stream turbulence does prevail. Up to now it has been assumed that a turbulent-free flow exists, but in most real systems such ideal conditions are not encountered.

The effects of free-stream turbulence have been studied mainly experimentally. After surveying many research contributions TOROBIN et al. (1960) conclude that the resistance due to these effects depends primarily on the magnitude of the relative turbulence intensity and on the particle Reynolds number, while acceleration plays apparently a minor role.

The following discussion focuses first on experimental data obtained in the transition region and then on the ones below the transition.

4.2.4.6.1 IN THE TRANSITION REGION. Interesting and famous is the influence of turbulence on the critical Reynolds number. SCHILLER (1932)

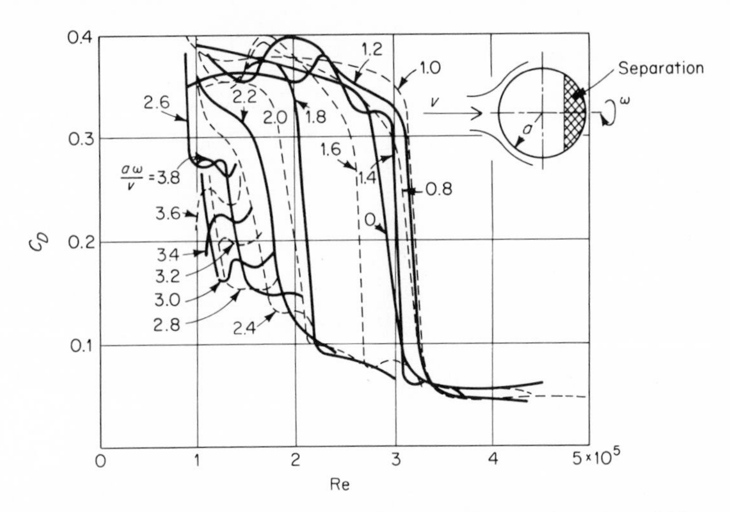

Fig. 4.13 Drag coefficient vs. Reynolds number for rotating sphere. [*After* SCHLICHTING (*1968*).]

gives an excellent account as to what might be referred to as the *Eiffel-Prandtl debate*. In short, Eiffel's and Prandtl's most carefully performed experiments showed serious disagreement for the value of the drag coefficient at the critical Reynolds number. However, Prandtl was eventually able to show that the data obtained by Eiffel were subject to free-stream turbulence, which causes the transition to occur at lower Reynolds numbers. This is presented in Fig. 4.14, which in some ways resembles Fig. 4.12 where the influence of the particle roughness was considered. Oversimplified,

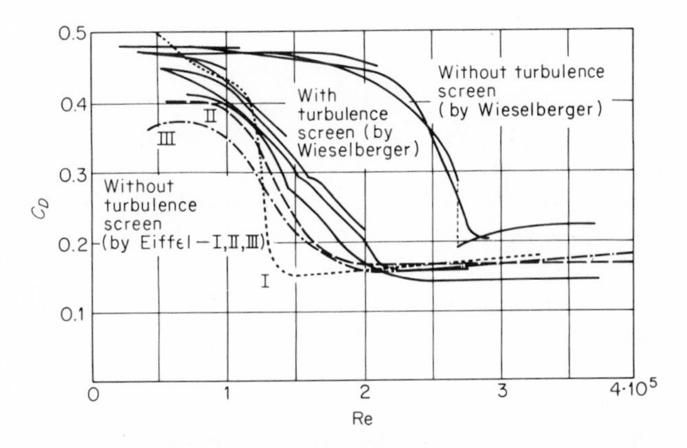

Fig. 4.14 Drag coefficient vs. Reynolds number of a spherical object as influenced by turbulence. [*After* SCHILLER (*1932*).]

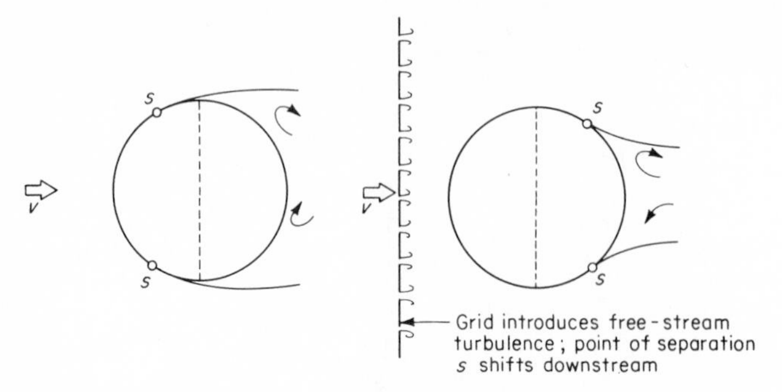

Fig. 4.15 Flow around a sphere below and above critical conditions.

the two phenomena can be explained such that, owing to increased roughness or turbulence, the point of separation changes from in front to behind the equator of the sphere and, thereby, less resistance is encountered. Photographs of these phenomena are presented in PRANDTL et al. (1934) for the roughness influence and in TOROBIN et al. (1960) for the turbulence influence, and are sketched in Fig. 4.15. Generally it was qualitatively concluded that increasing the turbulence level can be associated with decreasing the critical Reynolds number.

More recently a study by TOROBIN et al. (1960) with moving spheres shed some additional light on this problem. Not only did it show that the

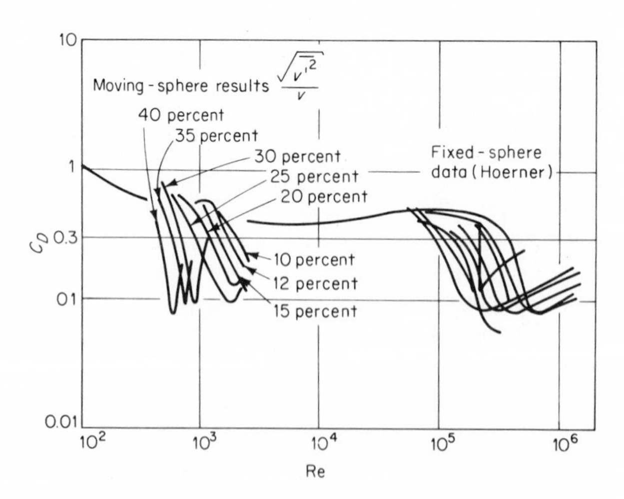

Fig. 4.16 Drag coefficient vs. Reynolds numbers for fixed and moving spheres as influenced by turbulence. [*After* TOROBIN *et al.* (*1960*).]

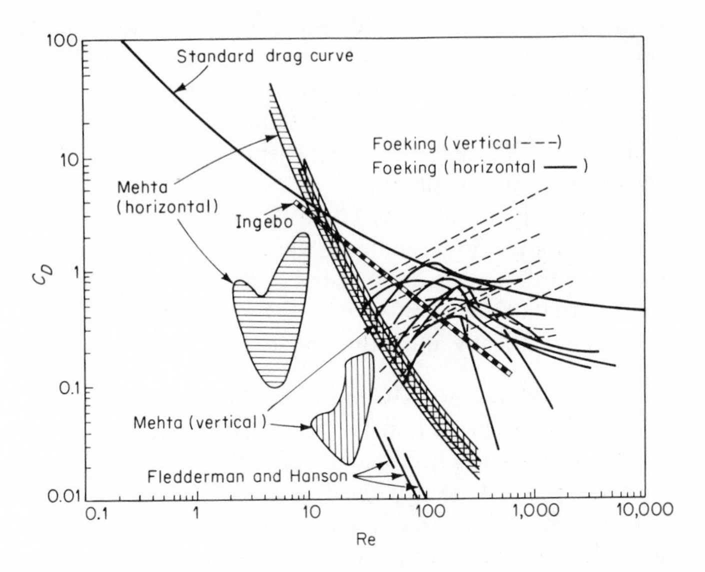

Fig. 4.17 Drag coefficients vs. Reynolds number, as influenced by turbulence; data obtained in concurrent solids-gas systems. [*After* TOROBIN *et al.* (*1960*).]

transition takes place at considerable lower Reynolds numbers altogether but also it presented quantitative results for the relative intensity of the turbulence, as can be seen in Fig. 4.16.

The foregoing research was extended to lower relative intensities and higher Reynolds numbers by CLAMEN et al. (1966). From these studies it becomes apparent that at low relative intensities the resistance does not alter appreciably, but as the relative turbulence intensities increase the resistance starts to differ considerably. Very high relative intensities, however, will cause new complications, because these might be responsible for a particle of small inertia to have a fluctuating motion of its own. TOROBIN et al. (1960) make an attempt to advance a theory for turbulence effects on the transition at high turbulence level. They suggest that the quotient of the critical Reynolds number and the relative turbulence intensity are constant.

4.2.4.6.2 BELOW THE TRANSITION REGION. From Fig. 4.16 we might tend to conclude that turbulence only affects the resistance in the transition region, while below and above no material effect is noticeable. However, TOROBIN et al. (1960) refer to quite a few studies where drag coefficients are reported to alter from the standard curve even below the transition region. These investigations are not at all conclusive, since some researchers report an increase, others a decrease of the drag coefficient, and again others no change at all. This confusing picture is summarized in Fig. 4.17, which shows a general tendency for the drag coefficient to be below the standard curve. In passing it should be mentioned that recently KRIEGEL et al. (1966) presented an analysis on this topic. They derived an equation which predicts a higher drag coefficient than the standard curve gives for Reynolds numbers under about 40 and lower coefficients above this range. No experiments accompany this speculation.

Recently, GRAF et al. (1970) have shown that test results for cylinders under turbulent flow conditions gave drag coefficients both above and below the standard C_D versus Re curve. Subsequently, it was demonstrated by KO (1970) that the turbulence intensity and, to a lesser degree, the turbulence scale systematically affect the drag coefficients of cylinders.

4.2.4.6.3 SMALL-PARTICLE MOTION IN TURBULENT ENVIRONMENT. It should be stressed that in the foregoing section it was assumed that the solid particle is large compared with the turbulent scale; the influence of the turbulence could be expressed, even if not too successfully, as a kind of "correction" to the steady-state drag coefficient. However, if the particle is small compared with the smallest scale of turbulence, it will tend to follow the turbulent fluid components, and the particle motion will be a nonsteady one. For a detailed discussion of the latter case, we refer to the extensive research done by TCHEN (1947), HINZE (1959), and SOO (1967). A theoretical analysis of this problem is possible only if it is assumed that the flow resistance of the particle with respect to the ambient fluid will be a solely viscous one.

4.2.4.7 Combinations of various effects. It was assumed up to now that only a single disturbance, namely another particle, solid or liquid boundaries, turbulence, etc., will influence the resistance to the moving spherical particle. It was far from being simple, at least in most cases, to express these causes with an appropriate correction factor. Even more complicated is the problem of two or more disturbances acting simultaneously, which we shall discuss in this section.

Consider a sphere falling steadily in the center of a circular cylinder under Stokes' conditions. This problem, like most of the problems falling in this category of combinations of various effects, can essentially be attacked in two ways. Assume that a combination of the various corrections as applied to the resistance of the sphere, i.e., the cylindrical boundary, the bottom of the cylinder, and the free surface, can be applied within its proper limits. The other way is to investigate the complete problem in itself.

4.2.4.7.1 COMBINATIONS OF CORRECTION COEFFICIENTS. BRENNER (1961) discussed the foregoing problem within the Stokes range—which was first investigated by Ladenburg—where it is assumed that individual corrections can be combined according to a Taylor series expansion, and we obtain

$$K = 1 + 2.1 \frac{a}{R_{cyl}} + \frac{9}{8}\frac{a}{s_1} + \frac{3}{4}\frac{a}{s_2} \tag{4.37}$$

This manipulation accounts only for the interactions between the particle and the boundaries, and ignores the interactions among the boundaries themselves. Although this technique of superimposing solutions is incorrect, no better one exists at present, but Brenner shows that the error is not too serious.

At larger Reynolds numbers a similar approach is frequently used, namely, to correct for one effect after the other with the appropriate drag coefficients.

4.2.4.7.2 INVESTIGATION OF THE COMPLETE PROBLEM. For problems in the viscous region quite a number of theoretical investigations are available which are concerned with these complicating effects on the uniform motion.

BRENNER (1962) and HAPPEL et al. (1965) study the translation of an arbitrary shaped particle in the proximity to container walls within the Stokes range. The theory advanced is confirmed with experimental data. Also studied is the wall effect on a particle rotating near the boundary.

SMYTHE (1964) and HAPPEL et al. (1965) present an extensive work of research on the motion in a viscous fluid of a spheroid in circular tubes and along plane walls. Cylindrical rods settling inside a cylinder are also discussed by HAPPEL et al. (1965). In the same reference the case of spheroids moving in a viscous liquid is treated. MORELAND (1963) studies the settling velocity of crushed coal particles in a mineral oil. It was concluded that an equation like Eq. (4.35) represents all the data. For the Reynolds-number range investigated, $10^{-4} < \mathrm{Re} < 10^{-2}$, the coefficient m varied from 6.5 to 7.5.

The influence of the roughness of various bodies, mainly at the larger Reynolds numbers, is presented by SCHILLER (1932). The effect is very similar to the one qualitatively discussed in an earlier section and represented in Fig. 4.12.

Limited to the viscous region HAPPEL et al. (1965) study the rotation of a deformed sphere and of a circular cylinder. In the study of the rotating cylinder it is found that due to rotation the critical Reynolds number changes its value, quite similar as shown in Fig. 4.13.

5
Effect of Particles on the Viscosity

Under various assumptions it seems reasonable to adopt the point of view that a fluid-solid particle suspension may be regarded as a continuum. In this chapter we shall focus our attention on the determination of viscosity of such a suspension. The basic variables involved in this problem include the nature of the fluid and the solid phase, the relative motion between them, and their concentration.

5.1 THEORETICAL CONSIDERATIONS

5.1.1 SUSPENSIONS OF SPHERICAL PARTICLES

For dilute suspensions of spheres Einstein developed a rather simple looking equation and obtained the following relationship:

$$\frac{\mu_{susp}}{\mu} = 1 + k_e C \qquad (5.1)$$

where

μ_{susp} = viscosity of the suspension
μ = viscosity of the liquid medium
k_e = Einstein's viscosity constant
C = volumetric concentration of the solid phase

In a review of Einstein's derivation HAPPEL et al. (1965) derived Eq. (5.1), which is based on fundamental equations of hydrodynamics. The following are the main assumptions: (1) The liquid is a newtonian, incompressible, and isotropic one; (2) the inertial terms in the Navier-Stokes equation either vanish or are negligibly small, or do not interact with the viscous term. This implies that each particle be sufficiently far away from the other particles or any other boundary, which is satisfied by having a dilute system; (3) the solid phase consists of rigid, smooth, and uniform spheres. Under these circumstances the Einstein viscosity constant was found to be

$$k_e = 2.5 \tag{5.2}$$

Considering first-order interaction effects, various investigators arrived at different equations, which may be conveniently summarized in the form of

$$\frac{\mu_{susp}}{\mu} = 1 + k_e C + k_2 C^2 + \cdots \tag{5.3}$$

Since collision effects—particle-particle and particle-boundary—become increasingly important, the proper physical and mathematical evaluation of Eq. (5.3) is very involved.

WARD (1955) discussed a solution for high volume concentrations as advanced by Hatschek, who made similar assumptions to those of Einstein, such as

$$\frac{\mu_{susp}}{\mu} = 1 + 4.5C \tag{5.4}$$

Another solution—its derivation is quite involved—is given by HAPPEL et al. (1965).

5.1.2 SUSPENSIONS OF NONSPHERICAL PARTICLES

Theories for prolated or oblated ellipsoidal particles have been advanced, but for dilute systems only. In such a case the viscosity of suspension could be expressed by

$$\frac{\mu_{susp}}{\mu} = 1 + k_J C \tag{5.5}$$

where k_J depends on the various axial ratios. The available work is summarized in a table by WARD (1955).

5.2 EXPERIMENTAL CONSIDERATIONS

All of the main assumptions made for deviation of Eq. (5.1) have to be fulfilled by the experiment in order to make a comparison with the theory possible.

5.2.1 SUSPENSIONS OF SPHERICAL PARTICLES

5.2.1.1 Stable suspensions. It becomes difficult to review in short the many equations proposed by researchers for the determination of the viscosity of suspension. WARD (1955) suggested that ready comparison may be made if the experimental results are expressed in form of

$$\frac{\mu_{susp}}{\mu} = 1 + k_e C + k_e^{\,2} C^2 + k_e^{\,3} C^3 + \cdots \tag{5.6}$$

where the k_e value is an experimentally determined Einstein viscosity constant. Table 5.1 gives values of k_e obtained by various researchers under various conditions.

The data in Table 5.1 show that the k_e values do not greatly differ from the ones theoretically derived, as given by Eq. (5.2). Furthermore, Eq. (5.6) reduces to Eq. (5.1) at extremely low concentrations. It should be noted

Table 5.1 Values for k_e from data of various researchers [*after* WARD (*1955*)]

Researcher	*Type of viscometer*	*Size ratio of suspended spheres*	k_e	*Minimum range of volumetric concentration over which equation applies, %*
Eirich, Bunzl, and Margaretha	Capillary	*	2.41	0–15
Broughton and Windebank	Rotating cylinder	2.0:1	2.57	0–15
Ward and Whitmore	Rising sphere	1.8:1	2.41	0–20
		1.2:1	2.77	0–20
Eveson	Rotating cylinder	1.4:1	2.56	0–20
Nandi	Capillary	1.4:1	2.42	0–35
Higginbotham	Capillary	1.2:1	2.38	0–25
Williams	Capillary	5.0:1	2.34	25–35
Oliver and Ward	Capillary	1.6:1	2.45	0–30

* Killed yeast cells, size ratio not exactly known. All other results refer to perfect spheres.

that Eq. (5.6) predicts the correct suspension viscosity for volume concentration up to 35 percent. Experimentally it was found that suspensions of spherical particles behaved as newtonian fluids at volume concentrations of 40 percent, the latter value being quoted by HIGGENBOTHAM et al. (1958). After examining many data WARD (1955) concluded that a suspension viscosity is independent of the absolute size of spheres but dependent on the size ratio of the spheres. EVESON (1957) shows that the first statement is true if the particle size for the spheres is above $d \approx 15\ \mu$. In a further investigation by EVESON (1958) on the effect of the size ratio—defined as the ratio of the diameter of the largest and smallest particles in the fraction—the results indicate only a random variation of viscosity with size ratio. Lastly, it should be noticed in Table 5.1 that the k_e value appears to depend to some extent on the type of viscometer.

5.2.1.2 Unstable suspensions. We refer to suspensions where the solid particles are either downward- or upward-settling as unstable suspensions. This, of course, complicates a comparison with the earlier advanced theory. Nevertheless, unstable suspensions are more common because of the frequent existence of density differences between the two phases. Experimental research reported by OLIVER et al. (1959) suggested that the data could be expressed as follows. For low concentrations Eq. (5.1) generally explains the data, with an Einstein viscosity constant ranging from 3.00 to 3.60, depending on the density difference. When $0.1 < C < 0.3$, the data may be represented by an equation such as

$$\frac{\mu_{susp}}{\mu} = (1 + k_1) + (1 + 2k_1)k_e C + (1 + 3k_1)k_e{}^2 C^2 + \cdots \tag{5.7}$$

being similar to Eq. (5.6). A relationship between k_1 and a density quotient was derived, such as

$$k_1 = 0.33\,\frac{\rho_s - \rho}{\mu} \tag{5.8}$$

which gives positive values of k_1 for downward-settling suspensions and negative ones for upward-settling ones. The available data are, however, very limited in extent and range.

Another type of equation is discussed in HAPPEL et al. (1965), which can be summarized as

$$\frac{\mu_{susp}}{\mu} = \frac{(1 - C)^r}{v_{susp}/v_s} \tag{5.9}$$

where v_{susp}/v_s is the ratio of the sedimentation velocity v_{susp} of the suspension to the individual particles' settling velocity v_s. The exponent r varies from 1 to 2.

5.2.2 SUSPENSIONS OF NONSPHERICAL PARTICLES

WARD et al. (1950) examined stable suspensions of irregularly shaped particles and found that their viscosity, quite in contrast to those of spheres, depended on the particle size. Similar results have been reported by MORELAND (1963*a*). It was concluded that differences in the behavior between suspensions of spherical and nonspherical particles is mainly due to layers of immobile liquid held to the surface of irregular particles. WHITMORE (1957) accommodates this by multiplying by an appropriate factor—the hydrodynamic volume factor k_ω—the "dry" concentration value, or

$$C' = k_\omega C \tag{5.10}$$

where C' = hydrodynamic volume concentration

C = concentration (dry)

k_ω = multiplication factor or hydrodynamic volume factor

Replacing C' for C in Eq. (5.1), it was found that all data could be explained with one equation, whereby Einstein's viscosity constant is given by $k_e = 2.0$.

The present knowledge about the viscosity of unstable suspension of nonspherical particles is very limited. Some information might be obtained in WARD (1955).

5.3 REMARKS ON NON-NEWTONIAN BEHAVIOR

Information is scarce on solid-liquid suspensions which behave as non-newtonian liquids. Even the available literature is often contradictory and no definite criteria are set as yet.

Many reports, however, for example, WARD (1955) or HAPPEL et al. (1965), refer to occurrence of non-newtonian behavior at higher volume concentrations; no value may be quoted, since this, again, depends on other factors.

EVESON (1957) suggest that at a given volume concentration small particles may behave non-newtonian, whereas large sizes do behave newtonian. There is also some evidence that a nonuniform distribution of the solids in the liquid phase causes non-newtonian behavior.

HAPPEL et al. (1965) discuss that high rates of shear indicate divergence of simple newtonian results.

Last but not least, it does depend on the type of viscometer; this clearly suggests as to how far we will still have to go in the proper evaluation of concentration effects in solid-liquid systems.

REFERENCES FOR PART TWO

ALBERTSON, M. (1953): Effects of Shape on the Fall Velocity of Gravel Particles, *Proc. 5th Iowa Hydraulics Conf.*, Iowa City, Iowa.

ANONYMOUS (1957): Measurement and Analysis of Sediment Loads in Streams, *Rept. 12, Subcom. on Sedimentation, Inter-Agency Committee on Water Resources*, Minneapolis, Minn.

BASSET, A. (1910): On the Descent of a Sphere in a Viscous Liquid, *Quart. J. Math.*, vol. 14.

——— (1961): "A Treatise of Hydrodynamics," chap. XXII, Dover, New York.

BOUSSINESQ, J. (1903): "Théorie Analytique de la Chaleur," vol. 2, Paris.

BRENNER, H. (1961): The Slow Motion of a Sphere Through a Viscous Fluid Towards a Plane Surface, *Chem. Eng. Sci.*, vol. 16.

——— (1962): Effects of Finite Boundaries on the Stokes Resistance of an Arbitrary Particle, *J. Fluid Mech.*, vol. 12/1.

BRUSH, L. M., H. W. HO, and B. C. YEN (1964): Accelerated Motion of a Sphere in a Viscous Fluid, *Proc. Am. Soc. Civil Engrs.*, vol. 90, no. HY1.

CLAMEN, A., and W. H. GAUVIN (1966): Effects of Turbulence on the Drag Coefficients of Spheres in a Super-critical Flow Regime, *Tech. Rept. 467, Pulp and Paper Inst. of Canada.*

CORRSIN, S., and J. LUMLEY (1956): On the Equation of Motion for a Particle in Turbulent Fluid, *Appl. Sci. Res.*, sec. A, vol. 6.

DALLAVALLE, J. (1943): "Micromeritics," Pitman, New York.

DUPLESSIS, M. P., and R. A. ANSLEY (1967): Settling Parameter in Solids Pipelining, *Proc. Am. Soc. Civil Engrs.*, vol. 93, no. PL2.

EVESON, G. F. (1957): The Rheological Properties of Stable Suspensions of Very Small Spheres at Low Rates of Shear, *J. Oil & Colour Chemists' Assoc.*, vol. 40, no. 2.

——— (1958): Stable Suspensions of Spherical Particles, *J. Oil & Colour Chemists' Assoc.*, vol. 41, no. 2.

———, E. W. HALL, and S. G. WARD (1959): Interaction between Two Equal-sized–Equal Settling Spheres Moving through a Viscous Liquid, *Brit. J. Appl. Phys.*, vol. 10.

FIDLERIS, V., and R. L. WHITMORE (1961): Experimental Determination of the Wall Effect for Spheres Falling Axially in Cylindrical Vessels, *Brit. J. Appl. Phys.*, vol. 12.

——— ——— (1961*a*): The Physical Interaction of Spherical Particles of Suspensions, *Rheol. Acta*, vol. 1, no. 4/6.

FROMM, J. E. (1963): A Method of Computing Nonsteady, Incompressible, Viscous Fluid Flow, *LA-2910, Los Alamos Sci. Lab., Univ. of Calif.*

GOLDSTEIN, S. (1929): The Steady Flow of Viscous Fluid past a Fixed Spherical Obstacle at Small Reynolds Numbers, *Proc. Roy. Soc.* (*London*), vol. 123A.

GRAF, W. H. (1965): The Preferred Orientation of an Elliptical Cylinder in a Potential Flow Field: A Discussion, *J. Geol.*, vol. 73, no. 3.

GRAF, W. H., and E. R. ACAROGLU (1966): Settling Velocities of Natural Grains, *Bull., Intern. Assoc. of Sci. Hydrology*, vol. XI, no. 4.

——— (1967): Remarks on the Rubey Equation for Computing Settling Velocities, *Proc. 7th Sedimentological Congr.*, Great Britain.

——— and S. C. KO (1970): Test of Cylinders in Turbulent Flow, *Proc. Am. Soc. Civil Engrs.* (submitted for publ.).

HAPPEL, J., and H. BRENNER (1965): "Low Reynolds Number Hydrodynamics," Prentice-Hall, Englewood Cliffs, N.J.

HEYWOOD, H. (1938): Measurement of the Fineness of Powdered Materials, *Proc. Inst. Mech. Engrs.*, vol. 140.

——— (1962): Uniform and Non-uniform Motion of Particles in Fluid, *Proc. 3d Congr. Europ. Fed. of Chem. Engrs.*, London.

HIGGINBOTHAM, G. H., D. R. OLIVER, and S. G. WARD (1958): Studies of the Viscosity and Sedimentation of Suspensions, pt. 4, *Brit. J. Appl. Phys.*, vol. 9.

HINZE, J. (1959): "Turbulence," McGraw-Hill, New York.

HJELMFELT, A. T., and L. F. MOCKROS (1967): Stokes Flow Behavior of an Accelerating Sphere, *Proc. Am. Soc. Civil Engrs.*, vol. 93, no. EM6.

HOERNER, S. F. (1958): "Fluid Dynamic Drag," Hoerner, Midland Park, New Jersey.

HUGHES, R. R., and E. R. GILLILAND (1952): The Mechanics of Drops, *Chem. Eng. Progr.*, vol. 48, no. 10.

IVERSEN, H. W., and R. BALENT (1951): A Correlating Modulus of Fluid Resistance in Accelerated Motion, *J. Appl. Phys.*, vol. 22, no. 3.

JENSEN, V. G. (1959): Viscous Flow Round a Sphere at Low Reynolds Numbers (40), *Proc. Roy. Soc.*, vol. 249A.

KO, S. C. (1970): "Effects of Turbulence on the Drag Coefficient of Cylinders. An Experimental Study," Ph.D. dissertation, Lehigh University, Bethlehem, Pa.

KOTSCHIN, N. J., I. A. KIBEL, and N. W. ROSE (1954): "Theoretische Hydromechanik," vol. 1, Akademie-Verlag, Berlin.

KRIEGEL, E., and H. BAUER (1966): Hydraulischer Transport körniger Feststoffe durch waagerechte Rohrleitungen, *VDI-Forschungsh.*, no. 515.

LAMB, H. (1945): "Hydrodynamics," Dover, New York.

MAMAK, W. (1964): "River Regulation," translated from Polish, USDI and NSF, Washington, D.C.

MARCHILLON, E. K., A. CLAMEN, and W. H. GAUVIN (1964): Oscillatory Motion of Freely Falling Disks, *Phys. Fluids*, vol. 7, no. 12.

MAUDE, A. D., and R. L. WHITMORE (1958): A Generalized Theory of Sedimentation, *Brit. J. Appl. Phys.*, vol. 9.

MCNOWN, J. S. (1951): Particles in Slow Motion, *La Houille Blanche*, vol. 6, no. 5.

———, H. M. LEE, M. B. MCPHERSON, and S. M. ENGEZ (1948): Influence of Boundary Proximity on the Drag of Spheres, *Proc. 7th Intern. Congr. Appl. Mech.*, London.

——— and J. MALAIKA (1950): Effects of Particle Shape on Settling Velocity at Low Reynolds Numbers, *Trans. Amer. Geophys. Union*, vol. 31.

McNown, J. S., J. Malaika, and H. Pramanik (1951): Particle Shape and Settling Velocity, *Proc. Intern. Assoc. Hydr. Res., 4th Meeting*, Bombay, India.

——— and J. Newlin (1951*a*): Drag of Spheres in Cylindrical Boundaries, *Proc. 1st Natl. Congr. Appl. Mech.*

Michael, P. (1966): Steady Motion of a Disk in a Viscous Fluid, *Phys. Fluids*, vol. 9, no. 3.

Milne-Thomson, L. M. (1963): "Theoretical Hydromechanics," Macmillan, New York.

Moreland, C. (1963): Settling Velocities of Coal Particles, *Canad. J. Chem. Eng.*, vol. 41, June.

——— (1963*a*): Viscosity of Suspensions of Coal in Mineral Oil, *Canad. J. Chem. Eng.*, vol. 41, February.

Oliver, D. R., and S. G. Ward (1959): Studies of the Viscosity and Sedimentation of Suspensions, Pt. 5, *Brit. J. Appl. Phys.*, vol. 10.

Olson, R. (1961): "Essentials of Engineering Fluid Mechanics," chap. 11, International Textbook, Scranton, Pa.

Oseen, C. (1927): "Hydrodynamik," chap. 10, Akademische Verlagsgesellschaft, Leipzig.

Pettyjohn, E. S., and E. B. Christiansen (1948): Effect of Particle Shape on Free-Settling Rates of Isometric Particles, *Chem. Eng. Progr.*, vol. 44, no. 2.

Prandtl, L. (1956): "Führer durch die Strömungslehre", Vieweg Verlag, Braunschweig.

——— and O. G. Tietjens (1934): "Applied Hyrdo & Aeromechanics," chap. V, Dover, New York.

Proudman, I., and J. Pearson (1957): Expansions of Small Reynolds Numbers for the Flow Past a Sphere and a Circular Cylinder, *J. Fluid Mech.*, vol. 2/3.

Rouse, H. (1938): "Fluid Mechanics for Hydraulic Engineers," chap. XI, Dover, New York.

Rubey, W. (1933): Settling Velocities of Gravel, Sand and Silt Particles, *Amer. J. Sci.*, vol. 225.

Schiller, L. (1932): Fallversuche mit Kugeln und Scheiben in "Handbuch der Experimental-Physik," vol. IV/2, Akademische Verlagsgesellschaft, Leipzig.

——— and A. Naumann (1933): Über die grundlegenden Berechnungen bei der Schwerkraftaufbereitung, *Z. VDI*, vol. 77.

Schlichting, H. (1968): "Boundary Layer Theory," McGraw-Hill, New York.

Shanks, D. (1955): Nonlinear Transformations of Divergent and Slowly Convergent Sequences, *J. Math. Phys.*, vol. 34, no. 1.

Smythe, W. R. (1964): Flow Around a Spheroid in a Circular Tube, *Phys. Fluids*, vol. 7, no. 5.

Soo, S. L. (1967): "Fluid Dynamics of Multiphase Systems," chap. 2, Blaisdell, Waltham, Mass.

Stewartson, K. (1955): On the Steady Flow Past a Sphere at High Reynolds Numbers, Using Oseen's Approximation, *The Phyl. Magaz.*, ser. 8, vol. 1/1.

Tchen, C. (1947): "Mean Value and Correlation Problems Connected with the Motion of Small Particles Suspended in a Turbulent Fluid," D.Sc. dissertation, Technische Hogeschool, Delft, Holland.

Torobin, L. B., and W. H. Gauvin (1959): Fundamental Aspects of Solid-Gas Flow, pt. I, *Canad. J. Chem. Eng.*, vol. 37.

——— (1960): Fundamental Aspects of Solid-Gas Flow, pt. V, *Canad. J. Chem. Eng.*, vol. 38.

——— (1960*a*): Fundamental Aspects of Solid-Gas Flow, pt. IV, *Canad. J. Chem. Eng.*, vol. 38.

Truesdell, C. (1960): A Program Toward Rediscovering the Rational Mechanics of the Age of Reason, *Arch. History Exact Sci.*, vol. 1, no. 1.

Valentik, L., and R. L. Whitmore (1965): The Terminal Velocity of Spheres in Bingham Plastics, *Brit. J. Appl. Phys.*, vol. 16.

Van Dyke, M. (1964): "Perturbation Methods in Fluid Mechanics," chap. VIII, Academic, New York.

Wadell, H. (1934): The Coefficient of Resistance as a Function of Reynolds Number for Solids of Various Shapes, *J. Franklin Inst.*, vol. 217.

Ward, S. G. (1955): Properties of Well-Defined Suspensions of Solids in Liquids, *J. Oil & Colour Chemists' Assoc.*, vol. 38, no. 9.

——— and R. L. Whitmore (1950): Studies of the Viscosity and Sedimentation of Suspensions, pt. 2, *Brit. J. Appl. Phys.*, vol. 1.

Whitmore, R. L. (1957): The Relationship of the Viscosity to the Settling Rate of Slurries, *J. Inst. Fuel*, May.

part three

Sediment Transport in Open Channels

CONTENTS

INTRODUCTION

LIST OF SYMBOLS

6 SCOUR CRITERIA AND RELATED PROBLEMS
- 6.1 Introductory Remarks
- 6.2 Critical Velocity Equations
  - 6.2.1 Theoretical Considerations
  - 6.2.2 Experimental Investigations
  - 6.2.3 Concluding Remarks
- 6.3 Critical Shear Stress Equations
  - 6.3.1 Theoretical Considerations
  - 6.3.2 Experimental Investigations
    - 6.3.2.1 Earlier Developments
    - 6.3.2.2 Modern Developments
  - 6.3.3 Concluding Remarks
- 6.4 Lift Force Mechanism
  - 6.4.1 Theoretical Considerations
  - 6.4.2 Experimental Investigations
  - 6.4.3 Concluding Remarks
- 6.5 Shear Stress Distribution
  - 6.5.1 Indirect Determinations
  - 6.5.2 Measurement of Shear Stress
    - 6.5.2.1 Shear Meter
    - 6.5.2.2 Preston Tube
  - 6.5.3 Direct Determinations
  - 6.5.4 Bank Scour
- 6.6 Design of Stable Channels
  - 6.6.1 Stable Cross Section
  - 6.6.2 Ideal, Stable Cross Section

7 THE BEDLOAD
- 7.1 Introductory Remarks
- 7.2 DuBoys-type Equations
  - 7.2.1 Theoretical Considerations
  - 7.2.2 Further Investigations
    - 7.2.2.1 Earlier Developments
    - 7.2.2.2 Modern Developments

7.3 Schoklitsch-type Equations
7.3.1 Theoretical Considerations
7.3.2 Experimental Investigations
7.3.2.1 Earlier Developments
7.3.2.2 Schoklitsch's Contributions
7.3.2.3 Contributions of the E.T.H.
7.4 Einstein's Bedload Equations
7.4.1 Introductory Remarks
7.4.2 Physical Model
7.4.3 Empirical Relation
7.4.4 Analytical Relation
7.4.5 Similar Kinds of Bedload Equations
7.4.5.1 Bagnold's Model
7.4.5.2 Yalin's Model
7.5 Equations Considering Bedform Motion
7.5.1 Theoretical Considerations
7.5.2 Experimental Investigations
7.6 Saltation as a Mode of Particle Transport
7.7 Concluding Remarks

8 THE SUSPENDED LOAD
8.1 Introductory Remarks
8.2 Earlier Studies
8.2.1 Earlier Developments
8.2.2 Earlier Turbulence Models
8.2.3 Concluding Remarks
8.3 Diffusion-Dispersion Model
8.3.1 Theoretical Considerations
8.3.2 Vertical Distribution of Suspended Matter
8.3.2.1 Uniform Turbulence Distribution at Steady-state Condition
8.3.2.2 Uniform Turbulence Distribution at Transient Condition
8.3.2.3 Nonuniform Turbulence Distribution
8.3.2.3.1 Theoretical Considerations
8.3.2.3.2 Experimental Investigations
8.3.2.3.3 Further Theoretical Considerations
8.3.2.3.4 Other Theoretical Contributions
8.3.2.4 Calculation of the Suspended Load
8.3.2.4.1 Lane's et al. Approach
8.3.2.4.2 Einstein's Approach
8.3.3 Longitudinal Distribution of Suspended Matter
8.3.3.1 For Neutrally Buoyant Dispersant
8.3.3.1.1 Fick's Law of Diffusion
8.3.3.1.2 Longitudinal Dispersion by Differential Convection
8.3.3.1.3 Concluding Remarks
8.3.3.2 For Suspended Sediment Particles
8.4 Concluding Remarks

9 THE TOTAL LOAD
9.1 Introductory Remarks
9.2 Indirect Determinations
9.2.1 Einstein's Bedload Functions
9.2.2 Modified Einstein Procedure

9.2.3 Bagnold's Approach
9.2.4 Chang's et al. Approach
9.3 Direct Determinations
9.3.1 Laursen's Approach
9.3.2 Bishop's et al. Approach
9.3.3 Graf's et al. Approach
9.4 Comparison and Application of Bed Material Load Equations
9.4.1 General Remarks
9.4.2 Comparative Sample Calculations
9.4.2.1 Description of the Test Reach
9.4.2.2 Hydraulic Calculations
9.4.2.3 Calculations of Bed Material Load
9.5 Hydrologic Effects on Sediment Transport
9.5.1 Stream Flow Effects
9.5.2 Temperature Effects
9.5.3 Effects of Rainfall
9.5.4 Dissolved Matter
9.6 Concluding Remarks

10 THE REGIME CONCEPT
10.1 Introductory Remarks
10.2 Canals in Regime
10.2.1 Kennedy's Study on the Prevention of Silting
10.2.2 Lindley's Study on Regime Channels
10.2.3 Lacey's Contributions
10.2.4 Blench's Contributions
10.2.5 Simons' et al. Contributions
10.2.6 Concluding Remarks
10.3 Rivers in Regime
10.3.1 Regime River vs. Regime Canal
10.3.2 Regime Equations for Width, Depth, and Velocity
10.3.3 Meandering of Rivers
10.3.4 Longitudinal River Profiles
10.4 Concluding Remarks

11 BEDFORM MECHANICS
11.1 Bedforms
11.1.1 Earlier Developments
11.1.2 Experimental Investigations
11.1.2.1 C.S.U.'s Contributions
11.1.2.2 Other Contributions
11.1.3 Theoretical Studies
11.1.3.1 Exner's Models
11.1.3.1.1 Models without Friction
11.1.3.1.2 Models with Friction
11.1.3.2 Potential Flow Models
11.1.3.2.1 General Considerations
11.1.3.2.2 The Sediment Transport Problem
11.1.3.3 Real Fluid – Sediment Models
11.1.3.4 Other Investigations
11.2 Bedforms and Flow Resistance
11.2.1 General Remarks

11.2.2 Flow Resistance in the Absence of Bedforms
11.2.3 Flow Resistance in the Presence of Bedforms
11.2.3.1 Flow Resistance Expressed with the Friction Velocity
11.2.3.2 Flow Resistance Expressed with the Friction Factor
11.2.4 Concluding Remarks

12 COHESIVE-MATERIAL CHANNELS
12.1 Introductory Remarks
12.2 Cohesive Materials
12.2.1 General Remarks
12.2.2 Clay Minerals and Their Properties
12.2.3 Cohesive Materials and Their Properties
12.2.3.1 Consistency
12.2.3.2 Soil Aggregates
12.3 Scour Criteria
12.3.1 Laboratory Investigations
12.3.2 Flume Investigations
12.3.3 Field Investigations
12.3.4 Design Criteria
12.3.5 Closing Remarks
12.4 Sedimentation-Flocculation Problem
12.4.1 General Remarks
12.4.2 Engineering Investigations
12.5 Transportation Problem

13 SEDIMENT MEASURING DEVICES
13.1 Introductory Remarks
13.2 Bedload Measuring Devices
13.2.1 General Remarks
13.2.2 Direct Measurements
13.2.2.1 Box- and Basket-type Samplers
13.2.2.2 Pan-type Samplers
13.2.2.3 Pit-type Samplers
13.2.2.4 Other Types of Samplers
13.2.3 Indirect Measurements
13.2.3.1 Calculation by Measuring the Bed Material
13.2.3.2 Sound Samplers
13.2.3.3 Tracking of Bedforms
13.2.3.4 Other Methods
13.3 Suspended-load Measuring Devices
13.3.1 General Remarks
13.3.2 Instantaneous Samplers
13.3.3 Integrating Samplers
13.3.3.1 Point-integrating Samplers
13.3.3.2 Depth-integrating Samplers
13.3.4 Continuously Recording Samplers
13.4 Total-load Measuring Devices
13.4.1 General Remarks
13.4.2 Indirect Measurements
13.4.2.1 Calculation and Measurement Procedure
13.4.2.2 Measurement Procedure
13.4.3 Tracing Methods

13.4.3.1 Types of Tracers
13.4.3.1.1 Radioactive Tracers
13.4.3.1.2 Paint and Fluorescent Tracers
13.4.3.1.3 Density Tracers
13.4.3.2 Measurement of Tracers
13.4.3.2.1 Direct Method
13.4.3.2.2 Indirect Method
13.4.3.3 Evaluation of Tracer Studies

14 MODEL LAWS
14.1 Similitude and Dimensionless Numbers
14.2 River and Channel Models
14.2.1 Fixed-bed Models
14.2.1.1 Undistorted Models
14.2.1.2 Distorted Models
14.2.2 Movable-bed Models
14.2.2.1 Empirical Approach
14.2.2.2 Rational Approach

REFERENCES FOR PART THREE

INTRODUCTION

A true watercourse, be it a stream, a channel, or a laboratory flume, represents usually an interface problem. There is the water which for a fixed volume is deformable and adjustable in any kind of boundary. There is loose and granular solid material upon which the water may act and be more or less successful in deforming and changing its boundaries. One responds to the other. Furthermore, water will infiltrate into the porous media of the loose bed, while loose, solid particles will become entrapped by the water. The change from one phase to the other will thus be a gradual one.

Part 3 of this book is restricted to the study of the entrainment of solid material into the water body of an open channel, its further effects and consequences, which are commonly referred to as the field of the *transport of sediment*. Despite the vast amount of research, being all too often empirical in order to obtain quick answers to specific questions, at present we may not consider all problems of this topic as solved. On the contrary, the field awaits in the years to come many a new impact for the eventual solution of this important area of water resources.

In Part 3, Chaps. 6 through 9 attempt to answer the following questions: (1) When and why does erosion and deposition begin to occur? How can one design a stable channel? (2) If erosion has started, how much solid material will be transported? (3) What is the mode of transport? The answers to question 1 are extensively discussed in Chap. 6. Question 2 is a quantitative one of considerable interest to the engineer. Answers to this question may be found in Chaps. 7 through 9 where also information pertaining to question 3 is given.

Chapter 10 considers the watercourse as an integral part of *geomorphology*. Over the years geologists and engineers have accumulated considerable material on various aspects of the geomorphology of canals and streams. Although most of the material is descriptive and frequently only applicable to specific problems and conditions, certain useful knowledge may be obtained. Chapter 11 deals with the occurrence, description, and effects of various bedforms. Chapter 12, covering the hydraulics of cohesive-material channels, becomes necessary because throughout the entire discussion cohesive material has been excluded. Sediment measuring devices will be discussed in Chap. 13. Chapter 14 deals with models and model laws for the transport of sediments in open channels.

LIST OF SYMBOLS

A	cross-sectional area
a	particle's radius ($2a = d$)
A'	sediment coefficient
A''	sediment coefficient
A_*	constants
B	width
C	Chezy's roughness coefficient
C	concentration; in weight percentage or volume percentage
C_D	drag coefficient
C_L	lift coefficient
C_o	coefficient of cohesion
C_x	coefficients
c_B	velocity of bedforms
d	$d = d_{50}$; diameter of particle or median diameter of particles in a mixture
D	depth of flow
d_p	pitot-tube-opening diameter
$\bar{d}$	average diameter of particles in mixture
F	force
F_B	bed factor, after BLENCH (1951)
F_C	cohesive force
F_D	drag force
F_G	force of gravity
F_I	force of inertia
F_L	lift force
F_P	force of pressure
F_S	side factor, after BLENCH (1951)
F_V	force of viscosity
f_L	silt factor, after LACEY (1929)
g	gravitational constant
g	rate of liquid flow in weight per unit time and unit width
g_s	bedload rate in weight per unit time and unit width
g_{ss}	suspended load rate in weight per unit time and unit width
g_{st}	total load rate in weight per unit time and unit width
G_s	bedload rate in weight per unit time
G_{ss}	suspended load rate in weight per unit time
G_{st}	total load rate in weight per unit time
ΔH	height of ridges (bedforms)
i_b	fraction of bed material in a given grain size
i_s	fraction of bedload in a given grain size
i_{ss}	fraction of suspended load in a given grain size

i_{st} fraction of total load in a given grain size
I_1 integral value
I_2 integral value
k Karman's constant; Strickler's roughness value
k_x constant
k_s Nikuradse's sand roughness
K_T virtual coefficient of diffusion (due to Taylor) or dispersion coefficient
L length dimension
l a length
m porosity of sand
M grain size distribution modulus, after KRAMER (1935)
M_L meander length
M_B meander width
M_R meander curvature
N_F Froude number
n Manning's roughness value
p probability of grain being eroded
P wetted perimeter
P_E parameter of total transport, due to EINSTEIN (1950)
p_s static pressure
p_t total pressure
q rate of liquid in volume per unit time and unit width
Q rate of liquid in volume per unit time; flow rate
q_s bedload rate in volume per unit time and unit width
q_{ss} suspended load rate in volume per unit time and unit width
q_{st} total load rate in volume per unit time and unit width
Q_s bedload rate in volume per unit time
Q_{ss} suspended load rate in volume per unit time
Q_{st} total load rate in volume per unit time
Re Reynolds number
R_h hydraulic radius
R_h' hydraulic radius, due to particles
R_h'' hydraulic radius due to channel irregularities
s specific density
S channel slope
t time
t_e exchange time of bed particles
u velocity (of liquid)
u_* shear velocity
V velocity dimension
v velocity
v_s velocity of sediment layers
v_{ss} settling velocity of particle
w velocity
W submerged weight of a sand grain
x distance (horizontal) from the reference level
y distance (vertical) from the reference level
z z value; exponent in the suspension distribution
α inclination of the bed
γ specific weight
δ thickness of laminar sublayer
ϵ_i diffusivity of fluid mass
ϵ_m molecular diffusivity
ϵ_M diffusivity of linear momentum
ϵ_s diffusivity of solid particles
ϵ_t turbulent diffusivity
η variability factor of lift; distance from reference level
θ inclination of bank from horizontal
λ wavelength of a bedform
λ_b single step of bedload, measured in diameter
μ coefficient of viscosity
ν kinematic viscosity
ρ density
τ shear stress, "drag," "tractive force"
φ angle of repose or specific abrasion
Φ intensity of bedload transport
Φ_* intensity of bedload transport for individual grain size
Φ_A transport parameter
ϕ velocity potential
χ characteristic sediment coefficient (due to duBoys)
Ψ intensity of shear on particles
Ψ_A shear intensity parameter
Ψ_* intensity of shear on individual grain size
ψ stream function

Subscripts
none liquid phase
b at bed or at bottom

cr	critical condition
H	denoting "horizontal"
max	maximum value
0	at boundary ($y = 0$) or at beginning ($t = 0$)
s	solid phase
V	denoting "vertical"
y	at a distance y
∞	at infinity
*	shear value

Superscripts

m	denoting "model"
p	denoting "prototype"
w	at sidewall
$'$	due to grain roughness; or turbulent fluctuation component
$''$	due to bedforms
—	average value

6
Scour Criteria and Related Problems

6.1 INTRODUCTORY REMARKS

Consider a plane stationary bed consisting of loose and cohesionless (mobile) solid particles of uniform size, and liquid flowing over it. This may exist in any kind of a conveyance system, an open channel, a closed conduit, or along a shoreline. Furthermore, it is assumed that the cross-sectional area of the conveyance system throughout a given distance is constant, and, through the same distance, liquid flow and bed are statistically in a steady state; in other words, average values are considered. As soon as liquid starts flowing, hydrodynamic forces are exerted upon the solid particles of the bed at the wetted perimeter of the conveyance system. A further increase in the flow intensity causes an increase in the magnitude of these forces. Hence, for a particular stationary bed, a condition is eventually reached at which particles in the movable bed are unable to resist the hydrodynamic forces and, thus, get first dislodged and eventually start to move. This movement is not an instantaneous one for all particles of a given size resting in the top layer. In fact, at any given hydraulic condition, some move and some do not move. This is owing to the statistical nature of the problem, which implicitly brings out the fact that the flow has to be a turbulent one. The condition of the "initial movement of the bed" is determined by observations; therefore, its definition is a very subjective one.

The initial movement of the bed, frequently called the *critical condition* or *initial scour*, will be explained in several ways:

1. With critical velocity equations; considering the impact of the liquid on the particles
2. With critical shear stress equations; considering the frictional drag of the flow on the particles
3. The lift force criteria; considering the pressure differences due to the gradient of the velocity

Hopefully, it can be shown that these three approaches, which might appear to be different, are not entirely different from each other. Since many conveyance systems, especially the natural ones, have a variable depth, a detailed discussion is included on the distribution of the shear stress and its measurement.

Information gained in these sections will be employed in the concluding section of this chapter to the design of stable channel cross sections. Chap. 6 deals exclusively with one particular conveyance system, the open channel.

6.2 CRITICAL VELOCITY EQUATIONS

6.2.1 THEORETICAL CONSIDERATIONS

The condition of incipient movement for an assembly of cohesionless, loose, and solid particles is described in terms of the forces acting on the particle by

$$\tan \varphi = \frac{F_t}{F_n} \tag{6.1}$$

where F_t and F_n are the forces parallel and normal to the angle of repose φ. In our study F_t and F_n are resultants of the hydrodynamic drag F_D, the lift force F_L, and the submerged weight W. The condition of incipient movement under the action of these three forces becomes, according to Eq. (6.1),

$$\tan \varphi = \frac{W \sin \alpha + F_D}{W \cos \alpha - F_L} \tag{6.2}$$

where angle α is the inclination of the bed from the horizontal at which incipient sediment movement takes place. This situation is illustrated in

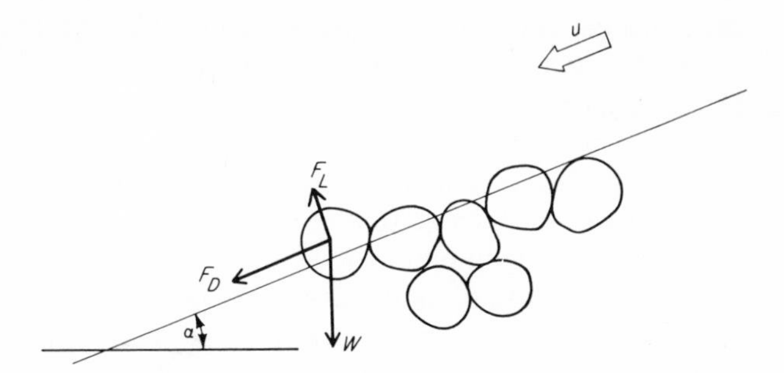

Fig. 6.1 Force diagram on particles in a cohesionless loose bed.

Fig. 6.1. In the usual way, drag and lift forces are expressed by

$$F_D = C_D k_1 d^2 \frac{\rho u_b^2}{2} \tag{6.3}$$

and

$$F_L = C_L k_2 d^2 \frac{\rho u_b^2}{2} \tag{6.4}$$

where u_b = fluid velocity at the bottom of the channel
C_D, C_L = drag and lift coefficient, respectively
d = particle diameter
k_1, k_2 = particle shape factors
ρ = liquid density

The submerged weight of the particle is expressed as

$$W = k_3(\rho_s - \rho)gd^3 \tag{6.5}$$

with k_3 being another shape factor and ρ_s being the solid-particle density.

Introducing Eqs. (6.3), (6.4), and (6.5) into Eq. (6.2) yields

$$\frac{(u_b^2)_{cr}}{(\rho_s/\rho - 1)gd} = \frac{2k_3(\tan\varphi\cos\alpha - \sin\alpha)}{C_D k_1 + C_L k_2 \tan\varphi} \tag{6.6}$$

with $(u_b)_{cr}$ as the critical bottom velocity at which, according to Eq. (6.1), incipient sediment motion takes place. The quantity of the right-hand side in Eq. (6.6) is referred to as the sediment coefficient A',

$$A' = \frac{2k_3(\tan\varphi\cos\alpha - \sin\alpha)}{C_D k_1 + C_L k_2 \tan\varphi} \tag{6.7}$$

The sediment coefficient A' depends on (1) the particles, their size, uniformity, shape, size distribution, texture, etc.; (2) the dynamics of the flow, since this determines the values of C_D and C_L; (3) the channel slope, which for most natural conditions amounts to $\cos \alpha \approx 1$; and (4) the angle of repose, which, however, is shown by LANE (1953) and MILLER et al. (1966) to depend on particle properties. (See Fig. 6.19.)

To test experimentally the extremely involved relationship of incipient motion as given by Eq. (6.6) has not been accomplished yet. However, under certain conditions, one or the other quantities might vanish, or at least become constant, and useful knowledge is thus obtained.

At this point it shall be said that FORCHHEIMER (1914) reports of an equation put forward by A. Brahms in 1753, which relates the bottom velocity with the weight of the particle. It was expressed as

$$(u_b)_{cr} = k_5 W^{1/6} \tag{6.8}$$

with W being the weight of a grain and k_5 being a constant. Equation (6.8) may be considered as a simple form of Eq. (6.6).

6.2.2 EXPERIMENTAL INVESTIGATIONS

The verification of Eq. (6.6) is particularly difficult because of lack of a good definition of the bottom velocity u_b and the difficulties encountered in measuring it accurately. The practical use of the experimental investigations is for these reasons very limited.

Important contributions of the last century are summarized in FORCHHEIMER (1914) who presents (on pp. 471 to 475) some results in table form and others in form of an equation, given by Sternberg as

$$(u_b)_{cr} = \xi\sqrt{d} \qquad \text{m/sec} \tag{6.9}$$

which resembles Eq. (6.6). The ξ value obtained with different experiments by various engineers is on the average $\xi \approx 4$. Forchheimer's discussion includes the research of duBuat (1786), Umpfenbach (1830), Telford (1838), Redtenbacher (1852), Sainjon (1871), Sternberg (1875), Blackwell (1872), Suchier (1883), Airy (1885), Franzius (1890), and others. It shall be remarked that some of these engineers use the surface or average velocity as criterion for incipient motion.

FORTIER et al. (1926) report on a most extensive study on "Permissible Canal Velocities," the maximum permissible value of the mean velocities. After a critical discussion of some laboratory data, they conclude:

> The best knowledge of scouring velocities comes from personal deductions, as to the performance of individual canals and not from direct experimental work. With this understanding in mind . . . (they) . . . submitted questionnaires to a number of irrigation engineers whose experience qualified them to form authoritative estimates of the maximum mean velocities allowable in canals of various materials.

ible canal velocities [*after* FORTIER *et al.* (1926)]

cavated for canal	*Velocity, fps, after aging, of canals carrying**		
	Clear water, no detritus (2)	*Water-transporting colloidal silts* (3)	*Water-transporting noncolloidal silts, sands, gravels, or rock fragments* (4)
idal)	1.50	2.50	1.50
loidal)	1.75	2.50	2.00
dal)	2.00	3.00	2.00
oncolloidal	2.00	3.50	2.00
	2.50	3.50	2.25
	2.50	3.50	2.00
	2.50	5.00	3.75
idal)	3.75	5.00	3.00
bles, when			
	3.75	5.00	5.00
olloidal	3.75	5.00	3.00
les, when colloidal	4.00	5.50	5.00
olloidal)	4.00	6.00	6.50
s	5.00	5.50	6.50
s	6.00	6.00	5.00

ess.

to these questionnaires together with data reported in the on engineering by Flynn (1892), Etcheverry (1916), and e summarized in a table which is reproduced as Table 6.1. the basis for canal design for many years. Since this study with original soil material, it seems almost impossible to in terms of Eq. (6.6), which really was established for an The excellent insight of the authors into the scour phenom- t by reading their conclusions, of which only a few are

of hydraulics governing the movement of loose silt and open channels are only distantly related to the laws the scouring of a canal bed and are not directly applicable. ial of seasoned canal beds is composed of particles of zes and when the interstices of the larger are filled by the e mass becomes more dense, stable, and less subject to the ion of water.

y required to ravel and scour a well bedded canal in any much greater than the velocity required to maintain movement of particles of that same material before becoming bedded or that have been raveled off by higher velocities than the bed would stand.

4. Colloids in either the material of the canal bed or the water conveyed by it, or in both, tend to cement particles of clay, silt, sand, and gravel in such a way as to resist erosive effects.
5. The grading of material running from fine to coarse coupled with the adhesion between particles brought about by colloids make possible high mean velocities without any appreciable scouring effect.

Erosion, transportation, and deposition are the subjects of a paper by HJULSTRÖM (1935). In an extensive analysis of data obtained "for monodisperse material on a bed of loose material of the same size of particles," Hjulström feels compelled to use the average flow velocity, because the more correct bottom velocity is seldom available. For this reason it was presumed that the average velocity is about 40 percent greater than the bottom velocity for a flow depth exceeding 1 m. The velocities, as given in Fig. 6.2, are for slightly varying depths. Figure 6.2 shows the limiting zone at which incipient motion starts and the line of demarcation between the sediment transport and sedimentation. The diagram also indicates that loose, fine

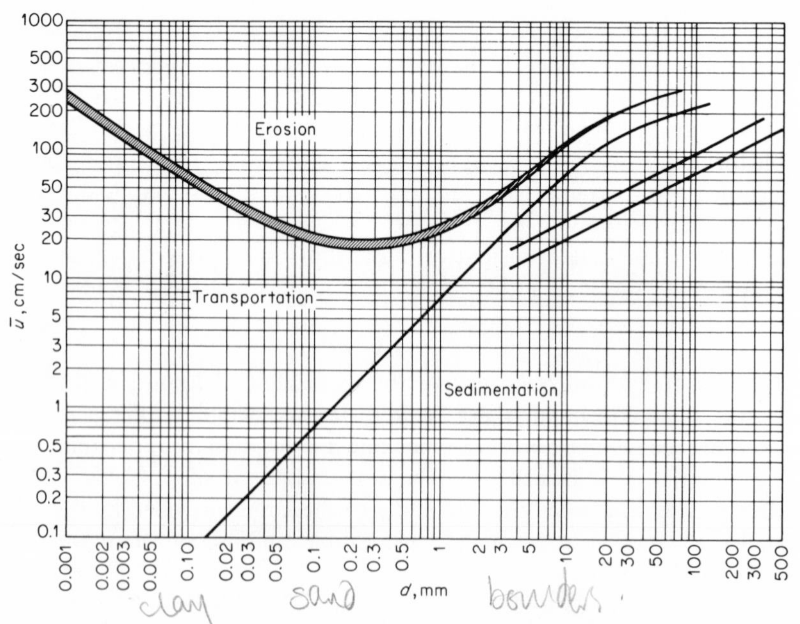

Fig. 6.2 Erosion-deposition criteria for uniform particles. [*After* HJULSTRÖM (*1935*).]

sand is the easiest to erode, and that the great resistance to erosion in the smallest particle range must depend on the cohesion and adhesion forces. Where applicable, the Hjulström diagram might serve as a useful concept. HELLEY (1969) has found agreement with Hjulström's diagram in a field study concerning large particles of $d \approx 33$ cm.

In a study by MAVIS et al. (1937) and its reappraisal by MAVIS et al. (1948), an equation for the *competent* bottom velocity was obtained. Data from about 400 tests were analyzed. The resulting equation of

$$(u_b)_{cr} = \frac{1}{2} d^{4/9} \sqrt{\frac{\rho_s}{\rho} - 1} \quad \text{fps} \tag{6.10}$$

with the grain diameter in millimeters, certainly does resemble Eq. (6.6).

RUBEY (1938) carefully analyzed Gilbert's data in the light of Eq. (6.7). It thereby was found that movement of coarse sand and gravel can be predicted with Eq. (6.7), but the smaller particles hiding in the laminar sublayer of the flow require much higher velocities than are indicated by this law.

Further research reported over the years is summarized by HEYNDRICKX (1948). This includes the work of Owens (1908), Gilbert (1914), Schaffernak (1922), Twenhofel (1932), Welikanoff (1932), Krey (1935), and Indri (1936). The suggested empirical equations are essentially of a type similar to Eq. (6.6). For natural sand with grain size distributions, some of the foregoing researchers introduce appropriate size distribution moduli.

Recently CARSTENS (1966) reported that a great many published data on incipient motion have been analyzed and could be represented by an equation such as

$$\frac{(u_b^2)_{cr}}{(\rho_s/\rho - 1)gd} \approx 3.61(\tan \varphi \cos \alpha - \sin \alpha) \tag{6.11}$$

Summarizing many Soviet data, BAREKYAN (1963) presents a graphical relationship between the average particle diameter and the *near-bottom current velocity*.

Numerous papers that have been published in the past report on studies using the average flow velocity $\bar{u}$ rather than the bottom velocity u_b as the initial motion criteria. Only a few recently developed ones shall be mentioned. JAROCKI (1963) discusses a formula deduced by Levi, which reads

$$(\bar{u})_{cr} = 1.4\sqrt{gd} \ln \frac{\bar{D}}{7d} \quad \text{m/sec} \tag{6.12}$$

applicable for a relative roughness value of $\bar{D}/d > 60$ only. Recently NEILL (1967) presented a "conservative design curve" for scour of coarse, uniform bed material with equation

$$\frac{\bar{u}_{cr}^2}{(\rho_s/\rho - 1)gd} = 2.50\left(\frac{d}{\bar{D}}\right)^{-0.20} \quad \text{cm/sec} \tag{6.13}$$

In the same paper a useful nomogram relating scour velocity $\bar{u}_{cr}$ to grain size, specific gravity, and depth of flow $\bar{D}$ is given. Equation (6.13) was established with data from various investigations.

6.2.3 CONCLUDING REMARKS

To use a velocity equation, such as discussed in the preceding section, as a criterion for incipient motion has been validly criticized by many researchers. The unanswered questions as to what is meant by bottom velocity u_b, and as to what is a proper relationship between bottom velocity u_b and average velocity $\bar{u}$, have led many hydraulicians to accept the more satisfactory quantity of the bottom shear stress $\tau_0 = \gamma S R_h$ as a scour criterion. For an elaborate discussion on this, the interested reader is referred to RUBEY (1938) and LANE (1953).

Finally a word of caution. As can be seen in Fig. 6.2, there exists for each grain size a certain velocity below which it will experience sedimentation, while above a certain velocity, called *critical scour velocity*, it will be eroded. These two phenomena are only slightly related, but are at times confused.

6.3 CRITICAL SHEAR STRESS EQUATIONS

6.3.1 THEORETICAL CONSIDERATIONS

According to FORCHHEIMER (1914), the relationship between the weight component of a water column and the friction force at the bottom can be given as

$$\gamma DS = k_4 u_b^2 \tag{6.14}$$

where D = water depth
S = slope of the energy grade line
k_4 = a constant

The expression γDS is the "tractive" force per unit surface. Other names currently in use are shear stress and drag force as well as the German expression *Schleppkraft* and the French one, *force d'entrainement*. The symbol frequently used is τ_0, or

$$\tau_0 = \gamma DS \tag{6.15}$$

This equation is more general if the depth is replaced by the hydraulic radius and reads

$$\tau_0 = \gamma R_h S \tag{6.16}$$

Already DUBUAT (1786) realized the advantage by using the "shear-resistance" concept, but it did not become popular prior to the work of SCHOKLITSCH (1914).

If the knowledge gained from Eqs. (6.14) and (6.15) is introduced into Eq. (6.6), we obtain[1]

$$\frac{(\tau_0)_{cr}}{(\gamma_s - \gamma)d} = A'' \tag{6.17}$$

where $(\tau_0)_{cr}$ is the critical shear stress or critical drag force at the point of incipient motion and A'' is a sediment coefficient.

6.3.2 EXPERIMENTAL INVESTIGATIONS

6.3.2.1 Earlier developments. Such experiments and their findings will be discussed which try to establish experimentally or empirically the kind of relationship given by Eq. (6.17).

SCHOKLITSCH (1914), acclaimed by Forchheimer, proposed an equation based on his own experiments, such as

$$(\tau_0)_{cr} = \sqrt{0.201\gamma(\gamma_s - \gamma)\lambda'\bar{d}^3} \qquad \text{kg/m}^2 \tag{6.18}$$

where $\bar{d}$ is the mean grain diameter and λ' is a shape coefficient, ranging from $\lambda' = 1$ for spheres to $\lambda' = 4.4$ for flat grains. At the same time, Schoklitsch proposes an equation of similar form as the previous one for sediment grains being of different size than the grain in the bed; this equation was checked with field data.

In the ensuing years, many laboratory and field investigations were carried out, notably by Gilbert, Engels, Schoklitsch, Schaffernak, and Krey. New information deduced from these tests led various investigators to fit their own correlation through a set of selected data. However, these investigations, like the ones suggested by KREY (1925), EISNER (1932), NEMENYI (1933), or O'BRIEN et al. (1934), do not differ considerably from the earlier Eq. (6.18), especially as far as its form is concerned.

An improvement over the foregoing studies was KRAMER's (1935) suggestion to describe the grain composition of the bed with the mean grain diameter $\bar{d}$ and a grain distribution modulus M, given by the ratio of areas F_A/F_B, as defined in Fig. 6.3.

KRAMER (1935) did suggest an equation which included his own data as well as those by other researchers. However, in a subsequent discussion, TIFFANY et al. (1935) present more experimental evidence, and it was felt that an equation of the form

$$(\tau_0)_{cr} = 29\sqrt{(\gamma_s - \gamma)\frac{\bar{d}}{M}} \qquad \text{g/m}^2 \tag{6.19}$$

is more representative, as can be seen in Fig. 6.3.

[1] In the presence of bedforms $(\tau_0)_{cr}$ is not dependent on the total hydraulic radius R_h, but rather on the portion of the hydraulic radius associated with the grain roughness R_h'.

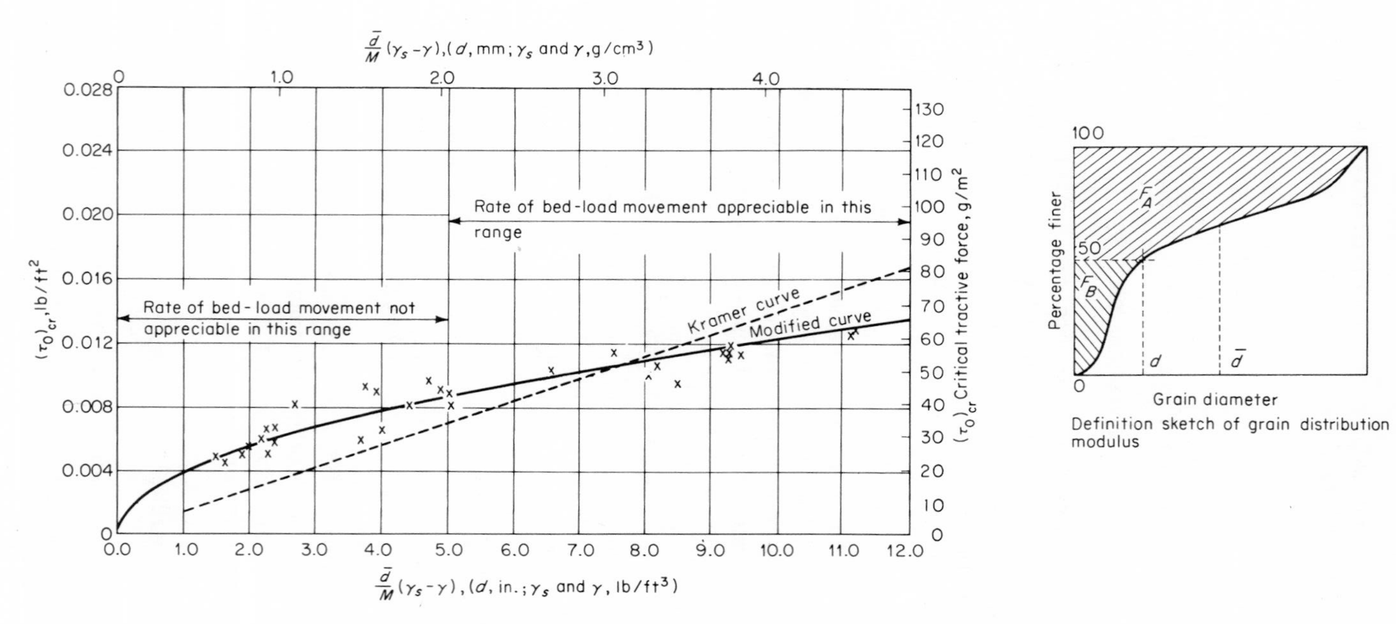

Fig. 6.3 Critical tractive force in relation to sand characteristics. [*After* TIFFANY *et al.* (*1935*).]

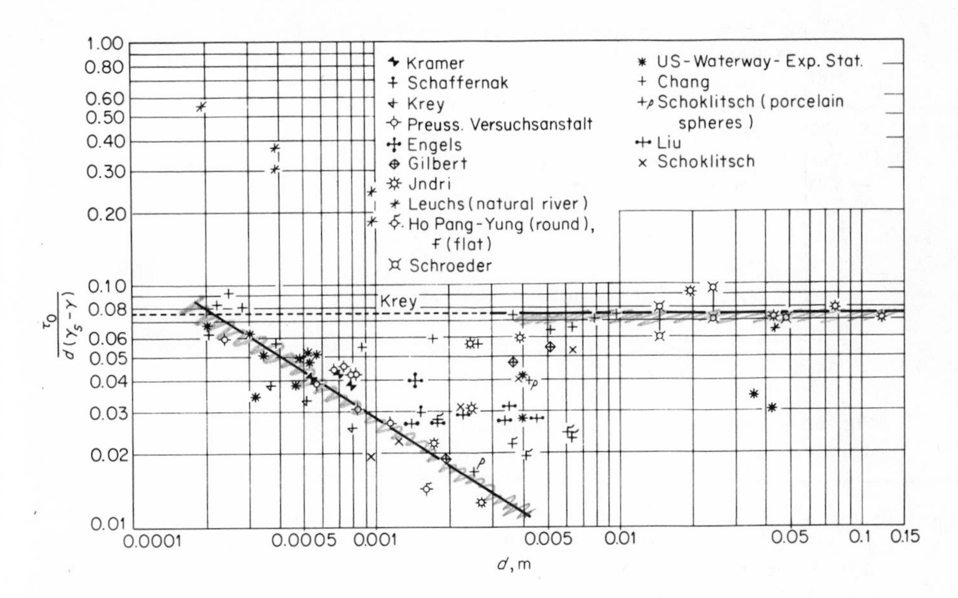

Fig. 6.4 Critical shear stress as function of grain diameter. [*After* SCHOKLITSCH (*1950*).]

Another set of laboratory experiments by CHANG (1939) presents data on the effect of the particle shape—elongated particles were studied—on the initial scour.

The complexity of the problem is probably nowhere better expressed than in SCHOKLITSCH (1950). Having been the prime mover in the field of sediment transport for 40 years, this researcher-engineer states: ". . . there is not too much known about the tractive force . . . ." Reanalyzing a great many data, two equations for the critical drag, *Grenzschleppspannung*, as he calls it, are given and plotted in Fig. 6.4.

For grains of $d \geq 0.006$ m, an equation already established by Krey is suggested, whereas for grains of $0.0001 \text{ m} < d < 0.003$ m, Schoklitsch suggests his own. If $d \geq 0.006$ m,

$$(\tau_0)_{cr} = 0.076(\gamma_s - \gamma)d \qquad \text{kg/m}^2 \tag{6.20}$$

and if $0.0001 \text{ m} < d < 0.003$ m,

$$(\tau_0)_{cr} = 0.000285(\gamma_s - \gamma)d^{1/3} \qquad \text{kg/m}^2 \tag{6.21}$$

with a transition between. The cause for the "strange shape" as given in Fig. 6.4—the shape of the Shields diagram comes into mind—according to Schoklitsch, cannot be explained. One wonders whether the research started by SHIELDS (1936) was intentionally or unintentionally ignored.

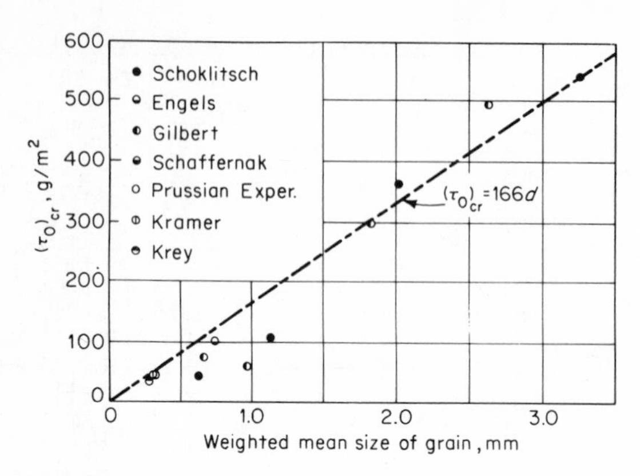

Fig. 6.5 Critical shear stress as function of grain diameter. [*After* LELIAVSKY (*1955*).]

Referring to the grain distribution modulus M, Schoklitsch attributes no advantage in using it.

LELIAVSKY (1955) feels that a convincing shear stress vs. grain size relationship, given in Fig. 6.5, can be represented with a simple relationship such as

$$(\tau_0)_{cr} = 166\bar{d} \qquad \text{g/m}^2 \tag{6.22}$$

where the mean particle diameter is given in millimeters. Equation (6.22) does apparently justice to most of the data, and is very simple in its use. That Schoklitsch's experiments fall below the line is explained by the fact that uniform material has been studied. The stabilization effect achieved with sediment mixtures was apparently not obtained. Equation (6.22) should not be applied if mean particle diameters are in excess of $\bar{d} = 3.4$ mm.

Considerable field data are used by LANE (1953) to establish the limiting tractive force diagram, as given by Fig. 6.6. This study, concerning itself with fine to coarse noncohesive materials, points out another seldomly mentioned fact. As can be clearly seen in Fig. 6.6, the critical shear stress for clear water is considerably lower than for water with a low or high content of sediments. The diagram given with Fig. 6.6 summarizes much of the important work done and, hence, should prove very helpful for the hydraulic engineer engaged in channel design.

6.3.2.2 Modern developments. Modern advancements in fluid mechanics suggested the desirability of expressing turbulent flow conditions with a quantity u_*. This term u_* is called *friction* or *shear velocity*, and represents a measure of the intensity of turbulent fluctuations. The relationship

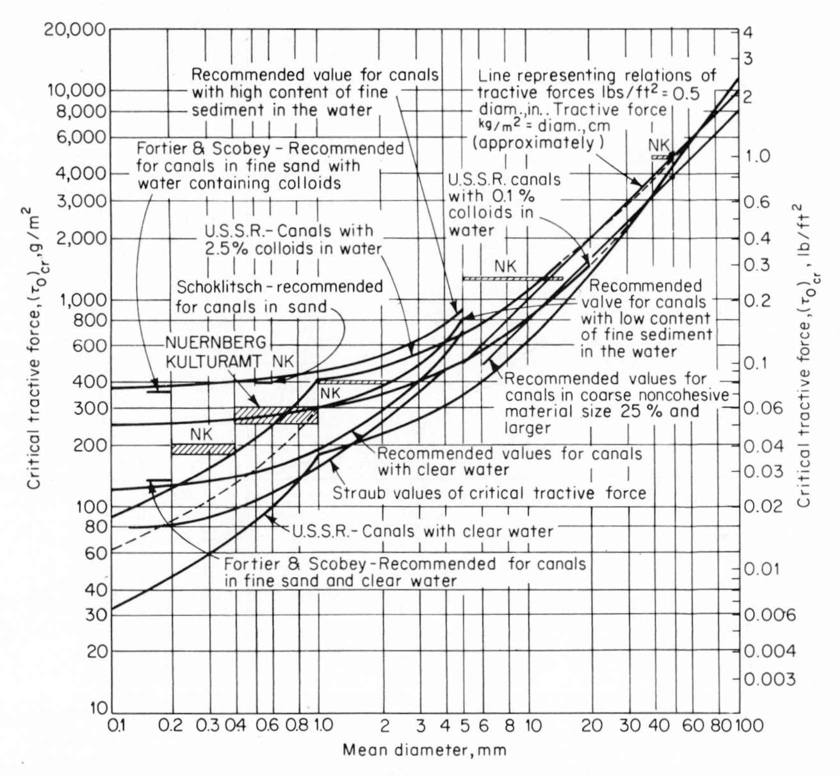

Fig. 6.6 Critical shear stress as function of grain diameter. [*After* LANE (*1953*).]

between the friction velocity and shearing stress is given by

$$u_* = \sqrt{\frac{\tau_0}{\rho}} \tag{6.23}$$

The first research in the mechanics of sediment transport, using the foregoing concept, was reported by SHIELDS (1936). For the special case of uniform grains on a flat bed, the sediment coefficient A'' in Eq. (6.17) becomes

$$\frac{(\tau_0)_{cr}}{(\gamma_s - \gamma)d} = \text{fct}\left(\frac{du_*}{\nu}\right) \tag{6.24}$$

where du_*/ν is referred to as the shear Reynolds number Re_*. In the establishment of Eq. (6.24), it becomes necessary to express the characteristic velocity with the laws of the logarithmic velocity distribution.[1] Physically

[1] A review of the logarithmic velocity distributions for pipes and channels is given by GRAF (1964).

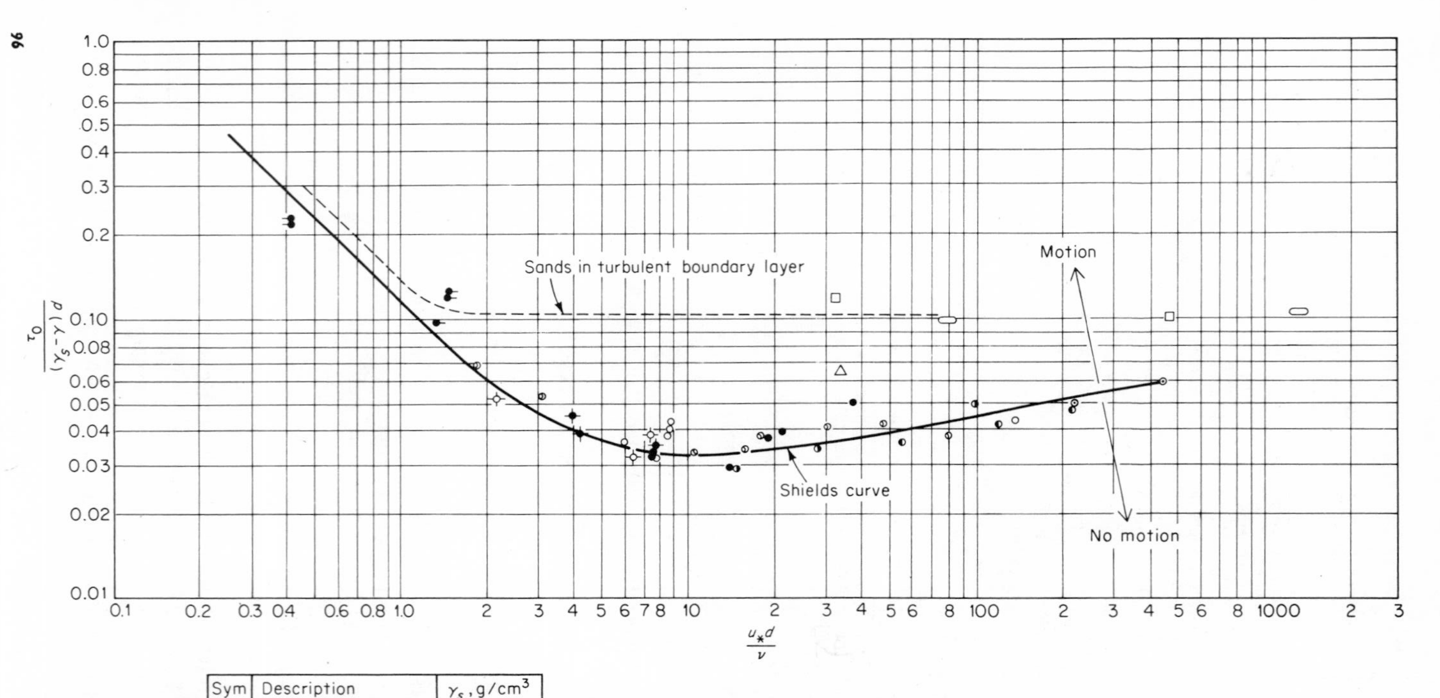

Fig. 6.7 Shields' diagram; dimensionless critical shear stress vs. shear Reynolds number. [*After* Vanoni (*1964*).]

it might be more helpful to express the shear Reynolds number in terms of the laminar sublayer δ or with the existing relationship of

$$\frac{du_*}{\nu} = 11.6\frac{d}{\delta} \tag{6.25}$$

Apparently it was independent research when Rubey (1938) suggested to use a *particle–laminar film ratio* as a scour criterion. The functional relationship, as given in Eq. (6.24), was established by Shields (1936) for a wide range of the variables involved, and is given in Fig. 6.7. On this graph a number of interesting facts can be observed. At once the similarity of the curves of incipient motion as suggested by Shields (1936) and those given by Hjulström (1935) (Fig. 6.2) and Schoklitsch (1950) (Fig. 6.4) are obvious. There is, however, no good reason to believe that this is a surprising fact. However, the shape of these curves is also very similar to the friction diagram for artificially roughened pipes and the drag coefficients relationship for a sphere or a cylinder. This perhaps does suggest that these phenomena are fluid mechanically related.

Three somehow distinct zones can be noticed in Fig. 6.7.

1. $d < \delta$: Up to $\mathrm{Re}_* \approx 2$, the particles are enclosed by a laminar film, their movement is mainly due to viscous action and independent of turbulence. The flow condition is similar to smooth boundary flow.
2. $d \gg \delta$: At large shear Reynolds numbers, the laminar sublayer gets interrupted by the existing bed roughness of the grains. The rough boundary is the source of the turbulence, and the dimensionless shear stress is independent of the shear Reynolds numbers. In Fig. 6.7, a value for $\mathrm{Re}_* \geq 400$ is obtained as

$$\frac{(\tau_0)_{cr}}{(\gamma_s - \gamma)d} = 0.06 \tag{6.26}$$

   Zeller (1963) finds this constant to be too high, and he obtains a value of 0.047.
3. $d \approx \delta$: At intermediate Reynolds numbers there lies a zone of transition when the laminar sublayer partially covers the particles and partially is interrupted. This transition curve has a minimum at $\mathrm{Re}_* \approx 10$. Shields' diagram indicates this as

$$\left.\frac{(\tau_0)_{cr}}{(\gamma_s - \gamma)d}\right|_{min} \approx 0.03 \tag{6.27}$$

   Below this value, no motion should ever occur.

Generally speaking, Shields' diagram, as Fig. 6.7 is usually referred to, has been checked by many experimentalists and is by now widely accepted.

Recently BOGARDI (1965) discussed Russian and Hungarian data, and finds the general trend, given by Shields' diagram, justified. However, the minimum of the dimensionless critical shear stress is shown around a value of 0.015.

Experimental evidence cited by TISON (1953) is such that within the laminar region $Re_* < 3.5$, neither a unique nor a straight-line relationship can be obtained. For the case of $Re_* > 3.5$, values are reported which fall below the curve in Shields' diagram. A study confined to a rather restricted phase of the problem is reported by IPPEN et al. (1953). Shields' relationship, as given by Eq. (6.24), was checked and found untenable where the sand of the bed and of the bedload is different in size and texture.

It should be mentioned that there exists evidence that Shields' relation does not explain data for extreme density differences between the fluid and sediment. Analysis of water-steel or water-plastics experiments which were reported by WHITE (1940), IPPEN et al. (1953), and WARD (1969) cannot readily be explained with the entrainment function of Fig. 6.7. A more general entrainment function was suggested and discussed by WARD (1969).

Since Shields' diagram was established for uniform sand, we would expect that nonuniform grain material as well as sticky or flocculant ones renders higher $(\tau_0)_{cr}$ values. Also, flat sand grains investigated by PANG (1939) gave larger values for incipient motion.

In Shields' diagram the grain diameter appears in the ordinate and abscissa. This grain diameter is the "representative" size d, and was and is taken as the median sieve size, the sieve size for which 50 percent per weight of material was finer or coarser. EGIAZAROFF (1965) proposes an equation for incipient motion for nonuniform mixtures, such as

$$\frac{(\tau_0)_{cr}}{(\gamma_s - \gamma)d} = \frac{0.1}{[\log 19(d/\bar{d})]^2} \tag{6.28}$$

where $\bar{d}$ is the average diameter of grain for both gradation curves, for grains in movement, and for total sediments. Previous Eq. (6.28) is only applicable when complete turbulence is assured. For a fine-graded mixture, $d < \bar{d}$, the resistance toward incipient motion is increased, while according to Eq. (6.28), the opposite is true for a coarse-graded mixture where $d > \bar{d}$.

EGIAZAROFF (1965) and PEZZOLI (1963) both try to establish a theoretical relationship for the Shields diagram as given by Eq. (6.24). Assuming that lift forces do not influence the problem, and applying the concept of logarithmic velocity distribution, a relation can be derived which, in the end, is still in need of experimental data. As such, it does not represent a great improvement over the functional relationship given by Eq. (6.24).

Also indicated in Fig. 6.7 are the experimental data obtained by WHITE (1940) and VANONI (1964). These relatively high shear stresses can be achieved if the frictional effects are limited to only a portion of the flow, i.e., a boundary layer which has not grown to the full channel depth. VANONI (1964) studied this particular problem, having in mind to improve the design of settling tanks. However, these findings have a limited value for the understanding of incipient motion in loose-boundary open channels.

The momentary rather than the averaged critical shear stress being considered in the study of incipient motion is suggested by KALINSKE (1942, 1947). Flow in open channels can be considered turbulent. Experimental evidence is such that velocity fluctuations are distributed according to the normal-error law, and it was shown that the standard deviation σ of a velocity u around its mean velocity $\bar{u}$ may be given for all practical purposes by

$$|u - \bar{u}|_{max} \approx 3\sigma \tag{6.29}$$

Data for velocity fluctuations near river bottoms were found to be

$$\frac{\sigma}{\bar{u}_b} \approx \frac{1}{4} \tag{6.30}$$

Combining Eqs. (6.29) and (6.30) renders a relationship of

$$u_b \approx 1.75\bar{u}_b \tag{6.31}$$

This implies that momentary velocities may exceed the average velocity by a factor of about 1.75. A relationship in terms of critical shear stress can be obtained, because the shear varies as the square of the velocity does. As such, momentary shear stresses can exceed the average ones by a factor of about 3. Shear stresses, as indicated in Shields' diagram, are the average ones and should be used as such. The work of KALINSKE (1947) was somehow extended by GESSLER (1965), who computed that the probability of a grain to be eroded, if the critical shear is taken from Shields' diagram, is 0.5. The median shear stress and the critical one for a given grain are equal if the probabilities of erosion and of nonerosion are equal. Similar results are reported by GRASS (1970).

6.3.3 CONCLUDING REMARKS

CHIEN (1954) has compared many of the tractive-force formulas; this comparison is reproduced in Fig. 6.8. Obviously, agreement among them is not very good; however, this is probably inherent in the nature of the sediment transport phenomenon. The reason for this is explained in the following.

Plotted on the vertical axis of Fig. 6.8 is the *critical tractive force.* Up to now, this definition was subject to interpretation of the individual researcher. SHIELDS (1936), for example, called it the *critical condition* at which zero sediment transport is obtained. KRAMER's (1935) definition of

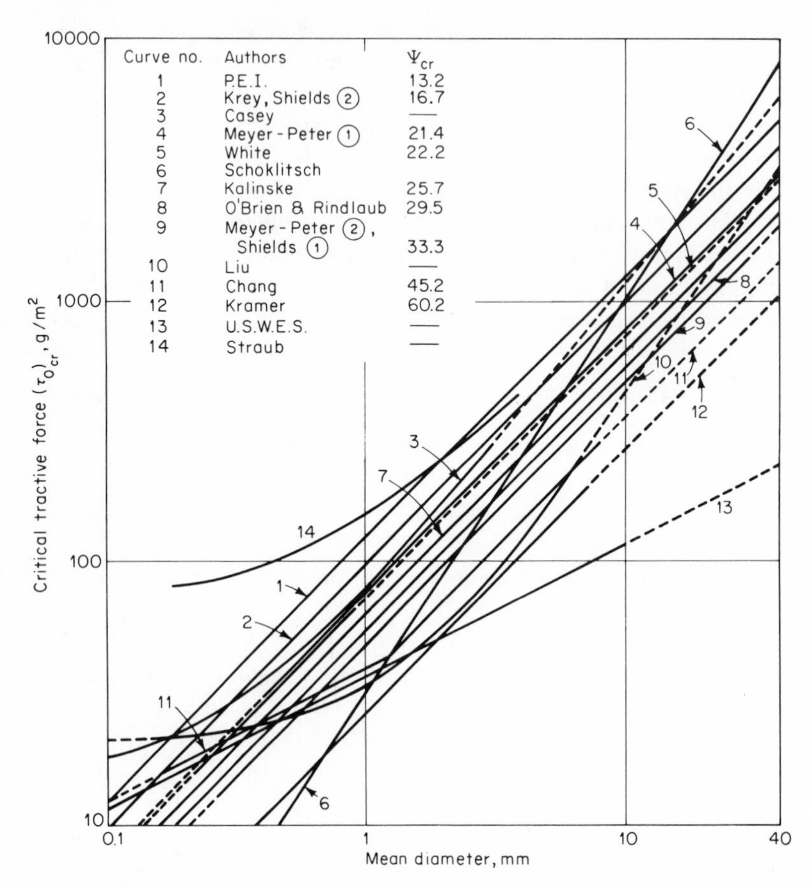

Fig. 6.8 Critical shear stress as function of grain diameter; a comparison. [*After* CHIEN (*1954*).]

weak movement seems to correspond to Shields' critical condition. It is understood that with weak movement few of the smallest sand particles are in motion in isolated spots. Most of the formulas are concerned with mean shear stresses; KALINSKE (1947) suggests that the maximum force should be considered which might, at times, exceed the average one by a factor of 3 or even 4. VANONI (1964) advances a criterion for incipient motion by counting the number of bursts per second; *critical* is defined with the burst frequency of $\frac{1}{3}$ to 1. GRASS (1970) employed high-speed cinephotography to trace sediment particles and fluid particles, the latter with the hydrogen bubble principle. A computer processed the film.

Plotted on the horizontal axis of Fig. 6.8 is the *mean diameter*. There is no evidence that the mean diameter represents most correctly the composition of a mixture. Parameters, such as the grain distribution modulus M, as introduced by KRAMER (1935), have not become very popular. The effects of small particles hiding behind larger ones or acting like a cement are some of the problems which certainly do exist, but are nowhere taken care of. Furthermore, in every kind of grain size mixture, the very small ones are most likely subject to cohesive forces, which are not represented in any of the models of the discussed formulas.

Considering the uncertainties involved in defining the arguments as plotted in Fig. 6.8, it should not be too astonishing to see the apparent spread. CHIEN (1954) has calculated the particular values of flow intensity Ψ', a parameter first used by EINSTEIN (1942), for the commencement of sediment motion; this is indicated in Fig. 6.8.

The design engineer facing the dilemma, as given by Fig. 6.8, should specify his problem most carefully, and then select a formula which suits his specifications best.

Nevertheless, all previously discussed critical shear stresses and velocity equations will predict that larger particles get less readily eroded than the smaller ones. As a result of this, the top layer (*Deckschicht*) is made up of coarser material than the original bed. This top layer will have a protective action on the underlying ones—an effect which is referred to as *armoring the bed.* Armoring (*Abpflästerung*) has already been described by DUBUAT (1796). Laboratory experiments[1] performed by GESSLER (1965) show the armoring effect very clearly. Figure 6.9*a* is a photograph of the granular bed in the test flume, made up of a mixture of particles before any flow is established. As water is allowed to flow, the bed configuration will change, and eventually a "fully developed" top layer is achieved. It can be seen in Fig. 6.9*b* that the mixture of particles has become considerably coarser. A special technique was developed to remove now this top layer. Figure 6.9*c* shows clearly that right beneath the top layer the original bed composition can be found. A longitudinal cross section through the bed is shown in Fig. 6.9*d*.

Finally a remark as to why the effects of lift forces on the scour problem apparently have been ignored. Strictly speaking, lift forces have been considered in deriving the basic scour criteria given with Eqs. (6.6) and (6.17). However, they never have been elaborated on in the discussed literature. This brings us to the conclusion that the lift is either ignored in the various investigations or that the sedimentation constants A' in Eqs. (6.6) and (6.7) and A'' in Eq. (6.17) absorb this effect among the other ones, like angle of repose, channel inclination, particle shape coefficients, etc. There exist some contributions to the scour problem which focus their attention on the effect of lift forces, which will be discussed subsequently.

6.4 LIFT FORCE MECHANISM

Although the importance of the lift forces was recognized in deriving the basic equation of scour, Eq. (6.2), the quantitative effect of the lift has not

[1] These experiments were done with clear-water flow.

(b)

Fig. 6.9 Granular bed in test flume [*after* GESSLER (*1965*)]. (*a*) Bed before water flow; (*b*) bed with *fully developed* top layer; (*c*) top layer, partially removed; (*d*) bed with *fully developed* top layer, longitudinal view.

(c)

(d)

Fig. 6.9 (*Continued.*)

been stressed in Secs. 6.2 and 6.3. This will be the subject of discussion in this section.

6.4.1 THEORETICAL CONSIDERATIONS

Lift forces may arise for at least two reasons. First, suppose that the particle under consideration rests on the bottom of a channel. This is the zone where velocity gradients are steepest; thus a pressure difference is set up which results in lifting the particle. Secondly, the same particle might experience lift because of the upward velocity components adjacent to the bed as a result of turbulence.

If the magnitude of the lift becomes equal to the weight, the smallest drag force would suffice to cause an initiation of motion. This can be seen quite clearly in Eq. (6.2) if a horizontal channel bed is assumed, i.e., $\alpha \approx 0$. Therefore we have to investigate as to what is the value of the lift force given by Eq. (6.4), or

$$F_L = C_L k_2 d^2 \frac{\rho u_b^2}{2} \tag{6.4}$$

and as to how it compares in magnitude to the drag force given by Eq. (6.3), and the particle's submerged weight given by Eq. (6.5).

JEFFREYS (1929) was apparently the first to show that classical hydrodynamics provides a simple explanation of lifting and carrying solid particles in fluids. Assuming a potential flow over a long, circular cylinder, with its major axis perpendicular to the flow, lift will take place if

$$\left(\tfrac{1}{3} + \tfrac{1}{9}\pi^2\right) u_\infty^2 > \frac{\rho_s - \rho}{\rho} ga \tag{6.32}$$

where u_∞ is the free-stream velocity. Attention is called to the similarity of Eq. (6.32), with Brahms' equation given by Eq. (6.8). An experimental check indicated that values obtained with Eq. (6.32) seem to be of the right order of magnitude. It was felt that modifying factors must be allowed for, since the two-dimensional model behaves differently from the flow past a grain, which is three-dimensional. The shortcoming of Jeffreys' model is that drag forces are altogether disregarded.

A similar idea was discussed by REITZ (1936), who suggested to express the beginning of sediment motion with a lift model. Circulation and viscosity are the important parameters in this investigation.

LANE et al. (1939) strongly emphasize the role of turbulence in the determination of the lift. The following assumptions became necessary: (1) Only particles having a settling velocity smaller than the instantaneous turbulent fluctuations at the bed will experience lift; (2) the velocity fluctuations vary according to the normal-error law; (3) the turbulent fluctuations and the shear velocity are related. A further development of this model by KALINSKE (1942, 1947) is found in Sec. 6.3.2.2.

The qualitative functional relationship in the Shields diagram, Fig. 6.7, can also be derived with a model for lift forces rather than for shear forces, as is shown by YALIN (1963). This, however, adds only further confidence to the Shields diagram.

6.4.2 EXPERIMENTAL INVESTIGATIONS

The average lift force on two sediment beds was measured directly as a pressure difference by EINSTEIN et al. (1949). Although these experiments did not aim to give a critical lift criterion, they ought to be considered as a step forward in understanding the lift mechanism. Used in the experiment were plastic spherical balls ($d = 0.225$ ft) and natural gravel of about the same average size ($d_{50} = 0.225$ ft), but with a considerable spread of grain size. All the data could be represented with an equation similar to Eq. (6.4), or

$$\Delta p = C_L \tfrac{1}{2} \rho u_y^2 \tag{6.33}$$

where Δp is the mean static-pressure difference in the beds at top and bottom of the grain. The velocity u_y is measured at a distance of 0.35 diameter (equivalent) from the theoretical wall; the lift coefficient C_L was found to have a constant value of 0.178. The results of this study were used by VANONI (1966) who calculated the ratio $\Delta p/\tau_0$, with τ_0 being the shear stress. This calculation gave values of $\Delta p/\tau_0$ of about 2.5, giving strong indication that lift forces are of considerable importance in the initial-motion mechanism. However, once a particle is displaced, lift forces tend to diminish and drag forces to increase, as pointed out by CHEPIL (1961).

Also investigated by EINSTEIN et al. (1949) were turbulent fluctuations of the lift. These experiments gave a constant average lift force with random fluctuations superimposed, following very accurately the normal-error law.

Lift forces on surface roughness elements in a windstream were studied by CHEPIL (1958) with direct and indirect measurements. Within the range of this study, lift was found to be equal to about 0.85 of drag for any size and fluid velocity; its Reynolds-number range is given with $47 < \text{Re} < 5{,}000$. Using an arbitrary scaling factor, the data are compared with the results given by EINSTEIN et al. (1949), which would indicate that the mechanism by which lift is produced is the same in any fluid-solid phase.

Only recently COLEMAN (1967) presented a preliminary study of drag and lift forces acting on a sphere resting on a hypothetical stream bed. Data for plastic and steel spheres were examined, and a lift-coefficient–vs.–Reynolds-number plot was obtained. Unexplained was, however, the fact why negative lift forces exist for Reynolds numbers less than 100.

An interesting entrainment mechanism was proposed by SUTHERLAND (1967). It was visualized that eddies exist within the turbulent boundary layers which disrupt the viscous sublayer, and then impinge onto the grains of the sediment bed. The motion within these eddies entrain the individual grains, provided it is strong enough. Qualitative experiments seem to support these arguments; however, no quantitative treatment is provided.

6.4.3 CONCLUDING REMARKS

Although lift forces obviously contribute to the incipient-motion problem, no "critical lift" criterion has been presented as yet which could be used as conveniently as, for example, Shields' diagram (Fig. 6.7) or Hjulström's diagram (Fig. 6.2). It was shown that, besides the lift forces, drag forces always exist. In other words, the model representing incipient motion, given by Eq. (6.2), seems to include all important forces. This model, which was the basis for the critical velocity equation, Eq. (6.6), and for the critical shear stress equation, Eq. (6.17), implicitly includes the lift. The discussion in this section should be considered a beginning to the explicit determination of the value of lift in Eq. (6.2).

6.5 SHEAR STRESS DISTRIBUTION

The shear stress, or unit tractive force, was earlier given by Eq. (6.15), or

$$\tau_0 = \gamma D S \tag{6.15}$$

For a two-dimensional flume, or in the special case of a very wide channel, Eq. (6.15) is quite correct. In this connection it should be remembered that the critical tractive force relations in Sec. 6.3 have been developed for cases where Eq. (6.15) holds true.

However, the more general shear stress equation includes the hydraulic radius R_h, and as such reads

$$\tau_0 = \gamma R_h S \tag{6.16}$$

Equation (6.16) represents the average value of the tractive force per unit wetted area. As will be shown subsequently, the shear stress in channels, except for a few cases, is not uniformly distributed along the wetted perimeter. Knowledge of the shear stress distribution is necessary for analyzing scour in channels on the basis of tractive force. In the following indirect and direct determinations of the shear distribution will be discussed.

6.5.1 INDIRECT DETERMINATIONS

Indirect determination of the shear stress distribution may be achieved with the knowledge of the velocity distribution. From this information the isovel pattern and its orthogonals may be obtained. Since the orthogonals are

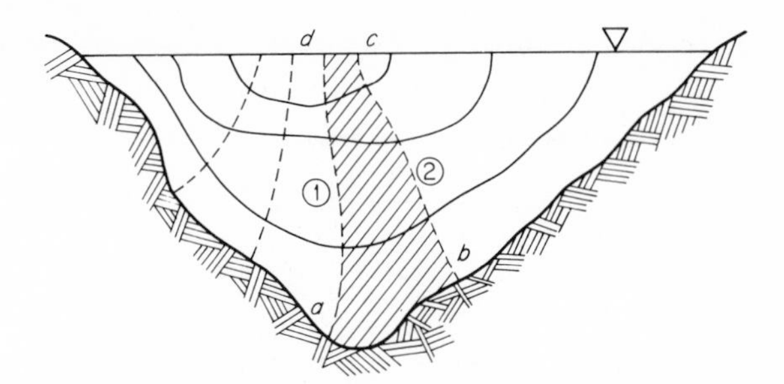

Fig. 6.10 Definition sketch for shear stress distribution; isovels (solid) and orthogonals (dashed).

assumed to be surfaces of zero shear, the entire weight component of the water between orthogonals and a boundary must be balanced by the shear stress. Figure 6.10 might help in understanding the problem; for example, the tractive force on the surface *ab* is calculated by multiplying the weight of the water enclosed by the two orthogonals 1 and 2, i.e., *abcd* with γS.

The first one to suggest this idea was LEIGHLY (1932) who obtained graphically the shear stress distribution for natural channels. LELIAVSKY (1955) suggested a similar method.

The shear stress distribution has been determined analytically for various cross sections, and is described by LANE (1953). Two methods were employed. One was based on an analysis of measured velocity distribution, as outlined by LEIGHLY (1932); but owing to a deficiency of data, this method did not render conclusive results. The other approach was a mathematical one, assuming a power law for the velocity distribution. By a membrane analogy and finite-difference methods the shear stress distribution for trapezoidal, triangular, and rectangular channels was obtained. In terms of maximum shear, the results are given in Fig. 6.11*a* for the sides and in Fig. 6.11*b* for the bottom; they are dependent on the width-depth ratio B/D and the side slopes, but are assumed to be independent of the actual size of the section. These results indicate that, within limits of the usual proportions of typical canal sections, the maximum shear stress is about equal to γSD and $0.75\gamma SD$ for bottom and sides of the channel, respectively, and zero shear stress exists in the corners.

The shear distribution for shallow channels with a plane bottom and a parabolic cross section was extensively studied by LUNDGREN et al. (1964). After a comparative discussion of the different analytical methods, a new one was developed, which is an extension of Prandtl's turbulence theory, taking into account the momentum transfer across normals to the bottom.

REPLOGLE et al. (1966) suggested a semianalytical approach, reducing the initial phases of the problem to a solution of laminar flow-type equations.

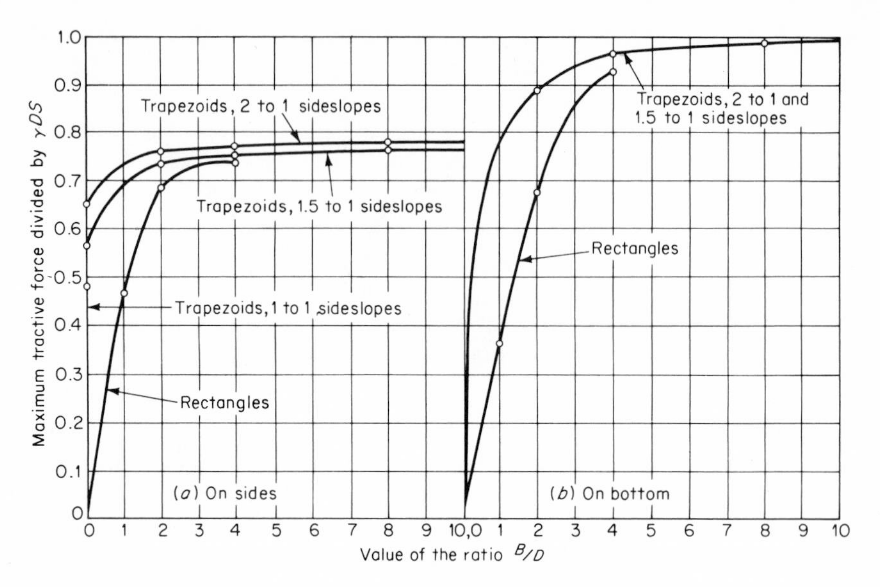

Fig. 6.11 Maximum shear stress in a channel. [*After* LANE (*1953*).]

This solution is then modified, not without assumptions, to simulate turbulent conditions. A digital computer was used to solve the equations by finite-difference methods. The procedure allowed for the prediction of the shear distribution in circular, trapezoidal, and rectangular open channels.

Indirect determination of the shear stress distribution requires either knowledge or assumption of the velocity distribution. Therefore the correctness of arriving at a proper shear stress is at times in question. Only in recent years instruments have been developed which measure the shear stress rather directly.

Some measurements have been published to allow comparison with data obtained from an indirect determination.

6.5.2 MEASUREMENT OF SHEAR STRESS

It appears that two kinds of instruments are at present available for the measurement of the stress; one is the shear meter or shear balance, the other is the Preston tube.

6.5.2.1 Shear meter. The principle of a shear meter is discussed in VENNARD (1961). A movable plate is built into a wall, which is mounted on elastic columns and fastened to a rigid support. As can be seen in Fig. 6.12, the shear stress created by the flow will minutely deflect the plate. The

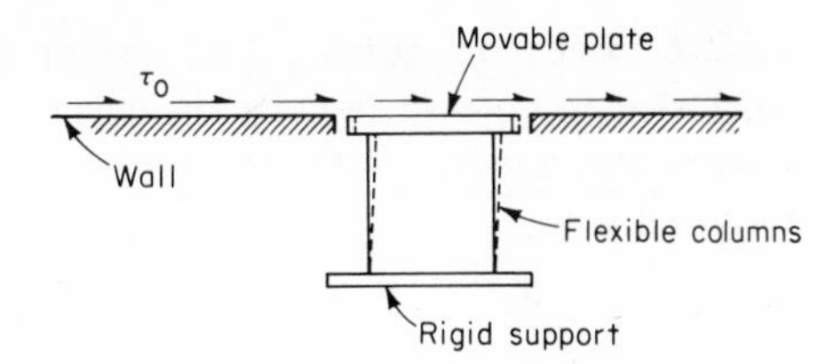

Fig. 6.12 Shear meter.

deflection of the plate, for example, measured by a strain gage after proper calibration, is an indication of the amount of shear.

An apparatus to measure shear stress over rather large areas, up to 0.015 m², was developed by STRAUSS (1962). A movable floating table is built into the wall which rests on a balance. Water movement will cause motion of table, which is recorded electronically. The instrument is limited to cases where the flow direction is known. Another measuring device was reported by STRAUSS (1964), which can be used for "point" shear stress measurement in stationary and nonstationary flow, and is given in Fig. 6.13. A deflection shaft attached to the table changes the resistance in the electronic pickup device, depending on the shear stress. The equipment, after calibration, is shown to pick up the turbulent shear stress fluctuations in stationary and nonstationary flow, but it does not indicate direction.

Disadvantages common to all shear meters are the required mechanical arrangements and disturbances of flow around the meter.

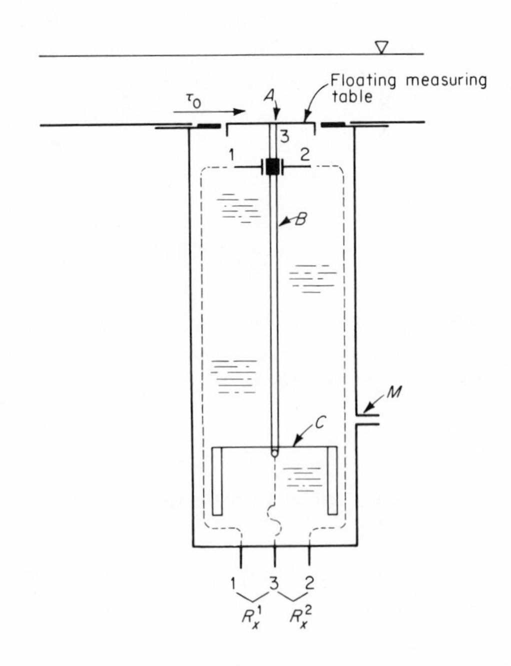

Fig. 6.13 Shear meter [*after* STRAUSS (*1964*)]. *A*—measuring table (10 cm²); *B*—deflection shaft; *C*—membrane; *M*—liquid opening; 1, 2, 3, R_x^1, R_x^2—electronic pickup device.

6.5.2.2 Preston tube. The use of a pitot tube at the wall has been suggested for measuring the shear stresses at the wall. Close to a smooth wall is the laminar sublayer, where the velocity distribution may be represented with

$$\frac{u_y}{u_*} = \frac{yu_*}{\nu} \tag{6.34}$$

where u_y is the velocity at a distance y, and $u_* = \sqrt{\tau_0/\rho}$ is the frictional velocity. A flattened pitot tube, suggested by STANTON et al. (1920), is immersed into the laminar sublayer of the flow, and it was shown that the dynamic pressure recorded by the pitot tube is in functional relationship with the shear stress. However it is very difficult to achieve a sufficiently accurate measurement of the velocity gradient in this layer, because in practice it was impossible to make the tube small enough to skim only the laminar sublayer material.

Near the surface of a wall there exists a region which includes the laminar sublayer, which is many times thicker than this region, and it can be represented by velocity distribution of

$$\frac{u_y}{u_*} = \text{fct}\left(\frac{yu_*}{\nu}\right) \tag{6.35}$$

The relationship of Eq. (6.35) is independent of pressure gradients and of upstream disturbances, and is the most general velocity distribution law for smooth boundaries. Larger pitot tubes can now be used, skimming more material. Their advantages are twofold: their mechanical robustness is greater and they record higher pressure readings.

A circular pitot tube with an outside diameter d_p resting on the wall reads the total pressure p_t. PRESTON (1954) showed that the total pressure p_t measured relative to the static pressure p_s, $(p_t - p_s)$, depends on the independent variables ρ, ν, τ_0, and d_p. Functional grouping yielded a relation of the form

$$\frac{p_t - p_s}{\rho\nu^2} d_p^{\,2} = \text{fct}\left(\frac{\tau_0 d_p^{\,2}}{\rho\nu^2}\right) \tag{6.36}$$

The functional relationship can be determined by direct calibration in pipe flow. For a circular pipe of a diameter r, the pressure drop Δp over the length l allows calculation of the boundary shear, such as

$$\tau_0 = \Delta p \frac{r}{2l} \tag{6.37}$$

Actual observations for four different pitot tubes could well be represented by

$$\log \frac{\tau_0 d_p^{\,2}}{4\rho\nu^2} = -2.604 + \frac{7}{8} \log \frac{(p_t - p_s)}{4\rho\nu^2} d_p^{\,2} \tag{6.38}$$

If the left-hand side becomes smaller than 4.5, Eq. (6.38) renders incorrect results; if larger than 6.5, no experimental data are available. PRESTON (1954) points out that, for the simple relationship given by Eq. (6.38) to hold true, two precautions are necessary: the pitot tubes must lie in the region given by Eq. (6.35), which is expected to hold for about one-tenth of the boundary-layer thickness, and the tube diameter must be small compared to the pipe diameter. The relationship of Eq. (6.38) can now be used to convert dynamic-pressure readings, $(p_t - p_s)$, into wall shear stress.

A pitot tube used in previously described manner is commonly referred to as a Preston tube. It is successfully employed in various branches of fluid mechanics. For example, LEUTHEUSSER (1963) used it in closed-conduit airflow, and IPPEN et al. (1962) in open-channel flow. The tube used in the latter study is shown in Fig. 6.14.

HWANG et al. (1963) report on a shear-measurement technique with a Preston tube for rough surfaces. The analytical relation between the dynamic pressure, $(p_t - p_s)$, and the shear stress was given as

$$\frac{p_t - p_s}{\tau_0} = 16.531\left\{\left(\log\frac{30h}{k_s}\right)^2 - \log\frac{30h}{k_s}\left[0.25\left(\frac{d'_p}{2h}\right)^2 + 0.0833\left(\frac{d'_p}{2h}\right)^4 + 0.00704\left(\frac{d'_p}{2h}\right)^6 + \cdots\right] + \left[0.25\left(\frac{d'_p}{2h}\right)^2 + 0.1146\left(\frac{d'_p}{2h}\right)^4 + 0.0586\left(\frac{d'_p}{2h}\right)^6 + \cdots\right]\right\} \tag{6.39}$$

where

d'_p = inside diameter of the Preston tube
k_s = size of roughness
h = position of the tube in relation to the datum

It was found that the experimental results agree quite well with Eq. (6.39).

6.5.3 DIRECT DETERMINATIONS

Using the Preston-tube technique, it became possible to check indirect determinations against the direct measurement. IPPEN et al. (1962) measured with an instrument, shown in Fig. 6.14, the peripheral distribution of local boundary shear stress in a straight trapezoidal channel.

With a rather similar technique, REPLOGLE et al. (1966) made direct measurements of the shear force in circular open channels. It was found that the shear stress distribution is dependent on the depth-to-channel-diameter ratio D/D_{ch}. A typical result is shown in Fig. 6.15. The maximum tractive force occurred close to the centerline of the channel, while the smallest values were observed near the water-surface edge. These facts and

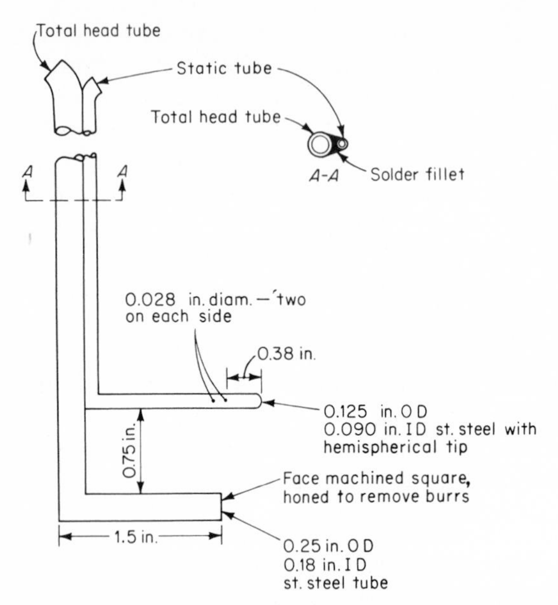

Fig. 6.14 Preston tube for measurement of shear stress. [*After* IPPEN *et al.* (*1962*).]

the independency of Reynolds and Froude numbers were found to be true in all cases tested. The semianalytical approach, suggested by REPLOGLE et al. (1966), predicted the shear stress distribution, actually measured with the Preston-tube technique, for circular channels (Fig. 6.15) and for trapezoidal ones (Fig. 6.16). Plotted in Fig. 6.16 are the experimental findings

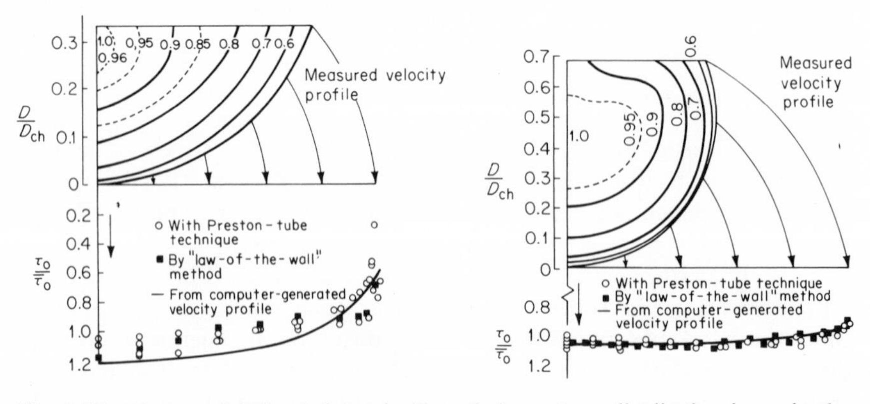

Fig. 6.15 Direct and indirect determination of shear stress distribution in a circular channel. [*After* REPLOGLE *et al.* (*1966*).]

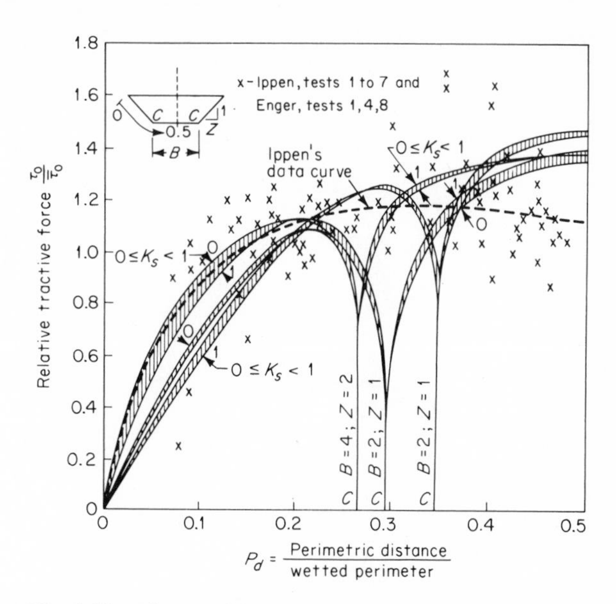

Fig. 6.16 Direct and indirect determination of shear stress distribution in trapezoidal channels, [*After* REPLOGLE *et al.* (*1966*).]

of IPPEN et al. (1962) and ENGER (1961). Experimental results on the distribution of shear stress in smooth and rough, rectangular, and trapezoidal channels have been reported by GOSH et al. (1970). Although the numerical solution predicts a zero tractive force at the corner of the channel, such a sharp intersection is not evident in the data. This brings out the fact that an ideal corner cannot be achieved in practice, and a finite radius of curvature must thus prevail. (The k_s value in Fig. 6.16 is a coefficient necessary for the numerical evaluation.) Detailed investigations on the shear distribution in corners are reported by LEUTHEUSSER (1963) and LIGGETT et al. (1965). It was shown that the turbulent shear distribution that depends on the Reynolds number has a greater uniformity than the laminar one. Leutheusser's results are plotted in Fig. 6.17 and certainly exhibit this trend; Liggett's et al. data are essentially a duplication of Leutheusser's research. The more even shear distribution can be explained by the existence of secondary currents.

6.5.4 BANK SCOUR

On an almost horizontal channel bed the initiation of particle motion can and was successfully expressed in terms of shear stress. The shear stress is a result of water forces which try to move the sediment particle down the channel in the direction of flow. However, on a channel bank another force

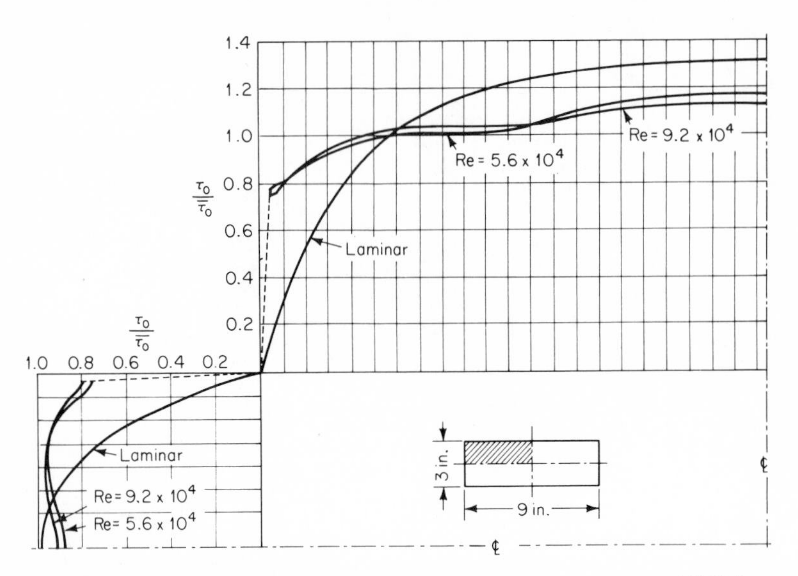

Fig. 6.17 Shear stress distribution in corners. [*After* LEUTHEUSSER (*1963*).]

must be considered—the gravity force that causes the particles to move down the sloping sides of the channel.

Earlier, the criterion for incipient motion was expressed as the ratio of forces parallel and normal to the angle of repose φ. With reference to Fig. 6.18, a relationship of incipient motion on banks was developed:

$$\tan \varphi = \frac{\sqrt{(W \sin \theta)^2 + 2F_D W \sin \theta \sin \beta + F_D{}^2}}{W \cos \theta - F_L} \tag{6.40}$$

where

F_D, F_L = drag and lift forces, respectively
θ = angle of inclination from the bank with the horizontal
β = angle of inclination of the shear stress as a result of secondary motion, which is especially pronounced in flow through channel curves

Equation (6.40) reduces to Eq. (6.2) under the appropriate assumptions, for $\theta = \alpha = 0$ and $\beta = 0$. A similar relationship, as given with Eq. (6.40), has been suggested by BROOKS (1963*a*) and CARSTENS (1966). Evaluations of Eq. (6.40) are hampered by the fact that nothing conclusive is yet available on the value of β.

Expressing the initiation of particle motion with the shear stress criterion, FORCHHEIMER (1914) implied and later LANE (1953) developed Eq.

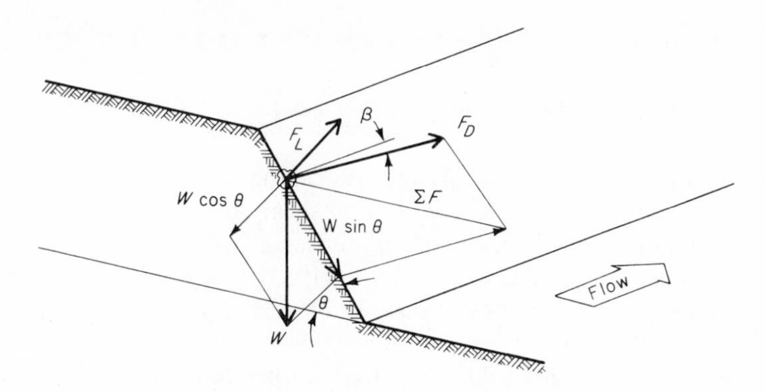

Fig. 6.18 Forces diagram on particle resting on a bank.

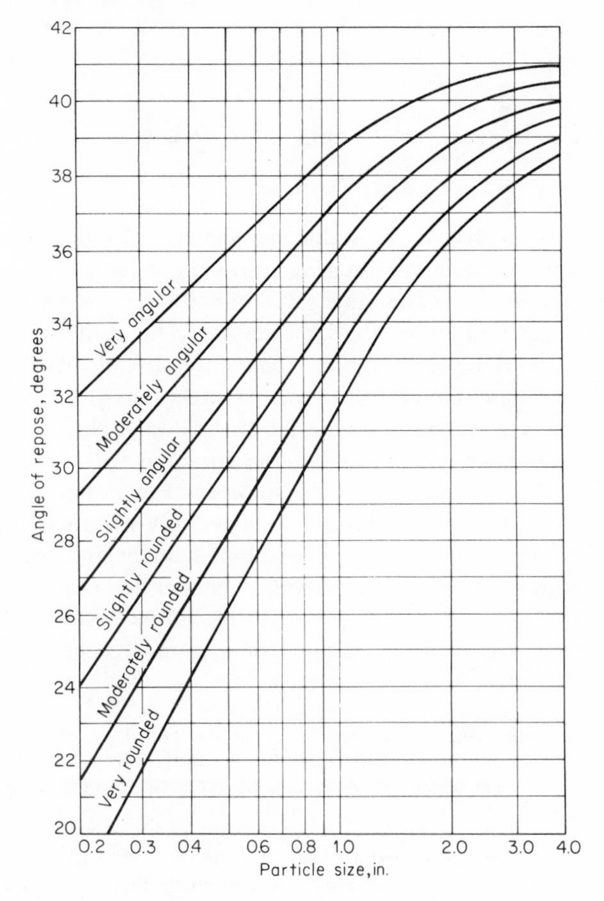

Fig. 6.19 Angles of repose of noncohesive material. [*After* LANE (*1953*).]

(6.41), where the critical wall shear stress $(\tau_0{}^w)_{cr}$ is given as[1]

$$(\tau_0{}^w)_{cr} = (\tau_0)_{cr}\left(\cos\theta\sqrt{1-\frac{\tan^2\theta}{\tan^2\varphi}}\right) \tag{6.41}$$

By knowing the critical shear stress on the bottom, for instance, with the aid of Shields' diagram the critical wall shear stress can be calculated, provided information on the angle of repose is available. Until more exact relations are available, a diagram by LANE (1953), given in Fig. 6.19, may be used, which gives values for the angle of repose for material above 0.2 in. in diameter for various degrees of roughness. Typical values for Eq. (6.41) are such that the allowable shear stress on banks is smaller than the one on the bed $(\tau_0{}^w)_{cr} < (\tau_0)_{cr}$. It is evident that, for a bank to be stable, the angle of the bank θ must be less than the angle of repose, φ or $\theta < \varphi$. Equation (6.41) has been suggested for the use in channel designs, but still lacks the experimental verification.

6.6 DESIGN OF STABLE CHANNELS[2]

6.6.1 STABLE CROSS SECTION

For the design of a stable channel cross section in noncohesive material, the various methods discussed throughout Secs. 6.2, 6.3, and 6.5 may be employed. For the same given conditions, there are almost no two equations which will render the same result. This could have been anticipated merely by looking at Fig. 6.8. Therefore special attention has to be paid when selecting among the various relations. A most careful comparative analysis of the soil material in the design channel with the one used by the researcher who proposed the equation will certainly be useful. Also the regional effects should be studied. For example, a designer in the western United States is probably better advised to make use of LANE's (1953) relationship, which was developed under and for these particular circumstances. On the other hand, channel design in central Europe might profit more from relations proposed by SCHOKLITSCH (1950).

Detailed calculations of a scour-free channel cross section are best described with an example.

Example 6.1 A rectangular canal shall be designed in a terrain where the slope is predetermined and given by $S = 0.01$. This canal, supposed to carry 30 m³/sec of clear water, should be free of scour for all purposes involved. The vertical banks

[1] Equation (6.41) can be derived from Eq. (6.40), assuming that $F_L \approx 0$ and $\beta = 0$; then F_D and W are substituted according to Eqs. (6.3) and (6.5), respectively; then the velocity vs. shear stress relation is introduced, given by Eqs. (6.14) and (6.15); and an equation of $(\tau_0{}^w)_{cr}$ is obtained. With an additional assumption to Eq. (6.41) of $\theta = 0$, $(\tau_0)_{cr}$ is obtained. The ratio of $(\tau_0{}^w)_{cr}/(\tau_0)_{cr}$ is the result given by Eq. (6.41).

[2] Nonscouring.

were made of wooden boards. A soil analysis gave this information: coarse quartz gravel with a representative size of $d = 50$ mm and a Manning's value of $n = 0.025$. What are the dimensions of the rectangular canal?

Approach 1 Use a "critical velocity equation" as the scour criterion; for example, Fig. 6.2.

For $d = 50$ mm $\rightarrow$ $(\bar{u})_{cr} \approx 2.5$ m/sec

a. With $\bar{u} = \dfrac{Q}{A}$:

$$\therefore \quad A = \frac{30}{2.5} = 12 \text{ m}^2$$

b. With $\bar{u} = \dfrac{1}{n} S^{1/2} R_h^{2/3}$:

$$\therefore \quad R_h = \left(\frac{\bar{u}n}{S^{1/2}}\right)^{3/2} = \left[\frac{(2.5)(0.025)}{\sqrt{0.01}}\right]^{3/2} = 0.49 \text{ m}$$

c. With $R_h = \dfrac{A}{P}$:

$$\therefore \quad P = \frac{A}{R_h} = \frac{12}{0.49} = 24.5 \text{ m}$$

With $A = DB = 12$ and $P = 2D + B = 24.5$:

$$\therefore \quad D = 0.51 \text{ m} \qquad B = 23.5 \text{ m}$$

Approach 2 Use a "critical shear stress equation" as the scour criterion; for example, Fig. 6.6.

For $d = 50$ mm $\rightarrow$ $(\tau_0)_{cr} \approx 4.5$ kg/m^2

a. With $\tau_0 = \gamma S R_h$:

$$\therefore \quad R_h = \frac{4.5}{(1{,}000)(0.01)} = 0.45 \text{ m} \approx D$$

b. With $\bar{u} = \dfrac{1}{n} R_h^{2/3} S^{1/2}$:

$$\therefore \quad \bar{u} = \frac{1}{0.025}(0.45)^{2/3}(0.01)^{1/2} = 2.35 \text{ m/sec}$$

c. With $Q = \bar{u}A$:

$$\therefore \quad A = \frac{Q}{\bar{u}} = \frac{30}{2.35} = 12.8 \text{ m}^2 \qquad B = \frac{A}{D} = \frac{12.8}{0.45} = 28.4 \text{ m}$$

The numerical differences obtained with the two approaches point to the fact that the choice of equation or figure used for incipient motion has to be made as carefully as possible. The conditions of the problem under investigation should be similar to those conditions under which the equation or graph had been established.

Example 6.2 A trapezoidal, completely unlined channel shall convey 57 m³/sec of clear water. The slope of 0.001 has been suggested; the bottom is made up of a mixture of slightly rounded quartz of $d_{50} = 37$ mm. An n value of 0.02 has been determined. What are the dimensions of the channel?

a. Angle of repose: From Fig. 6.19,

$$\varphi = 37^\circ$$

For stability reasons, $\varphi > \theta$
Assume a side slope of $1\frac{1}{2}$ horizontal and 1 vertical; $\theta = 33.6^\circ$

b. With $\dfrac{(\tau_0{}^w)_{cr}}{(\tau_0)_{cr}} = \cos\theta\sqrt{1 - \dfrac{\tan^2\theta}{\tan^2\varphi}}$: (6.41)

$$\frac{(\tau_0{}^w)_{cr}}{(\tau_0)_{cr}} = 0.83\sqrt{1 - \left(\frac{0.66}{0.75}\right)^2} = 0.40$$

c. Use critical shear stress criterion, for example, Fig. 6.7:

Assumption of rough zone $\dfrac{(\tau_0)_{cr}}{d(\gamma_s - \gamma)} \approx 0.06$

$\therefore$ bottom can take safely $(\tau_0)_{cr} = 3.7$ kg/m²

$\therefore$ side can take safely $(\tau_0{}^w)_{cr} = 0.4(\tau_0)_{cr} = 1.48$ kg/m²

d. With $(\tau_0{}^w)_{cr} = 0.75\gamma SD$ or with $(\tau_0)_{cr} = \gamma SD$ (from Fig. 6.11):

Solve for depth D

$$\therefore \quad D = \frac{1.48}{(0.75)(1{,}000)(0.001)} = 1.98 \text{ m}$$

$$\therefore \quad D = \frac{3.70}{(1{,}000)(0.001)} = 3.70 \text{ m}$$

With $D = 1.98$ m, Re_* is in the *rough* zone, thus assumption is correct.

$(\tau_0{}^w)_{cr}$ will be the controlling value.

e. With $Q = A\bar{u}$:

$$\bar{u} = \frac{1}{n} S^{1/2} R^{2/3} = \left(\frac{1}{n} S^{1/2}\right) f(D,B)$$

$$A = f'(D,B)$$

$$\therefore \quad B \approx 11 \text{ m (for } D = 1.98 \text{ m)}$$

f. The obtained values are on the safe side since after some time of flow the armoring effect will take place. Such that

d_{50} (before flow) $< d_{50}$ (due to armoring)

6.6.2 IDEAL, STABLE CROSS SECTION

In the design of a channel, as discussed in Sec. 6.6.1, the state of incipient motion occurs over a small portion of the wetted perimeter—usually somewhere on the sides of the channel—while the remaining part of the perimeter

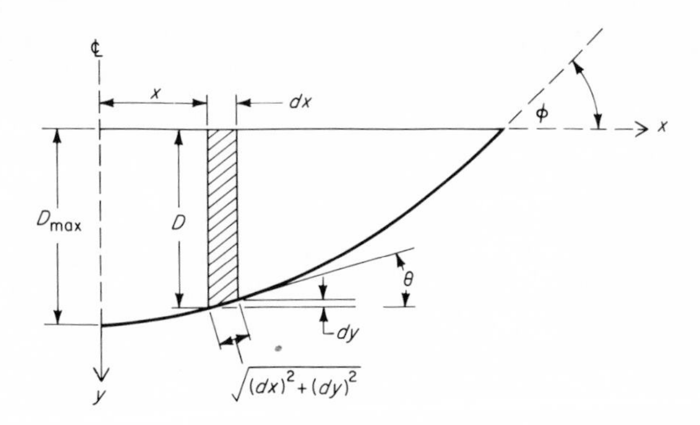

Fig. 6.20 Channel cross section; an ideal, stable one.

is not stressed to capacity. However, a channel cross section can be designed such that incipient motion occurs simultaneously at every point of the wetted perimeter. Using the tractive-force concept, GLOVER et al. (1951) developed theoretically this ideal, stable cross section for erodible channels. According to LANE (1953), the following five assumptions had to be made in the derivation:

> (1) at and above the water surface, the side slope is at the angle of repose of the material; (2) at points between the center and edge of the channel the particles are in a state of incipient motion, under the action of the resultant of the gravity component of the particles submerged weight acting down the side slope and the tractive force of the flowing water; (3) at the center of the channel the side slope is zero and the tractive force alone is sufficient to cause incipient motion; (4) the particle is held against the bed by the component of the submerged weight of the particle acting normal to the bed; and (5) the tractive force on any area is equal to the component of the weight of the water above the area in the direction of flow. Under Assumptions 1, 2, and 3 the particles on the entire perimeter of the canal cross section are in a state of impending motion.

Owing to assumption 5, the shear stress acting on an element of length

$$\sqrt{(dx)^2 + (dy)^2}$$

of the sloping-channel cross section can be expressed as

$$\tau_0{}^w = \gamma DS \cos \theta = \frac{\gamma DS\, dx}{\sqrt{(dx)^2 + (dy)^2}} \tag{6.42}$$

In Fig. 6.20 the continuously curved cross section is shown; its shape will be determined by subsequent analysis. If the shear stress, as given by

Eq. (6.42), is everywhere equal to the critical one, given by Eq. (6.41), a relationship is obtained such as

$$\gamma DS \cos\theta = \gamma D_{max} S\left(\cos\theta\sqrt{1 - \frac{\tan^2\theta}{\tan^2\varphi}}\right) \tag{6.43}$$

with the centerline depth of D_{max} at $x = 0$. After proper cancellations and replacing of $\tan\theta = dy/dx$, the following differential equation for the cross section is obtained:

$$\left(\frac{dy}{dx}\right)^2 + \left(\frac{D_{max}}{D}\right)^2 \tan^2\varphi - \tan^2\varphi = 0 \tag{6.44}$$

Separation of variables, integration, and introduction of a boundary condition ($D = D_{max}$ at $x = 0$) produce the solution of Eq. (6.44), or

$$D = D_{max}\cos\left(\frac{\tan\varphi}{D_{max}}x\right) \tag{6.45}$$

Equation (6.45) defines the geometry of an ideal, stable cross section; it is a simple cosine curve.[1] With further calculations, other properties of this stable cross section are obtained, such as the area of flow A, the top width B, and the average flow velocity $\bar{u}$, or

$$A = \frac{2D_{max}^2}{\tan\varphi} \tag{6.46}$$

$$\bar{u} = \frac{1}{n}\left[\frac{D_{max}\cos\varphi}{E(\sin\varphi)}\right]^{2/3} S^{1/2} \tag{6.47}$$

$$B = \frac{D_{max}\pi}{\tan\varphi} \tag{6.48}$$

where $D_{max} = \tau_0/\gamma S$ and $E(\sin\varphi)$ is an elliptic integral given by $E(\sin\varphi) \approx (\pi/2)(1 - \frac{1}{4}\sin^2\varphi)$. Equations similar to Eqs. (6.46) and (6.47) were given by MORRIS (1963).

It was found that the ideal, stable channel, as described by the foregoing equations, is one of minimum excavation, maximum hydraulic efficiency, minimum perimeter, and minimum top width for a given discharge.

The discharge for the theoretical section can now be computed with $Q = \bar{u}A$. Most likely the design discharge differs from this value. Suppose the design discharge is less than the theoretical one, $Q_D < Q$, then the width B has to be decreased according to $(B - B')$, with

$$B' - B = B\sqrt{\frac{Q_D}{Q}} \tag{6.49}$$

[1] Channels which approximate this geometry may be obtained in the laboratory, but also small channels of braided rivers have, at times, this shape. Most streams, however, are better approximated by a trapezoidal section.

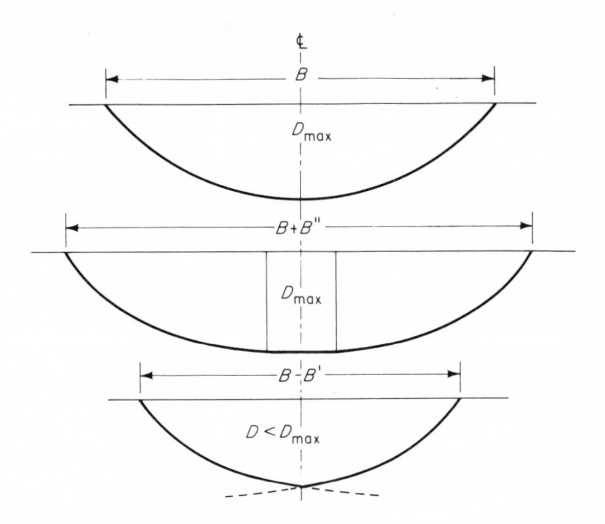

Fig. 6.21 Channel cross sections; ideal, stable one for changing widths.

If the opposite is true, i.e., $Q_D > Q$, then the width B can be increased to $(B + B'')$, with

$$B'' = \frac{n(Q_D - Q)}{D_{max}^{5/3} S^{1/2}} \tag{6.50}$$

The effect of the width change on the geometry of the channel is shown in Fig. 6.21. Development of Eq. (6.50) is reported by CHOW (1959).

It appears that FORCHHEIMER (1914) already realized the concept of an ideal, stable section and expressed it with the radius of curvature of the cross section. Equation (6.51),

$$r = \frac{D_{max}}{\sin \varphi \tan \varphi} \sqrt{\frac{D_{max}^2}{D^2} + \tan^2 \varphi} \tag{6.51}$$

indicates a radius of curvature of infinity at the water-surface intersection, and a minimum radius at the center of the channel cross section. In a further analysis, FORCHHEIMER (1930) included also the influence of the longitudinal slope of the canal. However, LANE (1953) seems to doubt whether this effect is significant.

Other contributions dealing with design formulas for ideal, stable cross sections have been made by GHETTI (1952) and BRETTING (1958). Some of the recent Russian work has been discussed by SCHEIDEGGER (1961).

We shall conclude this section by demonstrating the application of gained knowledge to a problem.

Example 6.3 The ideal, stable cross section for the same channel as given in Example 6.2 shall be determined.

a. Given are:

$$\varphi = 37^\circ \qquad S = 0.001 \qquad n = 0.02 \qquad Q = 57 \text{ m}^3/\text{sec} \qquad (\tau_0)_{cr} = 3.7 \text{ kg/m}^2$$

b. $D_{max} = \dfrac{\tau_0}{\gamma S} = \dfrac{3.7}{(1{,}000)(0.01)} = 3.7 \text{ m}$

c. Shape of cross section:

$$D = D_{max} \cos\left(\frac{\tan \varphi}{D_{max}} x\right) = 3.7 \cos 0.204x$$

d. $B = \dfrac{D_{max}\pi}{\tan \varphi} = 15.4 \text{ m}$

e. $A = \dfrac{2D_{max}^2}{\tan \varphi} = 37.1 \text{ m}^2$

f. $\bar{u} = \dfrac{1}{n}\left[\dfrac{D_{max} \cos \varphi}{E (\sin \varphi)}\right]^{2/3} S^{1/2} = 2.56 \text{ m/sec}$

g. $Q = \bar{u}A = 95 \text{ m}^3/\text{sec}$

$Q_D = 57 \text{ m}^3/\text{sec}$

$\therefore \quad Q_D < Q$

h. $B' = B\sqrt{\dfrac{Q_D}{Q}} = 15.4\sqrt{\dfrac{57}{95}} = 11.95 \text{ m}$

7
The Bedload

7.1 INTRODUCTORY REMARKS

Assume the hydraulic parameters in a channel with loose material to be such that a small change of one of the constituents will cause the *critical condition* of the bed to be exceeded, and as such it will be responsible for sediment transport. Suppose the motion of entrained particles is one of rolling, sliding, and sometimes jumping (saltating), and this particle motion takes place close to the bed of the channel. This kind of sediment transport is commonly referred as the transport of the *bedload* or the *contact load*.

In the past numerous bedload equations have been proposed, but some of them look very similar. Analogous to the three groups of *initial scour* criteria, there are essentially three slightly different approaches to the bedload problem. They are:

1. The *duBoys-type* equations, considering a shear stress relationship
2. The *Schoklitsch-type* equations, considering a discharge relationship

3. The *Einstein-type* equations, based upon statistical considerations of the lift forces.

Also presented is a bedload equation which considers the bedform motion.

A brief section on saltation as a possible mode of particle transport is included. An extensive discussion of the different bedload equations will conclude this chapter.

7.2 DUBOYS-TYPE EQUATIONS

7.2.1 THEORETICAL CONSIDERATIONS

Ninety years ago DuBoys (1879) advanced a model of sediment transport, which has since been the subject of many investigations and much criticism as well. Nevertheless, many a channel or canal have since been designed on the basis of this very model.

It is assumed that the sediment moves in layers, each of which has a thickness ϵ. These layers move because of the tractive force given by Eq. (6.15) or because $\tau_0 = \gamma DS$ is applied to them. The last layer, i.e., the first layer in Fig. 7.1, is the one where the tractive force balances the resistance force between these layers, such that

$$\tau_0 = \gamma DS = c_f n\epsilon(\gamma_s - \gamma) \tag{7.1}$$

where c_f is a frictional coefficient. The fastest moving layer is the one closest to the water, and moves with the velocity of $(n - 1)v_s$. If the layer between the first and nth layer moves according to a linear velocity distribution, then the amount of solid material per time and per unit width is given by

$$q_s = \epsilon v_s \frac{n(n-1)}{2} \qquad \text{m}^3/(\text{sec})(\text{m}) \tag{7.2}$$

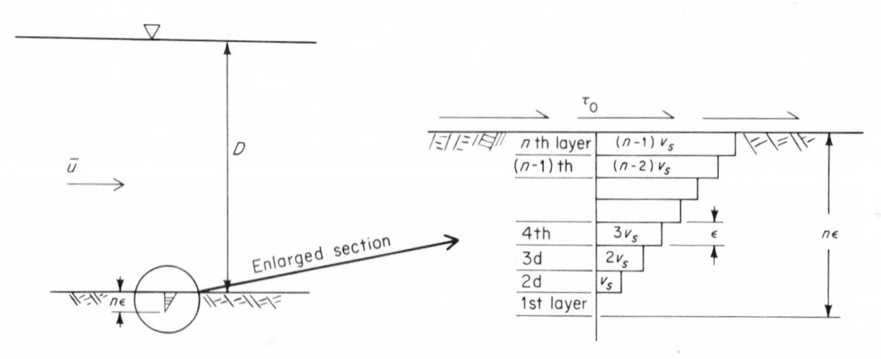

Fig. 7.1 Sketch of duBoys' bedload model.

In Eq. (7.2) the quantity $n\epsilon$ is the thickness of the sediment material moving with an average velocity of $[v_s(n-1)]/2$. The "critical condition" at which sediment motion is just about to begin is given by $n = 1$; then Eq. (7.1) becomes

$$(\tau_0)_{cr} = c_f \epsilon(\gamma_s - \gamma) \tag{7.3}$$

This, in turn, results in the relationship of

$$\tau_0 = n(\tau_0)_{cr} \tag{7.4}$$

which is introduced into Eq. (7.2), and the following is obtained:

$$q_s = \left[\frac{\epsilon v_s}{2(\tau_0)_{cr}^2}\right] \tau_0[\tau_0 - (\tau_0)_{cr}] \tag{7.5}$$

DuBoys (1879) referred to the quantity $\epsilon v_s/2(\tau_0)_{cr}^2$ as a characteristic sediment coefficient, and gave it the symbol χ. The DuBoys equation predicting the volume amount of bedload per unit width and time may thus be written as[1]

$$q_s = \chi \tau_0[\tau_0 - (\tau_0)_{cr}] \tag{7.6}$$

In effect, the total amount of solid material Q_s is given by $\int_{x_1}^{x_2} q_s \, dx$, or

$$Q_s = \chi \int_{x_1}^{x_2} \tau_0[\tau_0 - (\tau_0)_{cr}]\, dx \qquad \text{m}^3/\text{sec} \tag{7.7}$$

with the coordinate x indicating the width of the channel.

For the derivation of Eq. (7.6) certain assumptions had to be made, which have been shown to be in general disagreement with observations. The oversimplified model that the bedload moves as sliding layers is in strong disagreement with observations; this has been pointed out already by Schoklitsch (1914). However, the great simplicity of Eq. (7.6) and its—at least at times—surprising good agreement with field and laboratory data make it probably the widest used bedload equation.

However, it can be shown that Eq. (7.6) can also be derived without making such strong assumptions. The bedload transport is most likely related to the tractive force, such that

$$q_s = \text{fct}\,(\tau_0) \tag{7.8}$$

It was proposed by Donat (1929) to approximate the function with a power series and

$$q_s = k_1 + k_2(\tau_0) + k_3(\tau_0)^2 + \cdots \tag{7.9}$$

after neglecting higher order terms. Two boundary conditions are found easily; if the shear stress is zero, no bedload transport exists and $k_1 = 0$; if

[1] This is exactly the form as it can be found in DuBoys' (1879) original paper, on page 160.

the shear stress is finite and if the *critical condition* exists, the bedload transport is just about to start, but it is still little and $k_2 = -k_3(\tau_0)_{cr}$. With this knowledge, Eq. (7.9) becomes

$$q_s = k_3\tau_0[\tau_0 - (\tau_0)_{cr}] \tag{7.10}$$

where $k_3 = \chi$ is the characteristic sediment coefficient. Obviously, Eqs. (7.10) and (7.6) are identical.

In the following discussion, all bedload equations based on excess shear stress, i.e., $\tau_0 - (\tau_0)_{cr}$ or $\tau_0/(\tau_0)_{cr}$ will be classified as *duBoys-type* equations. Information on the shear stress or the critical shear stress may be obtained from Chap. 6. It thus remains that proper use of Eq. (7.6) depends on the correct evaluation of the characteristic sediment coefficient χ.

7.2.2 FURTHER INVESTIGATIONS

7.2.2.1 Earlier developments. The experiments of SCHOKLITSCH (1914) proved duBoys' model of sliding layers to be wrong, but they could be well represented by duBoys' equation. For uniform grains of various kinds of sand and of porcelain, the characteristic sediment coefficient could be determined empirically as

$$\chi = 0.54 \frac{1}{\gamma_s - \gamma} \tag{7.11}$$

Equation (7.11), which is written in metric units, has the shortcoming that only limited data are represented, and the experimental flume was of small dimensions. The excellent and extensive data reported by GILBERT (1914), who apparently was not aware of duBoys' work, were analyzed by DONAT (1929), and a definite relationship between the χ value and the mean diameter was found. Later STRAUB (1935) elaborated on this point by examining the research of various investigators. Averaged values of χ and $(\tau_0)_{cr}$ for sand sizes tested are given in Table 7.1, or by Eq. (7.12) as

$$\chi = \frac{0.173}{d^{3/4}} \tag{7.12}$$

Both Table 7.1 and Eq. 7.12 are restricted to the use in the British system of units. For the metric system ZELLER (1963) gives a graph, i.e., Fig. 7.2.

Table 7.1 Evaluation of parameters in Eq. (7.6) [*after* STRAUB (*1935*)]

d, mm	$\frac{1}{8}$	$\frac{1}{4}$	$\frac{1}{2}$	1	2	4
χ, ft^6/(lb^2)(sec)	0.81	0.48	0.29	0.17	0.10	0.06
$(\tau_0)_{cr}$, lb/ft^2	0.016	0.017	0.022	0.032	0.051	0.09

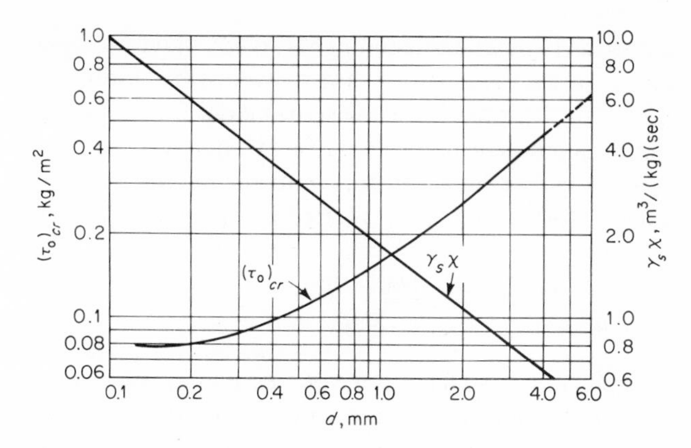

Fig. 7.2 Evaluation of parameters in Eq. (7.6) for metric units. [*After* ZELLER (*1963*).]

STRAUB's (1935) work has been criticized, mainly on two accounts: (1) All the data are obtained in laboratories having small flume dimensions, and were taken only over a small range of particle sizes, and (2) field measurements have apparently not clarified the applicability of Eq. (7.12). Further experiments reported by CHANG (1939) could be represented by duBoys' equation, and it was suggested to express the characteristic sediment coefficient χ as a function of Manning's roughness coefficient n. The latter relationship was useful in finding an empirical expression for a set of 275 observations of bedload movement by the U.S. WATERWAYS EXPERIMENT STATION (1935).

O'BRIEN et al. (1934) generalized duBoys' equation. They propose that in a state of dynamic equilibrium shear stresses must be constant on planes parallel to the bottom. Under various assumptions and by pointing out a certain analogy of sediment motion with *Bingham* fluids, the following relation was obtained:

$$q_s = \chi'[\tau_0 - (\tau_0)_{cr}]^m \tag{7.13}$$

Analysis of Gilbert's data showed that the new parameters χ' and m are in functional relationship with the median diameter. Quite independently the U.S. WATERWAYS EXPERIMENT STATION (1935) found an empirical equation of the general form of Eq. (7.13). For sand mixtures of $0.025 < d < 0.560$ mm, the values of the exponent m are confined to a narrow range of $1.5 < m < 1.8$.

7.2.2.2 Modern developments. The concepts of modern fluid mechanics were successfully applied by SHIELDS (1936) to the problem of critical shear stress, which was discussed in Sec. 6.3.2.2 and was given by Eq. (6.24) or Fig. 6.7. An excess of shear stress, excess in relation to the critical one, presented to SHIELDS (1936) a useful model for the sediment transport. It was not his intent to establish a universal equation, but to represent in an abbreviated form the many factors influencing the problem. The semiempirical tractive-force equation is plotted in Fig. 7.3 and given by Eq. (7.14), or

$$\frac{q_s \gamma_s}{q S \gamma} = 10 \frac{\tau_0 - (\tau_0)_{cr}}{(\gamma_s - \gamma)d} \tag{7.14}$$

where q is the water discharge per unit width. Equation (7.14) is dimensionally homogeneous, and can thus be used in any system of units. The range of scatter is equivalent to a factor of 10, which is by no means out of the ordinary for sediment studies. The inclusion of specific gravity into the equation and the variety of data, $1.06 < s_s < 4.25$ and $1.56 < d < 2.47$ mm, make this relationship of Eq. (7.14) a useful contribution. Needless to say, Eq. (7.14) is, however, nothing but a form of duBoys' equation with a more specified χ value.

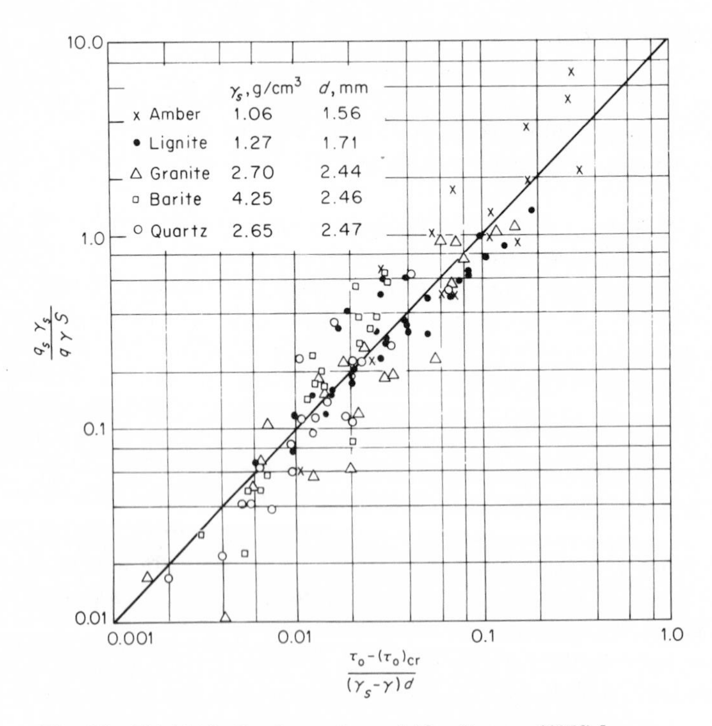

Fig. 7.3 Shields' bedload equation. [*After* BROWN (*1950*).]

That the turbulence mechanism in the flow above the bed does play an important role in the analysis of bedload motion was emphasized by KALINSKE (1947). It is assumed that the average rate of sediment motion can be expressed as

$$q_s = \left(\frac{\pi d^3}{6}\right)\left(\frac{p_k}{\pi d^2/4}\right)\overline{[u_b - (u_b)_{cr}]} \tag{7.15}$$

The first expression of Eq. (7.15) refers to the volume of a single particle, the second one to the number of particles participating in the motion, and the third one is the average particle velocity given approximately by the average difference of the instantaneous u_b and the critical fluid velocity $(u_b)_{cr}$. Simplifying Eq. (7.15) and multiplying with u_*, results in

$$q_s = \frac{2}{3} p_k u_* d \, \frac{\overline{[u_b - (u_b)_{cr}]}}{u_*} \tag{7.16}$$

Letting p_k equal 0.35 and expecting a functional relationship such as

$$\frac{\overline{[u_b - (u_b)_{cr}]}}{u_*} = \text{fct}\left[\frac{(\tau_0)_{cr}}{\tau_0}\right] \tag{7.17}$$

a dimensionless form of a bedload equation is given by

$$\frac{q_s}{u_* d} = \text{fct}_k\left[\frac{(\tau_0)_{cr}}{\tau_0}\right] \tag{7.18}$$

KALINSKE (1947) predicted analytically the shape of this function with knowledge of the turbulent behavior as given by Eqs. (6.29), (6.30), and (6.31).

Kalinske's bedload equation is similar to all the duBoys-type equations, and is compared with experiments in Fig. 7.4. It should be used with Kalinske's expression of critical shear stress, i.e., the average shear stress can be exceeded by the instantaneous one by a factor of 3, or $(\tau_0)_{cr} = 12d$ for turbulent rough flow, and with Eq. (6.30), i.e., $\sigma/\bar{u}_b \approx \frac{1}{4}$. While it was assumed that Eq. (7.18) holds only for uniform grains of spherical shape, in some cases mean values of the data were used, e.g., nonuniform sands are represented by the median diameters. Notice that at low $(\tau_0)_{cr}/\tau_0$ values the equation fails to predict accurately. This is attributed to the experimental difficulty in isolating the bedload from the suspended load.

Finally it should be remarked that the bedload equation by KALINSKE (1947) at least pays attention to many fluid-mechanical details, and as such could be considered more advanced than any of the other duBoys-type equations. However, judging from the scatter of data with Eq. (7.12), it is only another competitor with previously discussed equations.

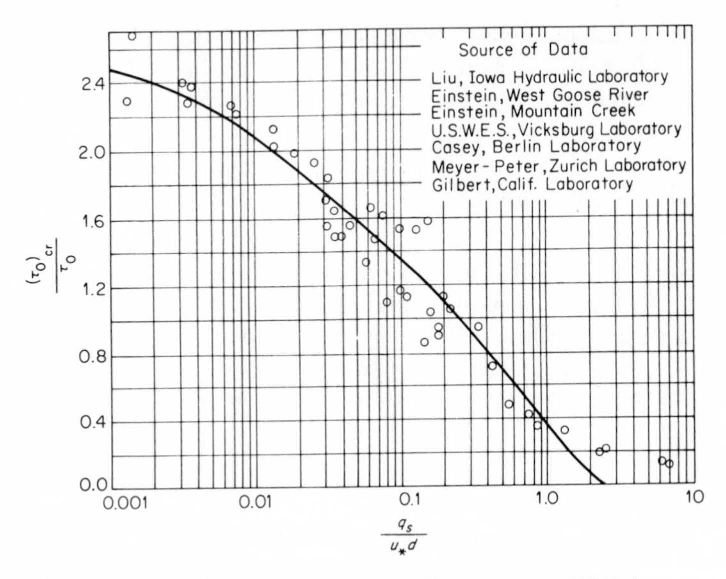

Fig. 7.4 Kalinske's bedload equation. [*After* KALINSKE (*1947*).]

7.3 SCHOKLITSCH-TYPE EQUATIONS

7.3.1 THEORETICAL CONSIDERATIONS

The basic duBoys-type equation relates the bedload transport with excess shear stress or

$$q_s = \chi \tau_0 [\tau_0 - (\tau_0)_{cr}] \tag{7.6}$$

In a further development, DUBOYS (1879) suggests the critical shear stress to be given by

$$(\tau^0)_{cr} = \gamma S D_{cr} \tag{7.19}$$

such that Eq. (7.6) becomes

$$q_s = \chi (\gamma S)^2 D(D - D_{cr}) \tag{7.20}$$

To our knowledge, Eq. (7.20) represents the working equation for duBoys. Sediment material moves if and when the water depth D exceeds the critical water depth D_{cr}. A further rearrangement was suggested by FORCHHEIMER (1914), such as replacing the quotient of DS with the representative average velocity $\bar{u}$, namely by $\bar{u}^2/C^2$, and Eq. (7.20) becomes

$$q_s = \chi \gamma^2 \left(\frac{\bar{u}^2}{C^2}\right) \left[\frac{\bar{u}^2}{C^2} - \left(\frac{\bar{u}}{C}\right)^2_{cr}\right] \tag{7.21}$$

where C is Chezy's roughness value. From Eq. (7.21) it may be deduced that the bedload motion starts at a certain velocity $(\bar{u})_{cr} = C\sqrt{(SD)_{cr}}$, and increases rapidly as the velocity exceeds this *critical condition* value. The layer movement, as visualized by DuBoys (1879), does, in fact, not take place. Donat (1929) went another step further and by rewriting Eq. (7.21) and making minor assumptions obtained

$$q_s = \chi\gamma^2 \frac{1}{C^4} \bar{u}^2[\bar{u}^2 - (\bar{u}^2)_{cr}] \tag{7.22}$$

with the average velocity expressed with Chezy's equation. Eventually, Schoklitsch (1930), independent from duBoys, but not in too strong a contradiction, suggested heuristically an equation based on laboratory experiments, such as

$$q_s = \chi'' S^k (q - q_{cr}) \tag{7.23}$$

where χ'' is a new characteristic sediment coefficient and q_{cr} is the water discharge at which the material begins to move.[1] Expressing the water depths in Eq. (7.20) with Manning's equation of average flow and assuming that the energy slope S and the roughness value n do not change with stage, the following is obtained:

$$q_s = \chi\gamma^2 S^{1.4} n^{1.2} q^{0.6}(q^{0.6} - q_{cr}^{0.6}) \tag{7.24}$$

If $q_{cr}^{1.2}$ is substituted for $q^{0.6}q_{cr}^{0.6}$, the exponent 1.2 may be compared with the exponent 1.0 as it appears in Eq. (7.23). Shulits (1935) has pointed out that this can be considered good agreement. A similar relationship, as given by Eq. (7.24), has been obtained by Rubey (1933), considering the energy balance in debris-laden streams. It appears, however, that the explanation of Eq. (7.23) advanced by Schoklitsch (1926) was that the rate of movement is proportional to the excess power.

Bedload equations with the general shape of Eq. (7.22), but mainly of Eq. (7.23), have been proposed by various researchers. Since Schoklitsch has been so instrumental in advancing this kind of an equation, they are referred to as *Schoklitsch-type* equations.

7.3.2 EXPERIMENTAL INVESTIGATIONS

7.3.2.1 Earlier developments. Gilbert (1914) carried on most important experiments which were designed systematically to determine "the transportation of debris" under changing conditions of water discharge, energy grade line, and sediment properties. These sets of experiments have

[1] According to Heyndrickx (1948), it was Hochenburger (1886) who put forward a similar relationship.

since proved to be invaluable for many researchers in checking out new theoretical advancements or in comparing them with other experimental findings.

Gilbert's measurements were done in three different flumes with lengths of 14, 31.5, and 150 ft, and a width varying within $0.23 < B < 1.96$ ft. The discharge of water could be varied from 0.019 to 1.19 cfs. Eight different kinds of sands were used, ranging in mean diameter from $0.305 < \bar{d} < 7.01$ mm, and in the density from $2.53 < \gamma_s < 2.69$ g/cm³. No attention was paid to the influence of particle form or density.

Empirical formulas were developed, which are, however, not recommended for immediate use to river problems. Each of these equations expresses the variation of the bedload with respect to one of the previously mentioned conditions, while the other ones are kept constant. For example, the bedload transport is proportional to the excess of discharge (slope or sediment property) above the critical—competent, as Gilbert called it—discharge (slope or sediment property). The mathematical relationship was given as

$$q_s = C_1(q - q_{cr})^{C_1^*} \tag{7.25}$$

$$q_s = C_2(S - S_{cr})^{C_2^*} \tag{7.26}$$

$$q_s = C_3\left(\frac{1}{d} - \frac{1}{d_{cr}}\right)^{C_3^*} \tag{7.27}$$

where C_1, C_2, and C_3 and C_1^*, C_2^*, and C_3^* are empirical constants. GILBERT (1914) indicates that it might be possible that a practical formula for the calculation of bedload may be of the form of Eq. (7.25), which, in turn, resembles Eq. (7.23). Equations (7.25), (7.26), and (7.27) are summarized together with Gilbert's experimental constants as follows:

$$q_s = C_4(q - q_{cr})^{0.81-1.24}(S - S_{cr})^{0.93-2.37}(d^{-1} - d_{cr}^{-1})^{0.50-0.62} \tag{7.28}$$

An extensive experimental program was also carried out by SCHAFFERNAK (1922), who found only partial agreement of his experiments with DUBOYS' (1879) bedload equation. However, DONAT (1929) reanalyzed the data, which were obtained for uniform sands and sand mixtures, and obtained remarkable agreement with Eq. (7.22). Minor deviations seemed to occur at the beginning of the sediment motion. The critical velocity $\bar{u}_{cr}$ and the characteristic sediment coefficient were tabulated by Donat, and are dependent on the sand itself and the grain size distribution. Field observations at the Tiroler Ache could be predicted with Eq. (7.22) and the parameters developed in the laboratory. A decade later, STRAUB (1939), apparently unaware of DONAT's (1929) work, studied the data by Schaffernak, supplemented with some of his own, and again suggested an equation of the type developed by DONAT (1929). The use of the average velocity has also been tried by O'BRIEN (1936), but the results scattered if there was a considerable variation in depth. A few empirical formulas which remotely resemble Eq. (7.22) are discussed in JAROCKI (1957); these formulas are due to the Polish investigators Debski and Jarocki, and to the Soviet investigators Goncharov, Levi, Lopatin, and Shamov.

A bedload equation using the average velocity was only recently suggested by BAREKYAN (1962). This equation, somehow similar to Eq. (7.22), was checked out with Soviet data and the data of Gilbert, and is given in the nondimensional form by

$$q_s = 0.187\gamma\left(\frac{\gamma_s}{\gamma_s - \gamma}\right)qS\left(\frac{\bar{u} - \bar{u}_{cr}}{\bar{u}_{cr}}\right) \tag{7.29}$$

This equation was subject to experimental investigation by SIMONS et al. (1965), but was found not to fit all the data.

7.3.2.2 Schoklitsch's contributions. If the total amount of the solid material Q_s is desired and only the sediment discharge per unit width q_s is given, a relationship such as

$$Q_s = \int_{x_1}^{x_2} q_s \, dx \tag{7.30}$$

with the coordinate x indicating the channel width, will provide information. In a rectangular channel the motion of the bedload is usually distributed over the entire bed. In natural channels bedload motion will occur only over that portion of the channel where the "critical condition" is exceeded. Let us now take a look at field observations, including bedload measurements, for the Tiroler Ache River. From Fig. 7.5 it can be seen that the critical

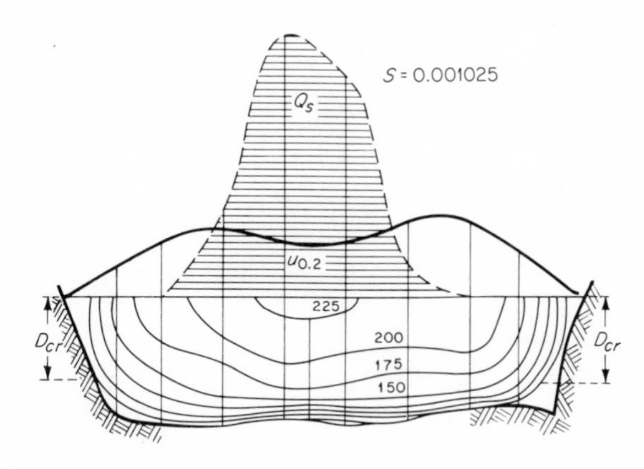

Fig. 7.5 Velocity and bedload distribution in the Tiroler Ache River. [*After* SCHOKLITSCH (*1930*).]

depth D_{cr} used in duBoys' working Eq. (7.20) is certainly a poor criterion. Also it does not seem that the bottom velocity, given here as $u_{0.2}$ is an admissible criterion. One might continue and argue that the shear stress $\tau_0 = \gamma SD$ is an equally poor choice, because the values of S and γ are constant and the depth D was found to be a poor criterion. This was apparently sufficient cause for SCHOKLITSCH (1930) to accept the water discharge q_{cr} as an incipient motion criteria, at least for engineering purpose, and to advance Eq. (7.23).

The total amount of bedload, Q_s—for lack of better information SCHOKLITSCH (1950) suggests to average it over the entire movable section B—can thus be expressed as

$$Q_s = \chi'' S^k (Q - Bq_{cr}) \tag{7.31}$$

That the general shape of Eq. (7.31) is sound was substantiated with field data of the Mur River, the Rhine, and the Danube. For the yearly amount of bedload $Q_s{}^y$, SCHOKLITSCH (1930) proposes

$$Q_s{}^y = \chi_s \sum (Q - Q_{cr}) \tag{7.32}$$

where Q_{cr} is the water discharge at which sediments begin to move. The sediment coefficient χ_s was calculated for the Mur River and Rhine, and is given as $\chi_s = 0.00019$ and $\chi_s = 0.00013$ for metric units, respectively. Figure 7.6 shows, as an example, the calculation of the bedload for the Mur River.

Owing to Eq. (7.31), the maximum yearly discharge transports the most material; it is, however, of short duration. According to SCHAFFERNAK (1922) there exists a "bedforming" discharge which lies between the maximum and the critical discharge of water. Because of its duration, the bedforming discharge is capable of transporting the most bedload during the year.

Meanwhile, MACDOUGALL (1933) found that a Schoklitsch-type equation, as given with Eq. (7.23), fitted his set of data best. The critical discharge q_{cr} was found from the experimental data at the zero bedload

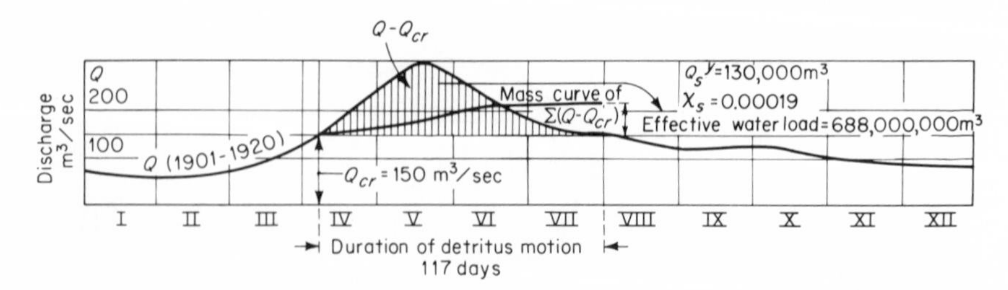

Fig. 7.6 Hydrograph of Mur River; determination of the yearly amount of bedload. [*After* SCHOKLITSCH (*1930*).]

point, but no relationship was suggested. MacDougall's experiments were extended by JORISSEN (1938). Both sets of experiments done with sand mixtures of $0.66 < d < 1.44$ mm and $0.354 < M < 0.654$ suggested that the characteristic sediment coefficient χ'' varied within the range of 100 to 1,000—for English units—and the power of the energy slope k was within 1.2 and 1.6.

CASEY (1935), analyzing his own laboratory data, found that a Schoklitsch-type equation, i.e., Eq. (7.23), was quite suitable.

The data of GILBERT (1914) were used again, this time by SCHOKLITSCH (1934), in a detailed study of the factors influencing Eq. (7.23), given by

$$q_s = \chi'' S^k (q - q_{cr}) \tag{7.23}$$

First Schoklitsch set out to determine the critical flow rate at which bedload starts to move. Extrapolation of the data with sediment transport to the point at which the flow rate of water produces no sediment transport gave the following relationship,

$$q_{cr} = \frac{(1.944)(10^{-5}d)}{S^{4/3}} \qquad \mathrm{m^3/(sec)(m)} \tag{7.33}$$

for a uniform sand grain measured in millimeters, ranging from 0.305 to 7.02 mm in diameter. Supplementing Gilbert's data with his own, the bedload equation is obtained as

$$g_s = \frac{7{,}000}{d^{1/2}} S^{3/2}(q - q_{cr}) \tag{7.34}$$

where g_s is the rate of bedload in weight per unit time and width. If B represents the width over which transport takes place, then the total amount of bedload going through a cross section becomes

$$G_s = \frac{7{,}000}{d^{1/2}} S^{3/2}(Q - Bq_{cr}) \tag{7.35}$$

Both Eqs. (7.34) and (7.35) are given in the metric unit system. These equations have been introduced to the English-speaking world by SHULITS (1935), who believes them to be the only ones with comprehensive evidence behind them. Although these equations are developed for uniform sands, they can be applied successfully to mixtures.

Suppose the grain size distribution of a sand mixture is given. The mixture is then divided into groups with a mean diameter of $d_a, d_b, d_c, \ldots,$ and a weight fraction of the total of $a, b, c, \ldots.$ The total bedload of the mixture is then calculated as

$$G_s = aG_a + bG_b + \cdots \tag{7.36}$$

where G_a is the bedload [according to Eq. (7.35)] for the fraction a, etc. SCHOKLITSCH (1934) reports encouraging agreement when Eq. (7.36) is compared with some of Gilbert's data. Furthermore, it is recommended that a representative diameter of a natural mixture is in the range of d_{40} to d_{60}.

Based on numerous experiments in laboratory flumes, but also on actual bedload measurements in the field, such as the Danube and the River Aare in central Europe, SCHOKLITSCH (1950) suggests a modification to Eq. (7.34) or (7.35). The new Eq. (7.39) is essentially of simpler form, but also incorporates implicitly the critical-shear-stress concept. The critical flow rate is given by the Gaukler-Strickler equation as

$$q_{cr} = \frac{1}{n} D_{cr}^{5/3} S^{1/2} \tag{7.37}$$

For grains of $d \geq 0.006$ m, the flow depth is given by Eq. (6.20), or

$$D_{cr} = 0.076 \frac{\gamma_s - \gamma}{\gamma} \frac{d}{S}$$

and the n value may be calculated according to

$$n = 0.0525 d^{1/6}$$

Introducing this information into Eq. (7.37), we obtain

$$q_{cr} = 0.26 \left(\frac{\gamma_s - \gamma}{\gamma} \right)^{5/3} \frac{d^{3/2}}{S^{7/6}} \tag{7.38}$$

where the grain diameter, measured in meters, for a sand mixture is given by the value of the 40 percent finer fraction d_{40}. The rate of bedload in weight per unit time and width is given as

$$g_s = 2{,}500 S^{3/2} (q - q_{cr}) \tag{7.39}$$

These foregoing equations are in metric units. Equations (7.38) and (7.39) represent the last information given to us by Schoklitsch; the problem of sedimentation was very important to this outstanding hydraulician of our century.

7.3.2.3 Contributions of the E.T.H. Extensive work on the problem of sedimentation and its application to engineering design has been undertaken by researchers at the *Eidgenössische Technische Hochschule*, called E.T.H., in Switzerland.

The first empirical relation was reported by MEYER-PETER et al. (1934) for uniform grain of sand, barite, and lignite. The E.T.H. experiments were performed in a laboratory flume with a cross section of 2 × 2 m and a total

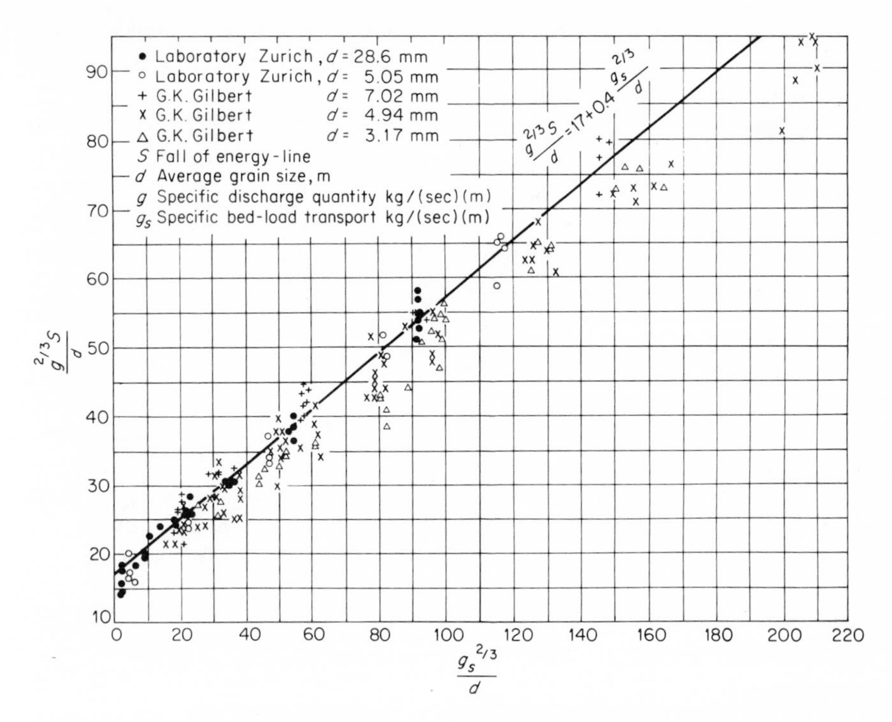

Fig. 7.7 The E.T.H. bedload equation. [*After* Meyer-Peter *et al.* (*1934*).]

length of 50 m. The water discharge could be varied up to about 5 m³/sec, and the sediment discharge up to 4.3 kg/(sec)(m). The sizes of the grain were on the large side, with one size of $d = 28.6$ mm and the other one of $d = 5.05$ mm; these were supplemented with only the largest-grain-size data of Gilbert (1914). Froude's law of similarity was used as a guide in the design of the experimental program. The validity of this can clearly be seen if both sides of the resulting bedload equation [Eq. (7.40)] are divided by the factor $\rho^{2/3}$.

For sand, Meyer-Peter's et al. effort resulted in the formulation of the following bedload equation, given in the metric system,

$$0.4\,\frac{g_s^{2/3}}{d} = \frac{g^{2/3}S}{d} - 17 \tag{7.40}$$

where g is the flow rate of water in weight per unit time and width. Equation (7.40) was developed with the E.T.H. data, but in Fig. 7.7 it is also compared with Gilbert's data of larger grain sizes. It should be pointed out that the strongest deviations from Eq. (7.40) occurred with data of small grain diameters. The authors speculate that, for small diameters of grains,

Froude's law might be a poor choice. Certainly this fact limits the applicability of Eq. (7.40) to coarse materials and excludes a priori alluvial streams.

Either from Fig. 7.7 or from Eq. (7.40) the critical value, $(g^{2/3}S/d)_{cr} = 17$, at which no sediment transport occurs can be found. Rearranging of Eq. (7.40) shows clearly its similarity to both of Schoklitsch's equations, namely Eqs. (7.34) and (7.39).

Further experiments with barite ($s_s = 4.2$) and lignite breeze ($s_s = 1.25$) confirmed the general form of Eq. (7.40). All the data, including the one with natural gravel, could be represented by Eq. (7.41), or

$$\frac{g^{2/3}S}{d} - 9.57(\gamma_s - \gamma)^{10/9} = 0.462(\gamma_s - \gamma)^{1/3}\frac{(g_s')^{2/3}}{d} \tag{7.41}$$

where g_s' is the bedload rate in under-water weight per unit time and width, or $g_s' = g_s(\gamma_s - \gamma)/\gamma_s$.

The E.T.H. experiments were extended to include data with particle mixtures. However, the first attempt to include a representative grain diameter into Eq. (7.40) was not at all successful. Therefore MEYER-PETER et al. (1948) questioned the general form of both Eqs. (7.40) and (7.41). After some data fitting, supported with a certain amount of hydraulic backing, the following equation fitted all of the data surprisingly well, as can be seen from Fig. 7.8, or

$$\frac{\gamma R_h(k/k')^{3/2}S}{d} - 0.047(\gamma_s - \gamma) = 0.25\sqrt[3]{\rho}\,\frac{(g_s')^{2/3}}{d} \tag{7.42}$$

where d is the median diameter of the mixture and R_h is the hydraulic radius which equals the depth of the flow D when bank resistance is negligible or does not exist. The quantity $(k/k')^{3/2}S$ is the kind of slope which is adjusted such that only a portion of the total energy loss S, the one due to the grain resistance S', is responsible for the bedload motion. The division of the bed resistance into its components, the one due to grain resistance S' and the other due to bedform resistance S'', is accomplished by holding the hydraulic

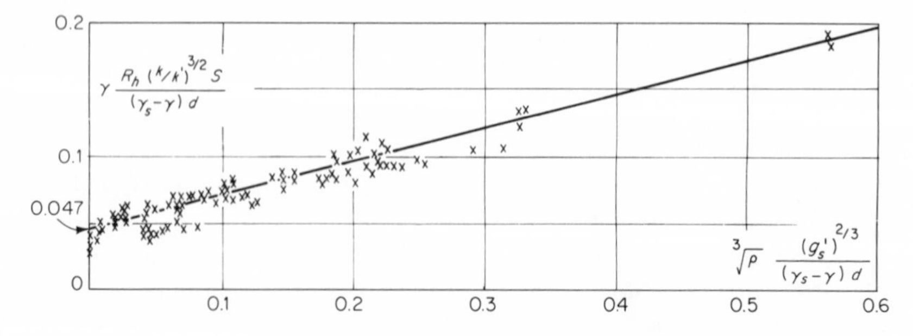

Fig. 7.8 The E.T.H. bedload equation. [*After* MEYER-PETER *et al. (1948)*.]

radius R constant and dividing the energy slope S. In Eq. (7.42) the energy slope has been expressed with Strickler's equation, with k being the roughness coefficient due to S, and k' the one due to S'. Correctly this results in a relationship such as

$$\frac{S'}{S} = \left(\frac{k}{k'}\right)^2 \tag{7.43}$$

but test results showed the relationship to be of the form of

$$\frac{S'}{S} = \left(\frac{k}{k'}\right)^{3/2} \tag{7.43a}$$

Equation (7.43*a*) was used for the establishment of Eq. (7.42). There is some difficulty in evaluating the ratio S'/S given by Eq. (7.43*a*). MÜLLER (1943) proposes to calculate the roughness coefficient k', which is a result of the friction of the top layer of the grains,

$$k' = \frac{26}{d_{90}^{1/6}} \qquad \mathrm{m^{1/3}/sec} \tag{7.44}$$

where d_{90} is the size of the sediment in the bed for which 90 percent of the material is finer. This is certainly a reasonable diameter, since the top layer, being made up of the largest grains, is armoring the bed—a fact which was discussed in Sec. 6.3.3 and is vividly shown in Fig. 6.9. It might be convenient to know that the ratio of k/k' varies within the limits of $0.5 < k/k' < 1.0$, where $k/k' = 1$ when no bedform exists and $k/k' = 0.5$ for strong bedforms; these figures are reported by ZELLER (1963).

Furthermore, notice that zero bedload transport transforms Eq. (7.42) into a shear stress relationship, such as given by Eq. (6.17), with a numerical value of 0.047, which compares favorably with other results plotted in Fig. 6.7. Recently CHIEN (1954) has shown that Eq. (7.42) can also be reduced to the form of a duBoys-type equation, but he also showed its similarity to the bedload equation by EINSTEIN (1950).

The E.T.H. contributions for uniform grain, given by Eq. (7.41), and for grain mixtures, given by Eq. (7.42), enjoy great popularity, especially in central Europe. The findings of the research have also been published in German by MEYER-PETER et al. (1949) and in French by MEYER-PETER (1949, 1951). The latter ones also include some special problems of alpine and subalpine rivers.

7.4 EINSTEIN'S BEDLOAD EQUATIONS

7.4.1 INTRODUCTORY REMARKS

The contributions by EINSTEIN (1942, 1950) to the problem of bedload transportation represent somehow a departure from the *duBoys-type* and

Schoklitsch-type equations. This is mainly so because Einstein's physical model makes ample use of the advancements in fluid mechanics.

There are at least two basic considerations which break with the past: (1) The definition of a *critical* value for the initiation of sediment motion is a difficult, if not impossible, proposition. Therefore, such a criterion was avoided. (2) It is suggested that the bedload transport is related to the fluctuations of the velocity rather than to the average value of the velocity. The beginning and the end of the particle motion have to be expressed with the concept of probability, which relates instantaneous hydrodynamic lift forces to the particle's weight.

Experimental evidence has shown that there exists an intimate relationship between the bed and the bedload:

1. A steady and intensive exchange of particles was observed to exist between the moving bedload and the bed.
2. The bedload moves slowly downstream; the motion of an individual particle is one of quick steps with comparatively long intermediate rest periods.
3. The average step made by any bedload particle appears to be independent of the flow condition, the transport rate, and the bed composition, and is always the same.
4. Different transport rates can be achieved by a change of the average time between two steps and of the thickness of the moving layer.

With these concepts EINSTEIN (1942, 1950) developed a bedload formula that relates the rate of bedload transport with properties of the grain and of the flow causing the movement. This relationship was presented empirically in 1942 and was replaced by an analytical function in 1950. The forthcoming analysis will exclude all particles finer than the bed—for practical purposes all particles finer than 10 percent of the bed material, filling the pores between larger ones, might be excluded as well—and all of the bed material moving in suspension.

7.4.2 PHYSICAL MODEL

The bedload equation expresses an equilibrium condition of the exchange of bed particles between the bed layer and the bed. This implies that the number of particles deposited per unit time and per unit bed area must be equal to the number of particles eroded per unit time and unit bed area.

Deposition. Each particle with a given diameter d will perform individual steps of the length $A_L d$, and must be assumed to be deposited over an area which is $A_L d$ long and has unit width. If g_s is the bedload rate in weight per unit of time and of width, and i_s is the fraction of bedload in a given grain size, then $g_s i_s$ is the rate at which the given size moves through the unit width per unit of time. The weight of a single particle is given by $\gamma_s k_2 d^3$; thus the number of particles of this fraction which is deposited per unit time and unit bed area is given by

$$\frac{g_s i_s}{(A_L d)(\gamma_s k_2 d^3)} = \frac{g_s i_s}{A_L k_2 d^4 \gamma_s} \tag{7.45}$$

where A_L is a constant of the bedload unit step and k_2 is the constant of the particle volume.

Erosion. Whether or not a particle of size d will be eroded depends on the availability of the particle and on the flow conditions, namely, on the turbulence level. If i_b is the fraction of bed material in a given grain size, then the number of particles d in a unit area of bed surface is given by $i_b/k_1 d^2$, with k_1 being the constant of the grain area. Let p/t_e be the probability of removal; then the number of particles which are eroded per unit time and unit area may be expressed as

$$\frac{i_b p}{k_1 d^2 t_e} \tag{7.46}$$

where t_e is the time consumed by each exchange. No direct method is available for the determination of the exchange time t_e; however, EINSTEIN (1942) suggests a dependency with the particle settling velocity, or

$$t_e \propto \frac{d}{v_{ss}} = k_3 \sqrt{\frac{d\rho}{g(\rho_s - \rho)}} \tag{7.47}$$

with k_3 being a constant of time scale.

Equilibrium. Since the rate of deposition balances the rate of erosion, Eqs. (7.45) and (7.46) will be equated and

$$\frac{g_s i_s}{A_L k_2 \gamma_s d^4} = \frac{i_b p}{k_1 k_3 d^2} \sqrt{\frac{g(\rho_s - \rho)}{d\rho}} \tag{7.48}$$

The resulting Eq. (7.48) is the bedload equation due to EINSTEIN (1942, 1950).

Exchange probability. The probability of erosion p is the fraction of time during which, at any one spot, the instantaneous lift exceeds the weight of the particle. At not too intensive sediment transport the probability of erosion p is small, and deposition is most everywhere possible. However, at strong sediment transport, p becomes larger, and deposition is not everywhere possible. EINSTEIN (1950) interprets that p may be used to calculate the distance $A_L d$, which a particle travels between consecutive places of rest. If p is small, the distance of travel is virtually a constant, and $A_L d = \lambda_b d$, with λ_b being a single step of bedload and having a value of about 100. Suppose p is larger such that only $(1 - p)$ particles find a chance of deposition after having traveled $\lambda_b d$ while p particles stay in motion or, better, would like to deposit, but find after $\lambda_b d$ that the lift forces again exceed the weight of the particles. Of these, $p(1 - p)$ particles are deposited after traveling $2\lambda_b d$ while p^2 particles are still not deposited, and so on. The total travel distance can be expressed in a series as

$$A_L d = \sum_{n=0}^{\infty} (1 - p)p^n(n + 1)\lambda_b d = \frac{\lambda_b d}{1 - p} \tag{7.49}$$

Bedload equation. Introducing the knowledge of Eq. (7.49) into the bedload Eq. (7.48) and separating p on one side of the equation, we obtain

$$\frac{p}{1 - p} = \left(\frac{k_1 k_3}{k_2 \lambda_b}\right)\left(\frac{i_s}{i_b}\right)\left(\frac{g_s}{\gamma_s}\sqrt{\frac{\rho}{\rho_s - \rho}}\sqrt{\frac{1}{gd^3}}\right) \tag{7.50}$$

The quantity on the right-hand side of Eq. (7.50),

$$\Phi = \frac{g_s}{\gamma_s}\sqrt{\frac{\rho}{\rho_s - \rho}\frac{1}{gd^3}} \tag{7.51}$$

is the dimensionless measure of the bedload transport and is called the *intensity* of bedload transport. Equation (7.50) can thus be written as

$$\frac{p}{1 - p} = A_*\left(\frac{i_s}{i_b}\right)\Phi = A_*\Phi_* \tag{7.52}$$

where A_* is a constant to be determined by experiments and Φ_* is the intensity of transport for an individual grain size. Φ and Φ_* are dimensionless parameters, and as such are invariant between model and prototype, and can be used in any consistent system of units.

7.4.3 EMPIRICAL RELATION

A bedload equation applicable for uniform sediment and mixtures acting like uniform sediment was advanced by Einstein (1942). A relationship quite similar to Eq. (7.52) was obtained, or

$$\frac{p}{1 - p} = A_*^{42}\Phi^{42} \tag{7.53}$$

where the superscripts of A_* and Φ merely indicate the year in which these particular symbols were proposed. The right-hand side of Eq. (7.53) was expressed as

$$A_*^{42}\Phi^{42} = \left(\frac{k_1 k_3}{k_2 \lambda_b}\right)^{42}\frac{1}{F}\frac{g_s}{\gamma_s - \gamma}\sqrt{\frac{\rho}{\rho_s - \rho}}\sqrt{\frac{1}{gd^3}} \tag{7.54}$$

where F is a dimensionless settling function, $F = 0.816$, for sand particles of $d \geq 1$ mm, and settling in water at normal temperature.

Probability determination. From the previous discussion, it is obvious that probability of erosion depends on hydrodynamic lift and particle weight. Mathematically expressed this means

$$p = \text{fct}\left(\frac{\text{effective weight of particle}}{\text{hydrodynamic lift}}\right) \tag{7.55a}$$

or $$p = \text{fct}\left[\frac{k_2(\rho_s - \rho)gd^3}{C_L \frac{1}{2}\rho k_1 d^2 u_b^2}\right] \tag{7.55b}$$

where C_L is the lift coefficient. The effective velocity u_b may be approximated by the velocity at the edge of the laminar sublayer if the wall is smooth, and

$$u_b \approx 11.6u_* \approx 11.6\sqrt{gR_h'S}$$

Equation (7.55*b*) can now be rewritten as

$$p = \text{fct}\left[\frac{k_2(\rho_s - \rho)gd^3}{C_L \frac{1}{2}\rho k_1 d^2 (135gR_h'S)}\right] \tag{7.56}$$

where R_h' is the hydraulic radius with respect to the grains,[1] which is the only important part of the total hydraulic radius as far as sediment transport is concerned. EINSTEIN (1942) has called the quantity

$$\Psi' = \frac{\rho_s - \rho}{\rho} \frac{d}{SR_h'} \tag{7.57}$$

the flow intensity and

$$B_*^{42} = \frac{k_2}{C_L \frac{1}{2} k_1 135}$$

a universal constant of the scale of Ψ' to be determined experimentally.

Now Eq. (7.56) with its appropriate abbreviations can be introduced into Eq. (7.53).

Weak sediment transport. For this special case which includes values of $\Phi^{42} < 0.4$, Eq. (7.53) reduces to

$$p = A_*^{42}\Phi^{42} \tag{7.58}$$

Combining the knowledge of Eqs. (7.58) and (7.56), a new form of the bedload equation is obtained, or

$$A_*^{42}\Phi^{42} = p = \text{fct}\,(B_*^{42}\Psi') \tag{7.59}$$

[1] For a detailed discussion of R_h', its further meaning and calculation, see Chap. 11 or EINSTEIN et al. (1952).

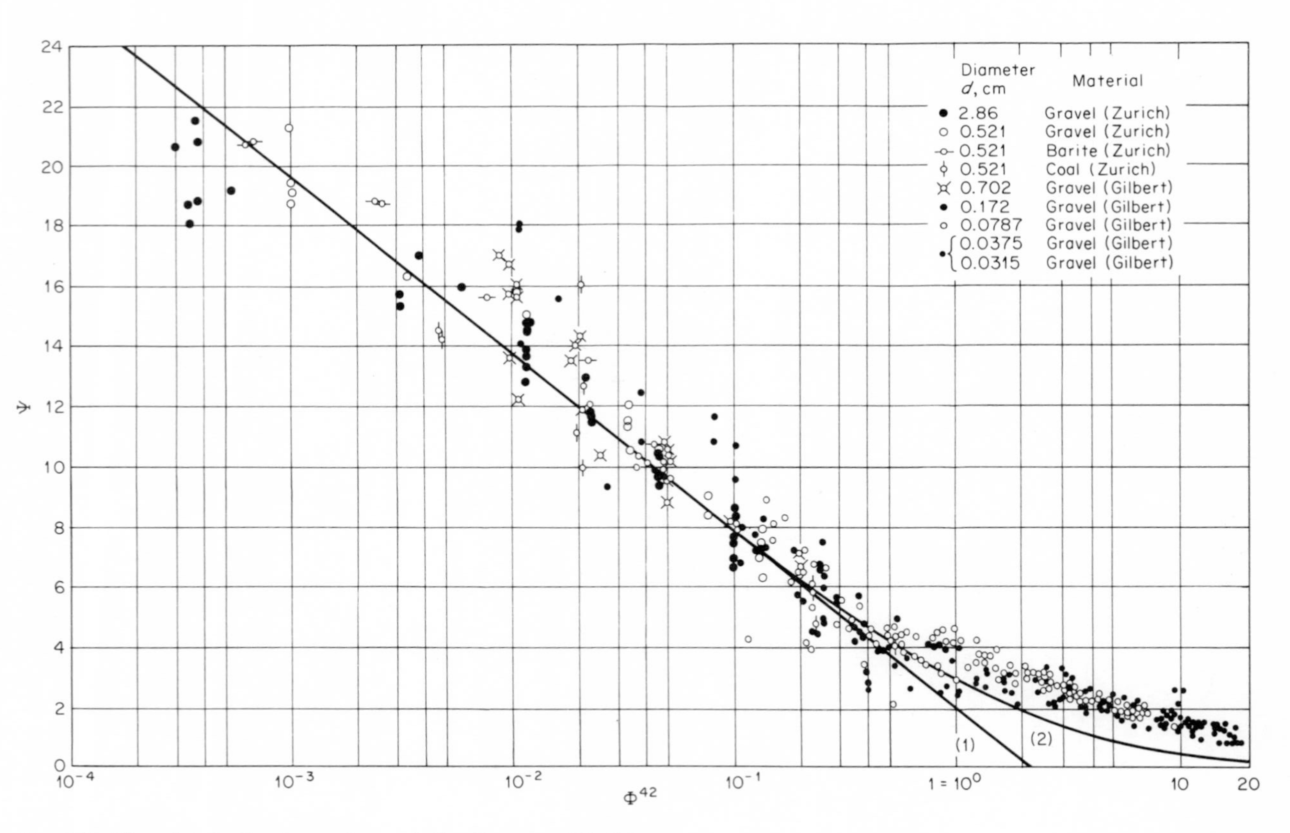

Fig. 7.9 Einstein's bedload equations. [*After* EINSTEIN (*1942*).]

The two constants A_*^{42} and B_*^{42} and the function fct must be determined empirically. This was done by using the data of GILBERT (1914) and of MEYER-PETER et al. (1934); the range of the data is indicated in Fig. 7.9, where values of Ψ' are plotted against values (on logarithmic scale) of Φ^{42}. From Fig. 7.9 it seems that all data with $\Phi < 0.4$ can be represented with a single curve, curve 1, the equation of which is

$$0.465\Phi^{42} = e^{-0.391\Psi'} \tag{7.60}$$

Strong sediment transport. For large values of Φ, i.e., $\Phi > 0.4$, the curve given by Eq. (7.60) deviates from the data. This is obviously to be expected because bedload motion is strong, and the approximation given by Eq. (7.58) must fail. However, the complete Eq. (7.53) was found to predict a curve, such as curve 2, which explains the data better, as can be seen in Fig. 7.9. The still-existing deviation may be attributed to the fact that the experimental data included suspended-load material.

Sand mixtures. Analyzing some data of the U.S. WATERWAYS EXPERIMENT STATION (1935), Einstein determined the effective diameter of sand mixtures. Experimentally it was found that the grain size of bed material of which 35 to 45 percent is finer is the most usable value.

Discussion. First of all, let us keep in mind that the Ψ' value as given in Eq. (7.57) resembles very much the critical-shear-stress relation given by Eq. (6.17). For the practicing engineer both concepts are useful, because they indicate the flow under which a natural or artificial bed will be stable. In terms of the flow intensity this *critical* value may be extrapolated and is $\Psi'^{-1} = 0.056$.

Revising the same sets of data, BROWN (1950) finds that a single linear function on a logarithmic plot predicts all the data, but at the lowest Φ values.

7.4.4 ANALYTICAL RELATION

As mentioned earlier, the empirical Φ vs. Ψ' relationship by EINSTEIN (1942) was replaced with an analytical one. The latter, also advanced by EINSTEIN (1950), is probably the most generally applicable, but also the most involved bedload equation.

Probability determination. The probability of erosion p may be expressed as the probability of the ratio of effective weight to instantaneous lift, which has to be smaller than unity:

$$1 > \frac{k_2(\rho_s - \rho)gd^3}{C_L \frac{1}{2}\rho k_1 d^2 u_b{}^2(1 + \eta)} \tag{7.61}$$

In an extensive investigation on the influence of the instantaneous lift, discussed in Sec. 6.4.2, EINSTEIN et al. (1949) have found that the lift coefficient C_L has a constant value of 0.178, and the random function η is distributed according to the normal-error law, where the standard deviation η_0 is a universal constant of $\eta_0 = \frac{1}{2}$. The velocity u_b was found to be at a distance of $0.35X$ from the theoretical bed, where X is the characteristic grain size of the mixture, with $X = 0.77\Delta$ if $\Delta/\delta > 1.80$ and $X = 1.39\delta$ if $\Delta/\delta < 1.80$. If u_b is expressed with the law of logarithmic velocity distribution, the following is obtained:

$$u_b = u_* 5.75 \log \left[\frac{(30.2)(0.35X)}{\Delta}\right] \tag{7.62}$$

an equation applicable for rough and smooth boundaries as well as for the transition between the two. The apparent roughness diameter Δ must be obtained from Fig. 7.10—a graph EINSTEIN (1950) derived from Nikuradse's experiments—with $k_s \approx d_{65}$ and the laminar sublayer $\delta = 11.5\nu/u_*$. Introducing this information into Eq. (7.61) results in

$$1 > \frac{1}{1+\eta}\left(\frac{\rho_s - \rho}{\rho}\frac{d}{R_h'S}\right)\left[\frac{2k_2}{(0.178k_1)(5.75)^2}\right]\left[\frac{1}{\log^2(10.6X/\Delta)}\right] \tag{7.63}$$

and, after introducing abbreviations,

$$1 > \frac{1}{1+\eta}\Psi' B\beta_x^{-2} \tag{7.64}$$

where Ψ' is the flow intensity, B is a constant of the scale Ψ', and β_x^{-2} is a value to be determined from Fig. 7.10. Furthermore, EINSTEIN (1950) suggests to introduce two correction factors ξ and Y. Small particles seem to hide between larger ones or in the laminar sublayer, such that their lift

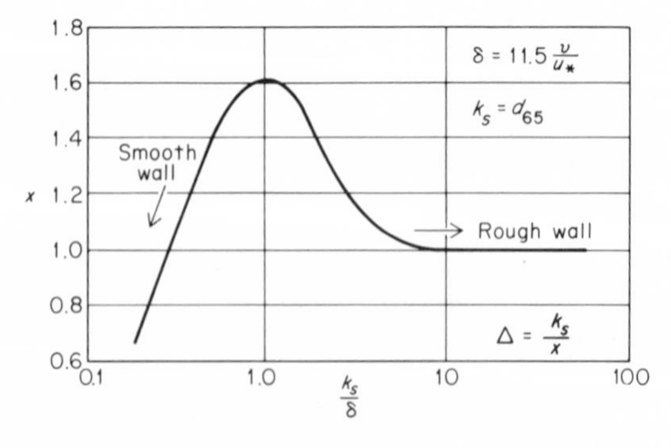

Fig. 7.10 Correction factor in the logarithmic velocity distribution. [*After* EINSTEIN (*1950*).]

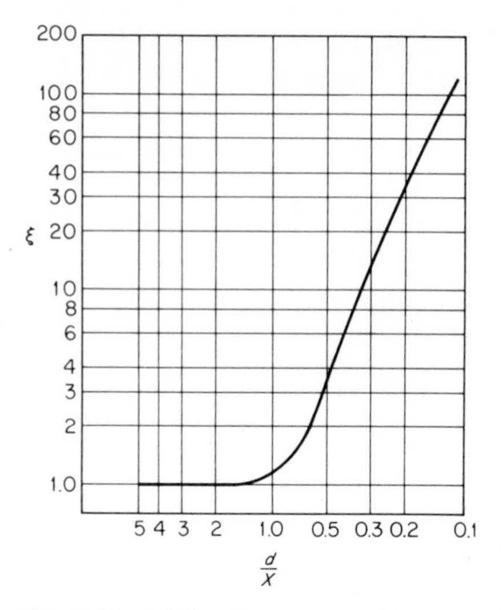

Fig. 7.11 Hiding factor, as used in Eq. (*7.65*). [*After* EINSTEIN (*1950*).]

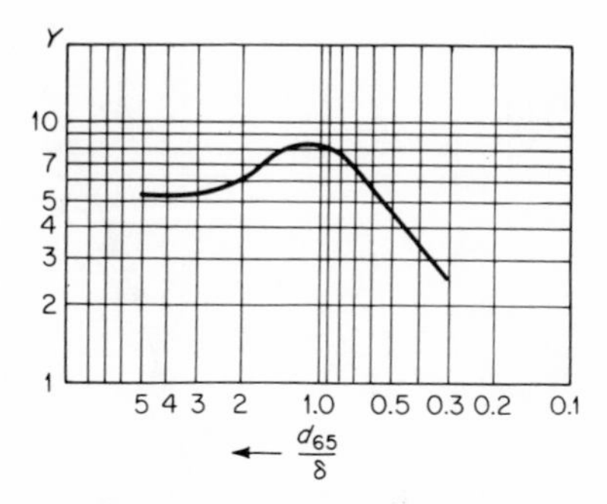

Fig. 7.12 Pressure correction, as used in Eq. (*7.65*). [*After* EINSTEIN (*1950*).]

must be corrected by ξ^{-1}. This *hiding factor* of grains in a mixture is in functional relation with d/X, and is given in Fig. 7.11. The other correction factor Y describes the change of the lift coefficient in mixtures with various roughnesses; it is a function of k_s/δ, and is plotted in Fig. 7.12. Both the correction and hiding factors have been determined experimentally by EINSTEIN (1950) but are subject to further improvement. For uniform grain they are unity. With this in mind, we can improve Eq. (7.64) and obtain

$$1 > \frac{1}{1+\eta}\,\xi Y B' \frac{\beta^2}{\beta_x^{\,2}}\Psi \tag{7.65}$$

having replaced $B' = B/\beta^2$ and $\beta = \log 10.6$. Or we can rewrite Eq. (7.65) such that

$$|(1+\eta)| > \xi Y B' \frac{\beta^2}{\beta_x^{\,2}}\Psi \tag{7.66}$$

Whereas η may be either positive or negative, the lift is always positive and the absolute-value sign on the left-hand side of Eq. (7.66) is in place. It is more convenient to square and divide by η_0, Eq. (7.66), and after expressing $\eta = \eta_0\eta_*$ we obtain

$$\left(\frac{1}{\eta_0} + \eta_*\right)^2 > \xi^2 Y^2 B_*^{\,2}\left(\frac{\beta^2}{\beta_x^{\,2}}\right)^2 \Psi^2 \tag{7.67}$$

or
$$\left(\frac{1}{\eta_0} + \eta_*\right)^2 > B_*^{\,2}\Psi_*^{\,2} \tag{7.67a}$$

where $B_* = B'/\eta_0$ and $\Psi_* = \xi Y(\beta^2/\beta_x^{\,2})\Psi$.

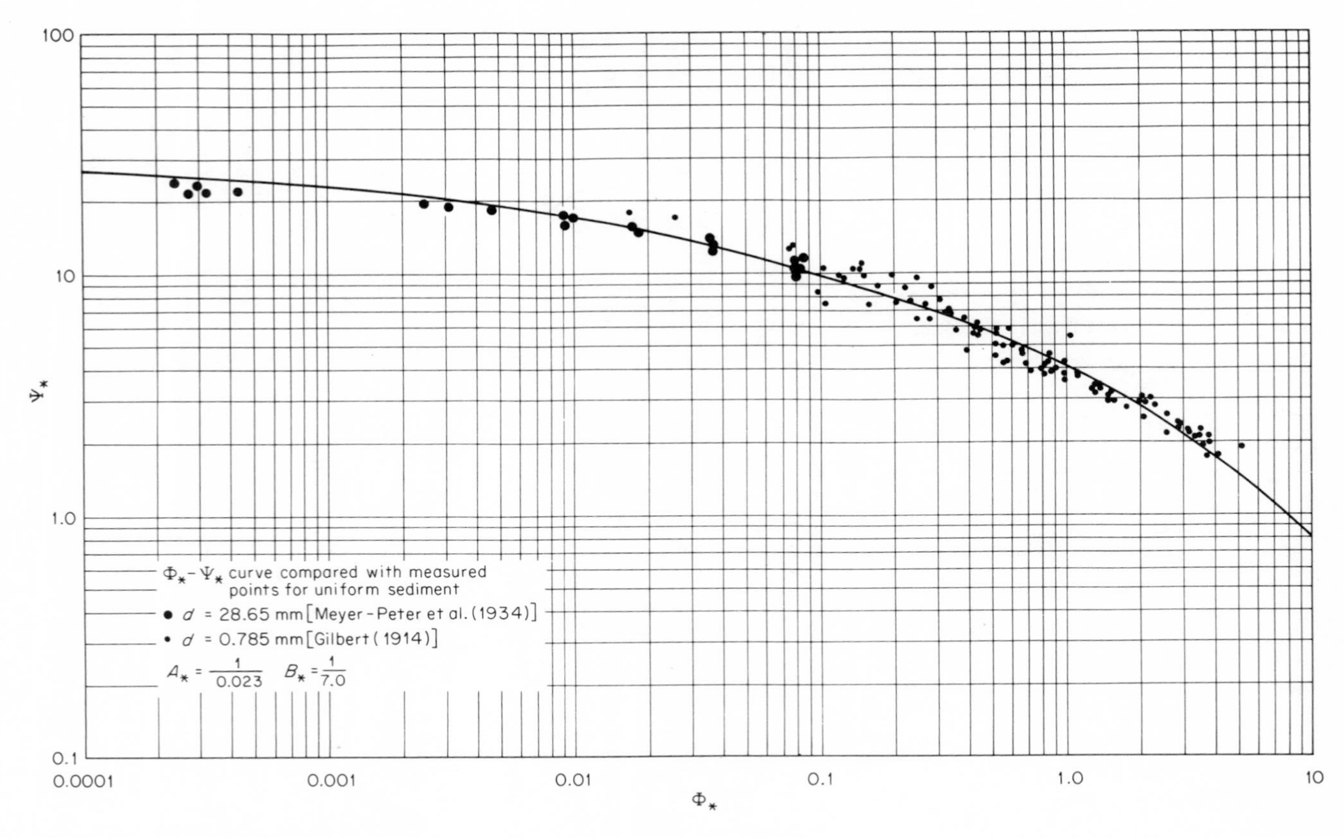

Fig. 7.13 Plot of Einstein's functions; Φ_* vs. Ψ_*. [*After* EINSTEIN (*1950*).]

From Eq. (7.67*a*) the limiting case of motion may be found if

$$\left(\frac{1}{\eta_0} + \eta_*\right)^2 = B_*^2 \Psi_*^2 \tag{7.68}$$

or $$(\eta_*)_{lim} = \pm B_* \Psi_* - \frac{1}{\eta_0} \tag{7.69}$$

It was previously mentioned that probability distribution is according to the normal-error law, such that the probability p for motion becomes

$$p = 1 - \frac{1}{\sqrt{\pi}} \int_{-B_*\Psi_* - 1/\eta_0}^{+B_*\Psi_* - 1/\eta_0} e^{-t^2}\, dt \tag{7.70}$$

where t is the only variable of integration.

Sediment transport. Combining Eq. (7.52) with Eq. (7.70) results in the final or second bedload equation suggested by EINSTEIN (1950), or

$$1 - \frac{1}{\sqrt{\pi}} \int_{-B_*\Psi_* - 1/\eta_0}^{+B_*\Psi_* - 1/\eta_0} e^{-t^2}\, dt = p = \frac{A_* \Phi_*}{1 + A_* \Phi_*} \tag{7.71}$$

where A_*, B_*, and η_0 are universal constants to be determined from experimental data. A Φ_* vs. Ψ_* relationship is shown in Fig. 7.13. The following constants were obtained from uniform sediments by using the data of GILBERT (1914) and of MEYER-PETER et al. (1934):

$$\begin{aligned} A_* &= \frac{1}{0.023} = 43.5 \\ B_* &= \tfrac{1}{7} = 0.143 \end{aligned} \tag{7.72}$$

η_0, the standard deviation of η, was given earlier as $\eta_0 = \frac{1}{2}$.

Discussion. Now that we know the transport rates of individual components i_s/i_b within a mixture, the total transport rate of the mixture as bedload can be determined by summing up these fractions. However, for mixtures with small size spread, the total transport of the mixture can be determined directly by using d_{35} as the effective diameter.

With the knowledge of the final bedload Eq. (7.71), Einstein proceeded to develop the relationship between the material traveling along the bed and that moving in suspension from the measured total sediment transport rate.

To be sure Einstein's derivations do not rely on any kind of *critical* value, and thus the constant in the different equations are free of *critical* values. The *critical* conditions for the initiation of bedload motion actually correspond to a particular Ψ value; these Ψ values have been computed for various investigations by CHIEN (1954) and are shown in Fig. 6.8.

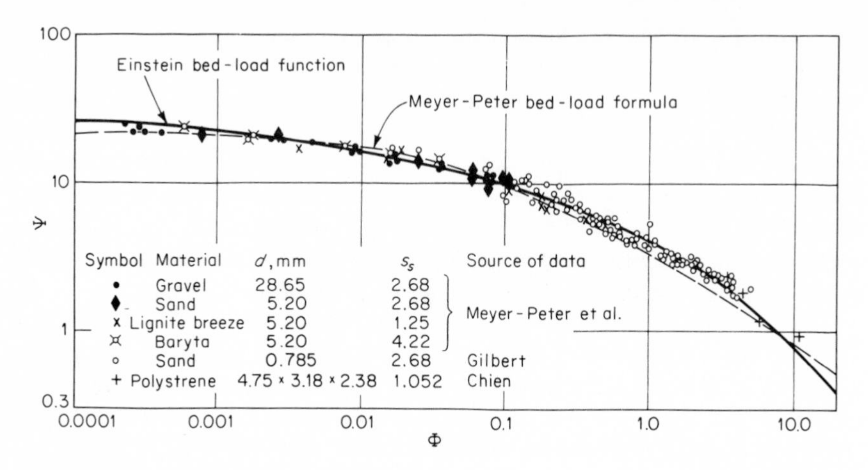

Fig. 7.14 Comparison of the bedload equation by Einstein and Meyer-Peter et al. [*After* CHEIN (*1954a*).]

Einstein's bedload Eq. (7.71) and the bedload Eq. (7.42) by MEYER-PETER et al. (1948) are considered to represent the most complete equations of all. Therefore, it is of particular interest to see where the two differ and how they do compare. This has been done by CHIEN (1954*a*). As far as hydraulics is concerned, the difference lies in how the total frictional resistance is divided. Einstein divides the hydraulic radius $R_h = R_h' + R_h''$ and keeps the slope constant, whereas Meyer-Peter et al. divide the slope $S = S' + S''$ and keep the hydraulic radius constant. Both assumptions facilitate the understanding of superposition of the effects due to grain roughness and bedform. Under these conditions, CHIEN (1954*a*) has shown that Eq. (7.42) by Meyer-Peter et al. can be written in terms of Φ and Ψ',

$$\Phi = \left(\frac{4}{\Psi'} - 0.188\right)^{3/2} \tag{7.73}$$

Equations (7.73) and (7.71) are presented in Fig. 7.14 for uniform material, and they show very good agreement. For sediment mixtures, by using d_{35} in the Einstein relation and d_{50} in the Meyer-Peter et al. relation, agreement was found to be equally good. Thus we may safely conclude that both methods demonstrate their merits.

7.4.5 SIMILAR KINDS OF BEDLOAD EQUATIONS

7.4.5.1 Bagnold's model. With an entirely different approach than the one used by EINSTEIN (1942, 1950), BAGNOLD (1956) developed a bedload equation that strongly resembles the one of Einstein. Bagnold's model implies that the sediment particles move in a thick suspension of a grain-fluid mixture. As far as hydraulics is concerned, Bagnold neither divides the hydraulic radius nor the energy slope, and thus makes no explicit provision for the effect of the bedforms. COLBY (1963) fears that Bagnold's equations may not be as generally applicable as the bedload equations by MEYER-PETER et al. (1948) and EINSTEIN (1950). Bagnold's method is involved and lengthy; for full details, the interested reader is advised to study the original contribution. However, BAGNOLD'S (1966) more recent approach to the sediment transport problem, which makes his "previous theory considerably simpler and more directly applicable," is discussed in some detail in Chap. 9.

7.4.5.2 Yalin's model. The average lift—not the instantaneous one—acting on a particle has been considered by YALIN (1963) in the development of a bedload equation. The particle motion is by saltation, a kind of particle movement that will be discussed in Sec. 7.6. However, it should be noted at this point that present thinking is such that saltation is unimportant as a separate mode of solid-liquid transport. Proceeding by way of dimensional theory, Yalin suggests that an increase of bedload discharge is due to the increased range of the particle's path, and not necessarily due to the number of particles participating in the motion. The resulting sediment transport equation is one of excess shear; its constant was determined by using, mainly, GILBERT'S (1914) data. The equation was compared with EINSTEIN'S (1942) relationship, which is based on statistical type of lift forces. The approach by YALIN (1963) was extensively reviewed by SCHMITT (1966), who finds that for the special case of a flat bed this relation yields better results than the methods due to EINSTEIN (1950), MEYER-PETER et al. (1948), or BAGNOLD (1956). YALIN'S (1963) paper was also critically discussed by NORDIN et al. (1964), who felt that neither bedload data from flume studies nor field data could be predicted with desired accuracy. This is probably in part true in regard to YALIN'S (1963) strong assumptions of a plane bed, uniform sand, and a depth range in which the ratio of grain size to actual depth is negligible.

7.5 EQUATIONS CONSIDERING BEDFORM MOTION

7.5.1 THEORETICAL CONSIDERATIONS

The concept of bedforms, as they exist in channels subject to sediment transport, will be extensively discussed in Chap. 11. There are, however, interesting contributions to the determination of bedload transport within the lower flow regime when ripples and dunes cover the bed.

It is assumed that all the material moving as bedload is in continuous contact with the bed. This bed, however, is not a flat one, but is made up of dune-shaped ridges. Particles move up the face of a ridge. The coarser ones drop down the steep slope to be deposited on the downstream face of the ridge or in the trough; the finer particles may be deposited just about everywhere, and might even be swept into suspension. The ridges, owing to erosion of the upstream and deposition on the downstream slope, move downstream.

The differential equation of transport for such a case is given by

$$\frac{\partial y}{\partial t} + \left(\frac{1}{1 - m}\right)\frac{\partial q_s}{\partial x} = 0 \tag{7.74}$$

where y = elevation of the sand bed above a horizontal datum
x = direction of the flow
m = porosity of the sand

Using a transformation given by

$$\delta = x - c_B t \tag{7.75}$$

where c_B is the velocity of the sand ridges, appropriate substitution results in

$$-c_B \frac{dy}{d\delta} + \left(\frac{1}{1 - m}\right)\frac{dq_s}{d\delta} = 0 \tag{7.76}$$

Rearranging Eq. (7.76),

$$q_s = (1 - m)c_B y + C \tag{7.77}$$

where C is the constant of integration. Assuming that ridges are triangular, of a height ΔH, Eq. (7.77) becomes

$$q_s = (1 - m)c_B \frac{\Delta H}{2} + C \tag{7.78}$$

At the initiation of movement, and when the bed is entirely covered with ridges, $C = 0$; for a flat bed at $\Delta H = 0$, $C = q_s$.

The erosion equation, as Eq. (7.74) might be called, was suggested by Exner (1925). Only recently it has been used again by Simons et al. (1965), Hansen (1966), and Crickmore (1970) to develop an equation for the bedload transport, such as Eq. (7.78).

7.5.2 EXPERIMENTAL INVESTIGATIONS

More than 100 observations in a recirculating laboratory flume were made by Simons et al. (1965*a*). Measurements were taken of the ridge height ΔH and its length λ. The ridge velocity c_B and the height were recorded also by a sonic depth recorder. The median sand diameter varied with $0.19 < d < 0.93$ mm. By comparing the measured with the computed bedload quantity, it was concluded that Eq. (7.78) is especially well suited for dune-bed configurations and for coarse bed material. SIMONS (1967) mentioned that, with more sophisticated sonic equipment for measuring the various ridge parameters, the theory, as given by Eq. (7.78), provides an excellent method for the determination of the bedload discharge.

HANSEN (1966) has found that the bedload discharge in a Danish watercourse could be well predicted with an equation similar to Eq. (7.78). Finally, SIMONS et al. (1965*a*) show that, if empirical relations for the ridge velocity and ridge height, as given by BAREKYAN (1962), are introduced into Eq. (7.78), a relationship similar to Eq. (7.29) may be obtained.

7.6 SALTATION AS A MODE OF PARTICLE TRANSPORT

A bounding motion of grains has been described by GILBERT (1914) and was called *saltation*. Later, BAGNOLD (1941) in the classic treatise on air-solid problems defined and described this kind of motion extensively. The characteristic path of a saltating particle was obtained photographically by Bagnold, as shown in Fig. 7.15. Initially, the grain has a very large upward velocity w_1 and a small forward velocity u_1; however, during the subsequent flight, its forward velocity is increased owing to a supply of energy from the fluid, the upward velocity diminishes and is in balance with the gravity forces at a certain level from the ground, and eventually the latter take over entirely. The equation of motion for saltating grains has been investigated by OWEN (1964), YALIN (1963), and TSUCHIYA (1969) not without their making certain quite restrictive assumptions.

The exact mechanism of the initial upward motion is still subject to investigations. BAGNOLD (1941) visualizes the grains in saltation moving like ping-pong balls. When a solid particle is airborne, it receives a supply of energy from the forward fluid pressure. Upon impact of the almost horizontally moving particle, the same particle and/or possibly other stricken ones are ejected almost vertically upward; thus another characteristic path of a grain begins. Bagnold presents sufficient experimental evidence that an

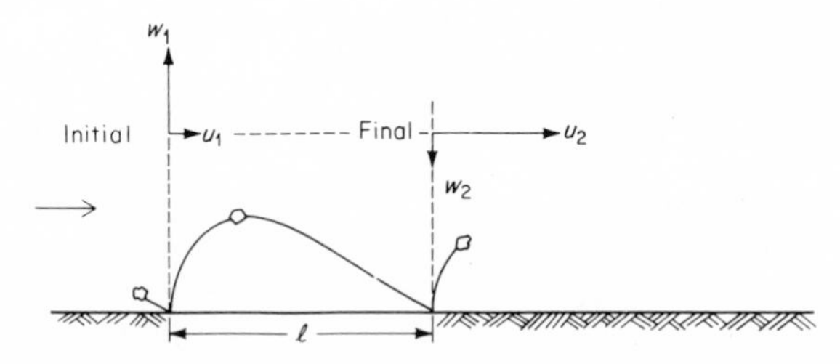

Fig. 7.15 Characteristic path of a saltating grain. [*After* BAGNOLD (*1941*).]

upward force exists, but the model of saltation being caused by impact has not been well accepted. From the point of fluid mechanics, it is more convincing to relate the initial upward motion to the lift experienced by the particle. The investigations of EINSTEIN et al. (1949) on the lift as well as the influences of turbulent fluctuations of the lift forces have been discussed in Sec. 6.4.2. Also significant is the contribution by CHEPIL (1958), who found, based upon experimental data, that lift and drag forces are of the same order of magnitude. In another contribution by CHEPIL (1961) the following was experimentally obtained and is shown in Fig. 7.16. The lift decreases with the height and virtually vanishes at a distance of a few grain diameters above the ground. Drag forces are generally greater than lift forces and increase quite rapidly with height because of the direct pressure of the wind. Chepil concludes that the vertical rise must be due to the presence of lift, but that lift alone could not be the sole factor. Looking for another *rising* mechanism, Chepil suggests that saltating grains experience a rebound action from the ground. EINSTEIN et al. (1949) have presented a convincing study showing that it is not the average lift force that is of importance, but rather the instantaneous force that has to be consulted when establishing a motion criterion. This was also the conclusion of a qualitative experiment reported by BISAL et al. (1962).

After the particle has experienced the initial upward force, its motion in water and air appears to be basically the same, and differs only in degree. In air the height of vertical rise is up to 1,000 grain diameters, and the downstream range is correspondingly large. In water, however, the vertical rise rarely exceeds a few grain diameters, and the whole characteristic path

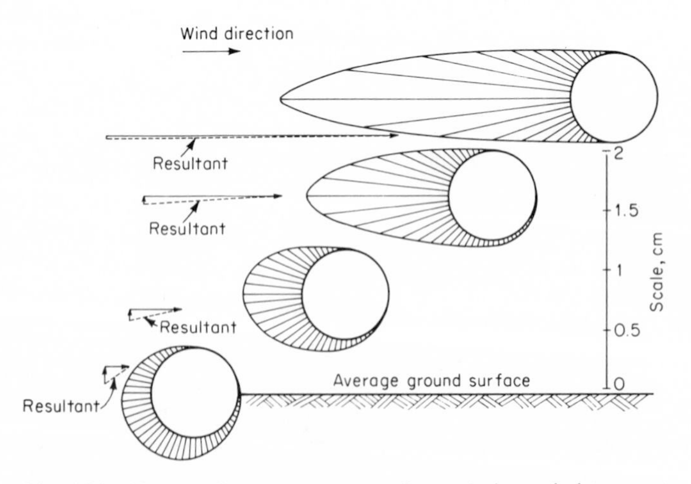

Fig. 7.16 Change of pressure pattern of a grain in a windstream at various positions above the ground. [*After* CHEPIL (*1961*).]

is extremely short. This has been shown by KALINSKE (1942) with a simple mathematical calculation. For the same condition of shear stress and of particle size, the height of bounce depends on the density ratio only; taking buoyancy into account, the geometric values of the characteristic path will be only about 1/1,200 of what they are in air. For all purposes, we may safely conclude that saltation as a separate mode of sediment transport in water is unimportant. Its effect, however, is included in the bedload motion, as was stated at the beginning of Sec. 7.1.

The reader interested in the qualitative and quantitative aspects of aeolian transport is well advised to study BAGNOLD (1941) and WOODRUFF (1965). It should be noted, however, that Bagnold found that of the total sand transport in air about three-fourths moves in saltation and the rest as surface creep. The total sand movement g_{st} under these circumstances is given by

$$g_{st} = k_A \bar{d}^{1/2} \frac{\rho_A}{g} \left(\frac{\tau_0}{\rho_A} \right)^{3/2} \tag{7.79}$$

where ρ_A is the air density and k_A is an empirical constant. Equation (7.79) may be compared in its form with Eq. (7.13).

7.7 CONCLUDING REMARKS

Initially, the bedload was defined as the sediment material the weight of which while in motion is supported by the nonmoving material; the individual particles are rolling, sliding, and occasionally jumping. Many equations have been developed to predict the amount of the bedload. Certain similarities among almost all of these equations are evident. The empirical or semiempirical character of most of them should not be overlooked. Therefore, the application of each equation should remain limited to similar hydraulic conditions and similar sedimentary material as originally used in the development of the equation; if the foregoing is not sufficiently satisfied, the equation should be used with extreme care.

The ultimate goal would be achieved if any or all of the bedload equations could be used to predict the amount of bedload in a natural watercourse. At the present state of the knowledge this is an extremely difficult task for at least two reasons:

1. Every single bedload equation relies on the experimental determination of its coefficients. Except for a few cases in which the bedload was actually measured in watercourses, the bedload motion has been studied in small-scale laboratory flumes.[1] Thus the applications of bedload equations to field studies remain limited and pose many unsolved problems.

[1] A compilation of published and unpublished data of laboratory investigations on bedload transport was made by JOHNSON (1943).

2. Bedload is hard to measure. Owing to the meager definition of the bedload itself, it becomes an exceeding task to build measuring equipment which produces reliable data. Published data from field measurements which could be used to reevaluate the coefficients in the bedload equation are scarce and also hard to work with.

With this in mind, an application of a bedload equation to field determination remains but an educated guess. However, even the available equations do not provide confident data; using a few different equations under the same general conditions, a disagreement of up to 100 percent among the data may be found as not too surprising.

An example is best suited to illustrate the application and the applicability of some bedload equations.

Example 7.1 What is the rate of bedload transport in the Danube canal which flows through the city of Vienna?

The following hydraulic data have actually been measured: A somewhat nonuniform water slope with an average of $S = 6.5 \times 10^{-4}$; the average depth $\bar{D} = 5.87$ m, fairly uniform over the entire width of the rectangular channel; the width $B = 46.52$ m; an average velocity $\bar{u} = 1.52$ m/sec was calculated from velocity-distribution measurements showing fairly uniform distribution; the following roughness values have been computed from the velocity-distribution data which, by the way, indicated that the flow is slightly accelerating, an effect which we shall ignore: $n = 0.0212$ or $C = 56$.

Since no information on the sediment material is available, the following was done: with the Strickler relationship, a grain diameter $d_{90} = 0.059$ m was calculated, a value that compares favorably with other sediment material in this area, and from size distribution curves at the Danube, $d_{50} \approx 0.012$ m was extrapolated.

Approach 1 Use a Meyer-Peter et al. equation, or

$$a.\quad \frac{\gamma R_h (k/k')^{3/2} S}{d(\gamma_s - \gamma)} - 0.047 = 0.25 \sqrt[3]{\rho}\, \frac{(g_s')^{2/3}}{d(\gamma_s - \gamma)} \tag{7.42}$$

where $k/k' = 1$ (no report or evidence on any bedforms).

$$\frac{(1.000)(5.87)(1)(6.5)(10^{-4})}{(1.2)(10^{-2})(1.650)} - 0.047 = 0.25 \sqrt{\frac{1{,}000}{9.8}}\, \frac{(g_s')^{2/3}}{(1.2)(10^{-2})(1.650)}$$

$$0.192 - 0.047 = \frac{(0.25)(4.67)}{19.8} (g_s')^{2/3}$$

$$\frac{0.145}{0.059} = 2.46 = (g_s')^{2/3}$$

$$b.\quad g_s' = (2.46)^{3/2} = 3.82$$

$$c.\quad g_s = g_s' \frac{\gamma_s}{\gamma_s - \gamma}$$

$$g_s = 3.82 \frac{2.65}{1.65} = 6.15 \text{ kg/(m)(sec)}$$

$$G_s = B g_s = 288.0 \text{ kg/sec}$$

The Meyer-Peter et al. equation should be used carefully at such high g_s; although experiments have been performed in this region, they were done with lignite breeze ($s_s = 1.25$). The equation has two major advantages: (1) It has been tested with large grains and (2) it is dimensionally homogeneous.

Approach 2 Use a Schoklitsch equation, or

$$a.\ g_s = 2{,}500S^{3/2}(q - q_{cr}) \tag{7.39}$$

$$b.\ q = \bar{D}\bar{u}$$

$$q = (5.87)(1.52) = 8.9\ \text{m}^3/(\text{sec})(\text{m})$$

$$c.\ q_{cr} = 0.6\,\frac{d^{3/2}}{S^{7/6}}$$

$$q_{cr} = 0.6\,\frac{[(1.2)(10^{-2})]^{3/2}}{[(6.5)(10^{-4})]^{7/6}} = 4.4$$

$$d.\ g_s = 2{,}500[(6.5)(10^{-4})]^{3/2}(8.9 - 4.4)$$

$$g_s = (1.85)(10^{-1}) = 0.185\ \text{kg}/(\text{sec})(\text{m})$$

$$G_s = Bg_s = 8.6\ \text{kg/sec}$$

The Schoklitsch equation must be used with extreme care. Equation (7.39) is dimensionally not homogeneous; this does, in our case, represent a distinct disadvantage. The equation was checked out for Gilbert's data, which neither include coarse materials nor gentle slopes. However, Schoklitsch quotes good agreement of this equation with field measurement at two watercourses.

Approach 3 Use the Kalinske equation, or

$$a.\ \frac{q_s}{u_* d} = \text{fct}_k\left[\frac{(\tau_0)_{cr}}{\tau_0}\right] \tag{7.18}$$

where fct_k is given in Fig. 7.4.

$$b.\ (\tau_0)_{cr} = 1.2\ \text{kg/m}^2, \text{ from Fig. 6.6.}$$

$$c.\ \tau_0 = \gamma RS = (1{,}000)(5.87)(6.5)(10^{-4}) = 3.8\ \text{kg/m}^2$$

$$\frac{q_s}{u_* d} = \text{fct}_k\left(\frac{1.2}{3.8}\right) = \text{fct}_k(0.32) = 1$$

$$q_s = u_* d$$

$$q_s = \sqrt{\frac{3.8}{100}}\,(1.2)(10^{-2}) = (2.34)(10^{-3})$$

$$d.\ g_s = q_s\gamma_s = (2.34)(10^{-3})(2.65)(10^{-3}) = 6.2\ \text{kg}/(\text{m})(\text{sec})$$

$$G_s = g_s B = 290\ \text{kg/sec}$$

The Kalinske equation should be used with care at such high g_s; only limited experiments have been checked in this region. Since the equation is dimensionally homogeneous, it may be consulted.

Approach 4 Use Einstein's equation, or

$$a.\ \Phi = \text{fct}(\Psi') \tag{7.71a}$$

Assume $R_h \approx R'_h$ and $d_{50} \approx d_{35}$

$$b.\ \Psi' = \frac{\rho_s - \rho}{\rho} \frac{d}{SR'_h}$$

$$\Psi' = \frac{1.65}{1} \frac{(12)(10^{-3})}{(6.5)(10^{-4})(5.87)} = (0.52)(10)$$

$$c.\ \Psi' = 5.2; \qquad \text{from Fig. 7.13:} \quad \Phi = 0.58$$

$$d.\ \Phi = \frac{g_s}{\gamma_s} \sqrt{\frac{\rho}{\rho_s - \rho} \frac{1}{gd^3}}$$

$$\Phi = \frac{g_s}{2{,}650} \sqrt{\frac{1}{1.65} \frac{1}{(10)[(1.2)(10^{-2})]^3}}$$

$$\frac{\Phi}{g_s} = \frac{1}{2{,}650} \frac{1}{1.28} \frac{10^3}{4.12} = \frac{1}{13.0}$$

$$e.\ g_s = (13.0)(0.58) = 7.52 \text{ kg/(m)(sec)}$$

$$G_s = Bg_s = 350.0 \text{ kg/sec}$$

Einstein's relationship has all the advantages of the Meyer-Peter et al. equation; in addition even more varied experimental data are used in its establishment.

Discussion Four bedload equations have been used to seek an answer to the given problem. The rate of bedload transport might be given with $G_s \approx 300$ kg/sec. Three of the

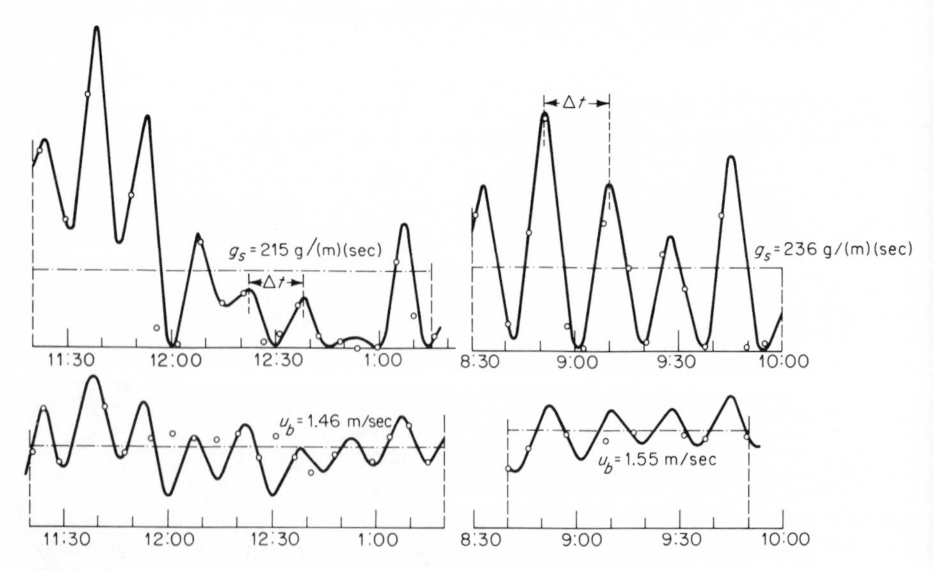

Fig. 7.17 Unsteady bedload transport. [*After* EHRENBERGER (*1931*).]

four bedload equations predict this very value within an accuracy of about 15 percent. The fourth equation is considerably below any of the other values. This equation was not tested under conditions applying to the present problem; since the equation is not even dimensionally homogeneous, it should be excluded from the equations used for our problem.

A further remark is in place here. All the bedload equations are derived such that they predict the maximum bedload that a stream in equilibrium can possibly carry at the given hydraulic and sedimentary condition. The *transporting capacity*, as it is frequently called, may or may not be equal to the actual load if the channel undergoes aggradation or degradation. The actual load depends among others on hydrologic and geologic factors. Because it is very difficult to determine when capacity is reached, it is understandingly even more difficult to make correct use of the field measurements. Furthermore, when using field measurements for checking or establishing bedload equations, special care should be taken that the bedload does not include parts of the suspended load and/or washload.

The problem whether the bedload transport is a steady one has received very little attention. Measurements in the Danube by EHRENBERGER (1931) gave some indications that the bedload transport represents a highly unsteady phenomenon. A typical example is given in Fig. 7.17. The bedload transport fluctuates with an almost constant period but quite variable amplitudes; the velocity at the bed was observed to be in phase. This phenomenon was explained with the migration of bedforms which pass the sediment-measuring device. It is also possible that the bedload moves in distinct clouds of denser and less dense concentrations. Similar field observations have been recorded by MÜHLHOFER (1933) for the Inn and by NESPER (1937) for the Rhine. A promising start to the understanding of the problem has been made by DEVRIES (1962, 1965).

Much has been done to understand the phenomenon of solid transport moving close to the bed; however, we hope that future research will shed more light on the problem.

8
The Suspended Load

8.1 INTRODUCTORY REMARKS

For a set of hydraulic parameters in an open channel with a movable bed, the *critical condition* of the bed may be exceeded and particle motion occurs. If and when the entire motion of the solid particles is such that they are surrounded by fluid, they are said to move in suspension. Owing to the weight of the particles, there is a tendency for settling, which, however, is counterbalanced by the irregular motion of the fluid particles, i.e., the turbulent velocity components. Thus the hydraulic conditions of a stream determine if and when a given size fraction will be in suspension. Furthermore, sediment particles being part of the suspended load at one time may, at another time, be part of the bedload, and vice versa. There exists an active interchange between the suspended load and bedload, but also between the bedload and the bed itself.

The upper limit, where suspension ceases, is the water surface; its lower limit is extremely poorly defined. Coming closer toward the bed, it becomes more difficult to fulfill the foregoing requirements for suspension.

The mode of sediment transport that exists in the immediate vicinity of the bed, in the so-called *bed layer*, is the bedload, extensively discussed in Chap. 7. Suspended load is always accompanied by bedload, and the transition between the two modes of transport is gradual.

Earlier studies concerned with the suspension problem offered very little conclusive or even encouraging results. However, the introduction of the diffusion-dispersion model made a considerable impact on the field. Laws of the vertical and longitudinal distribution of suspended matter were established and found in reasonably good agreement with data from flume experiments or field measurements.

8.2 EARLIER STUDIES

8.2.1 EARLIER DEVELOPMENTS

It was DUPUIT (1865) who put forward the first physical model in order to explain why flowing water was able to hold solid particles in suspension. The suspending agent was believed to be the differences in the magnitudes of the velocities of adjacent filaments; the opposing agent is the settling velocity of the solid particle. The power of suspension was suggested to be limited to a certain liquid-solid concentration, but becomes greater as the velocity increases. The interdependence of the suspension capacity of a stream and the velocity gradient is a correct observation; however, DUPUIT (1865), whose ideas had many a follower, was not successful in explaining it. An interesting discussion on Dupuit's model is given in LELIAVSKY (1955).

FORCHHEIMER (1930) suggested using a bedload equation like the one by SCHOKLITSCH (1914)—a combination of Eqs. (7.6) and (7.11)—or

$$q_s = 0.54 \frac{\gamma^2}{\gamma_s - \gamma} DS[DS - (DS)_{cr}] \qquad \text{m}^3/(\text{sec})(\text{m}) \tag{8.1}$$

and modified it such that it would be applicable for the determination of the rate of suspended material q_{ss} moving through a section in unit width and unit time. Assuming that the factor DS can be expressed with a velocity distribution equation, and that the critical value $(DS)_{cr}$ vanishes for small particles, the following has been obtained:

$$q_{ss} = k_F \frac{\gamma^2}{\gamma_s - \gamma} \frac{\bar{u}^5}{q} \tag{8.2}$$

where k_F is a constant and q is the unit water discharge. Realizing the strong assumptions necessary to obtain Eq. (8.2), FORCHHEIMER (1930) cites agreement with experimental data reported by Deacon (1894); this is especially noticeable in the proportionality of $q_{ss} \propto \bar{u}^5$.

8.2.2 EARLIER TURBULENCE MODELS

According to JAKUSCHOFF (1932), Partiot and Lechalas, in 1871, were the first to explain the mode of suspension as being caused by the eddies and vortices, both created by the bed roughness and the velocity fluctuations. Similar ideas are advanced by Boussinesq (1872) and McMath (1880). GILBERT (1914) believed that the process of suspension depended on the diversity in the direction of the strands of the currents. Quantitative answers were sought by KENNEDY (1895), who made observations at irrigation canals in India. He assumed that the suspended load was caused by vortices created at the bottom of the river, and the whole water body was thus kept in circulatory motion. The representative velocity component u_K causing this kind of flow was expressed by

$$u_K = k_K \frac{{}_s\bar{u}_s^{\,x}}{D^y} \tag{8.3}$$

where k_K is a coefficient and ${}_s\bar{u}_s$ is the velocity which neither silted nor scoured. Rearranging Eq. (8.3), we obtain

$${}_s\bar{u}_s = \left(\frac{u_K}{k_K}\right)^{1/x} D^{y/x} = C_K D^z \tag{8.4}$$

A large amount of data was statistically analyzed, and an empirical relation such as

$${}_s\bar{u}_s = 0.84 D^{0.64} \qquad \text{fps} \tag{8.5}$$

was suggested. If the actual velocity $\bar{u}$ is lower than ${}_s\bar{u}_s$, or $\bar{u} < {}_s\bar{u}_s$, silting will occur; if $\bar{u} > {}_s\bar{u}_s$, the canal will scour. If used at all, Eq. (8.5) should be used with extreme care; it lacks specific corroborative evidence.

Krey (1921) suggested computing the upward velocity component u_y; but since u_y and also the Reynolds number—expressed by $\bar{u}D/\nu$—are indicative of the turbulence level in the stream, a relation of

$$u_y = k_K \left(\frac{\bar{u}D}{\nu}\right)^x \tag{8.6}$$

is proposed. If Jasmund's (1918) data are used, Eq. (8.6) becomes, in metric units,

$$(u_y)_{max} = 0.17(\bar{u}D)^{0.46} \tag{8.7}$$

where $(u_y)_{max}$ is the maximum upward velocity component. Provided the particle diameter and its settling velocity are known, Eq. (8.7) gives an idea of whether the particle will settle out, stay in suspension, or will remain in suspension.

In many of these earlier approaches about suspension it was found that more suspended load moves closer to the bottom than to the water surface. This fact has been documented by field measurements in the Mississippi by Forshey (1851), in the Rhone by Surell (1882), in the Suttley in the Punjab by Buckley (1911), and in the Murg in Turkestan by Gluschkoff (1910). The distribution of the suspended load over the width of a stream is such that near the banks the concentration is lower than in the center. Field data seem to show no indication whether the sediment concentration increases, stays constant, or decreases along a given watercourse.

8.2.3 CONCLUDING REMARKS

It can be seen quite clearly that the suspension phenomenon was but poorly understood by the mid-thirties. None of the attempts to present quantitative informations were of any success. It was realized that turbulence and its associated consequences must be of considerable importance, but this could be expressed neither physically nor mathematically.

Looking back, it seems now that the work of SCHMIDT (1925) and, later, PRANDTL (1926) was indeed a relief and opened up new research avenues for years to come. As a matter of fact, the physical model of suspensions, developed in the second part of the 1930's, is still the backbone of our present-day thinking.

8.3 DIFFUSION-DISPERSION MODEL

Solid particles, but also other matter, are kept in suspension due to turbulence the effect of which on the particles may be assumed to be analogous to the diffusion-dispersion process. Although such a model does not account adequately for all influences, it has been found to explain satisfactorily many suspension problems.

Thus it seems appropriate to discuss and derive next a general diffusion-dispersion equation. Subsequently we shall focus on the vertical distribution of suspended matter, and later on the longitudinal one.

8.3.1 THEORETICAL CONSIDERATIONS

The differential equation of diffusion can be derived by considering an arbitrary control volume. Based on the principle of conservation of mass, equilibrium exists between the rate of change of a property in the control volume and the rate at which the property leaves the control volume, i.e., the rate of change of a property due to diffusion. Mathematically, it can be formulated—see, for example, FRANK et al. (1927, vol. II, p. 531)—such that

$$\frac{\partial C}{\partial t} = \epsilon_m \nabla^2 C \tag{8.8}$$

The diffusion coefficient ϵ_m was assumed not to vary with concentration, $\partial\epsilon_m/\partial C = 0$, a fact truly satisfied in dilute suspensions. Equation (8.8) is similar to the differential equation of heat conduction.

Equation (8.8) was derived for diffusion in an otherwise quiescent medium. Suppose that an influence of an external force exists such that an arbitrary flow with a velocity u is given. The left-hand side of Eq. (8.8) will only be balanced if the rate of change of a property due to diffusion and convection is considered, or

$$\frac{\partial C}{\partial t} = -\nabla \cdot (Cu) + \epsilon_m \nabla^2 C \tag{8.9}$$

Equation (8.9) reduces for incompressible flow to

$$\frac{\partial C}{\partial t} = -u \cdot \nabla C + \epsilon_m \nabla^2 C \tag{8.10}$$

Both Eqs. (8.8) and (8.10) are standard forms of the Eulerian diffusion equation, the latter one, however, in a convective flow field. A diffusion process is adequately described with the diffusion equation, given by Eq. (8.8), and a dispersion process by Eq. (8.10).

Since Eq. (8.8) is nothing but a special form of Eq. (8.10), we shall focus our discussion on the latter. In tensor form Eq. (8.10) becomes

$$\frac{\partial C}{\partial t} = -u_i \frac{\partial C}{\partial x_i} + \epsilon_m \frac{\partial^2 C}{\partial x_i \, \partial x_i} \tag{8.11}$$

in an x_1, x_2, x_3 rectangular coordinate system. Since ϵ_m is the molecular diffusion coefficient, Eq. (8.10) presents a physical model for laminar flows only. For dispersion in a turbulent flow field, given by $C = \bar{C} + C'$ and $u_i = \bar{u}_i + u_i'$, where $\bar{C}$ and $\bar{u}_i$ are the mean values of concentration and the velocity at a given position, and C' and u_i' represent their fluctuations, we obtain

$$\frac{\partial \bar{C}}{\partial t} = -\bar{u}_i \frac{\partial \bar{C}}{\partial x_i} - \frac{\partial}{\partial x_i}(\overline{C'u_i'}) + \epsilon_m \frac{\partial^2 \bar{C}}{\partial x_i \, \partial x_i} \tag{8.12}$$

Elder (1959) found it convenient to define a coefficient of turbulent diffusion such that[1]

$$\epsilon_{t_{ij}} \frac{\partial \bar{C}}{\partial x_j} = -\overline{C'u_i'} \tag{8.13}$$

Under the assumption that molecular and turbulent diffusions are independent and thus additive, we obtain

$$\epsilon_{ij}(x_i) = \epsilon_{t_{ij}} + \epsilon_m \tag{8.14}$$

[1] Further discussion may be based upon Reynolds' analogy for the equivalence of mass and momentum transfer; see Hinze (1958) or Sayre (1967).

where in open-channel flow the turbulent diffusivity is usually considerably larger than the molecular one. Introducing this knowledge into Eq. (8.12) results in

$$\frac{\partial C}{\partial t} = -u_i \frac{\partial C}{\partial x_i} + \frac{\partial}{\partial x_i}\left(\epsilon_i \frac{\partial C}{\partial x_i}\right) \tag{8.15}$$

In Eq. (8.15) the overbars denoting time averages are no longer needed and therefore dropped; the scalar ϵ_i has replaced ϵ_{ij}, commonly referred to as the diffusion tensor, by taking the axes of the coordinate system as the principal axes of the diffusion tensor. Performing the indicated operation in Eq. (8.15) for a three-dimensional problem, the equation may be written as

$$\frac{\partial C}{\partial t} = -u_1 \frac{\partial C}{\partial x_1} - u_2 \frac{\partial C}{\partial x_2} - u_3 \frac{\partial C}{\partial x_3} + \frac{\partial}{\partial x_1}\left(\epsilon_1 \frac{\partial C}{\partial x_1}\right)$$

$$+ \frac{\partial}{\partial x_2}\left(\epsilon_2 \frac{\partial C}{\partial x_2}\right) + \frac{\partial}{\partial x_3}\left(\epsilon_3 \frac{\partial C}{\partial x_3}\right) \tag{8.16}$$

Notice that u_i is the field velocity, or convective velocity, tending to move the suspended matter in the respective direction. Derivation of Eq. (8.16) or (8.15) is a most general one, applying equally well to solid particles or to dispersant in solution form. However, for solid particles, the field velocity must be considered that velocity at which particles are convected, call it u_{s_i}; and, similarly, the diffusivity of solid particles ϵ_{s_i} has to be used.

8.3.2 VERTICAL DISTRIBUTION OF SUSPENDED MATTER

Of considerable importance to the quantitative determination of the suspended matter is its vertical distribution under steady-state and under nonsteady-state conditions. First, we shall consider problems where the turbulence distribution is uniform over the entire vertical section. Then we shall focus on the more realistic and, therefore, very important problem of steady-state sediment distribution at a nonuniform turbulence distribution.

Before we start our discussion, a few words are in order on the general notion of the diffusivity as it appears in Eqs. (8.20) and (8.20*a*). The concept of an analogy between the process of mass and momentum transfer in turbulent flow has been known as the *Reynolds analogy*. Considering the transfer of momentum and mass in the x_2 direction, we can write

$$\text{Momentum flux} = \rho(\epsilon_M + \nu)\frac{\partial u_2}{\partial x_2} = \rho\epsilon_M \frac{\partial u_2}{\partial x_2}$$

$$\text{Mass flux} = (\epsilon_t + \epsilon_m)\frac{\partial C}{\partial x_2} = \epsilon_2 \frac{\partial C}{\partial x_2} \tag{8.17}$$

respectively. Under the assumptions that $\nu < \epsilon_M$ and $\epsilon_m < \epsilon_t$, the Reynolds analogy is valid if the mechanisms which control both transfers—the mass and momentum transfer—are actually identical. As this is most likely the case, we can use ϵ_M and ϵ_2 interchangeably in the x_2 direction, or

$$\epsilon_M = \epsilon_2 \tag{8.18}$$

Experimental data from KALINSKE et al. (1943) seem to substantiate Eq. (8.18). A mixture of hydrochloric acid and alcohol was injected into flowing water; the mixture was thus of a density equal to that of water. The data were analyzed by applying the statistical theory of turbulence, and were then used to predict the concentration distribution. Measurements of the concentration distribution showed agreement within the expected limits of experimental errors. This gives an indirect proof that the diffusivity of mass ϵ_2 is for all practical purposes identical to the diffusivity of linear momentum, ϵ_M; as we shall show subsequently, the latter coefficient predicts also the concentration distribution. JOBSON et al. (1970) have arrived at the same conclusion, namely, that the ratio ϵ_M/ϵ_2 can be approximated by unity. There remains to be shown the relationship of the diffusivity of fluid mass ϵ_2 and the diffusivity of solid particles ϵ_{s_2}. If and only if the solid particles follow the motion of the fluid particles, an equality such as $\epsilon_{s_2} = \epsilon_2$ exists. Nevertheless, we shall use a more general relationship, such that

$$\epsilon_{s_2} = \beta\epsilon_2 \tag{8.19}$$

where β is a factor of proportionality. A detailed discussion of β will follow, but here we may cite that in many cases good agreement between experiment and theory is obtained by taking β as unity.

8.3.2.1 Uniform turbulence distribution at the steady-state condition. Assume steady-state condition $\partial C/\partial t = 0$ and no variation of the concentration with either the x_1 direction, $\partial C/\partial x_1 = 0$, or the x_3 direction, $\partial C/\partial x_3 = 0$; ϵ_{s_2} is, furthermore, considered independent of elevation. Equation (8.16) can be expressed as:

$$0 = v_{ss}C + \epsilon_{s_2}\frac{\partial C}{\partial x_2} \tag{8.20}$$

where the subscript s indicates suspended solid matter and v_{ss} the settling velocity of the particle. Equation (8.20) represents the state of equilibrium between the upward rate of sediment motion due to turbulent diffusion and the downward volumetric rate of sediment transfer per unit area due to gravity. A solution of Eq. (8.20) can be obtained through integration, such that

$$\frac{C}{C_a} = e^{-[v_{ss}(y-a)/\epsilon_s]} \tag{8.21}$$

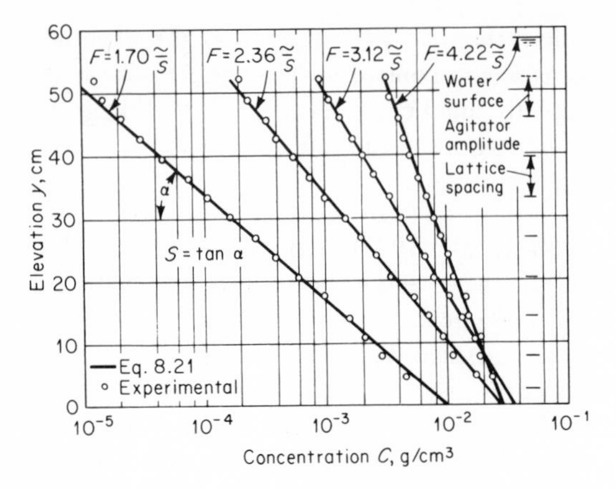

Fig. 8.1 Concentration distribution for sand, $d = \frac{1}{8}$ mm, with uniform turbulence distribution (F = frequency, the turbulence mechanism). [*After* ROUSE (*1938*).]

where C_a is a reference concentration at a distance a from the bed. Note that the more customary notation y and ϵ_s has replaced x_2 and ϵ_{s2}, respectively. According to Eq. (8.21), the concentration distribution is an exponential one; the concentration will be larger closer to the bed than further away from it. The steepness of the exponential curve is dependent on the diffusivity ϵ_s and the settling velocity v_{ss}.

Equation (8.20) and one of its assumptions, namely, that ϵ_s is independent of elevation, make an unrealistic problem, because it was observed that the diffusion coefficient is not constant in natural streams and laboratory flumes. Therefore, Eq. (8.21) should not be expected to predict field data. However, if this simple relation can be verified experimentally, it would give certain strength to Eq. (8.16) as a realistic model of a simple suspension problem.

Both HURST (1929) and ROUSE (1938) have investigated experimentally Eq. (8.20) and its solution. The apparatus, used by both researchers, was a vertical cylinder in which a lattice structure was either rotated or oscillated, and it produced the same degree of turbulence over the entire volume without any steady flow. The intensities of turbulence were achieved by variation of the speed or frequency. Sand with $0.03 < d < 0.9$ mm was used. The experiments by both investigators indicated convincingly that the general shape of the distribution Eq. (8.21) is correct, as can be seen in Fig. 8.1. For larger particles, agreement was not quite perfect. This led ROUSE (1938) to reason that mixing of larger particles is no longer governed solely by the turbulence of the fluid, and that the inertia of the particles will have to be considered.

8.3.2.2 Uniform turbulence distribution with transient conditions. Suppose that the problem of the suspended material transport under non-equilibrium conditions is restricted to two dimensions, the turbulence distribution is uniform and $\partial^2 C/\partial x_1^2 \ll \partial^2 C/\partial x_2^2$; then Eq. (8.16) becomes

$$\frac{\partial C}{\partial t} = -u_1 \frac{\partial C}{\partial x_1} + v_{ss} \frac{\partial C}{\partial x_2} + \epsilon_{s_2} \frac{\partial^2 C}{\partial x_2^2} \tag{8.22}$$

There are now two cases that are identical. One is when $u_1 = 0$, and Eq. (8.22) becomes

$$\frac{\partial C}{\partial t} = v_{ss} \frac{\partial C}{\partial y} + \epsilon_s \frac{\partial^2 C}{\partial y^2} \tag{8.23}$$

while the other is for $\partial C/\partial t = 0$, and Eq. (8.22) becomes

$$u \frac{\partial C}{\partial x} = v_{ss} \frac{\partial C}{\partial y} + \epsilon_s \frac{\partial^2 C}{\partial y^2} \tag{8.24}$$

where x, y, and ϵ_s replaced x_1, x_2, and ϵ_{s_2}, respectively. Equation (8.24) has been investigated by KALINSKE (1940) for the following boundary conditions: (1) The concentration at $x = 0$ is zero; (2) the concentration at $y = \infty$ is zero; and (3) the concentration at $y = a = 0$ remains constant. With these boundary conditions, which again represent certain simplifications, the following solution is obtained:

$$\frac{C}{C_a} = e^{-v_{ss}y/\epsilon_s} - \frac{2}{\pi} \int_0^\infty \frac{n}{-B} e^{[(B\epsilon_s x/\bar{u}) - v_{ss}y/(2\epsilon_s)]} \sin ny \, dn \tag{8.25}$$

where $B = -\left(n^2 + \dfrac{v_{ss}^2}{4\epsilon_s^2}\right)$

$\bar{u}$ = mean velocity in the x direction

x = distance downstream

and the value of the integral is obtained for all values of n from zero to infinity. Notice that Eq. (8.25) reduces to the equilibrium solution, given by Eq. (8.21), when $x = \infty$. As this is the only check on the possible validity of Eq. (8.25)—no experimental evidence has been reported on this problem—it has to be taken with extreme care. Nevertheless, Eq. (8.25) provides an insight into the factors determining the distance required to bring about a change in suspended load until, eventually, equilibrium conditions are attained.

DOBBINS (1943) investigated Eq. (8.23) for a somewhat different set of boundary conditions than KALINSKE (1940) had employed, and thus the solutions are, unfortunately not directly comparable. The two boundary conditions are: (1) no net transfer of solids through the water surface or at $y = D$, $\epsilon_s(\partial C/\partial y) = -v_{ss}C$; (2) the rate of pickup equals the rate of deposit at the bottom, or at $y = a = 0$, $\epsilon_s(\partial C/\partial y) = -v_{ss}C$. The initial condition was taken by a known concentration distribution, or $t = 0$, $C = f(y)$ for $0 < y < D$, whereas Eq. (8.23) requires only three boundary conditions. A fourth boundary condition was given and considered as well (!), namely, that the concentration distribution approaches the equilibrium condition at $t = \infty$. The following solution was given:

$$\frac{C}{C_a} = e^{-v_{ss}y/\epsilon_s} + \frac{C_0 - C_a}{C_a} e^{-v_{ss}y2/\epsilon_s} \sum_{n=1}^{\infty} \frac{2\alpha_n^2 v_{ss} e^{-\epsilon_s At}}{\epsilon_s A(AD + v_{ss}/\epsilon_s)} Y_n \tag{8.26}$$

where C_0 is the concentration at $t = 0$, $y = 0$, $A = \alpha_n^2 + v_{ss}^2/(4\epsilon_s^2)$, $Y_n = \cos \alpha_n y + v_{ss}/(2\epsilon_s\alpha_n) \sin \alpha_n y$, and α is a constant given by

$$2 \cot D\alpha = \frac{D\alpha}{v_{ss}D/(2\epsilon_s)} - \frac{D\alpha/(2\epsilon_s)}{D\alpha}$$

Note that Eq. (8.26) reduces, after steady-state is obtained, to Eq. (8.21). With $C_0 = 0$, Eq. (8.26) resembles Eq. (8.25) and is the solution for the case of scour, in which the original concentration is zero for all values of y. If the equilibrium concentration is zero, $C_a = 0$, and $t = \infty$, Eq. (8.26) is the

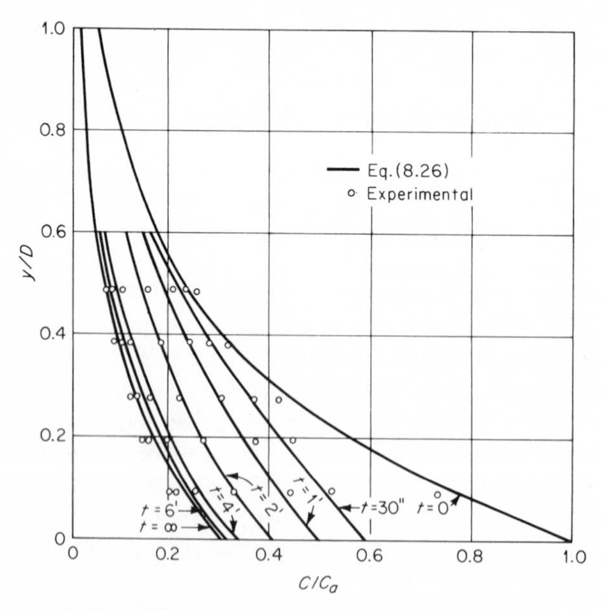

Fig. 8.2 Concentration changes with constant rate of pickup; $v_{ss}/\epsilon_s = 0.0638$ cm^{-1}; $D = 45.2$ cm. [*After* DOBBINS (*1943*).]

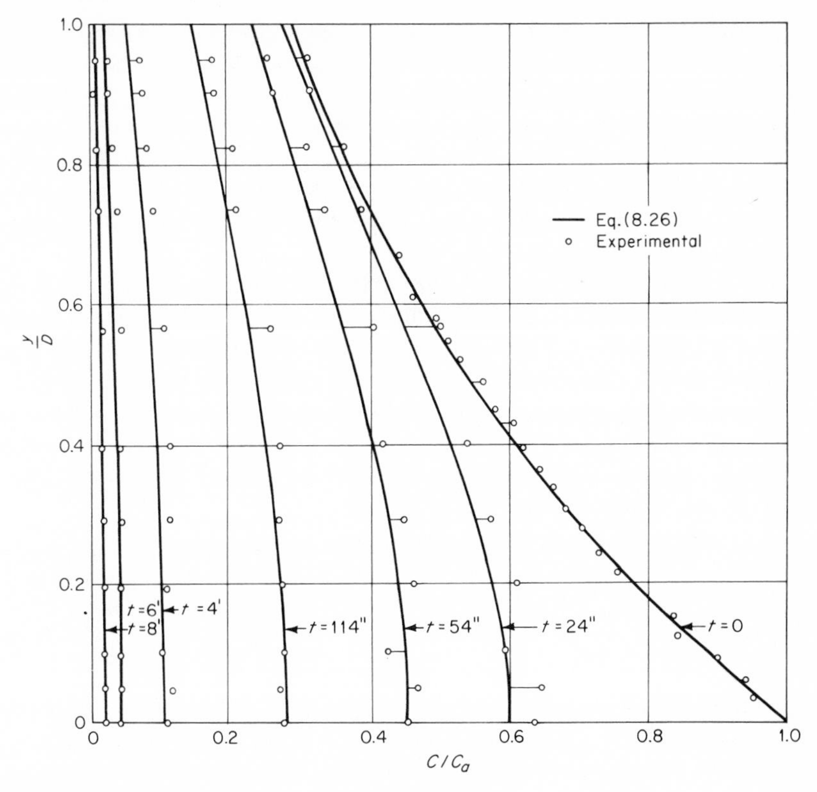

Fig. 8.3 Concentration changes with no pickup; $v_{ss}/\epsilon_s = 0.030$ cm^{-1}; $D = 41.6$ cm. [*After* DOBBINS (*1943*).]

solution for the case of a complete-settling-out-to-zero concentration at all depths.

Equation (8.26) has been investigated experimentally, and thus becomes the more interesting study. The experimental setup as used by DOBBINS (1943) was similar to one described earlier and used by ROUSE (1938). The mixing was achieved by a vertical oscillation; the rate of pickup was changed by lifting the lattice structure, and no pickup was obtained by suddenly removing the lattice structure from the sediment-covered floor. As solid matter, a resin called *lucite* was used, having a diameter of about 0.2 mm. Experimental data obtained are compared with values predicted with Eq. (8.26). From Fig. 8.2, for a constant rate of pickup, and from Fig. 8.3 for no pickup, we may safely conclude that the validity of the basic Eq. (8.22) and the assumptions regarding the boundary conditions are demonstrated. DOBBINS' (1943) contribution was successfully applied to the design of sedimentation basins by CAMP (1945).

Equation (8.24) with another set of boundary conditions was analytically solved by MEI (1969) and experimentally investigated by APMANN et al. (1970).

KALINSKE (1940) and DOBBINS (1943) considered the two-dimensional problem. However, GODA (1953) has presented a study on a three-dimensional problem including certain restrictions. Since the solutions are rather elaborate and experimental evidence is meager, this research must be followed in the original version.

8.3.2.3 Nonuniform turbulence distribution. In actual streams, the turbulence intensity and, thus, the diffusivity vary with the distance from the bottom. Since the problem of the suspended load in streams is of great interest, many contributions on this particular topic are available. We shall discuss first theoretical considerations, and then how well experimental data and theoretical ones agree; this will be followed by modified theories and various other contributions; in conclusion, we shall apply the knowledge to engineering calculations.

8.3.2.3.1 THEORETICAL CONSIDERATIONS. With Eq. (8.20) the diffusion relationship was given, or

$$0 = v_{ss}C + \epsilon_s \frac{\partial C}{\partial y} \tag{8.20a}$$

Equation (8.20*a*) was used by SCHMIDT (1925) to determine the distribution of dust particles caused by the turbulent mixing in the atmosphere. The diffusivity of solid particles ϵ_s was given as a function of the position, or $\epsilon_s = f(y)$. The application of Schmidt's model to the transport of solids in water was first realized by JAKUSCHOFF (1932) and LEIGHLY (1932, 1934).[1] The diffusivity distribution, however, has been advanced by O'BRIEN (1934). In a two-dimensional channel, the relative shear stress distribution is given by

$$\frac{\tau_y}{\tau_0} = \frac{D - y}{D} \tag{8.27}$$

and assuming the logarithmic velocity distribution, a further relation is written as

$$\frac{du}{dy} = \frac{\sqrt{\tau_0/\rho}}{ky} = \frac{u_*}{ky} \tag{8.28}$$

where k is *Karman's constant*. Due to Reynolds' analogy—see Eqs. (8.17) and (8.18)—the shear stress can be expressed as

$$\tau_y = \rho\epsilon \frac{du}{dy} \tag{8.29}$$

[1] According to JAROCKI (1957), the Soviet researchers, Makkaveev (1931) and Velikanov (1938) applied the principle of diffusion to the motion of suspended load.

By means of Eqs. (8.27), (8.28), and (8.29),

$$\epsilon = ku_* \frac{y}{D}(D - y) \tag{8.30}$$

and with the knowledge of Eq. (8.19) for $\beta = 1$,

$$\epsilon_s = ku_*(D - y)\frac{y}{D} \tag{8.31}$$

Introducing Eq. (8.31) into Eq. (8.20*a*) and separating variables, we obtain

$$\frac{dC}{C} = -\frac{v_{ss}}{ku_*}\frac{D}{y}\left(\frac{dy}{D - y}\right) \tag{8.32}$$

The quantity

$$\frac{v_{ss}}{ku_*} = z \tag{8.33}$$

is frequently used. Equation (8.32) will be integrated from a to y by resolution into partial fractions; then

$$\begin{aligned} \int_a^y \frac{dC}{C} &= \int_a^y d(\ln C) = \ln \frac{C}{C_a} \\ \int_a^y \frac{zD}{y(D - y)}\,dy &= \int_a^y d\left[\ln\left(\frac{D - y}{y}\right)^z\right] = \ln\left(\frac{D - y}{y}\frac{a}{D - a}\right)^z \end{aligned} \tag{8.34}$$

and the resulting relation may be written as

$$\frac{C}{C_a} = \left(\frac{D - y}{y}\frac{a}{D - a}\right)^z \tag{8.35}$$

The suspended-load-distribution equation was introduced by ROUSE (1937).[1] It may be used for calculation of the concentration of a given grain size v_{ss} at any distance y from the bed if a reference concentration C_a at a distance a is available. Integration of Eq. (8.35) over the region where suspended load occurs, say from a to D, is obtained such that

$$g_{ss} = \int_a^D Cu\,dy \tag{8.36}$$

where C and u are functions of y and g_{ss} is the suspended-load rate in weight per unit time and width. Equation (8.36) as well as Eq. (8.35) is limited because each can predict only the relative concentration.

[1] For a most enlightening historical remark on the derivation of Eq. (8.35), a note by ROUSE (1964) is recommended for reading.

The vertical concentration distribution according to Eq. (8.35) is shown in Fig. 8.4. The following should be noticed in Fig. 8.4. The concentration is smaller farther away from the bed than close to it. From a mathematical point, at the bed where $y = 0$, the concentration becomes $C = \infty$, which is impossible. That suspension does not exist close to the bed is physically sound; particles close to the bed are not embedded in water anymore, but rather represent part of the bedload. EINSTEIN (1940) has suggested that suspension is not possible in the so-called *bed layer*, which has a thickness of 2 grain diameters. For low z values, the sediment distribution is nearly uniform, whereas for large z values little sediment will be found at the surface of the channel. The particle size, expressed as the settling velocity v_{ss}, is directly responsible for the kind of distribution.

It appears worthwhile to mention that WILLIS (1969) proposed another model for the vertical distribution of suspended matter, which has the form of an error function. Certain advantages of such a function were pointed out and a comparison of field and laboratory data was made.

8.3.2.3.2 EXPERIMENTAL INVESTIGATIONS. *Concentration distribution.* Both in the laboratory and in the field, it was found that Eq. (8.35) describes

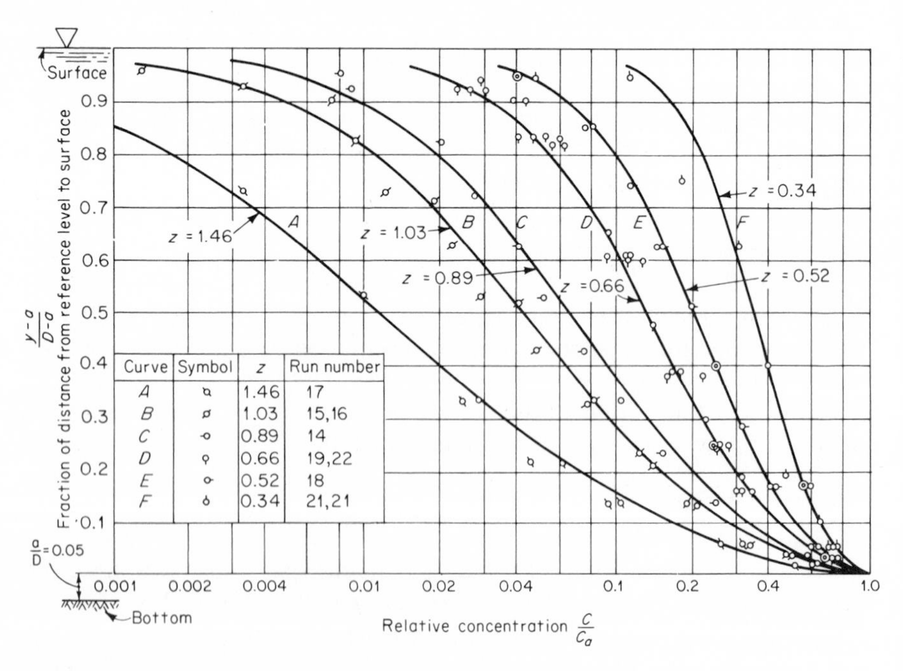

Fig. 8.4 Distribution of suspended sediment; comparison of experimental data with Eq. (*8.35*). [*After* VANONI (*1946*).]

the concentration distribution rather well. Actually, it is an agreement in form only, because the z values used are those giving the best fit instead of being determined independently.

Early laboratory studies were reported by RICHARDSON (1934). The results of carefully made laboratory experiments on the concentration of suspended material in a rectangular channel are reported by VANONI (1941, 1946). The experiments were performed in a closed-circuit system and adjustable slope flume, $33\frac{1}{4}$ in. wide by 60 ft long. The sides were kept smooth, while the bottom was made rougher by cementing sand to the steel structure. Three types of graded silica sand, $0.10 < d < 0.16$ mm, were investigated. The sediment distribution was determined from samples siphoned from the flume with a pipette, shaped much like a pitot tube. The data, plotted logarithmically according to Eq. (8.35), fell on a straight line with a slope of

$$z = \frac{\log (C/C_a)}{\log \{[(D - y)/y][a/(D - a)]\}}$$

This indicates at once that the concentration distribution follows the form of Eq. (8.35). A typical result, giving the relative concentration distribution as a function of the relative depth as measured upward from a reference level, is shown in Fig. 8.4. A theoretical distribution is shown as a solid line and the measurements are indicated by circles. ZELLER (1969) has shown that the distribution of clay suspension is also given with Eq. (8.35).

To check the theory with observed sediment distributions, CHRISTIANSEN (1935) analyzed the available data from the Colorado River and Imperial canals, as well as from the River Nile. In general, the agreement was good; however, the available data were not sufficiently complete to permit calculations without introducing certain estimated quantities. The vertical distribution of suspended sediment for one measuring station at the Missouri River is given by STRAUB (1936). The general trend, given by Eq. (8.35), seems to exist for all size fractions, from clay to coarse sand, of the sample. Most convincingly, ANDERSON (1942) has shown the existence of Eq. (8.35). A large number of suspended-load samples have been collected at the Enoree River, at a section reported as quite regular and fairly straight. The shape of the distribution curve agreed satisfactorily with the theory, even though some individual points scatter. Further evidence that the distribution function is of the correct form is given by LANE et al. (1941) for the Mississippi, by VANONI (1953) and by HARRISON et al. (1963) for the Missouri, by COLBY et al. (1955) for the Niobrara River, and by NORDIN et al. (1963) for the Rio Grande.

z value. Although the form of Eq. (8.35) has been found to apply to laboratory and field data, the exponent z that fits the observations is frequently different from Eq. (8.33). Notice that the z value has been obtained by making Eqs. (8.30) and (8.31) identical or putting $\beta = 1$ into Eq. (8.19). It thus appears that the relation given by

$$\epsilon_s = \beta\epsilon \tag{8.19}$$

will have to be examined. The relationship between the diffusivity of solid particles and the one of linear momentum is proportional and not necessarily identical. Some knowledge could be gained if we knew how willingly a solid particle followed its liquid environment. At the present stage of the research, it is sufficient to give a qualitative answer and to simplify the problem. Let us assume that the solid-particle motion is a nonuniform motion through a real fluid with linear(!) resistance and oscillatory(!) acceleration, and omit the gravitational force. If the liquid phase also oscillates, the slippage B/A is given by TCHEN (1947) as

$$\frac{B}{A} = 1 - \frac{2(\rho_s - \rho)}{(2\rho_s + \rho) - 9\rho[i\nu/a^2\omega - i(i\nu/a^2\omega)^{1/2}]} \tag{8.37}$$

for a solid particle $d = 2a$ and for a solid-particle velocity $Be^{i\omega t}$ and a liquid velocity of $Ae^{i\omega t}$. The amplitude ratio B/A and the ratio of the diffusivities are related according to CARSTENS (1952) such that

$$\left(\frac{B}{A}\right)^2 = \frac{\epsilon_s}{\epsilon} = \beta \tag{8.38}$$

From Eq. (8.37), where ω is the circular frequency and ν the kinematic viscosity, we may draw the following conclusions for $\rho_s > \rho$:

1. For ideal fluids,

$$\nu = 0 \qquad \frac{B}{A} = \frac{3\rho}{2\rho_s + \rho} \qquad \beta < 1$$

2. For real fluids,

$$\omega \to \infty \qquad \frac{B}{A} = \frac{3\rho}{2\rho_s + \rho} \qquad \beta < 1$$

$$\omega \to 0 \qquad \frac{B}{A} = 1 \qquad \beta = 1$$

$$a \to \infty \qquad \frac{B}{A} = \frac{3\rho}{2\rho_s + \rho} \qquad \beta < 1$$

$$a \to 0 \qquad \frac{B}{A} \Rightarrow 1 \qquad \beta \geq 1$$

Thus for sediment in water, it is not at all clear when β is equal to, smaller than, or possibly even larger than unity, or

$$\beta \gtreqless 1 \tag{8.39}$$

Certainly β depends on the frequency ω and on the particle size d, and the exact interrelationship is, even under our simplified assumptions, complicated. CARSTENS (1952) presented a mathematical expression for β, and concluded that β never exceeds unity but $\beta \leq 1$. For practical purposes, and if ω is about constant, we might conclude that:

For fine particles,

$$\beta \approx 1 \qquad \text{or} \qquad \epsilon_s \approx \epsilon$$

For coarse particles, (8.40)

$$\beta < 1 \qquad \text{or} \qquad \epsilon_s < \epsilon$$

In carefully conducted experiments, BRUSH et al. (1962), MATYUKHIN et al. (1966), and MAJUMDAR et al. (1967) provide sufficient experimental evidence for above-mentioned tendencies. The latter two experiments show clearly that increasing the particle size causes the β value to decrease. On the other hand, BRUSH's et al. (1962) study gives evidence to the fact that increases in the frequency ω cause the β value to decrease faster. Further ideas as to how the β value changes may be gotten from HOUSEHOLDER et al. (1969) and JOBSON et al. (1970).

The disagreement between theoretical z values, z, and the observed ones, z_1, are given by

$$\frac{v_{ss}}{\beta k u_*} = z_1 = \frac{z}{\beta} \tag{8.41}$$

It would be a mistake, however, to attribute all the disagreement between z and z_1 to the β value only, but rather the turbulence and concentration effects on the k value and on the settling velocity have to be considered as well. Experimental data from field and flume studies are presented by CHIEN (1954) and are reproduced in Fig. 8.5. The z_1 values are determined from the measured concentration distributions; the z values are calculated ones, but Karman's constant k has been determined from the measured velocity distribution, and the settling velocities v_{ss} are determined in still water.[1] It is evident from Fig. 8.5 that $z_1 < z$, and deviations between the two become consistently larger as z increases. A similar trend is noticeable in data from the Enoree River reported by ANDERSON (1942), from the Niobrara River by COLBY et al. (1955), and from the Rio Grande by NORDIN

[1] That z_1 values increase with concentration is reported in VANONI (1953).

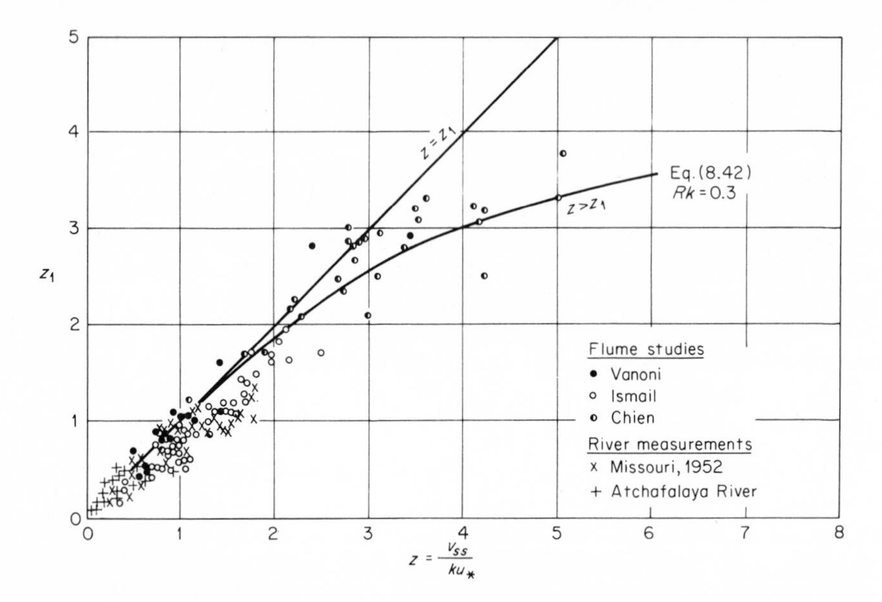

Fig. 8.5 Relationship between z and z_1. [*After* CHIEN (*1954*).]

et al. (1963). This cannot and must not be explained with Eq. (8.40). Also EINSTEIN et al. (1954) express doubt that this discrepancy can be attributed to the difference in the diffusivities alone, and offer a physical model to explain this. The derivation is based on the following considerations, and as such represents a *second approximation* to the solution of the suspended-load theory:

1. It considers higher-order terms of concentration gradients.
2. Fluctuation velocities of the upward and downward flow are the same.
3. $\epsilon_s = \epsilon$.
4. Fluctuation velocities follow the normal-error law.

The resulting z vs. z_1 relation was given by

$$z_1 = \frac{z}{e^{-L^2 z^2/\pi} + \dfrac{2zL}{\sqrt{2\pi}} \displaystyle\int_0^{\sqrt{2L^2 z^2/\pi}} e^{-x^2/2}\, dx} \tag{8.42}$$

where $x = \ln y$ and $L = \ln (1 + Rk)$. The best fitting curve is given by taking $Rk = 0.3$, and as such is shown in Fig. 8.5.

Although EINSTEIN et al. (1954) realized the uncertainties in using the settling velocity in clear, still water, they were compelled to believe that these do not seriously change the z vs. z_1 relation. A study by BRUSH et al. (1962) on settling velocities in oscillating fluids showed a substantial decrease of the settling velocities with increasing frequencies. The effect of concentration on the settling velocities was also discussed (see Sec. 4.2.4.3), but does not as yet render conclusive results. Finally, BRUSH et al. (1962) reanalyzed some data—the data by Ismail given in Fig. 8.5, which were interpreted wrongly for a justification of $\beta > 1$—and find $z \leq z_1$, a result which may now be assumed to be indicative of Eq. (8.40). However, owing to lack of a more extensive research on the settling of particles in a turbulent field, information on v_{ss}, as would be necessary in Eq. (8.41), is not as yet readily available. Until further knowledge is obtained, the z vs. z_1 relationship, as advanced by EINSTEIN et al. (1954) and given in Fig. 8.5, should be consulted.

k value. In open-channel flow without sediment, the *Karman k value*, the universal constant in logarithmic velocity law, is given by $k = 0.4$. In the presence of sediment the k values[1] were found to vary, with a tendency to decrease, $k < 0.4$, for increasing suspended loads. Early flume experiments by VANONI (1941, 1946) and field data discussed in EINSTEIN et al. (1954) exhibited this tendency. VANONI (1963) suggests that a reduction of k means that mixing is less effective and, apparently, the presence of sediment suppresses or damps the turbulence. EINSTEIN et al. (1954) offered the following explanation. The rate of frictional energy spent in supporting the suspended sediment per unit weight of fluid and unit time is given with

$$\sum \frac{\bar{C}v_{ss}}{\bar{u}S} \frac{\rho_s - \rho}{\rho} \tag{8.43}$$

where $\bar{C}$ is the average concentration by weight of a given grain size with a settling velocity v_{ss}, and the summation sign designates a summation over all the particles in suspension. This argument can be correlated with the k value, and is given in Fig. 8.6; scatter is evident but, nevertheless, a reasonable correlation is obtained with data from flume studies and river measurements. NORDIN et al. (1963) find the parameter of Eq. (8.43) useful, but report deviations when dunes form the bed. Also of interest are results obtained by ELATA et al. (1961) with particles of a relative density $\rho_s/\rho = 1.05$. Although it was found that the k value indeed decreased with a concentration increase, turbulence measurements indicated an increase of the intensity with an increase in concentration. Recently HINO (1963) offered a theory to explain this phenomenon. With two fundamental equations, an energy equation for flow with suspended particles and an acceleration balance equation of turbulent motion, the changes of the k values and of the

[1] The k values are determined from the semilogarithmic velocity distribution by $k = 2.3u_*J$, where J is the slope of the semilogarithmic velocity profile $d(\log y)/du$.

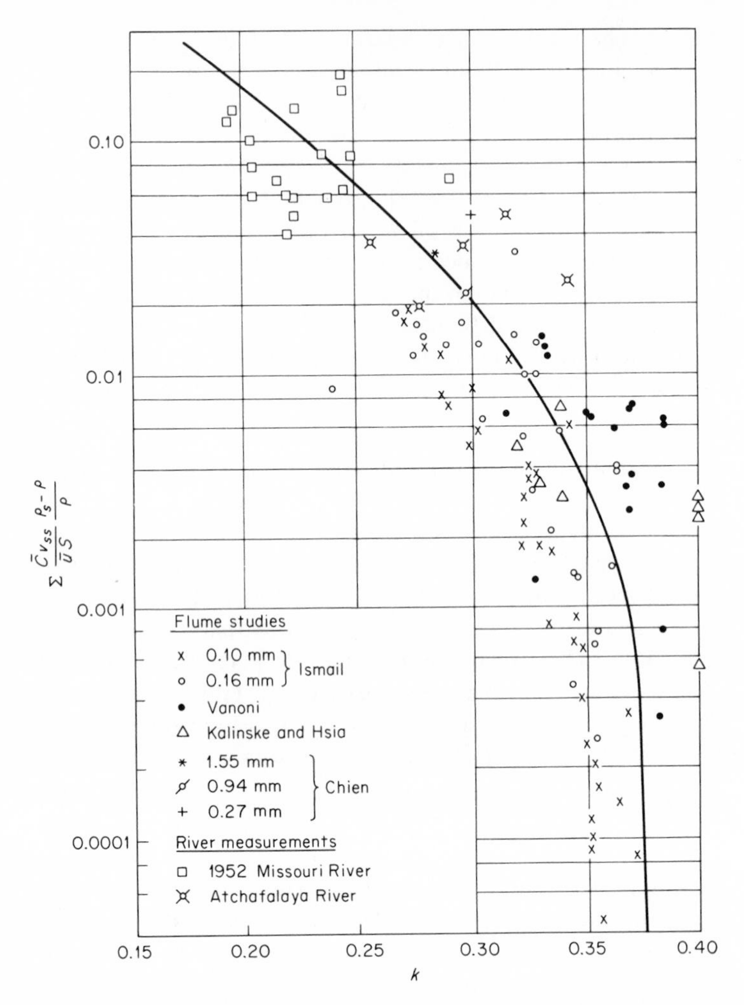

Fig. 8.6 Effect of suspended load on the k value. [*After* EINSTEIN et al. (*1954*).]

turbulence intensities could be predicted. For analyses the reader is referred to the original contribution, but good agreement between Hino's results and the experimental data was obtained.

However, a reduction in the k value does not necessarily mean that the average velocity of a flow with suspension increases proportionally to clear-water velocity. This is pointed out by CHIEN (1954), who suggests that the logarithmic velocity distribution equation for clear-water flow is no longer adequate. Ideas for the development of a new equation have been advanced by EINSTEIN et al. (1955) and are discussed in Sec. 8.3.2.3.3.

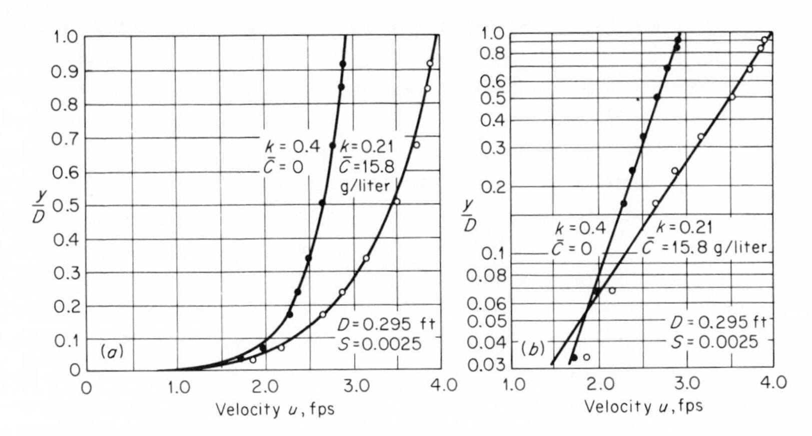

Fig. 8.7 Velocity profiles for clear-water and sediment-laden flow. [*After* VANONI *et al.* (*1960*).]

Experimental evidence is presented by VANONI (1941, 1946), which shows that for the same discharge the average velocity for sediment-laden flow is larger than for clear-water flow; an accompanying effect is an apparent reduction of the roughness coefficient. The velocity increase does not seem to give rise to an increase in the suspended load because depth decreases as well, and thus the bed shear stress $\tau_0 = \gamma DS$ is actually diminished. With Fig. 8.7 VANONI et al. (1960) present further evidence of this effect. Not only has the sediment-laden flow a larger cross-sectional velocity but also the velocity distribution becomes less uniform. The same laboratory study shows the effect of moving sediment reducing the frictional factor from 5 up to 28 percent. Another point has to be considered here. Sediment motion is responsible for the change of the loose bed, and various bedforms (ripples, dunes, etc.) are created. The change of the friction factor due to these bedforms is reported to be much larger than that due to the suspended load. As VANONI (1953) has pointed out, this implies that the friction factor of flow with suspension may increase, decrease, or not change at all, depending on magnitude of these factors.

Secondary flow. In VANONI's (1941, 1946) experiments, there are two observations that deserve special attention. Velocity distribution becomes less symmetrical and less uniform by addition of small amounts of sediment ($\bar{C} = 0.75$ g/liter) to a clear-water flow, and sediment transport appears to take place in longitudinal bands. It is reported that at "proper flow conditions" these bands are symmetrically spaced about the centerline, and are stable regardless of the flow rate. VANONI (1946) presents photographic evidence of one experimental run with three main streaks within the center portion of the channel and two weaker streaks at its walls occurring over the entire length of the flume. Thus VANONI (1941, 1946) believes that suspended load causes a flow to become unevenly distributed, which, in turn, causes secondary circulation. NEMENYI (1946) and LELIAVSKY (1955) advanced the following opinion. Since secondary currents are a normal rather than an exceptional occurrence in turbulent flow, they may be the cause of unevenly distributed sediments; the distribution of suspended matter and the secondary currents are mutually and inseparably interconnected. The latter statement will be even more appreciated when we realize that the common cause for secondary currents and for suspension is indeed the turbulence. It is not within the scope of this book to engage in a discussion on secondary currents. Earlier studies are reviewed in NEMENYI (1946); an outstanding contribution is by PRANDTL (1926). Recent research is reported by EINSTEIN et al. (1958), GESSNER et al. (1965), and TRACY (1965). In a rather limited semitheoretical study, CHIU et al. (1966) show secondary currents to have a remarkable effect on the distribution of suspended matter.

8.3.2.3.3 FURTHER THEORETICAL CONSIDERATIONS. Certain deficiencies in the suspended-load theory have been noticed when compared with field or laboratory data. This led some researchers to re-evaluate the available theoretical knowledge.

In a comprehensive experimental and theoretical investigation EINSTEIN et al. (1955) modify the velocity distribution equation, obviously influenced by the sediment motion and sediment distribution equation. They find it necessary to divide the sediment-laden flow into two zones. There is a zone close to the bed where the sediment is heavily concentrated; it is relatively shallow and may be referred to as *heavy-fluid zone*. The remaining part of the flow, where the sediment concentration is relatively small, does not alter the density of the fluid and is called the *light-fluid zone*. The heavy-fluid zone becomes an important physical concept. Turbulence is created at the boundary; the heavy-fluid zone acts like a filter, and it reduces the turbulence level because energy was spent to suspend the solid particles in this zone.

Velocity distribution. The clear-water flow follows the logarithmic law of velocity distribution, or

$$\frac{u}{u_*} = 5.75 \log 30.2 \frac{y}{k_s} \tag{8.44}$$

where k_s is the Nikuradse sand roughness. Equation (8.44) was derived by assuming that the shear stress can be expressed by Eq. (8.29), or

$$\tau_y = \rho\epsilon \frac{du}{dy} \tag{8.29}$$

which, in turn, is based on the exchange of water particles. For sediment-laden flow a more proper velocity distribution may be obtained if we include the fact that solid particles participate in the exchange mechanism. EINSTEIN et al. (1955) have derived the following relationship:

$$\tau_y = \left(1 + \frac{\rho_s - \rho}{\rho} C\right) \rho\epsilon \frac{du}{dy} \tag{8.45}$$

Within the light-fluid zone of the small concentration, Eq. (8.45) becomes Eq. (8.29). Under these circumstances an equation similar to the clear water equation (8.44), but with different numerical constants, might be expected to apply. Experimental research suggested the following relationship:

$$\frac{u}{u_*} = 17.66 + \frac{2.3}{k} \log \frac{y}{35.45k_s} \tag{8.46}$$

where the k value, as was explained earlier, is to be gotten from Fig. 8.6. Close to the bed, whenever the local sediment concentration reaches a value of 100 g/liter or if $y/D < 0.1$, Eq. (8.46) fails, as can be seen in Fig. 8.8. In this case, the shear stress given with Eq. (8.45) can be approximated by the bottom shear stress τ_0, or

$$\tau_0 = \int_0^D [\rho + (\rho_s - \rho)C]gS\,dy \tag{8.47}$$

The obtained velocity distribution is given by

$$\frac{u}{u_*} = 5.75 \frac{\sqrt{1 + \dfrac{\rho_s - \rho}{\rho} \dfrac{1}{D} \displaystyle\int_0^D C\,dy}}{\sqrt{1 + \dfrac{\rho_s - \rho}{\rho} C_0}} \log\left(A_e \frac{y}{k_s}\right) \tag{8.48}$$

where C_0 is the sediment concentration at the surface of the bed layer. However, we still have to determine the constant A_e and over what depth Eq. (8.48) applies.

As far as the determination of the average velocity is concerned, Eq. (8.46) may be used for the entire flow region without introducing any significant error, and

$$\frac{\bar{u}}{u_*} = 17.66 + \frac{2.3}{k} \log\left(\frac{D}{96.5k_s}\right) \tag{8.49}$$

In an experimental and theoretical study, YANO et al. (1969) could show that the "velocity distribution affected by the bedload deviated from the logarithmic formula near the bed." It was suggested that this deviation

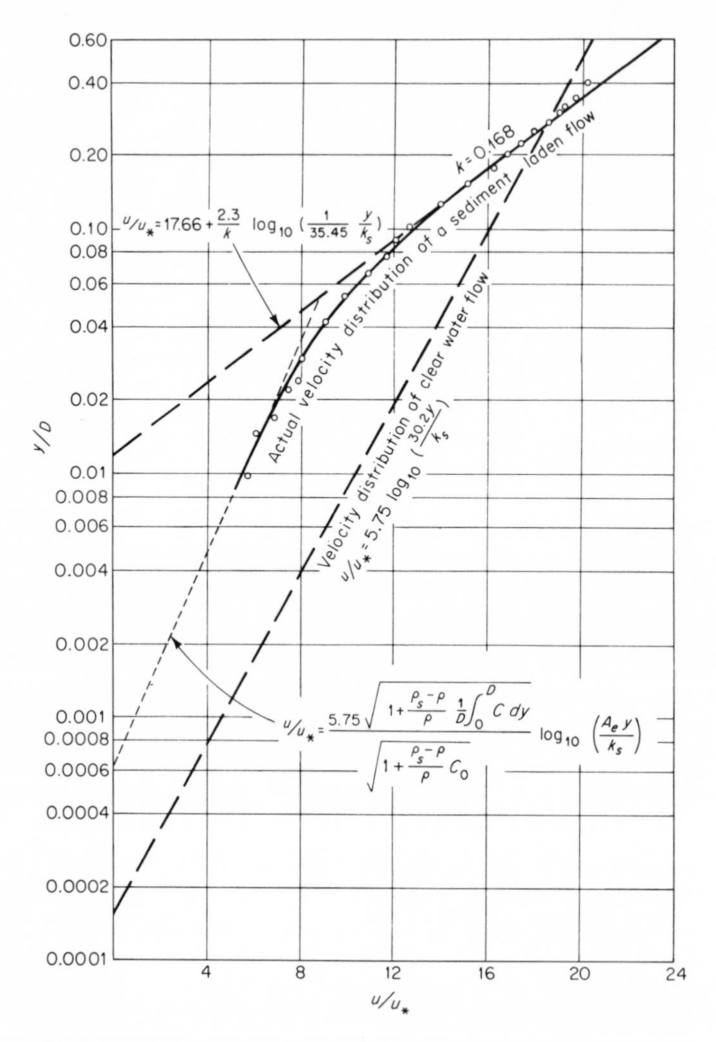

Fig. 8.8 Velocity distribution of clear-water and sediment-laden flow. [*After* EINSTEIN *et al.* (*1955*).]

is due to the shear stress which is generated by the rotation of the sediment particles.

Sediment distribution. Without lack of generality, Eq. (8.15) can be written as

$$\frac{\partial C}{\partial t} = -u_i \frac{\partial C}{\partial x_i} - C \frac{\partial u_i}{\partial x_i} + \frac{\partial}{\partial x_i}\left(\epsilon_i \frac{\partial C}{\partial x_i}\right) \tag{8.50}$$

For the special case of uniform flow in the x_1 direction and the concentration being constant with time, variations in the $x_2 = y$ component are considered, thus that Eq. (8.50) reduces to

$$0 = -u_y \frac{\partial C}{\partial y} - C \frac{\partial u_y}{\partial y} + \frac{\partial}{\partial y}\left(\epsilon_y \frac{\partial C}{\partial y}\right) \tag{8.51}$$

The rate of change of suspended matter is given by

$$0 = -u_s \frac{\partial C}{\partial y} - C \frac{\partial u_s}{\partial y} + \frac{\partial}{\partial y}\left(\epsilon_s \frac{\partial C}{\partial y}\right) \tag{8.52}$$

and, similarly, for water by

$$0 = -u_y \frac{\partial C}{\partial y} - (1 - C)\frac{\partial u_y}{\partial y} + \frac{\partial}{\partial y}\left(\epsilon \frac{\partial C}{\partial y}\right) \tag{8.53}$$

Owing to the relationship between the velocities of solid matter u_s and of liquid matter u_y, and the settling velocity of solid particles v_{ss}, we may write

$$u_s = u_y - v_{ss} \tag{8.54}$$

By eliminating u_s and u_y between Eqs. (8.52) and (8.53), we obtain

$$[\epsilon_s + C(\epsilon - \epsilon_s)] \frac{\partial C}{\partial y} + (1 - C)Cv_{ss} = 0 \tag{8.55}$$

where ϵ_s and ϵ represent the diffusivity of solid matter and liquid matter, respectively.[1] Note that Eq. (8.55) is the most general differential equation for the vertical distribution of suspended matter; following a rigorous mathematical derivation, it was first given by HUNT (1954) and later by EINSTEIN et al. (1955). Equation (8.55) becomes cumbersome, but the solution to it is greatly facilitated if the diffusion coefficients of solid and liquid matter are the same, or $\epsilon_s = \epsilon$; thus we obtain

$$\epsilon_s \frac{\partial C}{\partial y} + (1 - C)Cv_{ss} = 0 \tag{8.56}$$

a form of the diffusion equation first given by HALBRONN (1949). The solution to Eq. (8.56), as obtained by HUNT (1954), reads:

$$\left(\frac{C}{1 - C}\right)\left(\frac{1 - C_a}{C_a}\right) = \left[\left(\frac{1 - y/D}{1 - a/D}\right)^{1/2}\left(\frac{B_s - (1 - a/D)^{1/2}}{B_s - (1 - y/D)^{1/2}}\right)\right]^m \tag{8.57}$$

where $m = v_{ss}/k_s B_s u_*$, B_s is a constant of integration in the velocity distribution law ($B_s \leq 1$), and the k_s value is similar to the k value. The fact that Eq. (8.57) has two arbitrary constants, whereas Eq. (8.35) has only one, gives a slightly better fit of the data. This has been pointed out by VANONI (1963), who makes it clear that one reason why Eq. (8.57) is not used frequently is its complicated form, and the added difficulty of having two constants rather than one.

[1] As in Sec. 8.3.2.1, the assumptions are made that ϵ and ϵ_s are independent of elevation.

For small sediment concentrations, as encountered in the light-fluid zone, Eq. (8.56) reduces to Eq. (8.20), or

$$\epsilon_s \frac{dC}{dy} + Cv_{ss} = 0 \tag{8.20}$$

the solution of which was given earlier with Eq. (8.35).

For large sediment concentrations, Eq. (8.56) should be used. If the diffusivity of solid matter is given by Eq. (8.45), we obtain

$$\frac{dC}{dy} = -\left(1 + \frac{\rho_s - \rho}{\rho} C\right) C(1 - C) v_{ss} \frac{du}{dy} \frac{\rho}{\tau_y} \tag{8.58}$$

an equation where all variables can be measured. There is, however, some question as to what kind of settling velocity v_{ss} should be used. Equation (8.58) cannot be checked on its basic assumptions before the settling velocity of solid particles in a heavily concentrated zone is well known.

8.3.2.3.4 OTHER THEORETICAL CONTRIBUTIONS. JAROCKI (1957) and SCHEIDEGGER (1961) refer to it as the *gravitational* theory in order to distinguish it from the *diffusivity* theory, which was the subject of discussion until now. In the gravitational theory, the governing equations are derived separately for the movement of liquid and solid bodies.

To describe the turbulent motion mathematically, it is customary to distinguish between time-averaged values and fluctuation values; the instantaneous value for the velocity is given by

$$u_i = \bar{u}_i + u_i' \tag{8.59}$$

Replacing the foregoing equation into the Navier-Stokes equation, we arrive at the commonly referred to *Reynolds equations*. If the problem is restricted to two dimensions, these Reynolds equations are given—see SCHLICHTING (1968, p. 529)—by

$$\begin{aligned} \rho\left(\bar{u}\frac{\partial \bar{u}}{\partial x} + \bar{v}\frac{\partial \bar{u}}{\partial y}\right) &= \bar{X} - \frac{\partial p}{\partial x} + \mu\nabla^2\bar{u} - \rho\left(\frac{\partial \overline{u'^2}}{\partial x} + \frac{\partial \overline{u'v'}}{\partial y}\right) \\ \rho\left(\bar{u}\frac{\partial \bar{v}}{\partial x} + \bar{v}\frac{\partial \bar{v}}{\partial y}\right) &= \bar{Y} - \frac{\partial p}{\partial y} + \mu\nabla^2\bar{v} - \rho\left(\frac{\partial \overline{u'v'}}{\partial x} + \frac{\partial \overline{v'^2}}{\partial y}\right) \end{aligned} \tag{8.60}$$

For the very special case of uniform motion in the x direction, $\partial/\partial x = 0$, no time-averaged component in the y direction, $\bar{v} = 0$, and ignoring viscosity effects, we obtain a set of simplified equations, or

$$\begin{aligned} 0 &= g\rho S - \rho\frac{\partial}{\partial y}\overline{u'v'} \\ 0 &= g\rho - \frac{\partial p}{\partial y} - \rho\frac{\partial}{\partial y}\overline{v'^2} \end{aligned} \tag{8.61}$$

Velikanov's approach. According to KONDRAT'EV (1959),[1] who briefly discussed Velikanov's model, the amount of work per unit time and volume necessary to suspend solid particles is given by

$$(\rho_s - \rho)gv_{ss}C(1 - C) \tag{8.62}$$

Furthermore, it is suggested to bring the first of Eqs. (8.61) into dimensions of work per unit time and volume and consider in this equation the presence of sediment concentration, or

$$g\rho(1 - C)S\bar{u} - \frac{d}{dy}[\rho(1 - C)\overline{u'v'}\bar{u}] = -\rho(1 - C)\overline{u'v'}\frac{d\bar{u}}{dy} \tag{8.63}$$

The second term on the left-hand side in Eq. (8.63) can be differentiated term by term, and the following relation is obtained:

$$\rho(1 - C)\bar{u}\left(gS - \frac{d}{dy}\overline{u'v'}\right) = -\rho\bar{u}\overline{u'v'}\frac{dC}{dy} \tag{8.64}$$

The assumption of

$$gS = \frac{d}{dy}\overline{u'v'}$$

leads to the simple form of energy dissipation in the stream, or

$$0 = +\rho\overline{u'v'}\bar{u}\frac{dC}{dy} \tag{8.65}$$

Velikanov modifies Eq. (8.65) by equating it with Eq. (8.62) and therefore obtains

$$(\rho_s - \rho)gv_{ss}C(1 - C) = \rho\overline{u'v'}\bar{u}\frac{dC}{dy} \tag{8.66}$$

The mathematical appearance of Eq. (8.66) is akin to Eq. (8.56) or (8.20). Difficult as it is to attach proper physical character to the diffusivity of matter, it is equally difficult to get proper information on turbulent-fluctuation components. Although it is obvious that Eq. (8.66), upon integration, gives the concentration distribution as an exponential function, Kondrat'ev evades explaining coefficients and exponents in this equation. He stresses that an analytical comparison made by Velikanov shows that both theories, the gravitational and the diffusivity, give structurally akin results.

[1] Since the original contribution was not available, we follow the steps outlined by Kondrat'ev.

Ananian's et al. approach. Equations of biphase flow, representing each phase separately, are given by Frankle as follows:

$$\begin{aligned} -\frac{d}{dy}\,[\rho_s C(\overline{u_s' v_s'})] + \rho_s g C S - R_1 = 0 \\ -\frac{d}{dy}\,[\rho_s C(\overline{v_s'^2})] + \rho_s g C - C\frac{dp}{dy} - R_2 = 0 \end{aligned} \tag{8.67}$$

for the solid phase, and

$$\begin{aligned} -\frac{d}{dy}\,[\rho(1 - C)(\overline{u'v'})] + \rho(1 - C)gS + R_1 = 0 \\ -\frac{d}{dy}\,[\rho(1 - C)\overline{v'^2}] + \rho(1 - C)g - (1 - C)\frac{dp}{dy} + R_2 = 0 \end{aligned} \tag{8.68}$$

for the liquid phase, where R_1 and R_2 are components of the resistance on the solid-particle motion in the liquid. The subscript s indicates the solid phase. The four available Eqs. (8.67) and (8.68) contain eight unknown values, and it becomes necessary to establish further relationships. Determination of R_1 is possible by solving an equation of motion, i.e., a B.B.O equation (see Chap. 3). The pressure p may be calculated by using the hydrostatic law. Furthermore, the quantities $\overline{v_s'^2}$ and $\overline{v'^2}$ may be expressed with known constants and average velocities. Thus the system of four equations and four unknowns renders the following solution:

$$\frac{C}{C_a} = e^{-(y/D - a/D)/A_A} \tag{8.69}$$

where A_A is given by

$$A_A = \frac{0.0017\bar{u}^2}{gD}\left[\frac{\rho_s - (1 + k_A)\rho}{\rho_s - \rho}\right]$$

where k_A is an experimental coefficient. ANANIAN et al. (1965) present this derivation and report, but do not show it, that Eq. (8.69) was compared with experimental data and gave satisfactory results.

Similar to the foregoing two approaches is the one due to Barenblatt, as mentioned by SCHEIDEGGER (1961). Barenblatt's equations of transverse vertical energy balance were linearized by SINELTSHIKOV (1967), who obtained the following relation:

$$\overline{v'C'} = \frac{v_{ss}}{g}\,\overline{v'^2}\,\frac{dC}{dy} \tag{8.70}$$

The theories of Frankle and of Barenblatt are summarized and reviewed by VASILIEV (1969). Since all equations dealing with the gravitational theory are derivable from the Reynolds equation where the turbulent components are considered, only further research into the turbulence of flow in flumes and streams will give definite answers. The advances in turbulence research are such that we must conclude that the diffusivity theory presents a more satisfactory answer at present.

8.3.2.4 Calculation of the suspended load. So far this chapter has been devoted to the discussion of concepts and ideas. It seems appropriate to explore the possibilities available to obtain the load of sediment material which moves in suspension.

8.3.2.4.1 LANE'S ET AL. APPROACH. The general diffusion equation was given with Eq. (8.20a), or

$$0 = v_{ss}C + \epsilon_s \frac{dC}{dy} \tag{8.20a}$$

and under circumstances where ϵ_s is constant, a solution is obtained as

$$\frac{C}{C_a} = e^{-[v_{ss}(y-a)]/\epsilon_s} \tag{8.21}$$

In a stream the assumption of ϵ_s being constant is incorrect, but rather varies with depth, such as

$$\epsilon_s = ku_* \frac{y}{D}(D - y) \tag{8.31}$$

An average value $\bar{\epsilon}_s$ of the diffusivity can be given by

$$\bar{\epsilon}_s = \frac{\int_0^D \epsilon_s \, dy}{D} = \frac{ku_*}{D^2}\int_0^D (Dy - y^2)\, dy \tag{8.71}$$

which becomes for $k = 0.4$,

$$\bar{\epsilon}_s = \tfrac{1}{15} u_* D \tag{8.72}$$

LANE et al. (1941) suggested to use the relation in Eq. (8.72), and Eq. (8.21) becomes[1]

$$\frac{C}{C_a} = e^{-[15 v_{ss}(y-a)]/Du_*} \tag{8.73}$$

For a given sediment it is thus possible to determine the concentration at any and all points in a vertical section if the concentration is known at a single point in this vertical. This may be done with a simple calculation or a

[1] These equations are obtained from BROWN (1950). Introducing Eq. (8.72)—based on $\epsilon_s = f(y)$—into Eq. (8.21)—based on $\epsilon_s \neq f(y)$—is a violation of assumptions; for engineering purposes this may be considered permissible.

graphical construction. Comparison with field data—mainly for wide rivers—leads LANE et al. (1941) to the conclusion that Eq. (8.73), approximate as it may be, appears to be sufficiently accurate for practical purposes.

The suspended-load rate per unit time and width was given earlier as

$$g_{ss} = \int_0^D Cu\,dy \tag{8.36}$$

Introducing Eq. (8.73) leads to

$$g_{ss} = qC_a P_L e^{15v_{ss}a/(Du_*)} \tag{8.74}$$

with a factor P_L as a function of v_{ss}/u_* and of the relative roughness $n/D^{1/6}$, as shown in Fig. 8.9. Since Eq. (8.74) applies only to one grain size v_{ss}, such a relationship has to be solved for each grain-size range.

8.3.2.4.2 EINSTEIN'S APPROACH. The most advanced method for the computation of suspended load is due to EINSTEIN (1950). Again recall Eq. (8.36),

$$g_{ss} = \int_a^D Cu\,dy \tag{8.36}$$

where $y = a$ is the lower limit—i.e., where suspended load begins—defining the thickness of the bed layer. Introducing into Eq. (8.36) the suspension distribution Eq. (8.35), and expressing the velocity with the logarithmic velocity distribution, we obtain

$$g_{ss} = \int_a^D C_a \left(\frac{D-y}{y}\frac{a}{D-a}\right)^z 5.75u_*' \log\left(\frac{30.2y}{\Delta}\right) dy \tag{8.75}$$

where Δ is a corrective value, given in Fig. 7.10, and u_*' is the shear velocity due to grains only. Replacing a by a dimensionless argument, $A_E = a/D$,

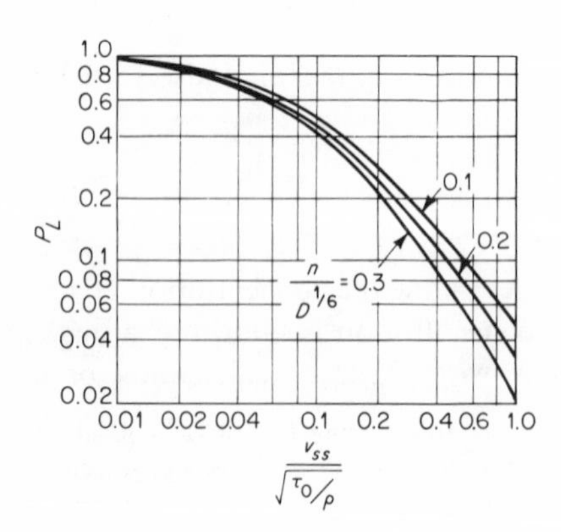

Fig. 8.9 Lane's et al. relationship of the factor P_L. [*After* LANE *et al.* (*1941*).]

and rearranging, the following is given if D is used as the unit for y:

$$g_{ss} = \int_{A_E}^{1} CuD\,dy$$
$$= Du'_* C_a \left(\frac{A_E}{1-A_E}\right)^z 5.75 \int_{A_E}^{1} \left(\frac{1-y}{y}\right)^z \log\left(\frac{30.2y}{\Delta/D}\right) dy \tag{8.76}$$

or

$$g_{ss} = 5.75 C_a u'_* D \left(\frac{A_E}{1-A_E}\right)^z \left[\log\left(\frac{30.2D}{\Delta}\right) \int_{A_E}^{1} \left(\frac{1-y}{y}\right)^z dy + 0.434 \int_{A_E}^{1} \left(\frac{1-y}{y}\right)^z \ln y\,dy\right] \tag{8.77}$$

The closed-form integration of this equation is impossible, but EINSTEIN (1950) has suggested and performed the numerical integration of the two integrals for various A_E and z values. The following arguments were evaluated:

$$I_1 = 0.216 \frac{A_E^{z-1}}{(1-A_E)^z} \int_{A_E}^{1} \left(\frac{1-y}{y}\right)^z dy$$
$$I_2 = 0.216 \frac{A_E^{z-1}}{(1-A_E)^z} \int_{A_E}^{1} \left(\frac{1-y}{y}\right)^z \ln y\,dy \tag{8.78}$$

and are graphically depicted in Figs. 8.10 and 8.11.

Equation (8.77) can now be rewritten as

$$g_{ss} = 11.6 C_a u'_* a \left[2.303 \log\left(\frac{30.2D}{\Delta}\right) I_1 + I_2\right] \tag{8.79}$$

Note again that Eq. (8.79) gives the suspended-load rate per unit time and width for a given size fraction v_{ss}.

Reference concentration C_a. According to Eq. (8.35) the concentration at $y = 0$ becomes infinite. Obviously this is in disagreement with the data. The sediment distribution equation does not apply right at the bed where $y = 0$. Furthermore, close to the bed the concept of a suspension, namely, solid particles being continuously surrounded by fluid particles, fails as well. The flow layer right on top of the bed, in which suspension becomes impossible, was designated by EINSTEIN (1950) as the *bed layer*, and was found to be of a thickness of $a' = 2d$. The material within this bed layer becomes the source of the suspended load, and the important determination of the lower limit or reference concentration C_a might thus be obtained.

From bedload theory, the bedload rate of a given size i_s is $g_s i_s$. If the velocity with which the bedload moves is u_B, then the weight of particles

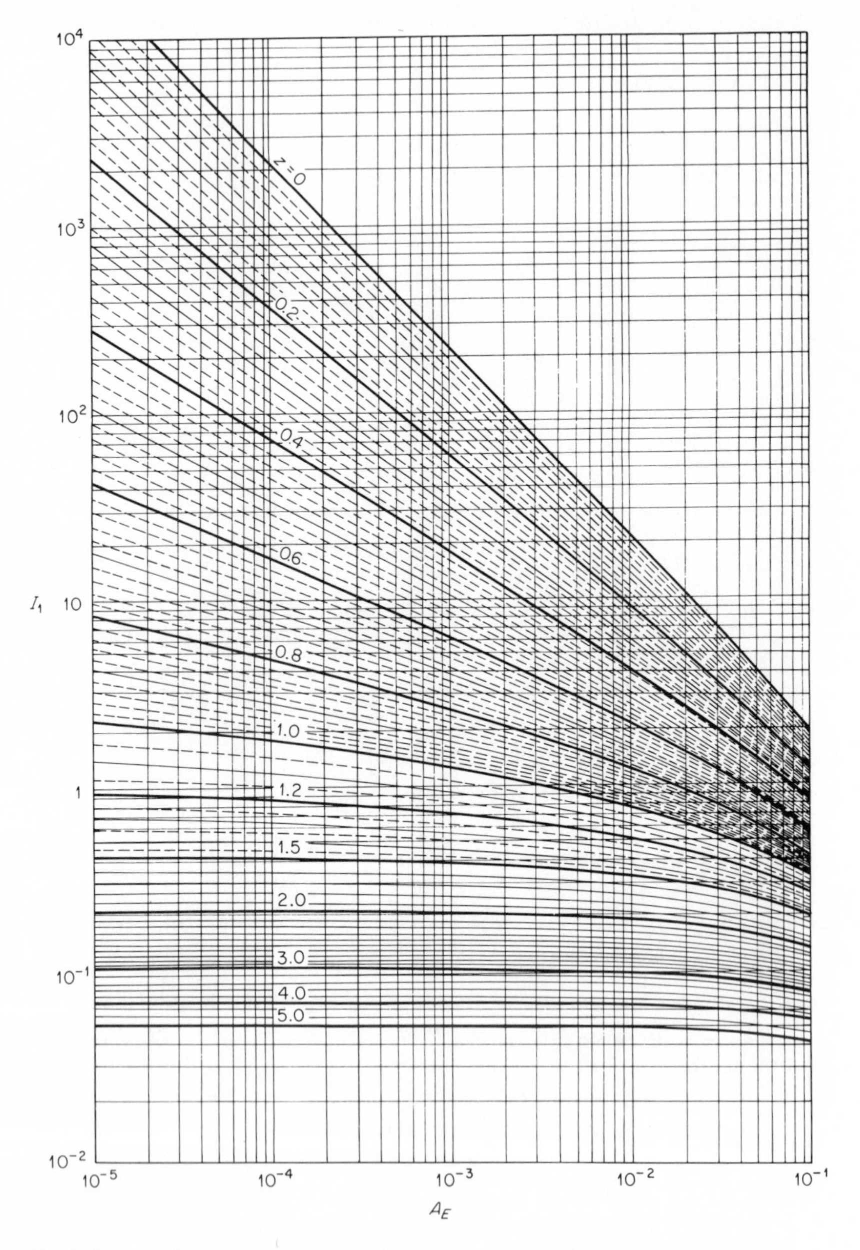

Fig. 8.10 Function I_1 in terms of A_E for values of z. [*After* EINSTEIN (*1950*).]

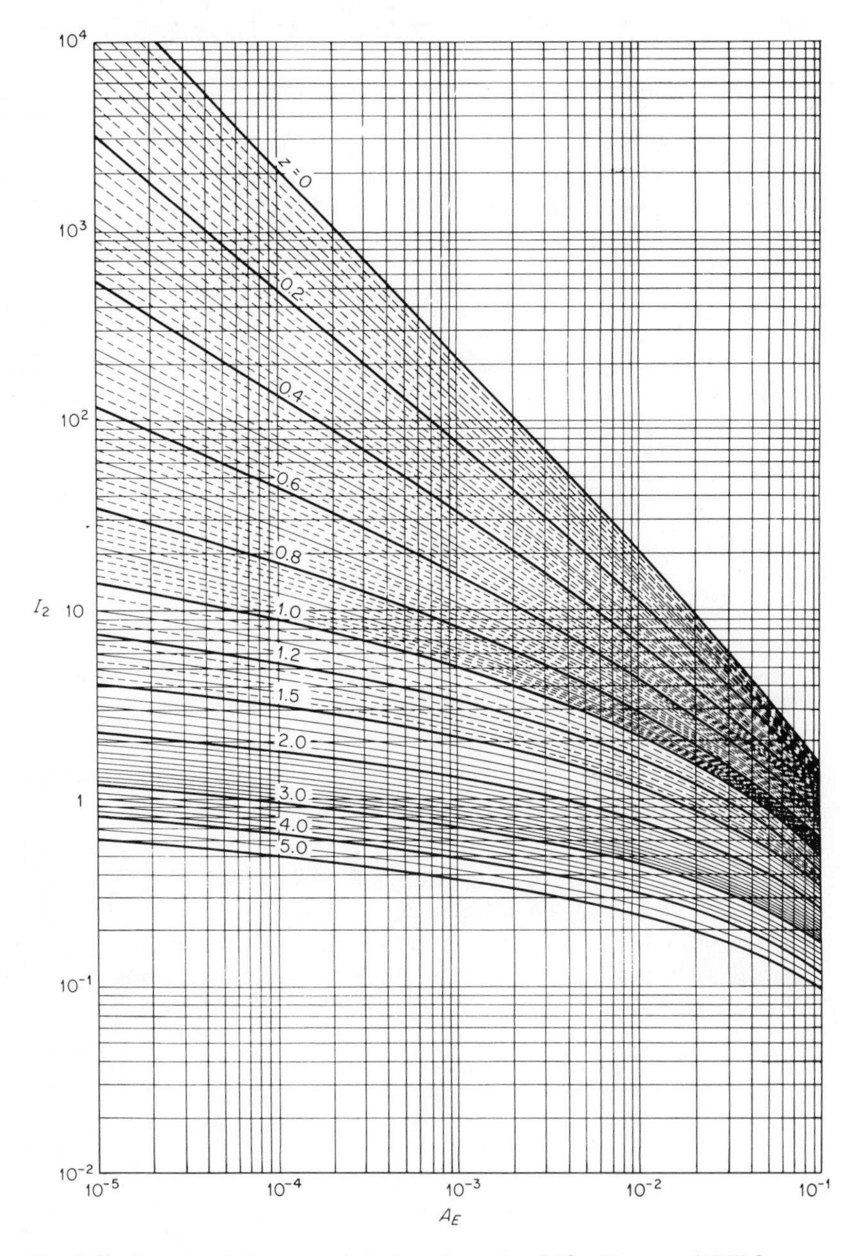

Fig. 8.11 Function I_2 in terms of A_E for values of z. [*After* EINSTEIN (*1950*).]

of a given size per unit area is $g_s i_s / u_B$. The average concentration in the layer is given by

$$C_a = A_5 \frac{i_s g_s}{a' u_B} \tag{8.80}$$

where A_5 is a constant of correction if the concentration over the entire bed layer is not constant. EINSTEIN (1950) was led by experimental results to rewrite Eq. (8.80), or

$$C_a = \frac{1}{11.6} \frac{i_s g_s}{u'_* a'} \tag{8.81}$$

by setting $u_B \propto u'_*$. The relation of the reference concentration, being known on behalf of the bedload movement, can be introduced into Eq. (8.79), and we obtain

$$i_{ss} g_{ss} = i_s g_s \left[2.303 \log \left(\frac{30.2 D}{\Delta} \right) I_1 + I_2 \right] \tag{8.82}$$

With P_E being a transport parameter,

$$P_E = 2.303 \log \left(\frac{30.2 D}{\Delta} \right) \tag{8.83}$$

we obtain a relationship of bedload transport to suspended-load transport for all size fractions for which a bedload function exists, and

$$i_{ss} g_{ss} = i_s g_s (P_E I_1 + I_2) \tag{8.84}$$

Equation (8.84) is dimensionally homogeneous and may be used in any consistent set of units, with g_{ss} being the suspended-load rate in weight per unit time and width.

With a slightly different approach BROOKS (1963) has developed an equation to determine the suspended load. Assuming the law of logarithmic velocity distribution, and with the concentration distribution, BROOKS (1963) obtained the following relationship, where g is the rate of liquid flow per unit time and width:

$$g_{ss} = C_{md} g \left[\left(1 + \frac{u_*}{k \bar{u}} \right) \int_{A_E}^{1} \left(\frac{1 - y}{y} \right)^z dy + \frac{u_*}{k \bar{u}} \int_{A_E}^{1} \left(\frac{1 - y}{y} \right)^z \ln y \, dy \right] \tag{8.85}$$

This formulation is more versatile than Eq. (8.77). The integrals can be evaluated, and Eq. (8.85) may thus be rewritten in terms of a transport function T_B,

$$\frac{g_{ss}}{g C_{md}} = T_B \left(\frac{k \bar{u}}{u_*}, z, A_E \right) \tag{8.86}$$

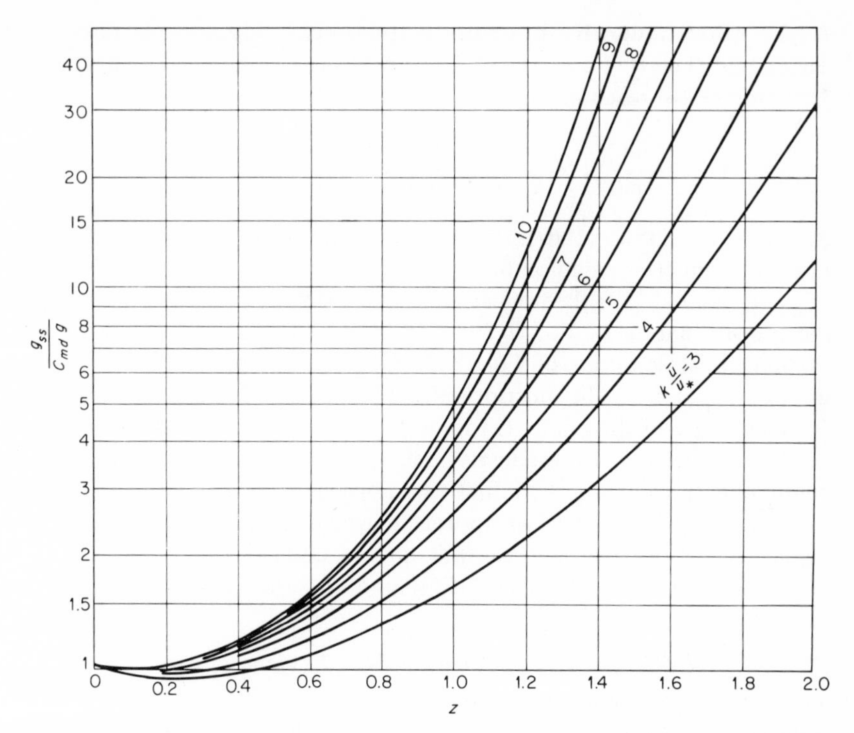

Fig. 8.12 Function g_{ss}/gC_{md} in terms of $k\bar{u}/u_*$ and of z values. [*After* Brooks (*1963*).]

where C_{md} is the reference concentration taken at $y = D/2$. The choice of the lower limit of integration is suggested to be at $u = 0$, and A_E becomes

$$A_E = e^{-(k\bar{u}/u_*)-1} \tag{8.87}$$

The functional relationship of Eq. (8.87) reduces to

$$\frac{g_{ss}}{gC_{md}} = T_B^*\left(\frac{k\bar{u}}{u_*},z\right) \tag{8.88}$$

and can be represented in a single chart, as given by Fig. 8.12.

8.3.3 LONGITUDINAL DISTRIBUTION OF SUSPENDED MATTER

Very little information is available for the dispersion process of solid particles having finite settling velocities. Therefore, we shall first discuss some research done on a neutrally buoyant dispersant with the hope that some knowledge can be gained. The little available knowledge on the dispersion theory, as applied to the transport of suspended sediment, will be covered subsequently.

8.3.3.1 For neutrally buoyant dispersant. Suppose the flow is unidirectional such that $u_2 = u_3 = 0$; furthermore, ϵ_1 should not vary with x_1. Equation (8.16) reads now

$$\frac{\partial C}{\partial t} = -u_1 \frac{\partial C}{\partial x_1} + \epsilon_1 \frac{\partial^2 C}{\partial x_1^2} + \frac{\partial}{\partial x_2}\left(\epsilon_2 \frac{\partial C}{\partial x_2}\right) + \frac{\partial}{\partial x_3}\left(\epsilon_3 \frac{\partial C}{\partial x_3}\right) \tag{8.89}$$

According to SAYRE (1967), the theoretical investigations of the dispersion in open channels focused on four problems, namely: (1) predicting the value of ϵ_1, ϵ_2, and ϵ_3; (2) theoretical analysis of simple cases of Eq. (8.89); (3) the transformation of Eq. (8.89) into forms more amenable for solution; (4) numerical solutions of Eq. (8.89) by finite-difference method with the use of computers. The most promising approaches have been found to be: (1) Fick's law of diffusion; (2) longitudinal dispersion by differential convection due to velocity gradients; and (3) Taylor's theory of diffusion by continuous movement. The first two approaches will be discussed in the following sections. The last approach, which is based on the actual turbulence (statistical) properties, will not be discussed here; the interested reader is referred to TAYLOR (1954) and to other contributions by this author.

8.3.3.1.1 FICK'S LAW OF DIFFUSION. TAYLOR (1954) described the longitudinal dispersion process by a bulk one-dimensional model, where $\bar{C}$ is the concentration distributed uniformly over the entire cross section and $\bar{u}$ is the cross-sectional average velocity. Thus Eq. (8.89) becomes

$$\frac{\partial \bar{C}}{\partial t} = -\bar{u}_1 \frac{\partial \bar{C}}{\partial x_1} + K_T \frac{\partial^2 \bar{C}}{\partial x_1^2} \tag{8.90}$$

The value K_T was referred to by Taylor as the virtual coefficient of diffusion, which represents the combined action of the velocity variation over the cross section and the turbulent diffusion. Since Eq. (8.90) describes the dispersion process and K_T is the only coefficient in the equation, it is customary to refer to K_T as the dispersion coefficient. The one-dimensional model, if applied to natural channels, was shown to be a valid description of the dispersion only after an initial period had passed. If t_R is the time from the release of the tracer, the model is good for

$$t_R > 1.5 \frac{B^2}{R_h u_*} \tag{8.91}$$

where B is the width and R_h is the hydraulic radius of the natural stream. Equation (8.91) has been suggested by FISCHER (1967*a*). Before the time [as given by Eq. (8.91)] has been reached, convective effects are of importance, and Eq. (8.90) is a poor model.

The dispersion coefficient K_T has been the subject of investigations for a few restricted problems. The original study was reported by TAYLOR (1954) for an infinitely long, straight pipe. ELDER (1959) extended this piece of research to the case of an infinitely wide, two-dimensional channel. Under the assumption of a logarithmic velocity distribution, the following relationship was given:

$$K_T = 5.9 u_* D \tag{8.92}$$

where D is the depth of the fluid and u_* is the friction velocity. SAYRE (1967) reports good agreement with Eq. (8.92) for laboratory flume experiments with no lateral velocity gradients. FISCHER (1967*b*) obtains values that are 50 percent higher than Eq. (8.92) would predict; this deviation is attributed to the lack of obtaining true two-dimensional flow in a flume. Nevertheless, FISCHER (1967*a*) pointed out that theory and actual field measurements do not agree. In natural streams there exist velocity distributions in the lateral and vertical directions; some portions of a stream move faster than adjacent portions. As a result of this, to omit the convective term, as was done in Eq. (8.90), is a violation imposed on the physical model. How the differential convection can be included will be described in the next section.

As an alternative to the dispersion coefficient in Eq. (8.92), THACKSTON et al. (1967) suggest the following relation:

$$K_T = 7.25 D u_* \left(\frac{\bar{u}}{u_*}\right)^{1/4} \tag{8.92a}$$

It was felt that such a relationship includes much better the form of the velocity profile and the bottom roughness. Experiments by the same authors are in good agreement if the effect of the lateral velocity profile is negligible.

A solution to Eq. (8.90) for diffusion of an initial concentration, $M = \int_{-\infty}^{\infty} \bar{C}\,dx$, is given by

$$\bar{C}(x_1,t) = \frac{M}{\sqrt{4\pi K_T t}} e^{-(x_1 - \bar{u}t)^2/(4K_T t)} \tag{8.93}$$

In conclusion it should be stated that the Fickian theory, given with Eqs. (8.90) and (8.92), and the solution, given by Eq. (8.93), are in good agreement with experiments by SAYRE (1967), SAYRE et al. (1968), and FISCHER (1967*b*), but only at considerable dispersion times and distances, as is given for a natural channel by Eq. (8.91).

8.3.3.1.2 LONGITUDINAL DISPERSION BY DIFFERENTIAL CONVECTION. Previously it has been discussed that the one-dimensional dispersion model, given with Eq. (8.90), is not valid if the dispersion time is small, i.e., smaller than indicated by Eq. (8.91). Thus Eq. (8.89) has to be reexamined. FISCHER (1967*a* and *b*) has presented the detailed analysis and obtained Eq. (8.90) for the dispersion coefficient as the coefficient in the bulk diffusion.

$$K_T = \frac{-1}{A}\int_0^B q'(x_3)\,dx_3 \int_0^{x_3} \frac{dx_3}{\epsilon_3 D(x_3)} \int_0^{x_3} q'(x_3)\,dx_3 \tag{8.94}$$

where A = cross-sectional area
B = width of the channel
ϵ_3 = lateral mixing coefficient
$D(x_3)$ = water depth as it varies over the channel width

Furthermore,

$$q'(x_3) = \int_0^{D(x_3)} u'(x_2,x_3)\,dx_2 \tag{8.95}$$

where u' is the deviation from the cross-sectional mean velocity $\bar{u}$. The lateral turbulent mixing coefficient ϵ_3 has been found experimentally by ELDER (1959) for a two-dimensional channel, and is given by

$$\epsilon_3 = 0.23 D u_* \tag{8.96}$$

In a flume study ORLOB (1959) and ENGELUND (1969) obtained a similar result, while FISCHER (1967*b*) reported the same relationship from field studies in an irrigation canal.

In the derivation of the relation of Eq. (8.94), FISCHER (1967*b*) restricted the analysis to cases where the vertical mixing coefficient becomes unimportant—a good assumption for wide channels. Remarks on the vertical mixing coefficient are found in FISCHER (1966).

Thus prediction of the dispersion coefficient, according to Eq. (8.94), requires knowledge of the channel geometry, given by B, A, and $D(x_3)$, of the shear velocity u_*, and of the value u'. For use in Eq. (8.94), ϵ_3 is assumed constant over the cross section.

It remains now to show how well Eq. (8.94) predicts the dispersion coefficient. Experimental data obtained in a laboratory flume with a salt solution as tracer were conducted by FISCHER (1967*a*). For six different flow conditions, the predicted [by Eq. (8.94)] and the measured dispersion coefficient K_T agreed within 20 percent. The *change-of-moment* method, outlined in FISCHER (1966), was also used for the computation of K_T, or

$$K_T = \frac{1}{2}\bar{u}^2 \frac{(\sigma_2^2 - \sigma_1^2)}{t_2 - t_1} \tag{8.97}$$

where σ_1^2 and σ_2^2 are the variances of the time-concentration curves at two stations, and t_1 and t_2 are the mean times of passages. Although the multiplier in Eq. (8.92), for strictly two-dimensional conditions, was given as 5.9, FISCHER (1967*a*) reported values up to 640, replacing the depth D by the hydraulic radius R_h. The highest values were obtained with a shallow water depth and channel banks covered with stones, conditions which demonstrate the importance of lateral differential velocities and, as such, the highly three-dimensional flow condition. According to an analytical study by SOOKY (1969), the multiplier in Eq. (8.92) is a function of the width–to–hydraulic-radius ratio, the cross-sectional shape, and the Reynolds number; the functional relationship has been provided.

In studying a natural stream, 70 ft wide and 6 ft deep, FISCHER (1968) found good agreement with his method. For a constant discharge of 310 cfs, the actual dispersion coefficient was between 70 and 90 ft²/sec; its predicted values were slightly higher.

8.3.3.1.3 CONCLUDING REMARKS. Dispersion of tracer particles, having no weight of their own, has been studied in laboratory and field conditions. A reasonable description of the mechanism of dispersion is available.

The process of dispersion can be described by a one-dimensional diffusion equation, given here by Eq. (8.90). Knowledge of the dispersion coefficient K_T, which appears in this equation, is paramount. Two cases have to be distinguished:

1. If the convective terms are of importance [see Eq. (8.89)], which is the case shortly after the release of the tracer, and whenever lateral or vertical velocity gradients dominate, the dispersion coefficient is given by Eq. (8.94).
2. If the one-dimensional diffusion equation describes adequately the physical picture, the dispersion coefficient can be calculated with Eq. (8.92).

The duration of the convective period and the beginning of the diffusion period is given by Eq. (8.91).

A typical example of the concentration variation at two different distances from the source is shown in Fig. 8.13. The measured distribution compares favorably with the routed one, for which Eq. (8.90) and a calculated dispersion coefficient were used. Notice that closer to the source, the distribution is more skewed than farther away. Eventually, or better, if Eq. (8.91) is satisfied, a normal or gaussian distribution is obtained; the solution to Fick's law, given with Eq. (8.93), can be written such that it represents the probability density function of the normal probability law. The initially skewed distribution is indicative for the existence of lateral and vertical differential convection.

Finally, it may be worthwhile to refer to a study by BUGLIARELLO et al. (1964) who presented a random-walk approach to the longitudinal-dispersion problem.

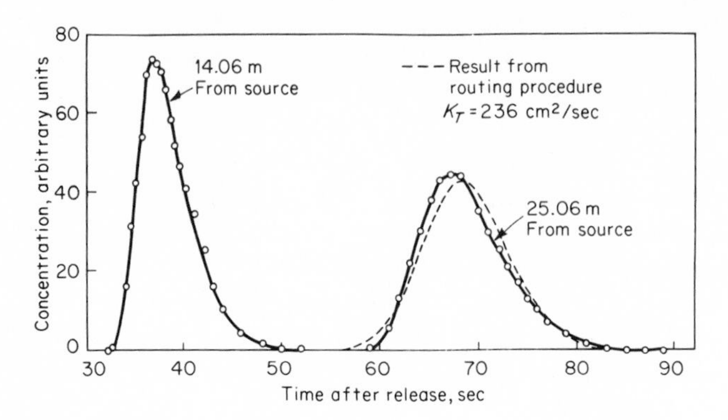

Fig. 8.13 Concentration traces, measured and computed (routed). [*After* FISCHER (*1967b*).]

8.3.3.2 For suspended sediment particles. Suppose the flow is uniform in the x_1 direction and ϵ_{s_1} does not vary with this x_1 direction, then Eq. (8.16) becomes

$$\frac{\partial C}{\partial t} = -u_s \frac{\partial C}{\partial x_1} + \epsilon_{s_1} \frac{\partial^2 C}{\partial x_1{}^2} + \frac{\partial}{\partial x_2}\left(\epsilon_{s_2} \frac{\partial C}{\partial x_2}\right) + \frac{\partial}{\partial x_3}\left(\epsilon_{s_3} \frac{\partial C}{\partial x_3}\right) + v_s \frac{\partial C}{\partial x_2} \tag{8.98}$$

where the subscript s stands for suspended particles, $v_s = v_{ss}$ is the settling velocity of the particle, and $u_s = u$ at a point. Equation (8.98) applies for cases where the volume of the particles can be ignored. Equation (8.98) in its complete form has never been solved, but certain simpler cases have been extensively investigated and were discussed in previous sections.

A rough idea of the variation of the dispersion coefficient K_{T_s} was obtained by ELDER (1959). Assumed was a parabolic velocity distribution, and ignored was the lateral velocity gradients effect; furthermore, for very small discrete particles, $d \approx 0.004D$, the dispersion coefficient for fluid particles and for solid particles was taken to be the same. Under these conditions ELDER (1959) obtained a relation to describe longitudinal dispersion of discrete particles, such as

$$K_{T_s}(z) = \frac{u_{max}^2 D^2}{7{,}560 k_m}(64 + 21z - 308z^2 - 210z^3 - 35z^4) \tag{8.99}$$

an expression valid only for large dispersal time. In its structure, Eq. (8.99) is similar to Eq. (8.94). The z value, defined earlier, is given by $z = v_{ss}/ku_*$, and k_m is the molecular diffusivity. The relative longitudinal dispersion

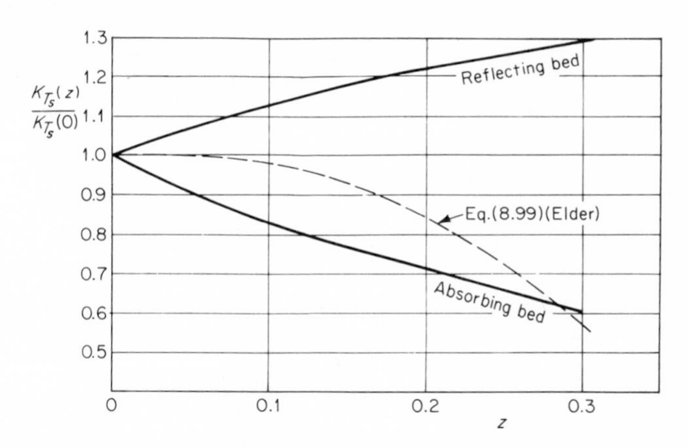

Fig. 8.14 Relative longitudinal dispersion coefficient for suspended sediment as a function of z. [*After* SAYRE (*1967*).]

coefficient is plotted as a function of z in Fig. 8.14, $K_{T_s}(z)$ rises by 0.55 percent above $K_T(0)$ to a weak maximum, and decreases quickly to zero near $z = \pm 0.5$.

Additional knowledge may be gained from SAYRE (1967) and SAYRE et al. (1968), who present in a detailed and exhaustive study further information on the dispersion problem in open-channel flow. Some of the results will be discussed briefly, and for further research the reader is referred to the original contribution.

Equation (8.98), even without the lateral-diffusion term, is difficult to solve. SAYRE (1967) and SAYRE (1969) brought Eq. (8.98) into a form more amenable for solution, and obtained numerical solutions to the transformed equation by means of finite-difference techniques using an electronic computer. Two boundary conditions imposed upon the equation are of interest: (1) The bed behaves as a completely absorbing boundary; and (2) the bed acts as a reflecting boundary such that no deposition is possible. Common to both cases is that no entrainment of sediment from the bed is permitted. For both cases the relative dispersion coefficients are plotted as a function of z in Fig. 8.14. The following conclusion might be drawn from this graph. The longitudinal dispersion coefficient for sediment particles resembles the one for dissolved matter, where $z = 0$. The differences, which increase with the z values, are due to particle settling and a vertical velocity gradient. ELDER'S (1959) results, also shown in Fig. 8.14, represent apparently the case where some particles are reflected and others are absorbed. In addition, SAYRE (1967) also considers the case when reentrainment of deposited particles is possible. This case, however, presents certain analytical and experimental difficulties.

8.4 CONCLUDING REMARKS[1]

Suspended load was said to be the material kept in suspension by the upward components of the turbulent velocities.

Much research, analytical and experimental, is available to give assurance that the vertical concentration distributions, based on a diffusion-dispersion model, do have some validity (at least until a better model is available). The concentration distribution in a stream is well represented by Eq. (8.35) and, to say it with LELIAVSKY (1955), "is believed to belong to the group of the most advanced achievements which the classical turbulence theory has produced in regard to various aspects of hydraulic flow."

[1] This chapter does not contain a sample calculation; however, Chap. 9, in the calculation of the total load procedure, includes the application of one suspended-load theory.

9
The Total Load

9.1 INTRODUCTORY REMARKS

The total-load rate g_{st} is obtained by addition of the bedload rate g_s and the suspended-load rate g_{ss}. The bedload rate may be obtained from an appropriate bedload equation, as discussed in Chap. 7; the suspended-load rate may be obtained from an equation discussed in Chap. 8. At low transport rates, where most of the sediment moves in contact with the bed, the bedload approximates, sufficiently well, the total load.

Besides this somehow indirect approach of the addition of the two fractions, there exist more direct approaches. In these cases, researchers establish a relationship which is immediately compared with measurements of the total load.

A more correct name for the total load is, actually, bed material load. The total load, predictable with previously mentioned equations, is made up of only those solid particles consisting of grain sizes represented in the bed. Bedload and suspended load equations were derived such that the particle supply is found within the bed material and, therefore, it appears appropriate to call it the transported fraction of the bed material, or the bed material load.

The name, total load, is thus erroneous and has given cause to much misunderstanding. The total load of a stream does not necessarily have to be identical with the bed material load. Besides the bed material load, there exists also the so-called *washload*, which is made up of grain sizes finer than the bulk of the bed material and, thus, is rarely found in the bed. This fact and an apparent lack of a definite relationship to the flow have made it difficult to advance an analytical method for the determination of the washload. The washload rate can be related to the available supply of solid particles within the watershed; it enters the watercourse by sheet wash, bank caving, etc., but is merely washed through the sections. Owing to its small size fractions, it moves readily in suspension and can thus be measured and calculated, provided the suspended bed material load is known. EINSTEIN (1950) suggested that the limiting sizes of washload and bed material load may be chosen quite arbitrarily from the mechanical size analyses as the grain diameter of which 10 percent of the bed mixture is finer. KRESSER (1964) reasons that the particle Froude number, $\bar{u}^2/gd$, is a useful criterion. When analyzing limited field data, a Froude number of $N_F = 360$ gave a surprisingly good correlation.

The foregoing is illustrated with two figures. The effect of the discharge, represented with the mean velocity, on the bed material load is clearly defined, as can be seen in Fig. 9.1. However, no definite relationships between the discharge and the washload can be established, as is obvious from Fig. 9.2. Washload usually is caused by land erosion and not by channel erosion; we may thus expect a certain relationship with the stream

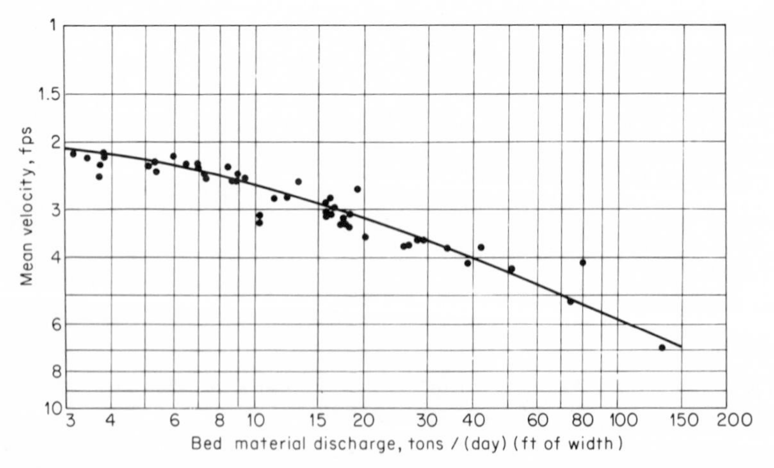

Fig. 9.1 Bed material load vs. mean velocity—Niobrara River. [*After* COLBY (*1963*).]

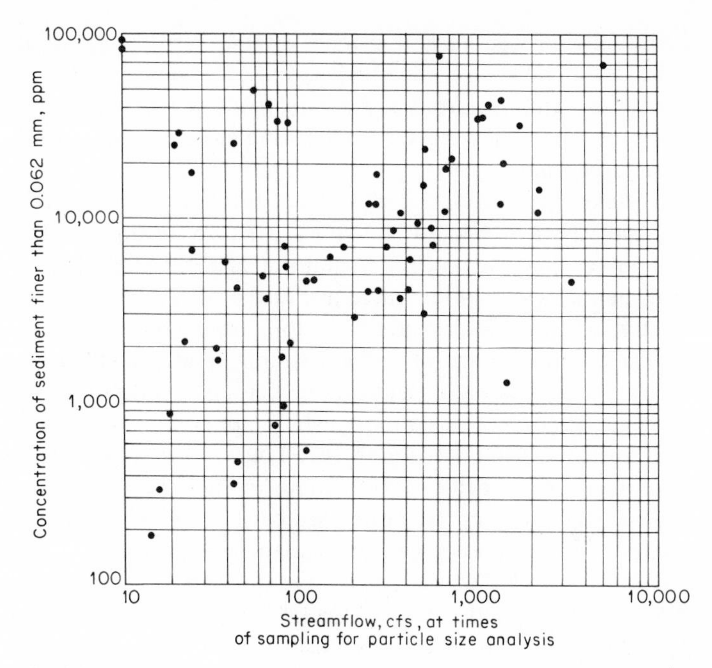

Fig. 9.2 Washload vs. streamflow—Powder River. [*After* COLBY (*1963*).]

flow. There is a bit of an indication in Fig. 9.2 that the washload is small if the stream flow is low, and the average concentration is much lower at low flows than at high ones.

It appears that, although the bed material load can be fairly well predicted with the hydraulic knowledge of the stream, the washload depends on the hydrological and geological conditions as well. Little is as yet known about the washload, but useful information pertaining to the estimation of the sediment erosion rate from watersheds is collected by agricultural engineers [see MEYER et al. (1965) and PIEST et al. (1970) as well as their references].

In this chapter the indirect determination of the total load will be discussed first, followed by various approaches for a direct determination. In a later section the comparison and an application of bed material load equations will be demonstrated, including a calculation of a sample problem. A subsequent section deals with the relation of the hydrology to the sediment problem, and is followed by general concluding remarks.

9.2 INDIRECT DETERMINATIONS

There exists one prominent approach for the determination of the bed material load by addition of its fractions, the bedload and the suspended load. This procedure is due to EINSTEIN (1950), making use of Einstein's bedload function. This method was later modified by COLBY et al. (1955), and is frequently referred as the *modified Einstein method*. More recently, BAGNOLD (1966) has made an interesting contribution, which should receive some attention.

Before we enter the discussion of the bed material determination by an indirect way, a few words of explanation are in order. There are at least two points that warrant differentiation of the bedload from the suspended load, one of which is the difference in the behavior of the movement of the different loads. Within a rather thin layer, the so-called *bed layer*, the particle motion is one of rolling, sliding, and sometimes saltating; this is defined as the bedload. Outside the bed layer is the suspended load, where the solid particles are continuously supported by the turbulent fluid. A clear line of division between the two kinds of loads does not exist. Particles being part of the bedload are sooner or later part of the suspended load. A general correlation between the bedload and the suspended load is not possible, as has been shown with many field data by SCHOKLITSCH (1930) and JAROCKI (1963). Generally, the quantity of suspended material (field investigations include the washload!) is larger than the one of the bedload, while the ratio of bedload to total load is lower in lowland streams than in mountain streams. Although there does not exist a sharp line of demarcation, the distinction between the two modes of transport becomes very important, because it serves to describe the sediment transport with two physically different models.

Secondly, it is convenient to distinguish between these two modes of transport, because the two loads are predominantly measured by different methods. The bedload is measured with suitable traps, whereas the suspended load is obtained from water samples carrying the sediment.

9.2.1 EINSTEIN'S BEDLOAD FUNCTIONS

EINSTEIN (1950) advanced the bedload and the suspended-load concept, having in mind that a summation of the two loads would lead to a proper determination of the bed material load. The bedload is given by

$$i_s g_s \tag{9.1}$$

for a size fraction of i_s, and the suspended load is given by

$$i_{ss} g_{ss} \tag{9.2}$$

for a size fraction of i_{ss}; thus the resulting bed material load for a size fraction i_{st} is

$$i_{st} g_{st} = i_s g_s + i_{ss} g_{ss} \tag{9.3}$$

where g_{st} is the bed material load rate in weight per unit time and unit width. Introducing the relation given with Eq. (8.84),

$$i_{st}g_{st} = i_s g_s(1 + P_E I_1 + I_2) \tag{9.4}$$

where P_E is a transport parameter, and I_1 and I_2 are integrals evaluated by EINSTEIN (1950). The bedload rate g_s can be obtained from a relationship of Φ_* vs. Ψ_*, given in Fig. 7.13 and extensively discussed in Sec. 7.4.

Equation (9.4) gives a stream's capacity as to how much bed material load it can transport under uniform and steady flow conditions; washload is not included in Eq. (9.4). In applying the method for a particular watercourse, EINSTEIN (1950) stresses the following points: (1) The length of a uniform reach should be such that the slope S may be determined accurately; (2) the channel geometry, the sediment composition, and all other factors influencing the roughness value n, such as vegetation, etc., should be uniform, so that an average representative cross section may be selected; and (3) the viscosity of the water has to be measured accurately.

The method presented by EINSTEIN (1950) is complicated and laborious for practical use. In the original paper (1950) a most detailed sample calculation is performed; we shall present one in a later section of this chapter. EINSTEIN'S (1950) method represents the most detailed and comprehensive treatment, from the point of fluid mechanics, that is presently available.

9.2.2 MODIFIED EINSTEIN PROCEDURE

EINSTEIN'S (1950) method of computing the bed material load is elegant and allows the calculation without measuring either the suspended or the bedload matter. This procedure was modified by COLBY et al. (1955) and adapted for computation of the total load. In addition to obtaining bed samples, this method requires, however, knowledge of depth-integrated samples of total suspended matter and stream flow measurements.

EINSTEIN (1964), in reviewing the new method, notices the following deviations for EINSTEIN'S (1950) approach: (1) Calculations are based on the actual measurement of the velocity, its local values and averaged ones, rather than on the slope S. In due course, the depth D replaces the hydraulic radius R_h; this facilitates the procedure, but may be responsible for introducing errors if $R_h'' + R_h' \neq R_h$, or $R_h'' > R_h'$ and $R_h'' = R_h$. The foregoing will result in a different numerical value of the transport parameter P_E, given with Eq. (8.83). (2) The z value used by EINSTEIN (1950) has been shown to be a good choice, but still subject to certain limitations. The z value used herein is determined by a trial-and-error method from the measurements for the predominant grain size. For other grain sizes, the z value is computed from the z value for the reference size, employing a proportionality of the 0.7 power of the settling velocity.

For some reason COLBY et al. (1955) and COLBY et al. (1961) present a highly involved relation for the total load, using a new set of integrals, but EINSTEIN (1964) has rewritten this relation, and it becomes

$$\frac{i_{st}g_{st}}{i_{ssm}g_{ssm}} = \left(\frac{E}{A_E}\right)^{z-1}\left(\frac{1-A_E}{1-E}\right)^{z}\frac{(1+P_E I_1 + I_2)_{A_E}}{(P_E I_1 + I_2)_E} \tag{9.5}$$

with the modified z value and P_E values as discussed previously. The integrals I_1 and I_2 have to be evaluated at A_E or at E as a replacement for A_E. Furthermore $i_{st}g_{st}$ represents the total load in a given size range, and $i_{ssm}g_{ssm}$ is the measured suspended load. The depth-integrating sampler cannot sample the zone close to the channel bottom, and E represents the ratio of unmeasured depth to water depth. Since this procedure relies on the measured suspended load, which apparently included at least part of the washload, it appears that Eq. (9.5) gives information on the total load (inclusive washload), rather than on the bed material load (exclusive washload). Equation (9.5) is dimensionally homogeneous, and can thus be used in any unit system. For a detailed discussion of the modified Einstein procedure, the reader is referred to COLBY et al. (1961), which also contains a sample calculation. Numerous graphs and nomograms facilitate a relatively fast computation. Although this method rendered very good predictions of the total load for the Niobrara and Loup Rivers, as was demonstrated by SCHROEDER et al. (1956), its general applicability to streams of different character must still be proved.

TOFFALETI (1969) proposed a method for the total-load calculation which is especially adaptable to computer programming. While, in general, the Einstein procedure was adopted the following departures were made: (1) Velocity distribution in the vertical; (2) a combining of several correction factors into one; and (3) a relating of stream parameters to sediment transport at other than the two grain diameters as bedload.

The comparisons of computed and measured sand discharge for 339 cases of major rivers of the United States and for 282 cases for different laboratory studies show good agreements.

9.2.3 BAGNOLD'S APPROACH[1]

Only recently has BAGNOLD (1966) confined his scope to problems of open-channel flow and, thus, has simplified BAGNOLD's (1956) earlier approach, so that it is more directly applicable.

From the point of general physics, BAGNOLD (1966) argues that the existence and maintenance of upward supporting stresses equal to the

[1] Symbols used throughout this section are mainly the ones used in BAGNOLD (1966), and do not appear in the list of symbols at the beginning of Part 3. They should not be confused with explanations found elsewhere in Part 3.

immersed weight of the solids represent the key issue in the sediment transport problem. The dry mass of transported solids m and the immersed weight $m'g$ are related by

$$m'g = \frac{\rho_s - \rho}{\rho_s} mg \tag{9.6}$$

Thus the bedload mass m'_b is defined as that part of the total load mass which is supported by a solid-transmitted stress $m'_b g$, while the suspended load mass m'_s is supported by the fluid-transmitted stress $m'_s g$. The transport rate of solids by immersed weight per unit width i is given as

$$i = i_b + i_s = \frac{\rho_s - \rho}{\rho_s} mg\bar{U} = m'_b g\bar{U}_b + m'_s g\bar{U}_s \tag{9.7}$$

where $\bar{U}$ is the mean transport velocity of solids, and $\bar{U}_b$ and $\bar{U}_s$ are the mean transport velocity of solids moving as bedload and suspended load, respectively. Equation (9.7) is actually useless unless changed into a different form. The important point made by BAGNOLD (1966), certainly subject to criticism, is the following: Dynamic transport rates, given by Eq. (9.7), have dimensions and quality of work rates, yet are not, in fact, work rates, since stress and velocity are not in the same direction. For the bedload fraction this may be achieved if multiplied by the coefficient of solid friction, tan α; and for the suspended load, by multiplying with $v_{ss}/\bar{U}_s$, where v_{ss} is the settling velocity. Thus the bedload work rate is given as

$$m'_b g\bar{U}_b(\tan\alpha) = i_b(\tan\alpha) \tag{9.8}$$

and the suspended-load work rate as

$$m'_s g\bar{U}_s\left(\frac{v_{ss}}{\bar{U}_s}\right) = i_s\left(\frac{v_{ss}}{\bar{U}_s}\right) \tag{9.9}$$

Furthermore, BAGNOLD (1966) introduces the general power equation by equating the rate of doing work with the available power times the efficiency. The available power, per unit length and unit width, is given by

$$\omega = \frac{\gamma QS}{B} = \gamma DS\bar{u} \tag{9.10}$$

where D = channel depth
B = width
$\bar{u}$ = average fluid velocity
S = energy slope.

The available power ω is the supply of energy for the transport of sediment. In this way BAGNOLD (1966) arrives at a relationship for bedload and suspended load, respectively, as follows:

$$i_b \tan \alpha = e_b \omega$$
$$i_s \frac{v_{ss}}{\bar{U}_s} = e_s \omega (1 - e_b) \tag{9.11}$$

where e_b and e_s represent the appropriate efficiency. Introducing Eqs. (9.11) into Eq. (9.7) leads to

$$i = i_b + i_s = \omega \left[\frac{e_b}{\tan \alpha} + \frac{e_s \bar{U}_s}{v_{ss}} (1 - e_b) \right] \tag{9.12}$$

The total load may thus be obtained with Eq. (9.12) if four parameters, namely, e_b, e_s, $\tan \alpha$, $\bar{U}_s$, are known. It is suggested that this equation is equally applicable to turbulent and laminar flow, but for the latter case, the second term in the equation disappears. For fully turbulent flow, Bagnold has shown that $e_b = \text{fct}(\bar{u}, d)$, a relationship reproduced in Fig. 9.3. From flume studies, ignoring some uncertainties, the quantity $e_s(1 - e_b)$ was given by 0.01. Furthermore, the solid-friction coefficient was expressed in terms of bed shear stress and grain size, and is given in Fig. 9.4, while the assumption was made that the mean velocities of fluid and of suspended solids are approximately equal. With this in mind, Eq. (9.12) reduces to

$$i = \omega \left(\frac{e_b}{\tan \alpha} + 0.01 \frac{\bar{u}}{v_{ss}} \right) \tag{9.13}$$

It should be stressed that Eq. (9.13) applies to fully turbulent flow conditions and flows with adequate depths. Using mainly GILBERT's (1914) data and plotting a relationship of i vs. ω, reasonable agreement was cited, a fact which is remarkable in view of the difficulties in the interpretation of the experimental data. Of course, agreement is confined to the larger transport rates, as would be expected owing to the previous limitation; it was found

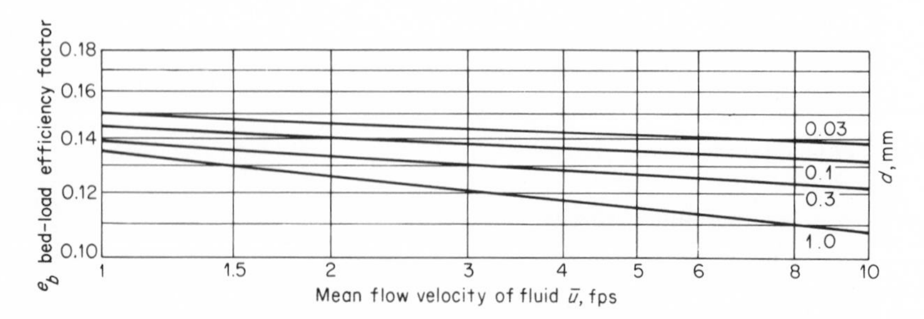

Fig. 9.3 The bedload efficiency factor. [*After* BAGNOLD (*1966*).]

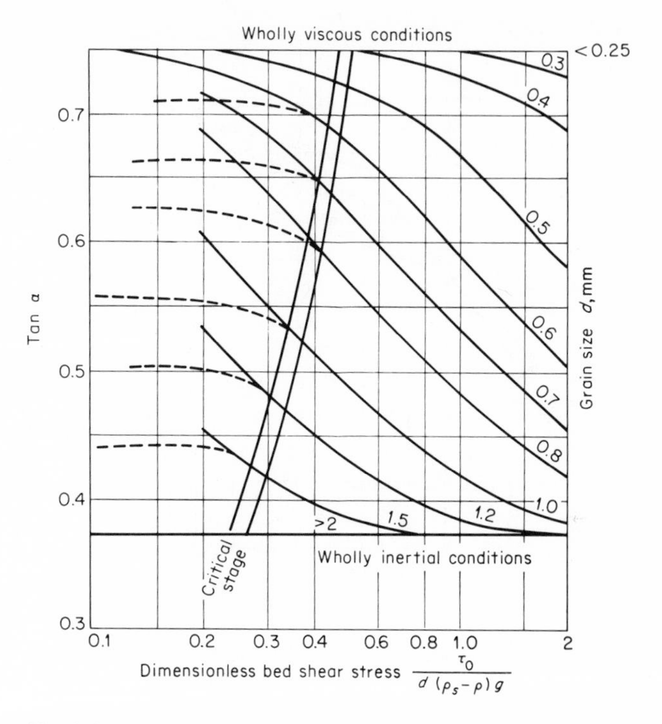

Fig. 9.4 The solid-friction coefficient. [*After* BAGNOLD (*1966*).]

that for $d \leq 0.015$ mm the theory becomes inapplicable. For available river data, a marked correlation was obtained.

9.2.4 CHANG'S ET AL. APPROACH

The bed material load may be given as

$$g_{st} = \int_0^a Cu_s\,dy + \int_a^D Cu_s\,dy \tag{9.14}$$

with the first term representing the bedload moving within the bed layer of a thickness a, and the second term representing the suspended load. To express the bedload, CHANG et al. (1967) employed the DUBOYS' (1879) relationship given by Eqs. (7.5) and (7.6), but modify it such that it becomes

$$g_s = K_T\bar{u}[\tau_0 - (\tau_0)_{cr}] \tag{9.15}$$

where K_T is the bed material discharge coefficient and $\bar{u}$ is the mean flow velocity. This coefficient was determined experimentally, and is reported to be in functional relation with bed material, bed configuration, and flow characteristic, and is given by Fig. 9.5 for flume data. Computed K_T values

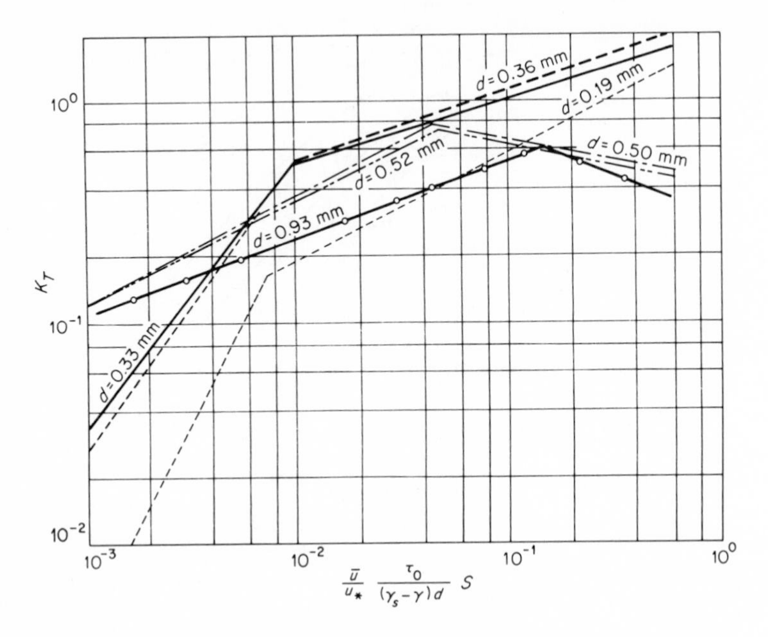

Fig. 9.5 Bed material discharge coefficient. [*After* CHANG *et al.* (*1967*).]

for each of the three natural rivers were constant, varying such as $0.27 < K_T < 1.10$ for different rivers.

The suspended load was expressed quite similar to EINSTEIN's (1950) approach, or

$$g_{ss} = g_s R_s \tag{9.16}$$

with R_s containing the two integrals, given by Eq. (8.78).

By substitution of Eqs. (9.15) and (9.16) into Eq. (9.14), we obtain

$$g_{st} = g_s + g_{ss} = K_T[\tau_0 - (\tau_0)_{cr}]\bar{u}(1 + R_s) \tag{9.17}$$

Equation (9.17) was tested with 184 flume data and 57 sets of natural river data, and agreement was found to be satisfactory.

An equation similar in its character to Eq. (9.17) had already been suggested by EGIAZAROFF (1965), who considered the bed material transport to be proportional to the excess of the mobility.

9.3 DIRECT DETERMINATIONS

Despite the fact that some researchers believe that bed material load has to be considered as the summation of fractional loads, there is a group of investigators who feel that no need exists to distinguish bedload from suspended load, because the hydrodynamic forces involved in lifting the bed material are the same. Considering the bed material load as such, it is thus unnecessary to define the thickness of the bed layer or the proper zone of demarcation of bedload and suspended load, a definition which has, for good reason, been subject to criticism.

9.3.1 LAURSEN'S APPROACH

An apparent functional relationship between a flow condition and the resulting sediment movement was sought by LAURSEN (1958), who advances parameters to explain this relation. One parameter is the shear velocity and settling velocity ratio, or

$$\frac{\sqrt{\tau_0/\rho}}{v_{ss}} \tag{9.18}$$

which expresses the effectiveness of the mixing action of the turbulence. The ratio, given by Eq. (9.18), has its importance in the suspended-load concept, while the other parameters somehow take care of the bedload concept. There is the critical tractive force $(\tau_0)_{cr}$, as discussed in Chap. 6 and given, for example, by Fig. 6.6, and there is the effective or available tractive force obtained by use of Manning's and Strickler's relation, and given by LAURSEN (1958) as

$$\tau_0' = \frac{\bar{u}^2 d^{1/3}}{30 D^{1/3}} \qquad \text{lb/ft}^2 \tag{9.19}$$

where τ_0' is the boundary shear associated only with the sediment particles. With this knowledge and "for reasons more intuitive than rational," the following empirical relationship was suggested:

$$\frac{\bar{C}}{\left(\frac{d}{D}\right)^{7/6}\left[\frac{\tau_0'}{(\tau_0)_{cr}} - 1\right]} = \text{fct}\left(\frac{\sqrt{\tau_0/\rho}}{v_{ss}}\right) \tag{9.20}$$

where $\bar{C}$ is the cross-sectional mean concentration per weight, and given for quartz sand as

$$\bar{C} = 265\,\frac{q_s}{q}$$

Data from flume studies, limited to sand of $d < 0.2$ mm almost exclusively, helped to establish the relationship of Eq. (9.20); it has been reproduced in Fig. 9.6 for total load—washload is excluded—and for bedload. Equation (9.20) can be rearranged thus:

$$\bar{C} = \sum i \left(\frac{d}{D}\right)^{7/6}\left[\frac{\tau_0'}{(\tau_0)_{cr}} - 1\right] \text{fct}\left(\frac{\sqrt{\tau_0/\rho}}{v_{ss}}\right) \tag{9.21}$$

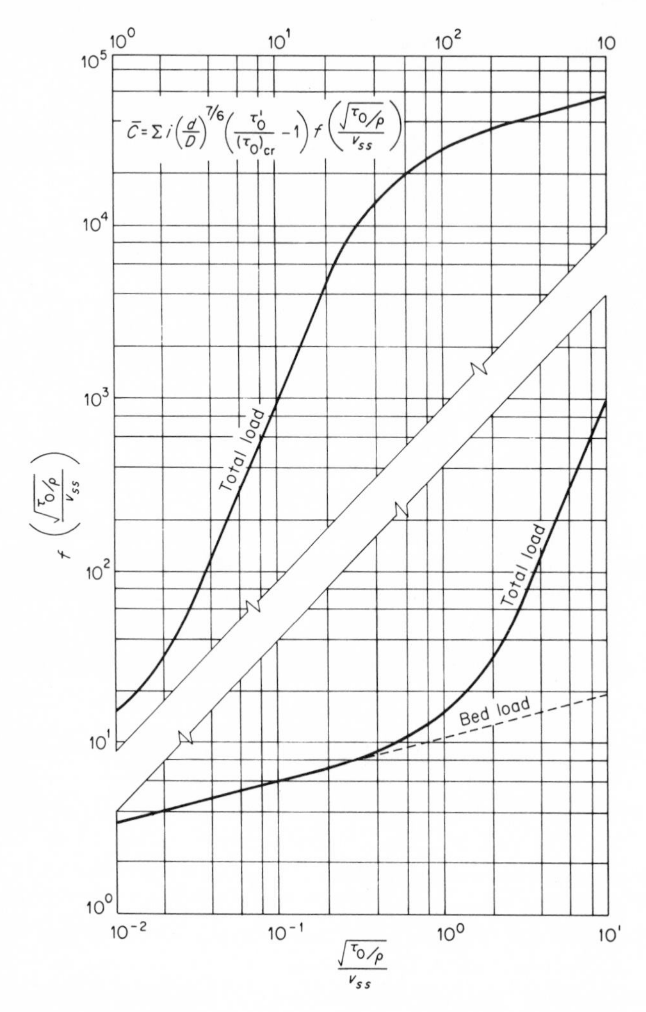

Fig. 9.6 Total-load relationship. [*After* LAURSEN (*1958*).]

where the contributions of each fraction i of a given grain size d are summed to obtain total mean concentration. The similarity of Eq. (9.21) with the duBoys equation and, in turn, with Eq. (9.17) should not be overlooked. Checking the empirical relation with field data of smaller rivers, LAURSEN (1958) obtained reasonable agreement. A vast amount of data, obtained at various reaches in the Missouri River, is presented and analyzed by BONDURANT (1958). However, these data did not agree well with Laursen's relationship. It was concluded, although the form of Eq. (9.20) can be very useful, there must be some revision, because all the data fall to the left of Laursen's graph, as given by Fig. 9.6.

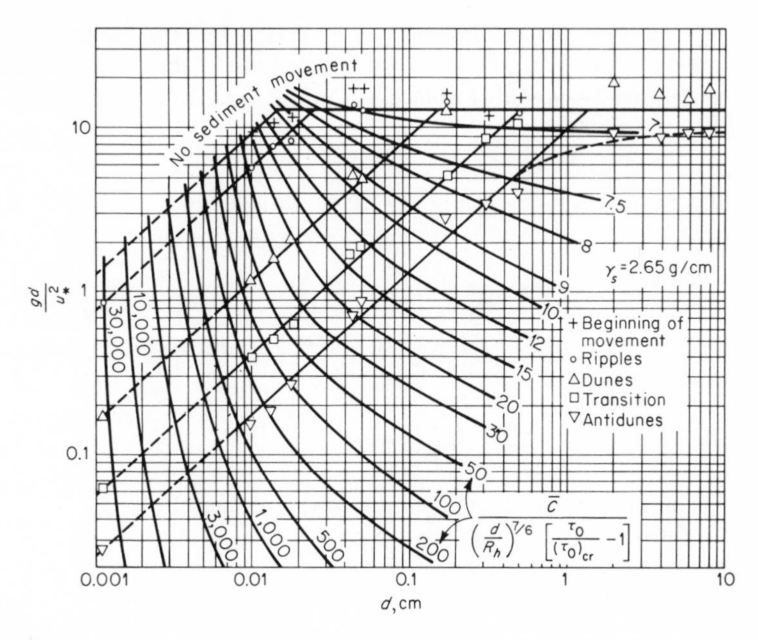

Fig. 9.7 Total-load relationship. [*After* BOGARDI (*1965*).]

It is worthwhile to mention the investigations by BOGARDI (1958, 1965), who developed a relation similar to Eq. (9.20), or

$$\frac{\bar{C}}{\left(\dfrac{d}{R_h}\right)^{7/6}\left[\dfrac{\tau_0}{(\tau_0)_{cr}} - 1\right]} = \text{fct}\left(\frac{gd}{u_*^2}, d\right) \tag{9.22}$$

where d and gd/u_*^2 offer information on sediment transport and on occurring bedforms. Equation (9.22) is plotted in Fig. 9.7.

In a discussion to LAURSEN's (1958) contribution, GARDE et al. (1958) suggest another empirical relationship, which was checked against numerous flume experiments, and is given as

$$\left(\frac{u_* D}{\nu} \frac{1}{\sqrt[3]{\bar{C}}}\right) = \left[\frac{D}{d} \frac{1}{\text{fct}\,(d)}\right]^{3/2} \tag{9.23}$$

This investigation was later extended by GARDE et al. (1963).

9.3.2 BISHOP'S ET AL. APPROACH

This approach makes use of the Φ_* vs. Ψ_* relation, as developed by EINSTEIN (1950), but rather than predicting the bedload transport, it is remodeled such that it predicts the bed material transport.

EINSTEIN (1950) has given an analytical relationship for the intensity of transport (bedload), Φ_*, and for the intensity of shear, Ψ_*; it was developed earlier and given with Eq. (7.71). In an abbreviated form it can be written as

$$\Phi_* = \text{fct}\,(\Psi_*) \tag{9.24}$$

and as such is given in Fig. 7.13. Knowledge of the intensity of shear Ψ_*, or

$$\Psi_* = \frac{\rho_s - \rho}{\rho}\frac{d}{SR_h'} \tag{7.57}$$

provides, according to Eq. (7.71) or (9.24), information on the intensity of transport, or

$$\Phi_* = \frac{g_s}{\gamma_s}\sqrt{\frac{\rho}{\rho_s - \rho}\frac{1}{gd^3}} \tag{7.51}$$

which, in turn, allows calculation of the bedload rate g_s.

BISHOP et al. (1965) reason that the shear intensity parameter Ψ_* may be used to predict immediately and directly the intensity of transport for the bed material discharge Φ_T, where Φ_T is given by

$$\Phi_T = \frac{g_{st}}{\gamma_s}\sqrt{\frac{\rho}{\rho_s - \rho}\frac{1}{gd^3}} \tag{9.25}$$

Using flume data for four different sands, the Φ_T vs. Ψ relationship was established and is reproduced in Fig. 9.8. Although the curves for each grain size exhibit the same general trend, they differ by a considerable degree. To remedy the latter effect, the scale constants A_* and B_*, which appear in Eq. (7.71), were adjusted and found to be in functional relationship with the median diameter of the sand, given in Fig. 9.9.

It is worthwhile to study the Φ_T vs. Ψ relationship, given by Fig. 9.10, for a specific sand. The authors imply that such a curve may be divided into three segments. The lower part of it represents data for flow regimes with bed figurations such as ripples and/or dunes. This part of the curve can be made fit to the Φ_* vs. Ψ_* relation by EINSTEIN (1950). The second segment is characterized by an inflection in the curve, with bedforms ranging from dunes to plane beds to antidunes. The uppermost part of the curve represents data with bedforms of plane beds and antidunes. Data in the second and third segment of the curve cannot be predicted with Einstein's relation, a fact which could be expected anyhow, because, most likely, the better part of the bed material is by now in suspension.

The computation of the bed material discharge is simplified by this empirical relation, and showed good agreement when used to predict actually observed bed material loads at streams, such as the Rio Grande, the Niobrara, the Loup, or the Colorado River.

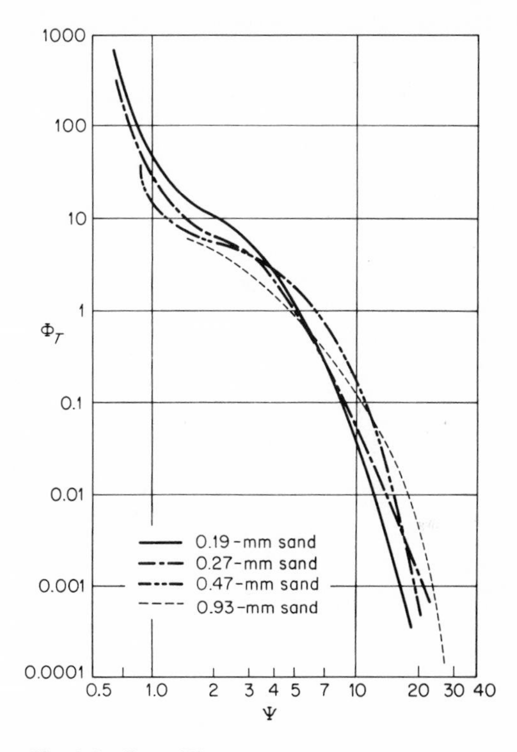

Fig. 9.8 Φ_T vs. Ψ relation for different sand sizes. [*After* BISHOP *et al.* (*1965*).]

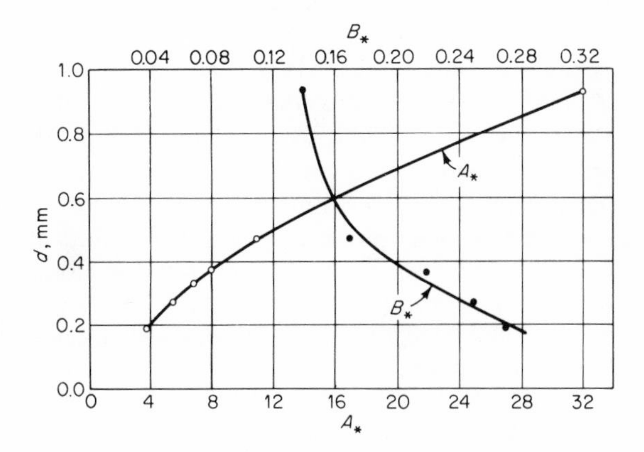

Fig. 9.9 Scale constants A_* and B_* vs. d. [*After* BISHOP *et al.* (*1965*).]

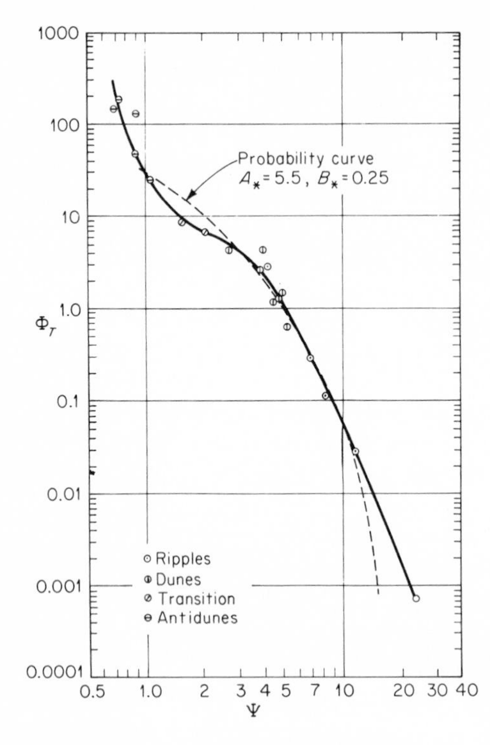

Fig. 9.10 Φ_T vs. Ψ' relation for a given sand of $d = 0.27$ mm. [*After* BISHOP *et al.* (*1965*).]

9.3.3 GRAF'S ET AL. APPROACH

The important feature of the model, proposed by GRAF et al. (1968), is that it was developed for the prediction of the bed material load in open-channel as well as in closed-conduit flow. Therefore some of the parameters are not, at least at this stage of the research, analyzed in such detail as could have been done when considering only one kind of flow. One of these parameters is the hydraulic radius. It was not possible to divide the hydraulic radius R_h into its two fractions, the one due to grain roughness R_h', and the other due to the bedforms R_h''. Secondly, it was not advisable that the bed material load be divided into the bedload and the suspended-load contribution. Data from closed-conduit experiments do not, at the present, permit such a distinction. The physical model for sediment transport in conveyance systems, both for open channel and for closed conduits, is discussed in detail in Part 4 of this book; here we shall merely present the findings.

A *shear intensity parameter* Ψ_A' was developed as a transport criterion, and is given by

$$\Psi_A' = \frac{[(\rho_s - \rho)/\rho]d}{SR_h} \tag{9.26}$$

The expression given by Eq. (9.26) is similar to EINSTEIN's (1950) parameter, called the *intensity of shear* on the particle, and was earlier expressed with Eq. (7.57). Whereas EINSTEIN (1950) used that part of the hydraulic radius associated with the grain roughness symbolized by R'_h, GRAF et al. (1968) use the entire hydraulic radius R_h. The hydraulic radius R_h is defined by the ratio of the net area of the cross section in which flow takes place to the wetted perimeter, a fact which can only be appreciated when considering closed-conduit flow.

Based on a work rate concept, a *transport parameter* was established, such as

$$\Phi_A = \frac{\bar{C}\bar{u}R_h}{\sqrt{[(\rho_s - \rho)/\rho]gd^3}} \tag{9.27}$$

where $\bar{C}$ is the volumetric concentration of the transported particles. Furthermore, it was shown that a functional relationship exists between the dimensionless transport parameter Φ_A and the shear intensity parameter Ψ_A. Since the mathematical expression of the function proved to be quite involved, the form of it was determined by using experimental data from laboratory and field measurements. Using flume data by GILBERT (1914), GUY et al. (1966), and ANSLEY (1963), and stream data reported in EINSTEIN (1944), the relationship was evaluated by a regression analysis as

$$\Phi_A = 10.39(\Psi_A)^{-2.52} \tag{9.28}$$

or

$$\frac{\bar{C}\bar{u}R_h}{\sqrt{[(\rho_s - \rho)/\rho]gd^3}} = 10.39\left\{\frac{[(\rho_s - \rho)/\rho]d}{SR_h}\right\}^{-2.52} \tag{9.28a}$$

Note that Eq. (9.28) was established from open-channel and closed-conduit data, and is shown in Fig. 9.11 with the open-channel data only. In spite of the scatter, which somehow depicts clearly the difficulty in obtaining sediment transport data, the Φ_A vs. Ψ_A relation explains the data rather well.

9.4 COMPARISON AND APPLICATION OF BED MATERIAL LOAD EQUATIONS

9.4.1 GENERAL REMARKS

The bed material load, being a sum of the bedload and the suspended load, cannot be predicted any better than each of its fractions. The difficulties involved in checking to a reliable degree the bedload and suspended-load equations have already been discussed. The various coefficients, not all of them known too accurately, used in each of these equations are probably responsible for errors of as much as 100 percent. This is especially so, if the bed material equation is used outside the range and/or condition for which

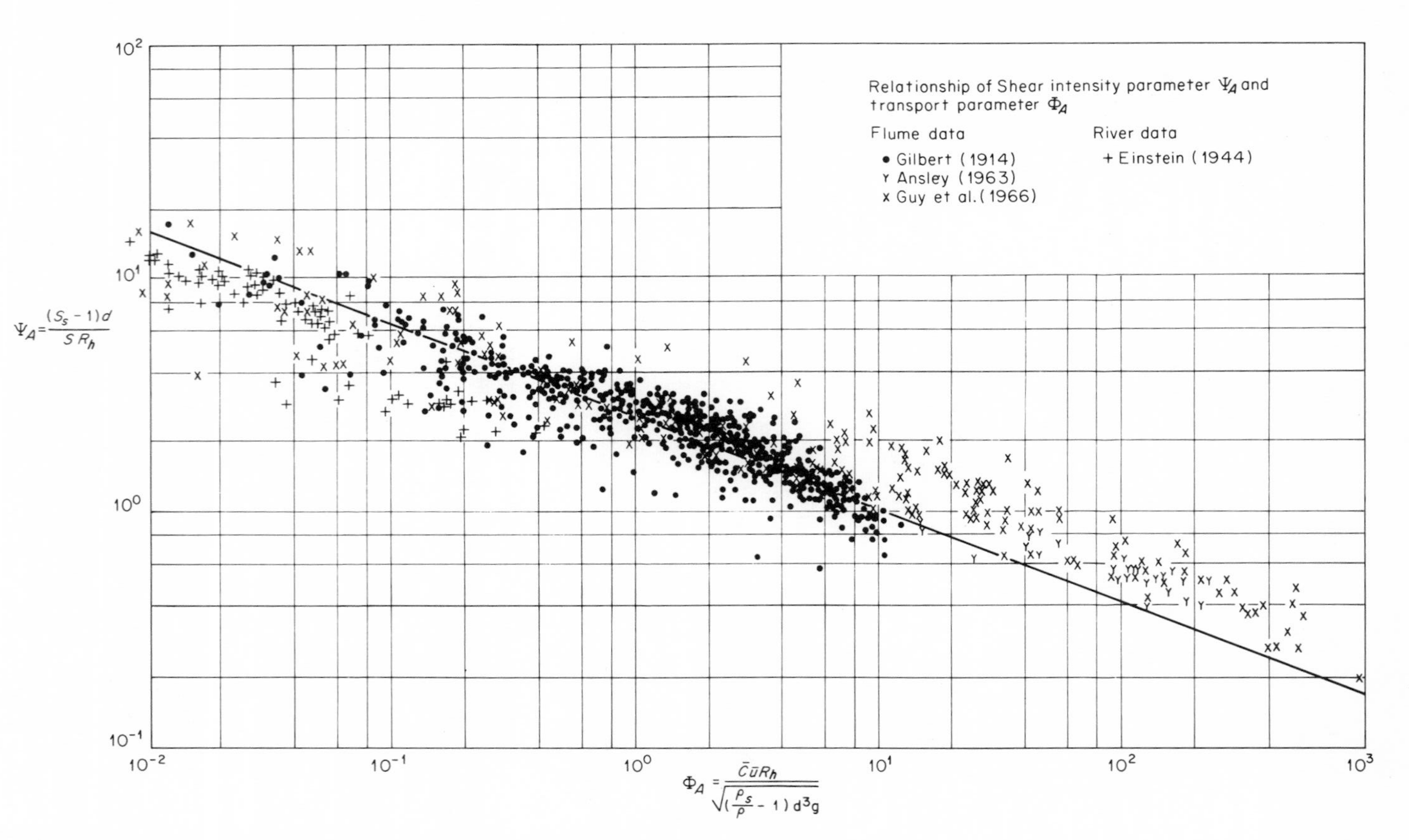

Fig. 9.11 Φ_A vs. Ψ_A relation; with open-channel data only. [*After* GRAF *et al.* (*1968*).]

it was established. Thus a meaningful comparison of measured and calculated bed material load is difficult. A few published comparisons are available and shall be discussed briefly. The sediment discharge for the Niobrara River was measured and is reported by COLBY et al. (1955). Values calculated with the approaches by DUBOYS (1879), SCHOKLITSCH (1934), EINSTEIN (1950), and Einstein's modified procedure were compared with the measured values. All of the formulas predicted—at particular stages—the sediment discharge reasonably well, but, with the possible exception of Einstein's bedload function, the formulas predicted—at certain stages—erroneous results of almost 100 percent. Another investigation by STALL et al. (1958) reports field measurements on the Money Creek. Approaches by EINSTEIN (1950), DUBOYS (1879), and SCHOKLITSCH (1934) were used to compute the sediment load. Whereas the latter method gave fairly good (within 30 percent) agreement, the former two gave values which were completely off (about 750 percent!). However, there is some question as to whether the reported measured load is not merely the bedload. (The total load was determined by letting an impounding reservoir act as a sediment collection basin.) This, in turn, would also explain why Schoklitsch's equation, a pure bedload equation, is in agreement, and Einstein's bedload function gives values which are apparently too high. Furthermore, comparisons for a few other streams are quoted by RAUDKIVI (1967), but present an equally dim picture.

Discrepancies, as exhibited by such studies, are not very encouraging and leave one to speculate that possibly not all the important factors governing the sediment problem are as yet fully understood.

9.4.2 COMPARATIVE SAMPLE CALCULATIONS

There exists no better way to demonstrate the usefulness of previously discussed relations than by illustration with a sample calculation. For one and the same stream, the bed material load will be calculated by three somewhat different methods.

9.4.2.1 Description of the test reach. To calculate or measure the flow and the sediment transport in a stream, a *test reach* has to be selected first. This test reach should be representative for the stream: It should be sufficiently long to determine rather accurately the slope of the channel; it should have a fairly uniform and stable channel geometry with uniform flow conditions and sediment composition; and it should have a minimum of outside effects, such as strong bends, islands, sills, or excessive vegetation. It must be stressed that uniformity of the flow requires no important tributaries to join the river, which might affect the test reach, nor that any sizable amount of flow leaves the test reach. Careful determination of the hydraulic and sediment parameters will help to select appropriate bed material load equations,

Table 9.1 Bed material information for sample problem

Grain size distribution, mm	*Average grain size* *mm*	*ft*	*%*	*Settling velocity* *mm/sec*	*fps*
$d > 0.589$	—	—	2.4	—	—
$0.589 > d > 0.417$	0.495	0.00162	17.8	5.20	0.205
$0.417 > d > 0.295$	0.351	0.00115	40.2	3.75	0.148
$0.295 > d > 0.208$	0.248	0.00081	32.0	2.70	0.106
$0.208 > d > 0.147$	0.175	0.00058	5.8	1.70	0.067
$0.147 > d$	—	—	1.8	—	—

$d_{35} = 0.29 \text{ mm} = 0.00094 \text{ ft}$
$d_{65} = 0.35 \text{ mm} = 0.00115 \text{ ft}$

or at least will help to exclude some of the equations which were derived for and under different conditions.

A test reach, representative of the watercourse to be investigated, has been selected. It was concluded that the channel can be represented by a trapezoidal cross section, with bank slopes of 1:1 and a bottom width of $B = 300$ ft. The channel slope was determined and given by $S = 0.0007$. Five samples, taken down to a depth of approximately 2 ft, were collected to obtain information on the grain size distribution of the entire wetted boundary areas (perimeter). The averaged values of the five samples are given in Table 9.1, and its grain size distribution in Fig. 9.12. 95.8 percent of the bed material falls between 0.589 and 0.147 mm, which is divided into four size fractions, for each of which the sediment transport will be determined, assuming the average grain diameter is the representative one for its size fraction.

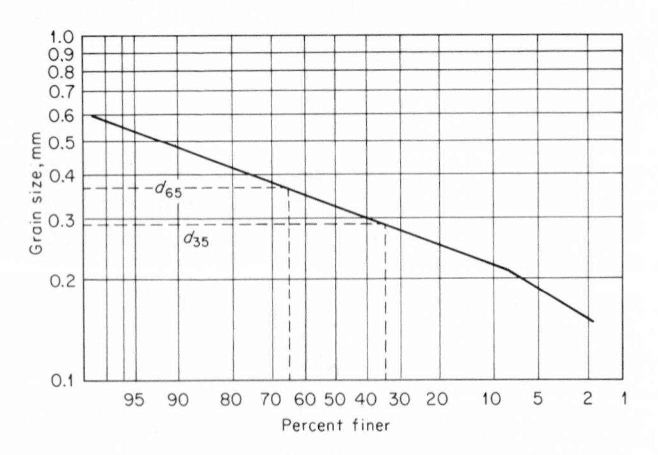

Fig. 9.12 Grain size distribution of bed material.

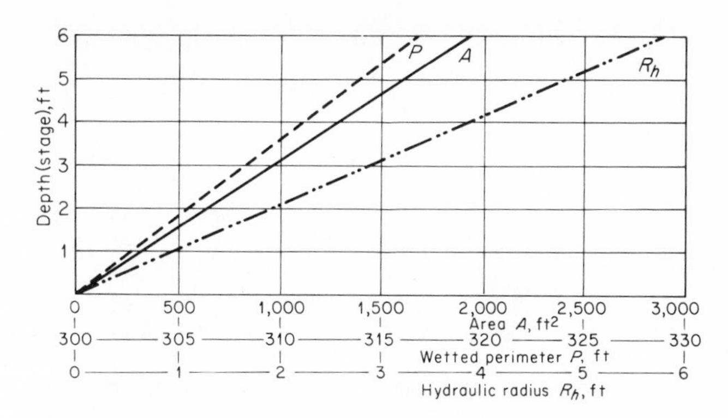

Fig. 9.13 Description of cross section for sample problem.

For the stream, the test reach of which was selected and investigated, a flow-discharge (in cfs) vs. sediment-discharge (in lb/sec) relation shall be found. The highest discharge of interest is $Q = 20{,}000$ cfs. The water viscosity is given by $\nu = 1.0 \times 10^{-5}$ ft²/sec, and the density of the sediment by $\rho_s/\rho = 2.65$.

The subsequent computation will be divided into (1) the hydraulic calculations for the stream and (2) the bed material load calculations.

9.4.2.2 Hydraulic calculations. The calculations of various important and relevant hydraulic informations precede any bed material load computation. It appears advisable to proceed with calculations in table form. The table heading, its meaning, and calculation are explained with footnotes. Each line in the table refers to a particular flow depth (stage) and discharge. The description of the cross section is given in Fig. 9.13; the resulting depth vs. discharge relationship is given in Fig. 9.14. The summary of the hydraulic calculations is found in Table 9.2.

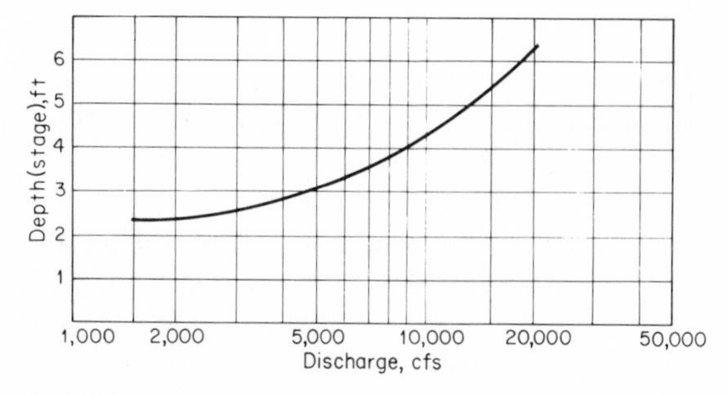

Fig. 9.14 Depth vs. discharge relationship for sample problem.

Table 9.2 Hydraulic calculations for sample problem*

R'_h	u'_*	δ	k_s/δ	x	Δ	$\bar{u}$	Ψ'	$\bar{u}/u''_*$	u''_*	R''_h
1	2	3	4	5	6	7	8	9	10	11
0.5	0.106	0.00110	1.05	1.61	0.00071	2.40	4.47	12.3	0.195	1.70
1.0	0.150	0.00077	1.50	1.53	0.00075	3.64	2.24	18.6	0.196	1.71
2.0	0.212	0.00055	2.10	1.35	0.00085	5.45	1.12	35.0	0.156	1.08
3.0	0.260	0.00045	2.56	1.25	0.00092	6.90	0.75	56.5	0.122	0.66
4.0	0.300	0.00039	2.95	1.19	0.00097	8.13	0.56	82.0	0.099	0.44
5.0	0.336	0.00035	3.29	1.15	0.00100	9.24	0.45	115.0	0.080	0.28
6.0	0.368	0.00032	3.60	1.12	0.00103	10.28	0.38	145.0	0.071	0.22

* See the following for explanation of symbols, column by column:

(1) R'_h, ft (hydraulic radius with respect to the grains); various values are assumed to cover the entire discharge range desired ($Q_{max} = 20{,}000$ cfs).

(2) $u'_* = \sqrt{gR'_hS}$, fps (shear velocity with respect to grain).

(3) $\delta = \dfrac{11.6\nu}{u'_*}$, ft (thickness of laminar sublayer).

(4) $k_s = d_{65}$ (roughness diameter).

(5) x (correction in the logarithmic velocity formula, distinguishing the smooth, transition, and rough regimes; $x = \text{fct}\,(k_s/\delta)$, given with Fig. 7.10.

(6) Δ, ft (apparent roughness diameter); given in Eq. (7.62); $\Delta = k_s/x$.

(7) $\bar{u}$, fps (average velocity); $\bar{u} = (u'_*)\,5.75 \log (12.27R'_h/\Delta)$.

(8) Ψ' (intensity of shear on representative particle); $\Psi' = \dfrac{\rho_s - \rho}{\rho}\dfrac{d_{35}}{R'_hS}$; similar to Eq. (7.57).

(9) $\bar{u}/u''_* = \text{fct}(\Psi')$; given with Fig. 11.16.

(10) u''_* fps (shear velocity due to channel irregularities).

(11) R''_h, ft (hydraulic radius with respect to channel irregularities); from $u''_* = \sqrt{gR''_hS}$.

9.4.2.3 Calculations of bed material load. The bed material transport is calculated for each grain fraction of the bed at each given flow depth. Since the particle size analysis showed that *fine* material is dominant, it was decided to select the following three approaches: (1) By EINSTEIN (1950) for the bedload and bed material load computation; (2) by LAURSEN (1958) for the bed material load computation and bedload estimation; (3) by GRAF et al. (1968) for the bed material load computation.

It is convenient to summarize the calculations in the form of tables; again, table headings and calculations are explained with footnotes in each table. Einstein's procedure is given in Table 9.3, Laursen's procedure in Table 9.4, and Graf's et al. procedure in Table 9.5. The results of the calculation for the bed material load are shown in Fig. 9.15, and are self-explanatory. The curves indicate that the sediment discharge increases rapidly with rising stage. For comparison's sake, the bedload and bed material load results have been

Table 9.2 (continued)

R_h	u_*	D	A	P	Q	X	Y	β_x	$(\beta/\beta_x)^2$	P_E
12	13	14	15	16	17	18	19	20	21	22
2.20	0.148	2.25	680	306.3	1,633	0.00153	0.84	1.36	0.57	11.49
2.71	0.165	2.77	845	307.8	3,075	0.00107	0.76	1.18	0.75	11.62
3.08	0.175	3.15	960	308.9	5,235	0.00076	0.60	0.98	1.10	11.65
3.66	0.192	3.75	1,140	310.5	7,860	0.00071	0.56	0.95	1.17	11.73
4.44	0.220	4.57	1,400	312.9	11,380	0.00075	0.54	0.91	1.27	11.90
5.28	0.230	5.45	1,670	315.4	15,420	0.00077	0.54	0.91	1.27	12.03
6.22	0.250	6.45	1,980	318.2	20,350	0.00079	0.54	0.91	1.27	12.18

(12) R_h, ft (hydraulic radius); $R_h = R'_h + R''_h$, with no additional friction from the banks, vegetation etc.; R_h represents the total hydraulic radius. For a detailed analysis, with additional friction, see EINSTEIN (1950).

(13) u_*, fps (shear velocity); from $u_* = \sqrt{gR_hS}$.

(14) D, ft (depth or stage); for wide channel $R_h \approx D$; see Fig. 9.13.

(15) A, ft² (cross-sectional area); obtained from description of the cross section; see Fig. 9.13.

(16) P, ft (wetted perimeter); obtained from description of the cross section; see Fig. 9.13.

(17) Q, cfs (flow discharge); $Q = \bar{u}A$; a stage-discharge relationship is plotted in Fig. 9.14.

(18) X, ft (characteristic distance); $X = 0.77\Delta$ for $\Delta/\delta > 1.80$ and $X = 1.39\delta$ for $\Delta/\delta < 1.80$.

(19) Y (pressure correction term); $Y =$ fct (k_s/δ); given with Fig. 7.12.

(20) β_x (logarithmic function); $\beta_x = \log(10.6X/\Delta)$.

(21) $(\beta/\beta_x)^2$; with $\beta = \log 10.6$.

(22) P_E (Einstein's transport parameter); $P_E = \frac{1}{0.434}\log\left(\frac{30.2D}{\Delta}\right)$; given by Eq. (8.83).

plotted in Fig. 9.16. These results merit a few words of discussion. The bedload prediction shows disagreement between Einstein's and Laursen's results, and becomes more pronounced at higher discharges than at lower ones. Since LAURSEN (1958) pays little attention to predicting the bedload and EINSTEIN (1950) considers this sediment fraction as an essential one in his calculation, the result of Laursen's procedure may be considered less reliable. For the bed material determination, all three approaches give surprisingly good results at larger flow rates. That the Graf et al. procedure deviates considerably at lower stages is not at all surprising. We must realize that the latter approach uses, for reasons explained earlier, the entire hydraulic radius, whereas the other investigations use only the fraction associated with grain roughness. This becomes even more evident if numerical values are compared. Take the lowest stage in the sample calculation; the ratio of hydraulic radius due to grains to the entire hydraulic radius is $R'_h/R_h = 0.50{:}2.20$; for the highest stage this ratio is $R'_h/R_h = 6.00{:}6.22$.

Table 9.3 Bed material load calculation for sample problem [*procedure by* EINSTEIN *(1950)*]*

10^3d	10^2i_b	R'_h	Ψ	d/X	ξ	Ψ_*	Φ_*	i_sg_s	i_sG_s	Σi_sG_s
1	2	3	4	5	6	7	8	9	10	11
1.62	17.8	0.5	7.64	1.06	1.12	4.10	1.00	0.01400	4.288	4.288
		1.0	3.86	1.51	1.00	2.20	2.90	0.04100	12.620	12.620
		2.0	1.93	2.13	1.00	1.27	5.80	0.08300	25.639	25.639
		3.0	1.29	2.28	1.00	0.85	9.10	0.13000	40.365	40.365
		4.0	0.97	2.16	1.00	0.67	11.70	0.16700	52.254	52.254
		5.0	0.78	2.10	1.00	0.54	14.70	0.21000	66.234	66.234
		6.0	0.66	2.05	1.00	0.46	17.30	0.24700	78.595	78.595
1.15	40.2	0.5	5.49	0.75	1.50	3.93	1.09	0.02000	6.126	10.414
		1.0	2.74	1.08	1.11	1.73	3.90	0.07200	22.162	34.782
		2.0	1.37	1.51	1.00	0.90	8.50	0.15600	48.188	73.827
		3.0	0.92	1.62	1.00	0.60	13.20	0.24200	75.141	115.506
		4.0	0.69	1.53	1.00	0.47	17.00	0.31200	97.625	149.879
		5.0	0.55	1.50	1.00	0.38	21.00	0.38600	121.744	187.978
		6.0	0.48	1.46	1.01	0.33	24.10	0.44200	140.644	219.239
0.81	32.0	0.5	3.86	0.53	3.20	5.91	0.44	0.00400	1.225	11.639
		1.0	1.93	0.76	1.50	1.65	4.15	0.03700	11.389	46.171
		2.0	0.97	1.07	1.12	0.72	10.80	0.09700	29.963	103.790
		3.0	0.65	1.14	1.08	0.46	17.30	0.15500	48.128	163.634
		4.0	0.48	1.08	1.01	0.34	23.50	0.21000	65.709	215.588
		5.0	0.39	1.05	1.13	0.30	26.80	0.24000	75.696	263.674
		6.0	0.33	1.03	1.14	0.26	31.00	0.27800	88.460	307.699
0.57	5.8	0.5	2.72	0.37	8.20	10.70	0.07	0.00003	0.009	11.648
		1.0	1.36	0.53	3.20	2.48	2.45	0.00100	0.308	46.479
		2.0	0.68	0.75	1.52	0.68	11.50	0.00460	1.421	105.211
		3.0	0.46	0.80	1.40	0.42	19.00	0.00760	2.360	165.994
		4.0	0.34	0.76	1.50	0.35	23.00	0.00930	2.910	218.498
		5.0	0.27	0.74	1.55	0.29	28.00	0.01130	3.564	267.238
		6.0	0.23	0.72	1.60	0.25	32.00	0.01290	4.105	311.804

* See the following for explanation of symbols, column by column:

(1) d, ft (grain size); the representative ones are taken from Table 9.1.

(2) i_b (fraction of bed material); taken from Table 9.1.

(3) R'_h, ft (hydraulic radius with respect to grains); taken from Table 9.2.

(4) Ψ (intensity of shear on particle); $\Psi = \dfrac{\rho_s - \rho}{\rho} \dfrac{d}{R'_h S}$; given in Eq. (7.57).

(5) d/X; for values of X see Table 9.2.

(6) ξ (hiding factor); $\xi = \text{fct}\,(d/X)$, given in Fig. 7.11.

(7) Ψ_* (intensity of shear on individual grain size); $\Psi_* = \xi Y(\beta^2/\beta_x^2)\Psi$; values are given in Table 9.2.

(8) Φ_* (intensity of transport for individual grain size); $\Phi_* = \text{fct}\,(\Psi_*)$ given with Fig. 7.13.

(9) i_sg_s, lb/(sec)(ft) (bedload rate in weight per unit width and time for a size fraction); $i_sg_s = i_b\Phi_*\rho_s g^{3/2} d^{3/2}\sqrt{(\rho_s/\rho - 1)}$; from Eqs. (7.51) and (7.52).

(10) i_sG_s, lb/sec (bedload rate in weight per unit time for a size fraction for entire cross section); $i_sG_s = (i_sg_s)P$.

(11) Σi_sG_s, lb/sec (bedload rate in weight per unit time for all size fractions for entire cross section).

Table 9.3 (continued)

$10^3 A_E$	z	I_1	$-I_2$	$P_E I_1 + I_2 + 1$	$i_{st}q_{st}$	$i_{st}G_{st}$	$\Sigma i_{st}G_{st}$
12	13	14	15	16	17	18	19
1.44	4.84	0.055	0.36	1.271	0.029	8.883	8.883
1.17	3.42	0.094	0.56	1.532	0.063	19.391	19.391
1.03	2.42	0.150	0.94	1.808	0.150	46.335	46.335
0.86	1.97	0.215	1.35	2.172	0.282	87.561	87.561
0.71	1.71	0.300	1.80	2.770	0.463	144.873	144.873
0.60	1.53	0.380	2.20	3.371	0.708	223.303	223.303
0.50	1.39	0.510	2.95	4.262	1.053	335.065	335.065
1.47	3.49	0.090	0.56	1.474	0.029	8.883	17.766
1.19	2.47	0.150	0.90	1.843	0.133	40.937	60.328
1.05	1.75	0.285	1.57	2.750	0.429	132.518	178.853
0.88	1.43	0.460	2.40	3.996	0.967	300.254	387.815
0.72	1.23	0.680	3.40	5.692	1.776	557.104	701.977
0.61	1.10	1.020	4.60	8.671	3.347	1,055.644	1,278.947
0.51	1.01	1.350	5.90	11.543	5.102	1,623.456	1,958.521
0.72	2.50	0.145	0.96	1.706	0.007	2.144	19.910
0.58	1.77	0.280	1.70	2.554	0.095	29.241	89.569
0.52	1.25	0.600	3.50	4.400	0.427	131.900	310.753
0.43	1.02	1.350	6.20	10.636	1.649	512.015	899.830
0.36	0.88	2.400	9.20	20.360	4.276	1,337.960	2,039.937
0.30	0.79	3.800	13.20	33.514	8.043	2,536.762	3,815.709
0.25	0.72	5.800	18.00	53.644	14.913	4,745.317	6,703.838
0.74	1.58	0.360	2.05	3.086	0.0001	0.028	19.938
0.60	1.12	0.960	4.20	8.352	0.008	2.462	92.031
0.53	0.79	3.300	10.80	27.645	0.127	39.230	349.983
0.44	0.64	7.300	19.50	67.129	0.510	158.355	1,058.185
0.36	0.56	12.500	30.00	119.750	1.114	348.571	2,388.508
0.31	0.50	19.000	42.00	187.570	2.120	668.648	4,484.357
0.26	0.46	27.000	56.00	273.860	3.533	1,124.201	7,828.039

(12) $A_E = a/D$ (ratio of bed layer to water depth).

(13) z (exponent for suspension distribution); $z = \dfrac{v_{ss}}{0.4u'_*}$; given by Eq. (8.33); all values are determined and given in Tables 9.1 and 9.2.

(14) I_1 (integral) read from Fig. 8.10; $I_1 = f(A_E,z)$; given by Eq. (8.78).

(15) I_2 (integral) read from Fig. 8.11; $I_2 = f(A_E,z)$; given by Eq. (8.78).

(16) $(P_E I_1 + I_2 + 1)$; given by Eq. (9.4).

(17) $i_{st}g_{st}$, lb/(sec)(ft) (bed material load in weight per unit width and time, for a size fraction); $i_{st}g_{st} = i_s g_s(P_E I_1 + I_2 + 1)$; given by Eq. (9.4).

(18) $i_{st}G_{st}$, lb/sec (bed material rate in weight per unit time for a size fraction for entire cross section); $i_{st}G_{st} = (i_{st}g_{st})P$.

(19) $\Sigma i_{st}G_{st}$, lb/sec (bed material rate in weight per unit time for all size fractions for entire cross section).

Table 9.4 Bed material load calculation for sample problem [*procedure by* LAURSEN (*1958*)]*

D	$10^3 d$	$\dfrac{\sqrt{\tau_0/\rho}}{v_{ss}}$	$f()_s$	$f()_{st}$	d/δ	$f(d/\delta)$	$(\tau_0)_{cr}$	$10^3 d/D$	$\left(\dfrac{d}{D}\right)^{1/3}$	$10^4\left(\dfrac{d}{D}\right)^{7/6}$
1	2	3	4	5	6	7	8	9	10	11
2.25	1.62	0.720	9.6	12.2	1.475	0.0335	0.00566	0.7200	0.08910	2.180
2.77		0.803	10.0	13.0	2.105	0.0350	0.00591	0.5850	0.08355	1.710
3.15		0.853	10.2	13.9	2.945	0.0370	0.00625	0.5150	0.08010	1.480
3.75		0.935	10.4	14.7	3.600	0.0380	0.00642	0.4320	0.07556	1.200
4.57		1.071	10.8	16.2	4.150	0.0392	0.00662	0.3540	0.07060	0.954
5.45		1.120	11.0	17.0	4.630	0.0400	0.00676	0.2970	0.06660	0.775
6.45		1.219	11.2	18.3	5.060	0.0420	0.00710	0.2510	0.06310	0.640
2.25	1.15	0.756	9.8	12.7	1.045	0.0330	0.00396	0.5110	0.08000	1.460
2.77		1.012	10.7	15.3	1.495	0.0335	0.00402	0.4150	0.07450	1.150
3.15		1.431	11.7	21.3	2.090	0.0350	0.00420	0.3660	0.07150	0.990
3.75		1.755	12.2	26.8	2.555	0.0360	0.00434	0.3070	0.06750	0.810
4.57		2.025	12.5	32.0	2.950	0.0370	0.00444	0.2520	0.06310	0.640
5.45		2.270	12.9	37.5	3.285	0.0375	0.00450	0.2110	0.05950	0.520
6.45		2.485	13.1	45.0	3.595	0.0381	0.00456	0.1780	0.05620	0.430
2.25	0.81	1.000	10.6	15.0	0.737	0.0335	0.00280	0.3600	0.07110	0.970
2.77		1.415	11.7	21.1	1.052	0.0330	0.00276	0.2920	0.06640	0.760
3.15		2.000	12.5	31.0	1.473	0.0334	0.00279	0.2570	0.06350	0.660
3.75		2.455	13.0	44.0	1.800	0.0340	0.00284	0.2160	0.06000	0.540
4.57		2.830	13.6	58.5	2.075	0.0350	0.00292	0.1770	0.05610	0.420
5.45		3.170	14.0	70.0	2.315	0.0353	0.00295	0.1490	0.05300	0.350
6.45		3.470	14.4	92.0	2.530	0.0360	0.00300	0.1260	0.05010	0.290
2.25	0.57	1.580	11.8	23.5	0.528	0.0354	0.00210	0.2530	0.06320	0.650
2.77		2.240	12.8	37.3	0.754	0.0333	0.00198	0.2060	0.05910	0.505
3.15		3.165	14.0	70.0	1.055	0.0330	0.00196	0.1810	0.05650	0.435
3.75		3.880	14.9	120.0	1.290	0.0331	0.00197	0.1520	0.05340	0.355
4.57		4.480	15.2	164.0	1.488	0.0335	0.00199	0.1250	0.05000	0.280
5.45		5.020	15.9	210.0	1.658	0.0340	0.00202	0.1045	0.04710	0.295
6.45		5.490	16.0	260.0	1.813	0.0341	0.00203	0.0885	0.04450	0.189

* See the following explanation of symbols, column by column:
(1) D, ft (depth or stage).
(2) d, ft (grain size); taken from Table 9.1.
(3) $\sqrt{\tau_0/\rho}/v_{ss}$ (shear velocity to settling velocity ratio); taken from Tables 9.1 and 9.2.
(4) $f(\sqrt{\tau_0/\rho}/v_{ss})_s$ (Laursen's bedload relationship); taken from Fig. 9.6.
(5) $f(\sqrt{\tau_0/\rho}/v_{ss})_{st}$ (Laursen's total-load relationship); taken from Fig. 9.6.
(6) d/δ (grain size to sublayer thickness ratio); taken from Table 9.2.
(7) $f(d/\delta)$ (Shields' shear stress relationship); taken from Fig. 6.7.
(8) $(\tau_0)_{cr}$, lb/ft² (critical tractive force).
(9) d/D (grain size to depth ratio).
(10) d/D (grain size to depth ratio).
(11) d/D (grain size to depth ratio).

Table 9.4 (continued)

$10^2\tau_0'$	$\dfrac{\tau_0'}{(\tau_0)_{cr}} - 1$	$10^4\bar{C}_s$	$10^4\bar{C}_{st}$	i_sG_s	$i_{st}G_{st}$	Σi_sG_s	$\Sigma i_{st}G_{st}$
12	13	14	15	16	17	18	19
1.710	2.02	0.422	0.537	4.300	5.472	4.300	5.472
3.690	5.24	0.895	1.162	17.173	22.297	17.173	22.297
7.940	11.72	1.772	2.415	57.885	78.889	57.885	78.889
12.000	17.70	2.210	3.120	108.392	153.025	108.392	153.025
15.550	22.50	2.230	3.475	158.355	246.764	158.355	246.764
18.900	26.95	2.290	3.560	220.346	342.546	220.346	342.546
22.200	30.30	2.050	3.345	260.317	424.762	260.317	424.762
1.535	2.88	0.411	0.534	4.239	5.441	8.539	10.913
3.290	7.19	0.876	1.263	16.809	24.234	33.982	46.531
7.080	15.85	1.835	3.340	59.943	109.106	117.828	187.995
10.700	23.80	2.355	5.170	115.504	253.570	223.896	406.595
13.900	30.32	2.420	6.200	171.847	440.269	330.202	687.033
16.900	36.60	2.460	7.150	236.703	637.979	457.049	980.525
19.750	42.25	2.380	8.180	302.222	1,038.729	562.539	1,463.491
1.365	3.88	0.398	0.564	4.056	5.747	12.595	16.660
2.930	9.06	0.801	1.450	15.370	27.823	49.352	74.354
6.290	21.50	1.775	4.400	57.983	143.732	175.811	331.727
9.520	32.55	2.280	7.720	111.826	378.638	335.722	785.233
12.350	41.30	2.360	10.050	167.586	713.663	497.788	1,400.696
15.050	50.00	2.450	12.250	235.741	1,178.705	692.790	2,159.230
17.600	57.60	2.405	15.380	305.397	1,953.014	867.936	3,416.505
1.212	4.78	0.366	0.730	3.730	7.439	16.325	24.099
2.610	12.19	0.789	2.290	15.139	43.941	64.491	118.295
5.600	27.57	1.783	8.380	58.244	273.744	234.055	605.471
8.460	42.00	2.200	17.900	107.902	877.931	443.624	1,663.164
11.000	54.25	2.310	24.900	164.036	1,768.179	661.824	3,168.875
13.390	65.20	3.060	40.400	294.436	3,887.320	987.226	6,046.550
15.630	76.00	2.300	37.400	292.063	4,749.202	1,159.999	8,165.707

(12) τ_0', lb/ft² (boundary shear associated with sediment particles); obtained with Eq. (9.19).
(13) $\tau_0'/(\tau_0)_{cr} - 1$.
(14) $\bar{C}_s$, % weight (mean concentration, bedload only).
(15) $\bar{C}_{st}$, % weight (mean concentration, total load).
(16) i_sG_s, lb/sec (bedload rate in weight per unit time for a size fraction for entire cross section); obtained with $\bar{C}_s = 265q_s/q$.
(17) $i_{st}G_{st}$, lb/sec (bed material load rate in weight per unit time for a size fraction for entire cross section); obtained with $\bar{C}_{st} = 265q_{st}/q$.
(18) Σi_sG_s, lb/sec (bedload rate in weight per unit time for all size fractions for entire cross section).
(19) $\Sigma i_{st}G_{st}$, lb/sec (bed material load rate in weight per unit time for all size fractions for entire cross section).

Table 9.5 Bed material load calculation for sample problem [*procedure by* GRAF *et al.* (*1968*)]*

$10^3 d$	R_h	Ψ'_A	Φ_A	$10^4 \bar{C}_v$	$i_{st}Q_{st} = \bar{C}_v Q$	$i_{st}G_{st}$	$\Sigma i_{st}G_{st}$
1	2	3	4	5	6	7	8
1.62	2.20	1.758	2.506	2.260	0.369	61.512	61.512
	2.71	1.428	4.241	2.045	0.629	104.854	104.854
	3.08	1.257	5.837	1.655	0.866	144.276	144.276
	3.66	1.057	9.035	1.705	1.340	223.244	223.244
	4.44	0.873	14.634	1.935	2.202	366.853	366.853
	5.28	0.733	22.735	2.120	3.269	544.615	544.615
	6.22	0.622	34.403	2.565	5.220	869.652	869.652
1.15	2.20	1.249	5.937	3.215	0.525	87.465	149.157
	2.71	1.013	10.057	3.060	0.941	156.771	261.625
	3.08	0.893	13.816	2.360	1.235	205.751	349.977
	3.66	0.751	21.379	2.420	1.902	316.873	540.117
	4.44	0.620	34.663	2.750	3.130	521.458	888.311
	5.28	0.520	54.115	3.180	4.904	817.006	1,361.621
	6.22	0.441	82.134	3.680	7.489	1,249.667	2,119.319
0.81	2.20	0.879	14.391	4.595	0.750	124.950	274.107
	2.71	0.714	24.276	4.150	1.276	212.582	474.207
	3.08	0.629	33.408	3.360	1.759	293.049	643.026
	3.66	0.529	51.563	3.440	2.704	450.486	990.603
	4.44	0.437	83.790	3.920	4.461	743.203	1,631.514
	5.28	0.367	129.875	4.480	6.908	1,150.873	2,512.494
	6.20	0.311	187.207	4.960	10.094	1,681.660	3,800.979
0.57	2.20	0.625	33.954	8.010	1.308	217.913	492.020
	2.71	0.507	57.722	7.300	2.245	374.017	848.224
	3.08	0.447	79.011	5.860	3.068	511.129	1,154.115
	3.66	0.376	122.235	6.112	4.804	800.346	1,790.949
	4.44	0.310	199.808	6.900	7.852	1,308.143	2,939.657
	5.28	0.260	277.066	7.070	10.902	1,816.273	4,328.767
	6.20	0.220	432.917	8.475	17.247	2,873.350	6,674.329

* See the following for explanation of symbols, column by column:

(1) d, ft (grain size); taken from Table 9.1.

(2) R_h, ft (hydraulic radius); given in Table 9.2.

(3) Ψ'_A (shear intensity parameter); defined in Eq. (9.26).

(4) Φ_A (transport parameter); defined in Eq. (9.27).

(5) $\bar{C}_v$ (average volume concentration); given by Eq. (9.27).

(6) $i_{st}Q_{st}$, lb/sec (bed material load rate in volume per unit time for a size fraction for entire cross section).

(7) $i_{st}G_{st}$, lb/sec (bed material load rate in weight per unit time for a size fraction for entire cross section).

(8) $\Sigma i_{st}G_{st}$, lb/sec (bed material load rate in weight per unit time for all size fractions for entire cross section).

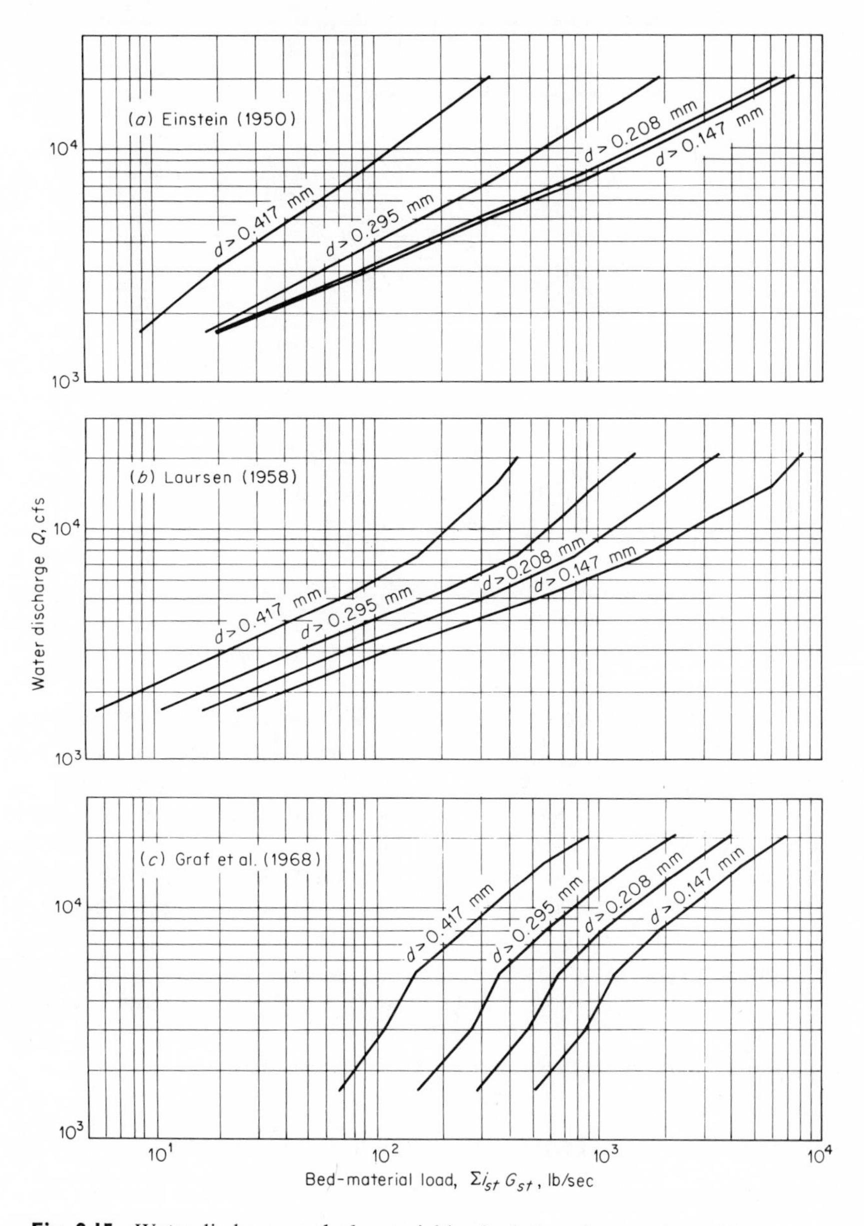

Fig. 9.15 Water discharge vs. bed material load relations for sample problem.

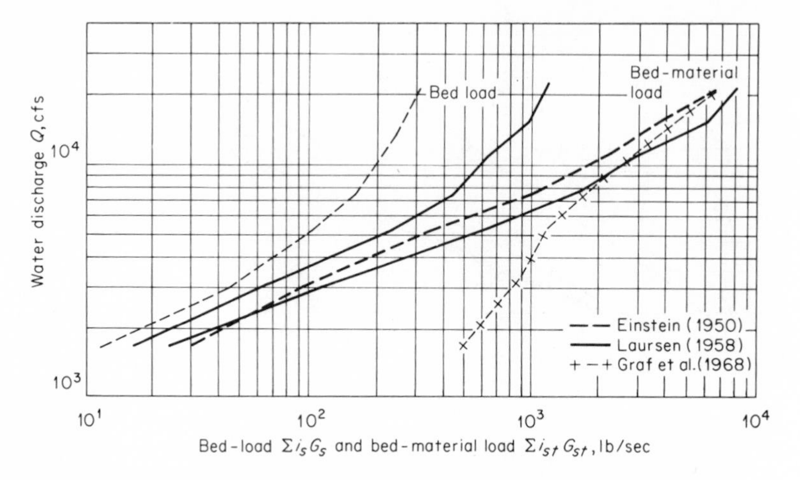

Fig. 9.16 Water discharge vs. bedload and bed material load for grains coarser than $d = 0.147$ mm for sample problem; a comparison.

Thus the assumption of using the total hydraulic radius in the computation accounts for deviations observed in Fig. 9.16, for the disagreement at low stages, and good agreement at high stages. On the other hand, we must ask as to how realistic are the measurements and, therefore, the calculations in a bedform-covered stream being 300 ft wide and 3 ft deep. Nevertheless, Einstein's and Laursen's approaches, essentially based on the same basic assumptions, render extremely encouraging and almost identical results.

It might be interesting to include a comparative study, by use of EINSTEIN's (1950) approach only, on the effects of varying channel width, channel slope, and viscosity on the bed material load. The basic data are identical to the ones for the present sample problem. The calculations are summarized in Fig. 9.17. Everything else being constant, and by varying the channel width results in an increase of bed material load with decreasing width. Increasing the slope causes more bed material discharge to move, while an increase of viscosity exhibits a tendency, pronounced at larger stages more than at smaller ones, of an increase in sediment discharge.

9.5 HYDROLOGIC EFFECTS ON SEDIMENT TRANSPORT

Besides the hydraulic factors there exist also hydrologic factors, which are of importance and strongly influence the transport of sediment. The discussion in Part 3 focused almost exclusively on the hydraulic factors, i.e., it considered hydrodynamical force relations. In this way it was possible to predict the bed material load. For certain streams, however, the total load is considerably larger than the bed material load. For example, GUY (1964), investigating streams on the Atlantic coast of the United States finds that the coarse-sediment discharge (approximately identical with the bed material load) is generally much less than half the total load (bed material load plus washload); COLBY (1963), having apparently in mind alluvial watercourses of the United States, states that most sediment transported by streams is washload and, thus, the hydraulic theories of bed material transportation are helpful but inadequate for the calculation of the total sediment transport.

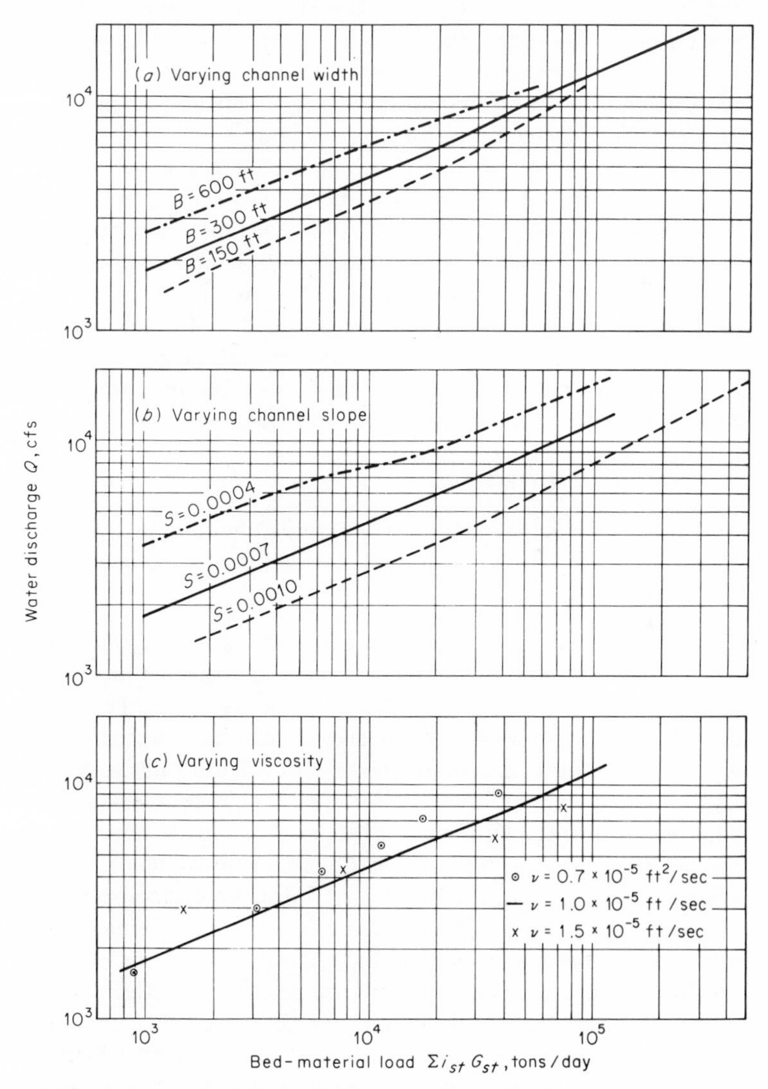

Fig. 9.17 Water discharge vs. bed material load for grains coarser than $d =$ 0.147 mm for sample problem: a comparison.

Such a dilemma has led researchers to search for more factors which might influence the sediment transport phenomenon. Certainly, meteorological, climatological, geological, geographical, biological, etc., factors, as they affect the hydrology of a watershed or river basin, must be studied. An important step into this direction is a study by GUY (1964). Data from seven streams in the Atlantic coast region of the United States are analyzed to determine the cause of variation in sediment transport. The independent variables were assumed to be: (1) season; (2) net surface runoff; (3) groundwater runoff; (4) long-term mean air temperature; (5) peak rate of water discharge; (6) a measure of storm intensity; (7) aerial mean precipitation; and (8) aerial mean precipitation intensity. The data analysis was achieved by a graphical correlation—for a preliminary study of the data and to note unusual events—and an analytical multiple-regression method. The vast amount of available data allowed merely some indication of trends. For example, it was found that the quantity of sediment discharge increases with water discharge, peak flow, rainfall quantity, and rainfall intensity. The method is described in detail by GUY (1964) and is illustrated with examples. Its more general applicability must be proved by further studies of streams, watersheds, and their hydrology of different character.

Nevertheless, a better understanding of the importance of hydrology may be gained by interpreting sediment discharge data.

9.5.1 STREAM FLOW EFFECTS

Since the stream is the carrier of the sediment material, when and if it reaches the channel, the stream flow and its relation to the transport should be investigated.

Sediment-rating curves. These are curves relating the water discharge and sediment discharge for a given stream. If the bed material discharge is understood to be identical with the sediment discharge, the two are functionally related, and a reasonable correlation may be expected. As a matter of fact, the correlation can be predicted, provided a satisfactory bed material load equation can be found. For the sample problem discussed in Sec. 9.4.2, sediment-rating curves for bed material loads have been established and are given in Figs. 9.15 and 9.16. In an investigation by COLBY (1964) it was pointed out that the bed material discharge can also be correlated, at least for some situations, with the mean velocity $\bar{u}$, the mean power $\gamma R_h S \bar{u}$, and/or with the $R_h' S$ value; total shear $\gamma R_h S$ is not considered a generally satisfactory measurement. (Based on such a knowledge, COLBY (1964) also established empirically graphs for the practical computations of bed material discharge.) Recently MADDOCK (1969) advanced similar ideas.

Hydrologic investigations on various streams have been carried out to establish sediment-rating curves with measured data. However, such

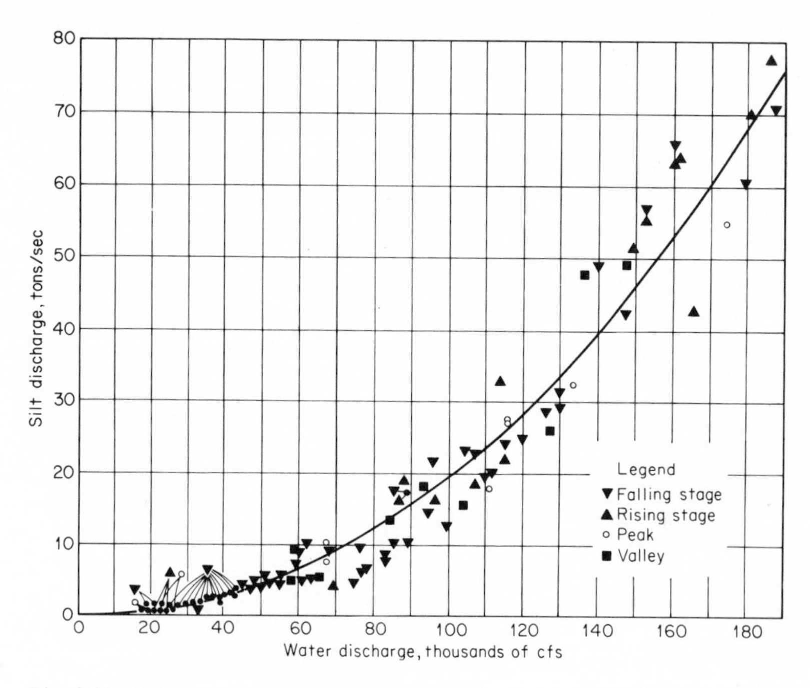

Fig. 9.18 Silt-discharge rating curve for the Missouri. [*After* STRAUB (*1936*).]

relations for streams are influenced by the limitations and proper identifications of these measurements. Since no commonly convincing method for bedload sampling exists, most of the field studies report only the suspended material. Furthermore, it appears that what is reported as suspended material is essentially the suspended fraction of the bed material load plus the washload. Present thinking is such that bed material discharge is truly dependent on the flow rate, whereas the washload does not exhibit a clear functional relationship with the flow rate, as shown by Figs. 9.1 and 9.2. Why should, therefore, the suspended material be a function of the flow rate?

Despite the uncertainty of obtaining a useful relation, some investigations are worth discussing. A *silt-discharge* rating curve is given for the Missouri River by STRAUB (1936). The results of 2 years of daily measurements of suspended material (apparently the suspended-load fraction of the bed material load plus the washload) could be well correlated by a relation first suggested by KENNEDY (1895), such as

$$G_{ssm} = pQ^j \tag{9.29}$$

which is given in Fig. 9.18, where G_{ssm} is the measured suspended material in tons per second, and Q is the water discharge in cubic feet per second; the exponent was found to be $j = 2.16$.

However, it was realized that for any particular day the suspended material may deviate greatly from the average, depending on various hydrologic factors, but that monthly values are in good agreement. Also, JAKUSCHOFF (1932) finds it possible to obtain a relationship similar to Eq. (9.29) for some streams in Turkestan.

The relation of suspended-sediment load to discharge for the Powder River is given in Fig. 9.19, which LEOPOLD et al. (1953) consider a typical relation. Figure 9.19 clearly demonstrates, what was to be expected, that there apparently does not exist a simple relationship between suspended sediment and discharge. A given flow rate may be a result of different hydrologic events; it might be caused by snow melt, by a rain of moderate

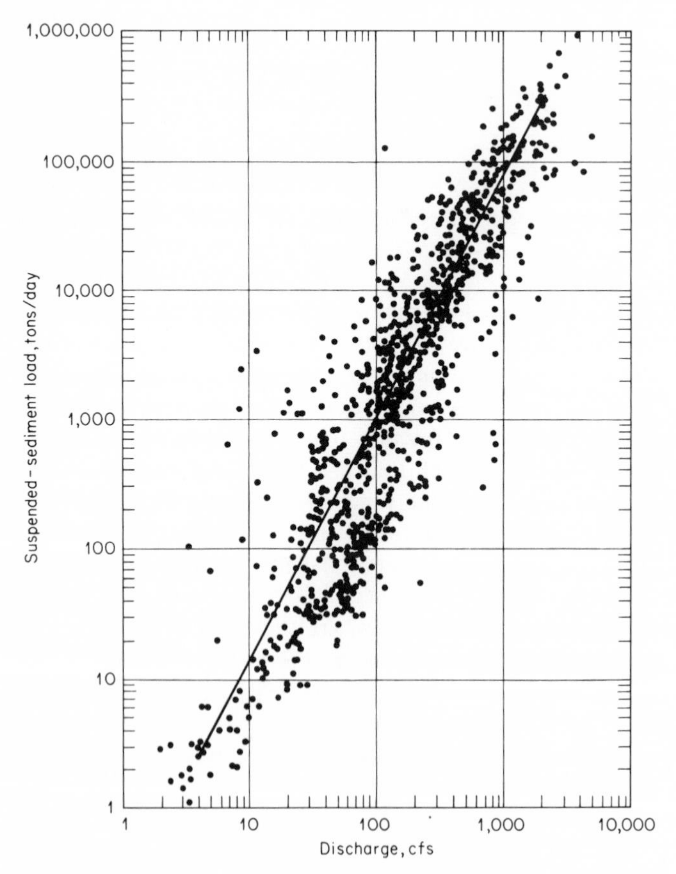

Fig. 9.19 Suspended-material-rating curve for the Powder River. [*After* LEOPOLD *et al.* (*1953*).]

or intense magnitude, or by various geographic dimensions and durations, etc., and a different suspended-sediment load would be the result in each case. Nevertheless, LEOPOLD et al. (1953) feel that a relationship, as written with Eq. (9.29), gives a rough correlation. For streams in the western parts of the United States, typical values for the exponent j lie in the range of $2 < j < 3$. Furthermore, it was found typical that suspended-sediment load increases more rapidly than does water discharge at a given station.

It should be mentioned here that JAROCKI (1963) discussed in some length the sediment-rating curves, mainly for lowland streams in Poland and in the Soviet Union. Sometimes it was found difficult and even impossible to deduce a meaningful correlation.

The foregoing discussion leads to the obvious conclusion that sediment-rating curves, where available at all, should be used with caution. They can provide fairly satisfactory results for the prediction of mean monthly or mean annual sediment rates, but may also give good results for small and homogeneous watersheds for a day-by-day estimate. Finally, it is worthwhile to refer to a hypothesis by NORGAARD (1968), who states that bedload movement is increased by stream flow fluctuations.

Sediment discharge and hydrographs. It is frequently observed that sediment discharge and water discharge do not always increase or decrease simultaneously. JAROCKI (1963) suggests that this is owing to the difference in the cause of the two effects. Intensive sediment transport from the river basin does not, necessarily, coincide with the occurrence of maximum flow rates. For small and homogeneous watersheds, the two peaks usually coincide, since the runoff or rain is responsible for both of them. For large rivers, it is often reported that the peak sediment discharge precedes the peak water discharge, depending on the hydrologic system of the watershed and the water velocity. The observations by EINSTEIN et al. (1940) for the Enoree River, by JAROCKI (1963) for the Vistula and Volga, and by NORDIN et al. (1963) for the Rio Grande, all exhibit this trend. In the Enoree River, a two-day flood was observed, and in the Volga the seasonal effect is exhibited; both are given in Fig. 9.20. Common to all observations is the fact that suspended-sediment concentration (inclusive washload) attenuates more slowly than the water discharge.

These investigations shed also some light on the sediment-rating curves. Consider, for example, a flood (see Fig. 9.20). During the passage of a flood the suspended material executes a hysteresis loop and, at the same flow rate, the quantity of the suspended material at the rising water stage is considerably larger than the one at the falling water stage. This was clearly observed by JAROCKI (1963) for the Vistula and by LEOPOLD et al. (1953) for the San Juan River, the Rio Grande, and the Colorado River, and it is shown in Fig. 9.21. For sake of completeness and comparison, Fig. 9.21 also indicates changes in velocity, width, depth, stream bed elevation, and

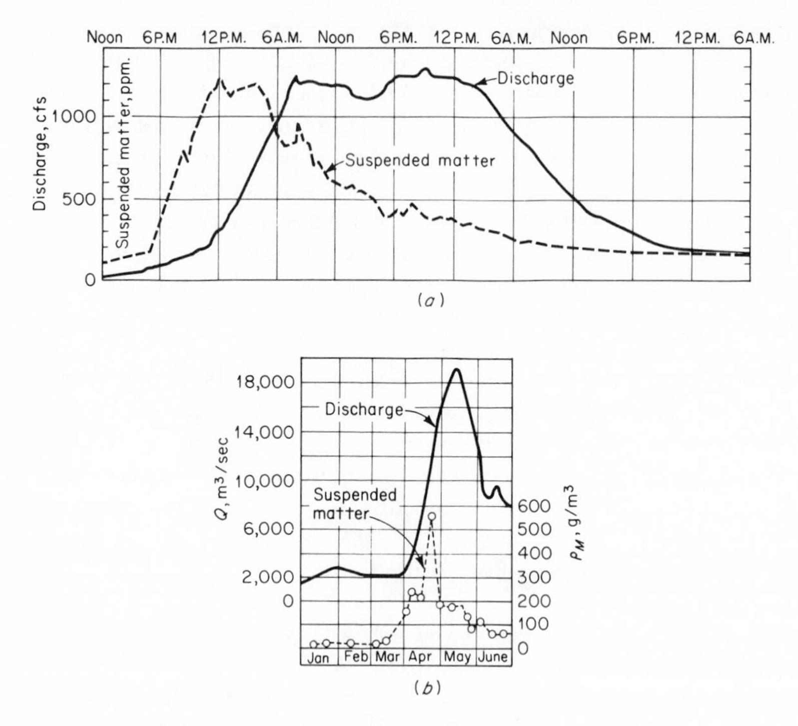

Fig. 9.20 Hydrograph and suspended-sediment curves; (*a*) for Enoree River [*After* EINSTEIN *et al.* (*1940*)] and (*b*) for Volga [*after* JAROCKI (*1963*)].

water surface, all of which show a hysteresis loop. These changes, but especially the river bed scour during flood, are extensively discussed by LEOPOLD et al. (1964).

Frequent events of flow rates and their relation to the transport of sediment have recently received some attention. However, before any general conclusion can be drawn, more research has to be conducted. A summary of current research, discussing exclusively streams of the United States, is given by LEOPOLD et al. (1964). In short, the sediment load at flood stages is larger than at moderate flow. Although excessive flows are very effective in transporting the sedimentary material, they do occur rather infrequently. The major work of transportation is thus not accomplished during periods of larger floods, but rather during the more modest but frequently occurring ones.

9.5.2 TEMPERATURE EFFECTS

Since LANE et al. (1949) and COLBY et al. (1965) reported an increase of sediment transport for the Colorado River and for the Niobrara, respectively, which was attributed to a decrease in water temperature, researchers paid attention to this effect. The same conclusion was made by FRANCO (1968) from a movable-bed-model study.

Perhaps the most striking indication of this phenomenon was reported for the Colorado River, where the flow rate remained fairly constant over seasons and over years. The considerable variation of sediment loads could be well correlated with the variations in the water temperature, as can be clearly seen from Fig. 9.22. It was reasoned that most of the effects of temperature are due to the pickup rate of sediment from the river bed.

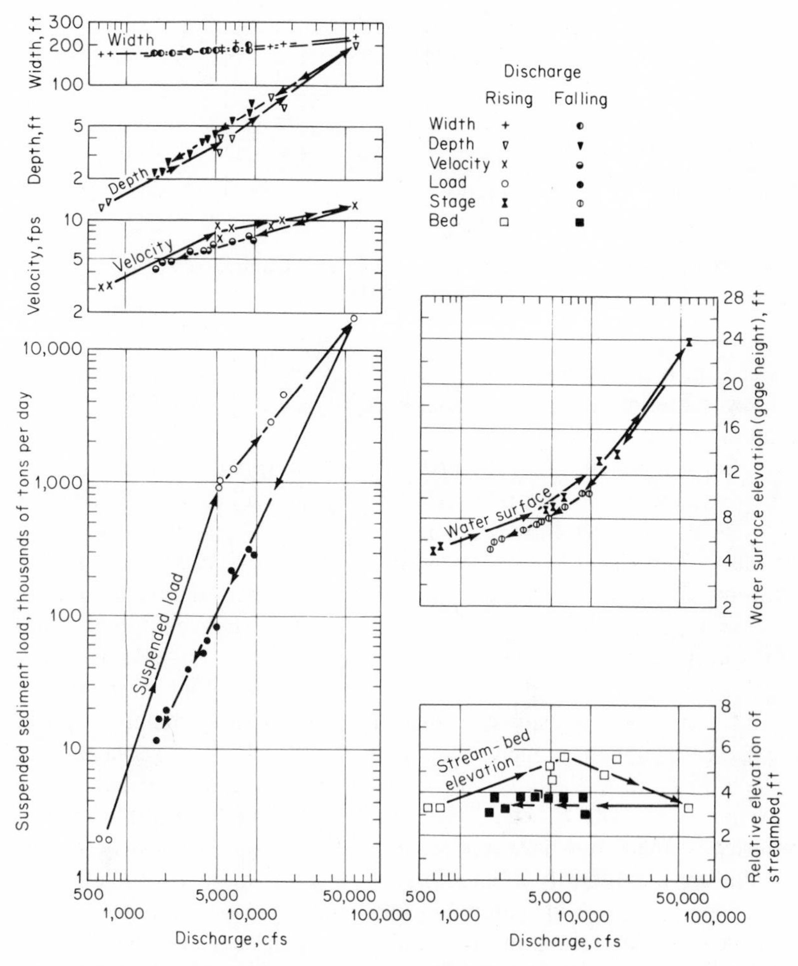

Fig. 9.21 Effect of a flood on the suspended-sediment load and other hydraulic parameters for San Juan River, Utah. [*After* LEOPOLD *et al.* (*1953*).]

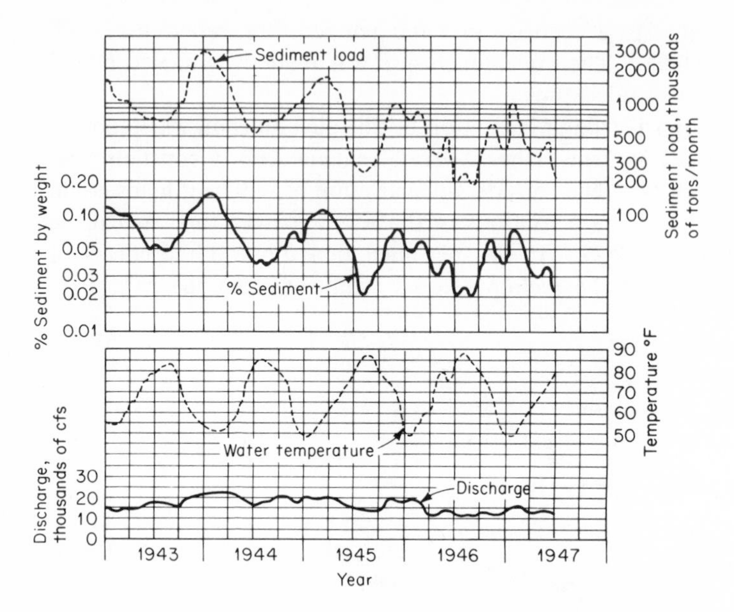

Fig. 9.22 Variation of sediment load, water temperature, and discharge—Colorado River. [*After* LANE *et al.* (*1949*).]

Indeed, in a recent investigation by COLBY et al. (1965) it was argued that the bed material discharge is roughly twice as large at a temperature of 40°F than at 80°F.

Since the analysis of the bed material computation by EINSTEIN (1950) represents the most comprehensive investigation up to now, we shall use this analysis to find out how temperature affects its results. Both the water density ρ and the water's kinematic viscosity ν are dependent on the temperature T. For all practical purposes under consideration, it is safe to assume that the density ρ remains unaffected; this definitely cannot be said for the kinematic viscosity ν. (A temperature–vs.–kinematic-viscosity relationship may be found in any standard book on fluid mechanics.)

Changes in viscosity will affect at least two parameters important in the present problem, namely, the thickness of the laminar sublayer δ and the particle's settling velocity v_{ss}; it may also affect other parameters such as the Karman's constant k, etc. The effect upon the sublayer's thickness is reportedly small, but may not be ignored. As such, it will be responsible for altering the intensity of shear in two ways, namely, through Ψ' and Ψ^*, and changes in the bed material results can thus be expected, accompanied by a possible alteration of the bedforms. Temperature changes are also exhibited by the exponent for the suspension distribution z, because it is dependent on the settling velocity v_{ss}. The obvious cause is a change in the distribution of the suspended matter. If the bed material load is calculated in a comparative way such that temperature effects can be observed—this has been done for the sample problem in Fig. 9.17—it can be seen that at larger flow rates, at least, there exists a tendency for the bed material load to increase with a decrease in temperature. In a comprehensive investigation by COLBY et al. (1965), it was found that the effect is small for shallow channels, but becomes larger for deep, natural streams. It is also suggested that sediments with particle sizes of $d > \frac{1}{2}$ mm experience little effect due to temperature variation on the bed material load, while for sizes of $d < \frac{1}{2}$ mm the effect may be considerable owing to the viscous forces which control the settling process. Studying laboratory channels, FRANCO (1968) concluded that "the effects of water temperature on the bedload appear to be mostly in the formation of the bed and bed roughness."

Thus it may be concluded that careful temperature measurements must be obtained for the investigation of sediment transport problems.

9.5.3 EFFECTS OF RAINFALL

It seems quite reasonable to expect that rainfall influences the hydraulics of the flow and, in turn, has an influence on the initial movement and, subsequently, on the sediment transport.

In studying shallow channel flow—in small channels flow occurs frequently during a period of extensive rainfall—SMERDON (1964) concluded that the mean flow velocity is reduced because of rainfall. The critical

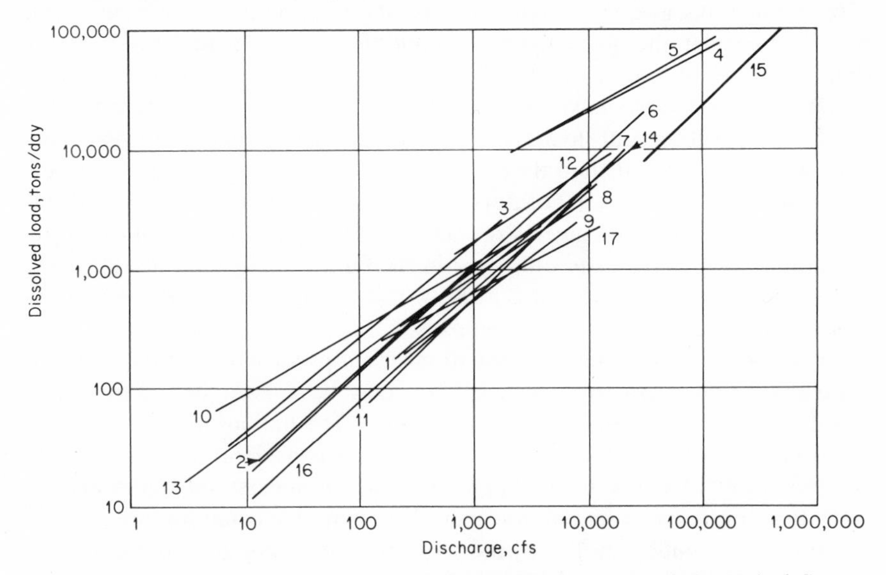

Fig. 9.23 Discharge vs. dissolved load relationship for streams of the United States. [*After* LEOPOLD *et al*. (*1964*).] (From Luna B. Leopold, M. Gordon Wolman, and John P. Miller, "Fluvial Processes in Geomorphology," W. H. Freeman and Company. Copyright *1964*.)

tractive force, however, increases slightly, which implies that rainfall reduces the tendency of flow to erode the channel bed, and results in reduced suspended-sediment transport. The reduction in shear is attributed to the momentum exchange from the flow to the raindrops in the zone affected by raindrop penetration.

9.5.4 DISSOLVED MATTER

Dissolved load, the material which has been eroded or corroded, but is transported in form of a solution, may, at times, be of considerable importance. The amount of the dissolved matter going through a cross section depends not only on the water discharge but also on various other hydrologic and geologic parameters. For a selected number of watercourses in the United States, LEOPOLD et al. (1964) have established a discharge-dissolved load relationship, as reproduced in Fig. 9.23. In all cases there exists a trend of decreasing concentration as the discharge increases. It was also shown that in zones of dry climate only a small part (approximately 9 percent) of the total load is dissolved, while this percentage increases (approximately 37 percent) with increase in annual runoff.

9.6 CONCLUDING REMARKS

In this chapter, equations to determine the bed material load were discussed. These equations give the maximum capacity that the flume or watercourse can carry under the given hydraulic conditions. The actual bed material load may be, at times, different from the transporting capacity. It may happen that a flood removes most of the sedimentary material from the bed and, under subsequent hydraulic conditions, the streams transport considerably less material than they could carry if the supply were not partially exhausted. Consider a laboratory flume with a sediment feeder. If the amount of sediment of a given size fraction is increased, there exists a certain limit above which deposits will occur in the flume. This limiting rate is the sediment-transporting capacity. Below this limiting rate the channel experiences no deposits; above it, deposits will occur.

In using the bed material equations for a particular problem, extreme care should be taken to select the ones that have been developed under conditions similar to the problem presented. The limitations of each set of equations ought to be realized and taken into consideration and the subsequent results viewed in this light. It must be understood that wherever washload plays an essential role, the bed material equations are merely helpful for the understanding of the problem, but do not give correct results.

Researchers have only recently paid attention to the importance of hydrological influences on the sediment problem. The beginning is encouraging and further fruitful results are to be expected.

10
The Regime Concept

10.1 INTRODUCTORY REMARKS

Uniform flow in an open channel with rigid boundaries is sufficiently described by a single equation, such as the Kutter or Manning formula. For any given discharge conveyed through a given channel, one and only one flow depth—the uniform one—will establish itself. An open channel with rigid boundaries has one degree of freedom.

Uniform flow in an open channel with loose and movable boundaries can only be described with a set of three independent equations. For any given discharge conveyed through a canal cut into movable material (soil), a flow depth will establish itself which depends on the adjusted slope and width, two quantities which in themselves are dependent on the discharge. An open channel with loose boundaries has thus three degrees of freedom.

According to BLENCH (1961) a fourth degree of freedom develops if the canal—and especially a river—is left all by itself. Artificially straight sections are found to be unstable, and erosive attacks at the sides will increase and ultimately develop into meanders.

A uniform system with one degree of freedom will find itself in a steady condition in a relatively short time. However, any system with more than one degree of freedom will take considerable time—depending on the number of degrees—until equilibrium is reached. Researchers in this field have chosen to replace the word *equilibrium* with *regime.* Some controversies have existed and, at times, still exist about the proper definition of *regime.* INGLIS (1949) gives this definition:

> Channels which do not alter appreciably from year to year—though they may vary during the year—are said to be in regime. . . .

BLENCH (1961, p. 437) says:

> *Regime* suggests considerable freedom of individual behavior within a framework of laws and has no short-period connotation . . . the term *regime channel* will be used, meaning that it is capable of acquiring regime, or equilibrium eventually by self-adjustment of its nonfluid boundaries, if the imposed conditions do not change on a long-term average . . . .

The relationship between the discharge (of water and of sediments) to be conveyed and the channel geometry to be established in the soil material must be subject to investigation. Such a study should provide a set of three equations, such that the problem is completely defined and can thus be solved. In search for the three equations, we could explore the possibility of obtaining them analytically in a physical or rational way, or empirically by analyzing available data with the laws of statistics. The first approach, the rational one, was the subject of discussion throughout Chaps. 6 through 9. This chapter is concerned with the empirical approach. Three empirical equations, known as the *regime formulas*, have been put forward. This set of formulas is frequently referred to as the *regime theory* or as the *Indian approach* throughout the technical literature.[1] The regime concept has enjoyed great popularity among Anglo-Indian engineers because it provides rather simple design criteria for silt-laden channels, but it also has been subject to attack for its lack of rationale and physical rigor.

In the following sections we shall first discuss the regime concept of canals and then the regime concept for rivers. At this point, we shall pay some attention to the meander problem and to the longitudinal profile of watercourses.

[1] The *regime theory* is at best an empirical theory. It is also called the *Indian approach*, because most of the field studies were done on the Indian subcontinent.

10.2 CANALS IN REGIME

In a canal the three degrees of freedom are given by the breadth, depth, and slope. For any given discharge, the geometry of the canal could be defined if three independent equations were known. The regime concept uses data of canals in regime to establish the three independent relations. To deduce these relations from natural channels, such as gulleys or rivers, is hopeless; since any natural watercourse has one additional degree of freedom, it meanders. Straight-river stretches are rare, especially those remaining in equilibrium. The possibility to obtain these relations in laboratory channels is difficult, because the usual spatial and time limitations of models and the meager knowledge of sedimentary model laws will not allow proper or adequate interpretation of the results. Compelled by the need for a better understanding of sediment-bearing channels and having irrigation canals available for deduction, British engineers were the first to advance regime formulas. The following was done according to BLENCH (1957, p. 12):

> . . . laws of self-formation were found by many years of observation in the field on a vast assemblage of canals that were virtually regime rivers simplified by controlled discharges, maintenance (but not forcing) of self-formed banks, and elimination of meandering . . . .

The majority of field observations occurred in Pakistan and India, while some data have been obtained in Egypt. The entire *Indian* canal system is discussed in some detail by BLENCH (1957). A typical canal section is trapezoidal with a side slope of about two to one. The bottom is made up of particles of $0.10 < d < 0.60$ mm, and the sides are of claylike material. The breadth-depth ratio was $4 < B/D < 30$, and the discharge was $1 < Q < 10{,}000$ cfs.

10.2.1 KENNEDY'S STUDY ON THE PREVENTION OF SILTING

In the 1880s the irrigation canals in the Punjab (India, Pakistan) were designed and excavated according to information obtained with a flow formula by Kutter, but breadths and slopes were used rather arbitrarily. After self-adjusting of the three degrees of freedom, namely, the breadth, slope, and depth, had taken place, excessive sedimentation was observed. This shortcoming of a proper design warranted KENNEDY (1895) to study the problem. Twenty-two channels of the Upper Bari Doab were selected, which had permanent and established cross sections, and neither silting nor scouring of the bed occurred. With these data KENNEDY (1895) proposed an empirical relationship between the mean velocity $\bar{u}$ and the depth $\bar{D}$, such as[1]

$$\bar{u}_{cr} = 0.84\bar{D}^{0.64} \tag{10.1}$$

[1] Unless mentioned otherwise, the equations in this chapter are in British units.

The velocity $\bar{u}_{cr}$ was referred to as the "critical velocity" at which, for a given depth $\bar{D}$, silting is just prevented; but what apparently is meant by "silting is just prevented" is "maintaining movement of the entire sediment load without eroding the channel."

Together with the flow formula of Kutter, Eq. (10.1) represents a set of two equations, still short of one equation to satisfy the three degrees of freedom. In passing, it should be remarked that KENNEDY (1895) implied that the multiplier and exponent in Eq. (10.1) may vary slightly, but any such variation will be small. It appears, however, that KENNEDY (1895) did not think it necessary to introduce a further equation to define properly the entire problem of a self-adjusting channel. Subsequently, Kennedy reassessed his research, which then made it permissible to design either a narrow, deep canal or a wide, shallow one to carry the same discharge (which is, of course, incorrect), and he gave a "rough-rule" relation for the ratio of depth to breadth. This presented, indeed, the lacking third equation and the proper design was at least basically assured.

Kennedy's equation, given with Eq. (10.1), represented to many a very prosperous approach to the sediment problem. Engineers in South India, Burma, Egypt, and in other countries adopted its general form and modified its constants to make it suitable for their own locality.

10.2.2 LINDLEY'S STUDY ON REGIME CHANNELS

An important step toward the establishment of the three regime equations was made by LINDLEY (1919). Investigations of 786 observations in the Lower Chanab Canal suggested new coefficients in the Kennedy equation, or

$$\bar{u}_{cr} = 0.95\bar{D}^{0.57} \tag{10.2}$$

It was also recognized that the bed as well as the banks may scour or fill, and an equation defining the breadth was suggested,

$$\bar{u}_{cr} = 0.57B^{0.355} \tag{10.3}$$

where B is the averaged breadth between banks. Equations (10.2) and (10.3) together with Kutter's flow formula prove the existence of a single solution for each channel in regime.

Research on Egyptian canals was presented by Molesworth et al. (1917), by Buckley (1919?), and by Kinder (1919), and is discussed in LELIAVSKY (1955). All of them have in common a Kennedy-type relationship and a depth-breadth-slope relationship. Further contributions to the regime concept have been reported by Woods (1927), Griffith (1927), and Bottomley (1928); they are also briefly discussed by LELIAVSKY (1955). The next important contribution was introduced by LACEY (1929).

10.2.3 LACEY'S CONTRIBUTIONS

Rather than producing further data, Lacey was placed on special research duty to bring some order into the mass of available data. Lacey set out to prove Lindley's theorem "that the dimensions, width, depth, and gradient of a channel to carry a given supply loaded with a given silt charge were all fixed by nature" and, thus, can be uniquely determined.

LACEY (1929) reanalyzed many data in the light of Eq. (10.1), but found that by changing the constant and exponent all data could be reasonably well represented with the following relation:

$$\bar{u}_{cr} = 1.17\sqrt{f_L R_h} \tag{10.4}$$

where R_h is the hydraulic radius and f_L is a defined silt factor. The expression[1]

$$Q f_L^2 = 3.8(\bar{u}_{cr})^6 \tag{10.5}$$

is the second relation and fits the data with remarkable accuracy. Taking Manning's flow formula and solving it for the slope in terms of previously defined quantities, the following was given:

$$S = \frac{f_L^{3/2}}{2587Q^{1/9}} \tag{10.6}$$

Equation (10.6) was altered in a reply to the discussion of the paper thus:

$$S = \frac{f_L^{5/3}}{1788Q^{1/6}} \tag{10.6a}$$

With Eqs. (10.4), (10.5), and (10.6),[2] it is possible, provided discharge and silt factor are known, to find the hydraulic variables for a stable channel. LACEY (1929) introduced a relationship for the determination of the silt factor, or

$$d = f_L^2/64 \qquad \text{in.} \tag{10.7}$$

where d is the diameter of the predominant type of sediment transported. A rather remarkable relation is obtained between Eqs. (10.4) and (10.5), since

$$P = 2.67\sqrt{Q} \tag{10.8}$$

where P is the wetted perimeter. Thus for a given discharge the wetted perimeter P of the channel is constant and independent of the slope and the sediment; however, the latter does influence the shape of the channel.

In subsequent contributions by LACEY (1935), LANE (1937), PETTIS (1937), etc., it was found that neither coefficients nor exponents are true constants, but depend on the locality of the canal. This must be expected, because the regime equations are not based on rational deductions but

[1] Q is the design discharge of the canal. In ordinary canal design and practice, major variations of discharge may be avoided and, indeed, are rather unusual.

[2] Many regime equations have been introduced, even by LACEY (1929, 1935, 1946) himself. Although they may appear on first sight to be quite different, all of them can usually be traced back to the general form of LACEY's (1929) original set of equations, given by Eqs. (10.4) through (10.6).

rather are a set of equations of an empirical nature. However, it should be mentioned that LACEY (1937) made an effort to apply the method of dimensional analysis in order to derive the regime equations. With the use of the Froude and Reynolds numbers, the regime equations were given as

$$f_L \propto \frac{\bar{u}^2}{gR_h}$$

$$\frac{P}{R_h} \propto \frac{\bar{u}}{(\nu g)^{1/3}} \tag{10.9}$$

$$\bar{u} \propto \left(\frac{\bar{u}R_h}{\nu}\right)^{1/6}\left(\frac{\bar{u}^2}{gR_h}\right)^{-1/3}\sqrt{gR_hS}$$

where ν is the kinematic viscosity and g is the acceleration of gravity.

CHIEN (1955) has investigated why the sediment load was omitted as an explicit variable in the regime equation. As to be expected, it was found that the silt factor f_L does include implicitly the sediment load. The functional relationship between the silt factor f_L and the sediment loads is given by CHIEN (1955).

Only recently ACKERS (1964) presented results of an experimental endeavor on model streams in alluvial material. The experiments, done with sands of median diameter of 0.16 and 0.34 mm, covered a range of discharge of $0.4 < Q < 5.4$ cfs. It was concluded that area and velocity relations are similar to Lacey's equations. However, the slope vs. discharge plot showed strong scatter and no correlation was established; plotting Eq. (10.6*a*) with $f_L = 1$, all the data were found to lie above this equation.[1]

A further confirmation to Lacey's equation relating discharge and wetted perimeter is reported by STEBBINS (1963). This is remarkable, because the scale of the test was small, with a discharge range of $3.1 < Q < 254.5$ in.^{3}/sec.

In conclusion we should stress that the three regime equations, as developed by Lacey, are empirical relationships; the many data used cover a relatively narrow range of the parameters involved (in a rather limited geographic location, the Punjab). The general application of these equations to conditions outside the range of the original data is questionable and often dangerous.

10.2.4 BLENCH'S CONTRIBUTIONS

Over the period of the last 25 years, Blench has been the staunch supporter of the regime concept. By way of an introduction, let us mention the conditions that must be fulfilled for either Lacey's or Blench's equation to apply.

[1] Simons (1969) is of the opinion that "in systems this small there is an undefined effect which seems to change the relation. This scale effect needs further discussion and probably further study."

According to BLENCH (1957, p. 28) the regime equations are applicable to those canals:

(*a*) That are straight.
(*b*) Where sides behave as if hydraulically "smooth." Technically, this means that the roughness constant of the sides depends only on the nature of the water-sediment complex.
(*c*) Where bed width exceeds three times the depth.
(*d*) Where sides stand at slopes approximating those found for cohesive sides in nature.
(*e*) Where discharges are steady.
(*f*) Where sediment load, however complex, is steady.
(*g*) That move a non-cohesive load, however small, along their bed in dune formation.
(*h*) That run at speed less than critical, i.e., less than the speed of small gravity waves.
(*i*) Where sediment size is small compared with depth of flow.
(*j*) That have adjusted their width, depth, and slope to final (or regime) values. (The implication here is that, except by the remotest chance, a channel of the type contemplated will not be designed to carry its sediment load and discharge without some addition to, or removal from, its boundaries; therefore self-adjustments will occur till regime is attained.)

The earlier contributions by Blench (1941) and King (1943) focused on a detailed study of Lacey's silt factor f_L. The effects of sides and bed were separated out. The original regime equations became thus generalized, and according to BLENCH (1961) may be expressed as follows:

The first equation is

$$F_B = \frac{\bar{u}^2}{D} \tag{10.10}$$

defining the bed (sediment) factor F_B, and known as the bed factor equation. BLENCH (1957) has remarked that a dimensionless form, obtained by dividing both sides of Eq. (10.10) by the gravitational constant g, is a Froude number in terms of depth; this in turn may be compared with the first equation in Eqs. (10.9).

The second equation is

$$F_S = \frac{\bar{u}^3}{A/D} \tag{10.11}$$

defining the side factor F_S, and known as the side factor equation. It was suggested by BLENCH (1951) that multiplying with $\rho^2\nu$, the right side appears as the square of shear stress.

The third equation is

$$\frac{\bar{u}^2}{gDS} = 3.63(1 + a_B\bar{C}_s)\left(\frac{\bar{u}A/D}{\nu_m}\right)^{1/4} \tag{10.12}$$

and is frequently referred to as the regime slope equation,[1] where

S = slope of the canal
$\bar{C}_s$ = bedload charge, in parts per 100,000 of weight
ν_m = viscosity of the fluid-solid mixture

Since Eq. (10.12) was originally given for small values of $\bar{C}_s$, the expression within the parentheses was unity. The present form of Eq. (10.12) was introduced by BLENCH et al. (1957), who recommended that $a_B = \frac{1}{233}$, a result which remains purely empirical due to lack of appropriate data. The flow formula, as Eq. (10.12) may be called, is according to BLENCH (1957) a generalization of the flow formula for smooth conduits (after Blasius), and has been subject of much discussion. A remote similarity between the third equation in Eqs. (10.9) and the present form of Eq. (10.12) is noticeable.

Equations (10.10) through (10.12) determine the problem of three degrees of freedom completely; however, in their forms, they are a bit cumbersome to use for design problems. Presented in a more practical form, we obtain for the average width,

$$\frac{A}{D} = \sqrt{\frac{F_B}{F_S}Q} \tag{10.13}$$

For the mean depth, we obtain

$$D = \sqrt{\frac{F_S}{F_B^2}Q} \tag{10.14}$$

and for the slope,

$$S = \frac{F_B^{5/6}F_S^{1/12}\nu^{1/4}}{3.63(1 + a_B\bar{C}_s)gQ^{1/6}} \tag{10.15}$$

From these equations we obtain for the flow velocity,

$$\bar{u} = \sqrt[6]{F_BF_SQ} \tag{10.16}$$

BLENCH (1957, 1966) elaborated on the use of the regime equation for design purposes by performing many sample calculations; the design engineer should by all means consult these pages to gain a better feeling for the regime concept.

[1] That the sediment discharge be included into the regime equations was pointed out by LANE (1937) and introduced by INGLIS (1948).

What remains is a discussion of estimating the side factor F_S and the bed factor F_B. Only rough guide lines are available to estimate the side and bed factors. BLENCH (1957, 1964) has suggested the following formula for the bed factor:

$$F_B = 1.9\sqrt{d}(1 + 0.12\bar{C}_s) \tag{10.17}$$

where d is the median bed material size by weight in millimeters. Notice that, for small bedload transport, the bed factor depends on the grain size only. Further research on this and related issues is reported by BLENCH et al. (1964). As far as the side factor is concerned, BLENCH (1957) recommends values of 0.1, 0.2, and 0.3 for loams of slight to high cohesiveness. In addition, the engineer might be well advised to select a test stretch of the canal and investigate both the side factor F_S and the bed factor F_B all by himself.

The regime equations by Blench seem to have greater generality than those by Lacey, because they are based on (limited) field data (Indian canals) as well as on laboratory data (Gilbert's experiments). Furthermore, interesting contributions were made by INGLIS (1948) who derived a similar set of regime equations.

10.2.5 SIMONS' ET AL. CONTRIBUTIONS

An important step toward a further generalization of the regime concept was the study of SIMONS et al. (1963). Based on the lack of application of the regime equations in the United States, a field study was undertaken, and its results together with a wealth of other published results were then analyzed.

The range of conditions of the data analyzed by SIMONS et al. (1963) is larger than was ever considered before, and is given in Table 10.1. The analysis of the data confirmed the general shape of the regime equations, as advanced in previous sections; however, as far as coefficients and exponents are concerned, it was found that both depend on the types of canal beds and

Table 10.1 Data from different canals [*after* SIMONS *et al.* (*1963*)]

Data	*Indian canals*	*San Luis Valley, Col.*	*Imperial Valley, Cal.*	*Irrig. canals, Wy., Col., Neb.*
Number of reaches	70	15	4	24
Q, cfs	5–9,057	17–1,500	—	43–1,039
S, ‰	0.06 –0.34	0.97–9.7	—	0.058–3.87
d, mm	0.035–0.43	20.0 –80.0	—	0.028–7.60
Sediment concr. ppm	156–3,590	—	2,500–8,000	—

banks. In SIMONS et al. (1963) investigation, five different types of geotechnical channel conditions were differentiated; a separate equation was fitted to each of these types, first graphically and then analytically, by HENDERSON (1966). The equations of regime are given in a most general form in the following:

For the average width,

$$\frac{A}{D} = 0.9K_1\sqrt{Q} \tag{10.18}$$

For the average depth,

$$\begin{aligned} D &= 1.21K_2Q^{0.36} \qquad \text{for } R_h \leq 7 \text{ ft} \\ D &= 2.0 + 0.93K_2Q^{0.36} \qquad \text{for } R_h \geq 7 \text{ ft} \end{aligned} \tag{10.19}$$

For a slope relation,

$$\frac{\bar{u}^2}{gDS} = K_4\left(\frac{\bar{u}A/D}{\nu}\right)^{0.37} \tag{10.20}$$

Needless to say, the similarity to Eqs. (10.13), (10.14), and (10.12), respectively, is noticeable. The coefficients K_1, K_2, and K_4 are summarized in Table 10.2; they are presented in HENDERSON (1966), and are given with slightly rounded-off values.

It must be agreed upon that Eqs. (10.18) through (10.20) provide a useful design procedure. With the known discharge and the geotechnical channel conditions, all other values can be obtained in a straightforward manner. Quite correctly, SIMONS et al. (1963) point out that regime canals can carry only a limited (total) bed material load (< 500 ppm), but are capable of carrying a considerable washload. The design of canals with large bed material loads may be accomplished with any of the total load equations discussed in Chap. 9.

Table 10.2 Coefficients for Eqs. (10.18), (10.19), and (10.20) [*after* HENDERSON (*1966*)]

Types of material	K_1	K_2	K_4
Sand bed and banks	3.5	0.52	0.33
Sand bed and cohesive banks	2.6	0.44	0.54
Cohesive bed and banks	2.2	0.37	0.87
Coarse noncohesive material	1.75	0.23	—
Sand bed and cohesive banks with heavy sediment load	1.70	0.34	—

10.2.6 CONCLUDING REMARKS

It seems appropriate to quote BLENCH (1966, p. 79) when he speaks of the regime concept as one which has

> . . . come to imply an inductive theory of channel self-formation that has grown systematically since about 1890, in terms of appreciation of canal self-adjustment and through measurements by engineers in the field (mainly in the Indian continent), has a quantitative orientation, and uses the functional forms of the equations of LACEY (1929, 1933) in terms of directly measurable variables . . . .

Of all the available regime equations, those developed by BLENCH (1961) and by SIMONS et al. (1963) have the greatest generality and are, therefore, recommended for use. The applicability of the regime equations is limited to canals only. The various points cited in Sec. 10.2.4 should be fulfilled for all practical purposes; for strong bed material load the regime concept is not well suited.

10.3 RIVERS IN REGIME

10.3.1 REGIME RIVER VS. REGIME CANAL

It has been said that a canal has three degrees of freedom while a river has four degrees of freedom. However, in reality, any channel with erodible sides, whether canal or river, has four degrees of freedom. The fourth degree of freedom is caused by the channel's tendency to meander. The fact that a canal is straight or only mildly curved, and thus has only three degrees of freedom is due to continuous engineering maintenance and/or gentle control of the canal's alignment. Straightness of channels is transient, abnormal, and maintained artificially. Furthermore, there exists another important difference between a river and a canal; the canal discharge can be kept fairly constant, but the river discharge is subject to fluctuations which may be considerable.

A canal may be considered, at least for many purposes, a special or simple kind of river.[1] Therefore any knowledge gained in an experience with canals must prove helpful for the river engineer. A set of three equations was derived for the regime canals. According to BLENCH (1957), direct river observations due to changes of discharge and meandering are not very suitable for establishing laws for rivers in regimes, although they give, at times, confirmations of those regime equations developed for canals.

A qualitative discussion of rivers as a result or cause of a fluvial process in geomorphology is beyond the scope of this treatise. Many books on geomorphology are entirely or in part devoted to this issue. LEOPOLD et al.

[1] Scale effects may be of importance especially in large rivers.

(1964) and SCHEIDEGGER (1961), two of the most recent contributions, and references cited in these might help the interested reader to get started within this fascinating field.

10.3.2 REGIME EQUATIONS FOR WIDTH, DEPTH, AND VELOCITY

The regime equations in the most general form for width, depth, and velocity are given by

$$B = C_B Q^{\alpha} \tag{10.21}$$

$$D = C_D Q^{\beta} \tag{10.22}$$

$$\bar{u} = C_{\bar{u}} Q^{\gamma} \tag{10.23}$$

Notice the similarity of Eqs. (10.21), (10.22), and (10.23) with Eqs. (10.13), (10.14), and (10.16), as given by BLENCH (1957). It should be stated in advance that no unique relationship may be given for these equations, or for the constants C_B, C_D, and $C_{\bar{u}}$ as well as for the exponents α, β, and γ which depend, at least up to the present state of research, on the stream and the location of the gaging station where the data are obtained. Not many investigations have been made to establish these relations, and it appears worthwhile to collect these data in form of a table. Table 10.3 includes the numerical values of the exponents α, β, and γ, the location where the data were obtained, the reporter of data, and other pertinent remarks. Also indicated are, for sake of comparison, the exponents of the regime equations for canals due to BLENCH (1957), Eqs. (10.13), (10.14), and (10.16).

Due to the principle of continuity, it follows that

$$(BD)\bar{u} = Q = (C_B Q^{\alpha} C_D Q^{\beta}) C_{\bar{u}} Q^{\gamma} \tag{10.24}$$

and

$$\alpha + \beta + \gamma = 1$$

$$C_B C_D C_{\bar{u}} = 1 \tag{10.25}$$

It remains to test the validity of these relations with data obtained in the field.

One of the most important investigations was reported by LEOPOLD et al. (1953). A large variety of rivers in the Great Plains and the Southwest of the United States, where semiarid conditions predominate, were analyzed. Seeking the average values of α, β, and γ for 20 river cross sections, the following was obtained:

$$\alpha = 0.26 \qquad \beta = 0.40 \qquad \gamma = 0.34$$

This is tabulated in Table 10.3. A rather typical example of such relations is shown in Fig. 10.1. It must be remembered that these exponents depend on the geometry and the geotechnical parameters of the stream channel, and thus are bound to vary from stream to stream. LEOPOLD et al. (1964) studied the geographic distribution of these exponents within the United

Table 10.3 Summary of regime relations for hydraulic geometry of rivers

Reporter	*B*	*D*	$\bar{u}$	*Location*	*Remarks*
LEOPOLD et al. (1953)	$Q^{0.26}$	$Q^{0.40}$	$Q^{0.34}$	20 river cross sections representing a large variety of rivers in the Great Plains and the Southwest of the U.S. (semiarid conditions)	Variations of hydraulic characteristics in a particular cross section
LEOPOLD et al. (1964)	$Q^{0.04}$	$Q^{0.41}$	$Q^{0.55}$	Brandywine Creek, Pennsylvania (humid eastern U.S.)	Variation of hydraulic characteristics in a particular cross section
LEOPOLD et al. (1964)	$Q^{0.12}$	$Q^{0.45}$	$Q^{0.43}$	Average of 158 gaging stations in the U.S.	Variation of hydraulic characteristics in a particular cross section
LEOPOLD et al. (1953)	$Q^{0.5}$	$Q^{0.4}$	$Q^{0.1}$	20 river cross sections representing a large variety of rivers in the Great Plains and the Southwest of the U.S. (semiarid conditions)	Variation of hydraulic characteristics in a downstream direction; for mean annual discharge
LEOPOLD et al. (1964)	$Q^{0.13}$	$Q^{0.41}$	$Q^{0.43}$	10 gaging stations on the Rhine River	Average at a station relation
LEOPOLD et al. (1956)	$Q^{0.5}$	$Q^{0.3}$	$Q^{0.2}$	Ephemeral streams in semiarid region of New Mexico, U.S.	Variation of hydraulic characteristics in a downstream direction; for mean annual discharge
NASH (1959)	$Q^{0.53}$	$Q^{0.27}$	$Q^{0.23}$	British and U.S. data from NIXON (1959)	For bankfull discharge
Rybkin (1947)	$Q^{0.57}$	$Q^{0.22}$	$Q^{0.21}$	Rivers in the Upper Volga and Oka Basin	Mean long-term discharge
LANGBEIN (1964)	$Q^{0.53}$	$Q^{0.37}$	$Q^{0.10}$	—	Theoretically
BLENCH (1957)	$Q^{0.50}$	$Q^{0.33}$	$Q^{0.17}$	—	Regime equations for canals

States. Considering the width relationship and its exponent α, it was concluded that humid eastern and wet mountain areas have lower α values than semiarid areas (see Table 10.3). Average values for the exponent for 158 gaging stations in the United States are given in LEOPOLD et al. (1964), and may be found in Table 10.3.

The set of regime equations, Eqs. (10.21) through (10.23), may also be used in a study of the variation of hydraulic characteristics in a downstream direction. Yet a comparison between two or more cross sections will only be meaningful if we can agree to investigate a flow of a given recurrence interval, e.g., peak flood, bankfull flow, mean annual flow, etc. LEOPOLD et al. (1953) have used the mean annual discharge in the investigation of a great number of streams of the midwestern United States. The values of the exponents in the regime equations (in a downstream direction) are summarized in Table 10.3. A typical plot of these relations from measuring stations within a river basin is given in Fig. 10.2. From Fig. 10.2 it becomes clear that, proceeding downstream in a given river, the discharge increases

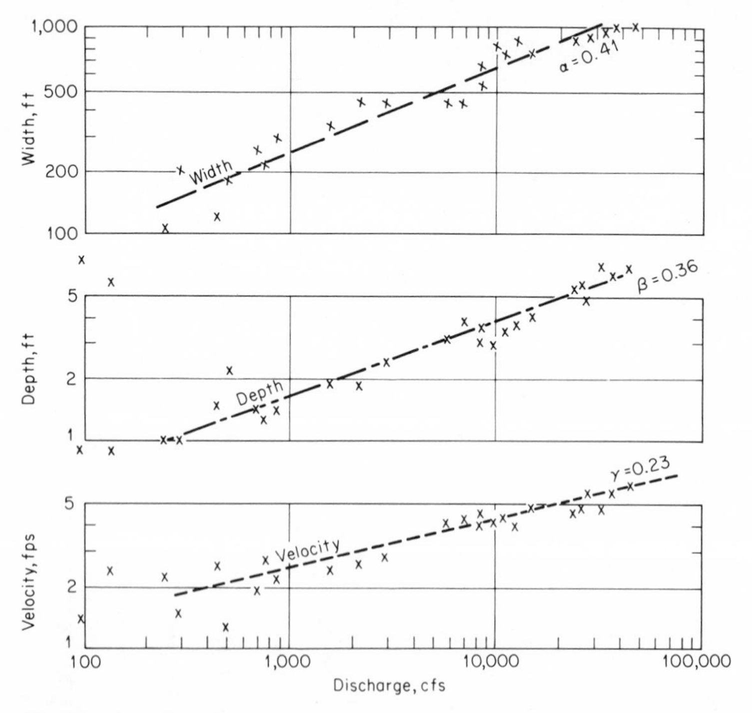

Fig. 10-1 Width, depth, and velocity vs. discharge at a gaging station, Cheyenne River near Eagle Battle, North Dakota. [*After* LEOPOLD *et al. (1953)*.]

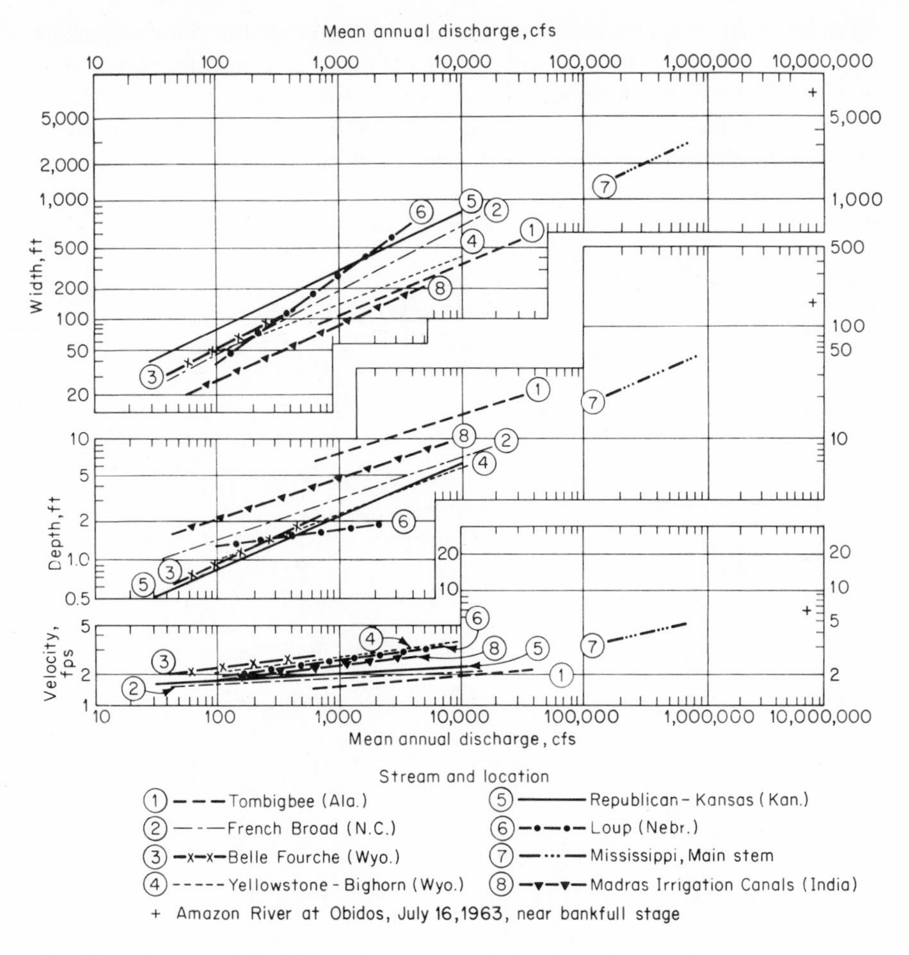

Fig. 10.2 Width, depth, and velocity vs. mean annual discharge, with the discharge increasing in downstream direction for various rivers. [*After* LEOPOLD *et al.* (*1953*).]

due to an increase in the drainage area; the way this increase takes place is remarkably similar for most streams. Furthermore, whereas the depth and width increase considerably, the velocity increases but slightly, a fact which is explained by the decrease of the slope as one proceeds from the headwater to the lower reaches.

In addition to the regime equations for depth, width, and velocity, various attempts have been made to correlate the sediment load in a similar way. This was briefly discussed in Sec. 9.5.1, and a relation such as

$$G_{ss} = pQ^j \tag{9.29}$$

was given for the suspended sediment load G_{ss} at a river cross section. Very little information on the exponent is available, but limited field investigations for rivers in the western United States indicate the following range of j, namely, $2 < j < 3$. A typical plot is given in Fig. 9.19.

The relations between the results obtained at a cross section and along a river are summarized in Fig. 10.3, which represents the average values of

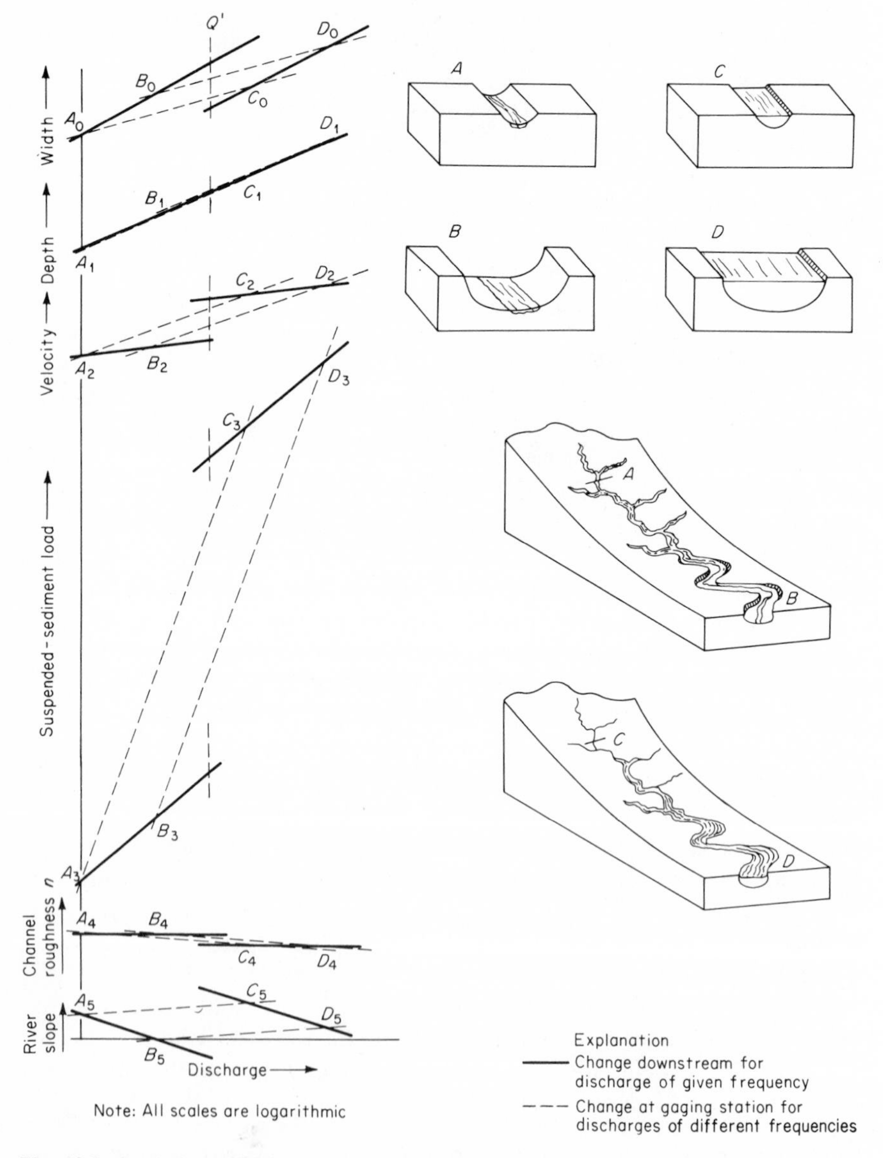

Fig. 10.3 Variation of hydraulic characteristics at a cross section of a river and in downstream direction. [*After* LEOPOLD *et al.* (*1953*).]

the exponents in the study by LEOPOLD et al. (1953). Sections A and C are at the headwater at low and high flow, respectively, while sections B and D are downstream from the headwater at low and high flow, respectively; this is indicated at the right-hand side of the diagram. In addition to the width, depth, and velocity relations, there are plots of the relations for suspended load, roughness coefficient, and channel slope. A detailed discussion and its usefulness for practical applications of Fig. 10.3 was given by LEOPOLD et al. (1953), to which the interested reader is now referred.

In another investigation, LEOPOLD et al. (1956) examine the regime relation for ephemeral streams, i.e., which carry water during storms only. The data are given in Table 10.3, and indicate that in ephemeral streams velocity increases downstream at a faster rate than in perennial streams.

A study on the bankfull discharges of rivers in England and Wales is reported by NIXON (1959). It was assumed a priori that the discharge affects the geometry of the channel, as given by Eqs. (10.13), (10.14), and (10.16). The exponents were thus given and the resulting coefficients were determined, which are, of course, of limited use, applicable only to rivers in England and Wales. However, in a subsequent discussion by NASH (1959) it was correctly pointed out that a least-square fit to the data results in different exponents than the original paper suggested; these exponents are included in Table 10.3.

Some Soviet contributions are summarized and discussed in KONDRAT'EV (1962). Rivers in the Upper Volga and Oka Basin have been studied by Rybkin (1947), and regime equations, similar to Eqs. (10.21) to (10.23), were derived. The numerical values of the exponents are given in Table 10.3, with the discharge value as the mean long-term discharge. The other Soviet studies discussed by KONDRAT'EV (1962) appear to be of rather limited value due to lack of complete description of the data used for the analysis and improper description of its symbols.

In an interesting contribution, LANGBEIN (1964) makes use of the probabilistic approach to the study of river geometry. Three physical equations, namely, a continuity, a friction, and a sediment equation, are written, and their exponents are compared. Under the assumptions that the rate of work in the whole system is extremely small, and that the rate of work per unit bed area is uniform, a minimal principle of least squares of the exponents is applied, and numerical values for the exponents are found. The agreement between the results given in Table 10.3 and those for regime canals by BLENCH (1957) and for regime rivers by LEOPOLD et al. (1953) is very good.

Finally, it is noteworthy to report the suggestions by BLENCH (1961). As far as the discharge is concerned, BLENCH (1961) used the peak flood Q_{max}, and computed the *flood breadth* and the *flood depth* by replacing the discharge Q by the peak flood Q_{max} in Eqs. (10.13) and (10.14), respectively. The regime equations for the slope, Eq. (10.15), may be used for rivers, but allowance must be made for the fact that meandering changes the slope to some extent. In general, however, it must be stressed that extreme care must be taken when the canal equations are applied to river problems. The regime equations contain various coefficients such as F_S, F_B, and $\bar{C}_s$, which are applicable only for canals and under rather limited conditions. More recently, BLENCH (1966) has suggested to apply a meander coefficient to the slope equations.

10.3.3 MEANDERING OF RIVERS[1]

The foregoing sections focused on the discussion of the *longitudinal* movement of streams. *Transverse* movement accompanies the *longitudinal* movement, and this is expressed by the channel pattern which is everything but straight.

LEOPOLD et al. (1964) have recognized three somewhat different kinds of channel patterns—*meandering*, *braided*, and *sinuous*. They were able to distinguish among these patterns by introducing a ratio of channel length to valley length. If this ratio, called *sinuosity*, is equal or greater than 1.5, the river is said to meander, otherwise the river is sinuous. Meandering and sinuous rivers flow in a rather well-defined channel. The braided river, however, does not at all have a single or well-defined channel but a network of interlacing streams. According to BLENCH (1961), these numerous streams gradually reduce in number and develop into a sinuous or meandering river; braiding is considered as an incipient form of meandering. In an extensive study on river channel patterns, LEOPOLD et al. (1957) have arrived at some worthwhile conclusions. It was found that braided-river reaches are generally steeper, wider, and shallower when compared with undivided reaches carrying the same discharge. Furthermore, a rather useful but preliminary criterion between channel slope S and bankfull discharge Q has been advanced, such as

$$S = 0.06Q^{-0.44} \tag{10.26}$$

which distinguishes between meandering and braiding rivers. For any given discharge, meanders occur at smaller slopes; at larger slopes accompanied by strong sediment transport, braided streams are encountered. SHEN et al. (1969) suggested that due to strong sediment transport "the channel may widen so much and the depth may become so shallow that the flow cannot take place in a single channel," which in turn may result in a braiding channel. Equation (10.26), which was obtained from stream data of the midwestern United States, is not a criterion for sinuous rivers. A classification of river patterns different from LEOPOLD et al. (1964) was suggested by CHITALE (1970).

[1] The name *meander* stems from the Maeander River in Phrygia (Turkey), proverbial for its winding.

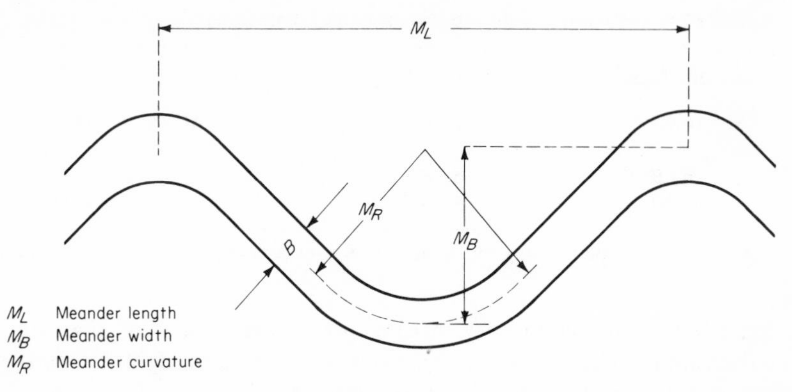

Fig. 10.4 Definition sketch for meanders.

As far as the geometry of meanders is concerned, important contributions have been made by JEFFERSON (1902), INGLIS (1938, 1947), and LEOPOLD et al. (1960). Three dimensions have been introduced to describe meanders. They are the length, width, and curvature of a meander, shown in Fig. 10.4.

A relationship between the width B of a river and the meander width was suggested by JEFFERSON (1902), who also pointed to the fact that limited data suggest that a ratio between the meander length and the meander width is of a constant value. The measurements of this study were subsequently used by INGLIS (1938) and BATES (1939). The analysis by INGLIS (1938) confirmed that both meander length and meander width vary with the discharge, and the meander length was given by

$$M_L = C_i\sqrt{Q} \tag{10.27}$$

where C_i is $15 < C_i < 30$. Furthermore, it was found that the ratio M_B/M_L is of the same order, or

$$\frac{M_B}{M_L} \approx 2.5 \tag{10.28}$$

for both incised[1] and flood plain meanders. Thus a discharge increase is accompanied by an increase in the meander length, and bends of meanders move downstream. Working with models, INGLIS (1947) reported a coefficient of C_i as $30 < C_i < 37$ and the ratio of M_B/M_L of the same order as given by Eq. (10.28).

LEOPOLD et al. (1960) found that the best-fitting empirical relations may be established between meander length or curvature and the channel width. They arrived at this conclusion after investigating all sizes of

[1] An incised meander is one whose free movement is restricted by the narrowness or absence of a flood plain. Incised meanders are sometimes called *fixed meanders*, while flood plain meanders are referred to as *free ones*.

Table 10.4 Meander geometry—empirical relations*

Meander length, M_L	*Meander width, M_B*	*Curvature, M_R*	*Reporter*
$10.9B^{1.01}$	$2.7B^{1.1}$	$M_L = 4.7M_R^{0.98}$	LEOPOLD et al. (1960)
$6.6B^{0.99}$	$18.6B^{0.99}$	—	INGLIS (1947)
$10.0B^{1.025}$	$4.5B^{1.00}$	—	ZELLER (1967)

* A summary supplementing this table with Soviet data may be found in ZELLER (1967*a*).

meanders, from laboratory streams to natural rivers of the size of the Mississippi. The important relations are summarized in Table 10.4, and compared with the results by INGLIS (1947).

The quality of the equations when compared with data may be examined in Fig. 10.5. Of considerable interest is that the foregoing relations are not limited to natural sediment-bearing rivers, but apparently also relate to the geometry of meltwater streams occurring on glacier surfaces and to the Gulf Stream, which is not confined by any boundaries. Recently, ZELLER (1967) investigated meandering channels in alluvial material, solid rock, glacier ice, and limestone lapies. The regression equations are given in Fig. 10.6 and are summarized in Table 10.4. This study is particularly worthwhile,

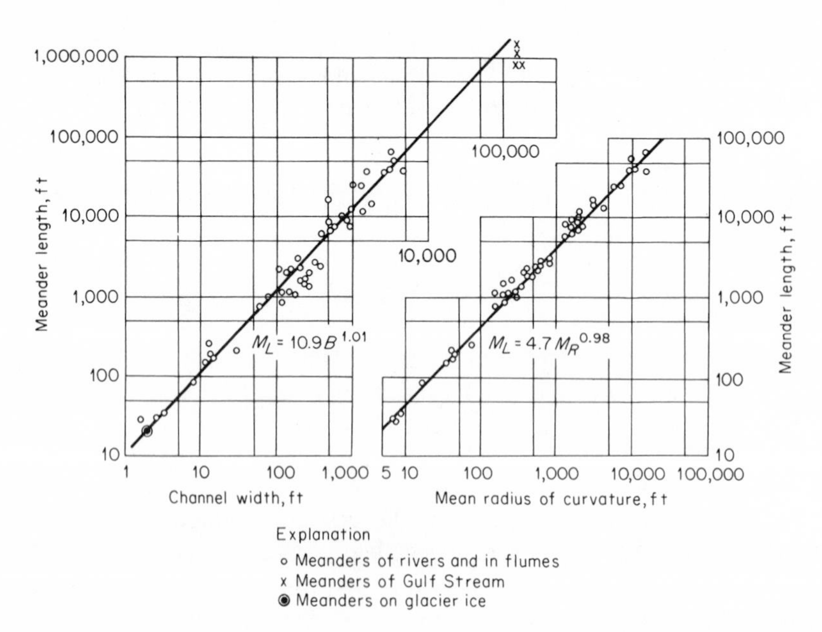

Fig. 10.5 Empirical relation for the geometry of meanders. [*After* LEOPOLD *et al. (1960)*.]

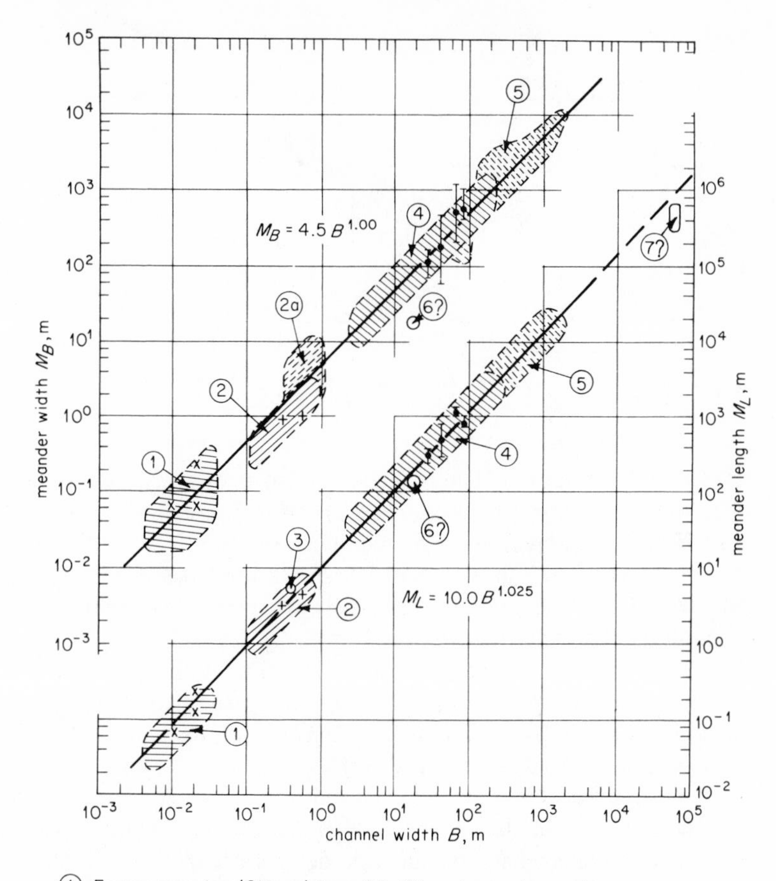

(1) Furrow meanders (Silbern) from 24 different meander reaches
(2) Ice meanders (Morteratsch – Glacier), 14 reaches
(2a) Ice meanders (East – Greenland), 12 reaches
(3) Alluvial meanders (VAWE – model – tests), 5 different tests in sand
(4) Meanders of Swiss Rivers, over 50 reaches
(5) Alluvial meanders, Leopold and Wolman's additional data, 46 reaches
(6) Meandering flow of density currents in a storage basin of River Reuss (basin width approx. 2 M_B), model tests
(7) Meandering flow of the Gulf Stream (Leopold and Wolman)

Fig. 10.6 Empirical relation for the geometry of meanders. [*After* ZELLER (*1967a*).]

because it not only includes further evidence of the existence of ice meanders and *free* meanders—exhibited by a density current flow in a storage basin—but also investigates the problem of miniature meanders. In a conclusion, ZELLER (1967) suggests that meandering may occur in any Froude- and Reynolds-number range, and may be accompanied by existence or absence

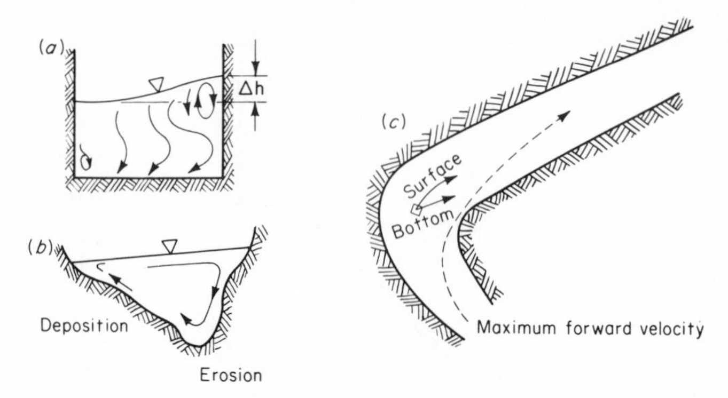

Fig. 10.7 Definition sketch of flow through a bend.

of bedload transport. The latter statement is of special interest, since both MATTHES (1941) and SCHUMM (1963) stressed the importance of sediment transport for the creation of a meander.

CHITALE (1970) has shown that the ratio of meander length to thalweg length and less the ratio of meander curvature to channel width are both dominant characteristics of river pattern. Data for 42 rivers, mainly Indian ones, were utilized to determine a statistical relationship.

The empirical relations give an idea as to what the geometry of any meander may be like. The field of mechanics, or hydraulics, of meanders, however, has not advanced very far.

Since meanders may be visualized as a sequence of bends, some useful information may be derived by a short discussion of bends. Flow of water in bends is discussed in considerable detail in ROZOVSKII (1957), LELIAVSKY (1955), RZHANITSYN (1960), SHUKRY (1949), and in textbooks on channel flow. We shall limit our discussion to available facts only. It is known that flow through a bend is accompanied by:

1. A superelevation of the water at the outside of the bend (Fig. 10.7*a*).
2. A strong downward current at the outside of the bend (Fig. 10.7*a*), which will cause erosion if at all possible (Fig. 10.7*b*).
3. A scour action at the outside of the bend and deposition at the inside of the bend, forming the characteristic bend cross section (Fig. 10.7*b*).
4. A maximum forward velocity being close to the inside of the bend of a rectangular unerodible channel (Fig. 10.7*c*).
5. A tendency that, at any location within the bend, a water or sand particle at the bed (bottom) moves usually stronger toward the center of curvature than a water or sand particle at the top (surface) (Fig. 10.7*c*).[1]

[1] As a consequence of this, it is advisable to construct sediment-free takeoff canals on the outside of a bend.

6. An extremely distorted and thus complicated velocity distribution throughout the bend.

According to the previously mentioned, a water or sand particle will move with a circulatory or spiral motion through the bend. Placing one bend after another, a very complicated flow pattern results. A typical meander may look like the one given in Fig. 10.8. The line of maximum depth, the thalweg, wanders from a deep pool at the outside of one bend over a shallow shoal (crossover) to another deep pool at the outside of the next bend, and so on. As a matter of fact, LELIAVSKY (1955) refers to this tendency of deepening and shoaling as the very tendency of meandering. Although a channel's alignment might be straight or curved, the thalweg migrates in a zigzag fashion over shoals and pools down the valley. ENGELS (1905) pointed this out when investigating German rivers; for United States streams, this is demonstrated by LEOPOLD et al. (1957). Figure 10.9 shows a sketch of a straight and a curved river stretch; both exhibit the meandering thalweg.

We shall now attempt to answer the inevitable question, Why does a straight river start to meander? Many opinions have been expressed, some still valid, others outdated, but still no definite answer is available. We list a few of the causes considered as the origin of a meander: (1) Local disturbances; (2) earth rotation; (3) excessive energy; (4) change in river stages; (5) forced oscillations; etc. However, most people will agree that due to superelevation a spiral flow is introduced and, thus, a nonsymmetric mass transport is produced. We now quote some of the foremost engineers engaged in the meandering problem.

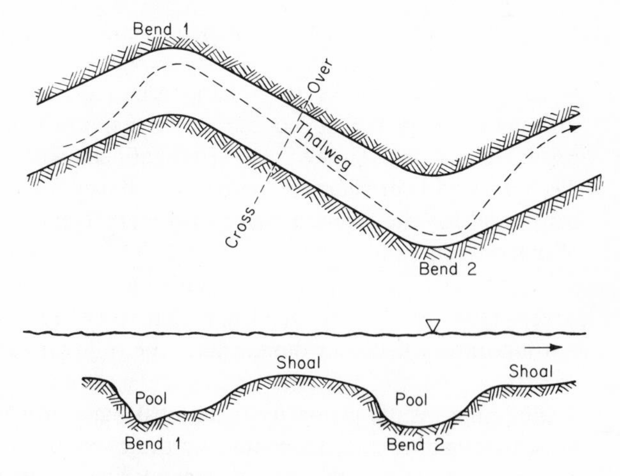

Fig. 10.8 Definition sketch of shoals and pools as they occur in a meander.

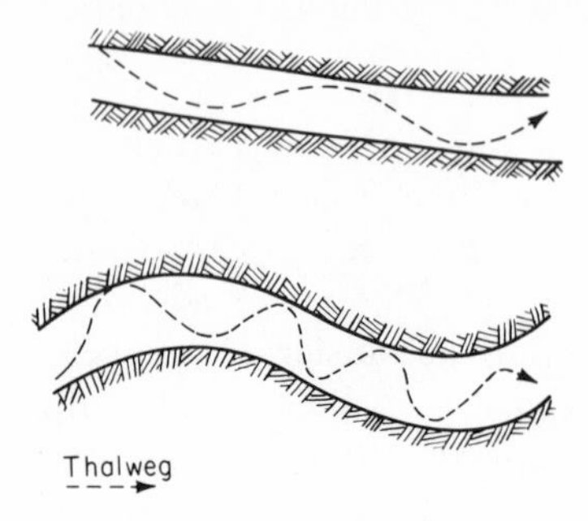

Fig. 10.9 Sketch of a straight and a curved river stretch.

LELIAVSKY (1955, p. 122) summarizes the European ideas—the major advocates of which were Fargue, deLeliavsky, and Engels—in his outstanding book on fluvial hydraulics:

> This centrifugal effect [which causes the superelevation] may possibly be visualized as the fundamental principle of the meandering theory; for it represents the main cause of the helicoidal cross-currents which removes the soil from the concave banks, transports the eroded material across the channel, and deposits it on the convex bank; thus intensifying the tendency towards meandering. It follows, therefore, that the slightest accidental irregularity in channel formation, turning as it does, the stream lines from their straight course may, under certain circumstances, constitute the focal point for the erosion process which leads ultimately to a meander.

INGLIS (1947), having had a vast experience in river hydraulics, says:

> Where, however, banks are not tough enough to withstand the excess turbulent energy developed during floods, the banks erode, and the river widens and shoals . . . . In channels with widely fluctuating discharges and silt charges, there is a tendency for silt to deposit at one bank, and for the river to move to the other bank. This is the origin of meandering . . . .

FRIEDKIN (1945, p. 3), who reported what some consider the most complete sets of laboratory meander data, makes the following statement:

> Meandering results primarily from local bank erosion and consequent local overloading and deposition by the river of the heavier sediments which move along the bed. Meandering is essentially a natural trading process of sediments from banks to bars . . . .

Once, for some reason, a local disturbance in the bank is initiated, it will start to grow into a bend. Further disturbances will develop as a result of impingement and deflection of the current, and these also will grow into a bend. This keeps on going until a sequence of bends exists; this is referred to as a stream meander. The meander, as such, has a tendency to move gradually downstream, a fact which can be observed with many meanders, given a certain time interval.

Since the origin of a meander is still subject to speculation, there can be no sufficiently convincing theory. Nonetheless, scientists have attempted to advance theories for meanders, which are rather limited in their scope. One contribution is due to EINSTEIN (1926). It was suggested that in straight sections the Coriolis forces play an important role. Since the friction of the river bed is responsible for a gradual change of these forces (toward the bottom), a circulation (spiral flow) is introduced. The effect of the earth rotation is thus expressed. This is known as the *Baer law*, i.e., rivers on the Northern Hemisphere have a tendency to erode their right bank, and on the Southern Hemisphere the left bank. The theory has been reviewed by KABELAC (1957). It should be mentioned that much pro-and-con discussion may be found in the literature.

WERNER (1951) has presented an analytical approach to the local-disturbance hypothesis. It was suggested that the transverse waves play an important part in the formation of meanders. At the slightest disturbance a gravity wave is created, which is swept downstream under the characteristic standing-wave pattern. The meander length was calculated as

$$M_L = \frac{2B\bar{u}}{\sqrt{C_w g\bar{D} - \bar{u}^2}} \tag{10.29}$$

where C_w is an empirical coefficient. Qualitatively this formula was compared with the laboratory data reported by FRIEDKIN (1945). Recently ANDERSON (1967) developed a quite similar model when basing the concept on transverse oscillations. The conclusion of this study was a relationship between the meander length and a Froude number, or

$$M_L = C_A\left(\frac{\bar{u}}{\sqrt{g\bar{D}}}A\right)^{1/2} = C_A\left(\frac{Q}{\sqrt{g\bar{D}}}\right)^{1/2} \tag{10.30}$$

where A is the cross-sectional area and C_A is a numerical constant. Experimental results show good agreement with Eq. (10.30) over a reasonable range of Froude numbers. It is not at all surprising that Eq. (10.29), given by WERNER (1951), and Eq. (10.30) are similar. However, the concept that meanders are the result of crosscurrents was advanced much earlier by EXNER (1919) who derived the following relation:

$$M_L = 2M_B\frac{\bar{u}}{\sqrt{g\bar{D}}} \tag{10.31}$$

The limited field data justified, at least qualitatively Eq. (10.31). One fact is clear, namely, the advancement over the last 50 years (1919–1967) measured by Eqs. (10.30) and (10.31) is minimal.

Correctly realizing the many deterministic influences on the meander phenomenon, LANGBEIN et al. (1966) suggest to treat it as a stochastic process. Random-walk techniques were used to generate meanderlike curves. River meanders could be best approximated by a sine-generated curve, a curve which has the smallest variation in changes of direction.[1] This important contribution has shown that the meandering stretch is more stable than the straight one, and that total energy losses are reduced to a minimum rate.

EXNER (1925) suggested to consider the water-sediment interface as a problem of stability. The mathematical model proposed is similar to the one discussed in Sec. 11.1.3.1. A linearized stability analysis for water flow in a channel with a loose bed and straight banks was reported by CALLANDER (1969). It was shown that such flows are usually unstable.

It appears, at present, that the most prosperous analytical approach is one for treating the spiral or secondary flow in the light of the Navier-Stokes equation. Consider the second and third Reynolds equations for turbulent flow, or

$$\begin{aligned}
\frac{\partial \bar{v}}{\partial t} + \bar{u}\frac{\partial \bar{v}}{\partial x} + \bar{v}\frac{\partial \bar{v}}{\partial y} + \bar{w}\frac{\partial \bar{v}}{\partial z} &= -\frac{1}{\rho}\frac{\partial \bar{p}}{\partial y} + \nu\nabla^2\bar{v} + \frac{\partial}{\partial x}\overline{(-u'v')} \\
&\quad + \frac{\partial}{\partial y}\overline{(-v'^2)} + \frac{\partial}{\partial z}\overline{(-v'w')} \\
\frac{\partial \bar{w}}{\partial t} + \bar{u}\frac{\partial \bar{w}}{\partial x} + \bar{v}\frac{\partial \bar{w}}{\partial y} + \bar{w}\frac{\partial \bar{w}}{\partial z} &= -\frac{1}{\rho}\frac{\partial \bar{p}}{\partial z} + \nu\nabla^2\bar{w} + \frac{\partial}{\partial x}\overline{(-u'w')} \\
&\quad + \frac{\partial}{\partial y}\overline{(-v'w')} + \frac{\partial}{\partial z}\overline{(-w'^2)}
\end{aligned} \tag{10.32}$$

Assume that the flow is uniform ($\partial/\partial x = 0$), and that convective and viscous terms can be neglected, then

$$\begin{aligned}
\frac{\partial \bar{v}}{\partial t} &= -\frac{1}{\rho}\frac{\partial \bar{p}}{\partial y} + \frac{\partial}{\partial y}\overline{(-v'^2)} + \frac{\partial}{\partial z}\overline{(-v'w')} \\
\frac{\partial \bar{w}}{\partial t} &= -\frac{1}{\rho}\frac{\partial \bar{p}}{\partial z} + \frac{\partial}{\partial y}\overline{(-v'w')} + \frac{\partial}{\partial z}\overline{(-w'^2)}
\end{aligned} \tag{10.33}$$

[1] LEOPOLD et al. (1966) have remarked that the sine-generated curve is also the minimum total work in bending. (A metal strip held between two hands and then made to bend will avoid any concentration of bending and, thus, assume the sine-generated curve.)

After differentiating Eqs. (10.33) by z and y, respectively, the difference between the two resulting equations gives

$$\frac{\partial}{\partial t}\left(\frac{\partial \bar{v}}{\partial z} - \frac{\partial \bar{w}}{\partial y}\right) = \frac{\partial^2}{\partial y\,\partial z}(\overline{w'^2} - \overline{v'^2}) + \frac{\partial^2}{\partial y^2}(\overline{v'w'}) + \frac{\partial^2}{\partial z^2}(\overline{v'w'}) \qquad (10.34)$$

Equation (10.34) describes the change of vorticity with time, since

$$\xi = \frac{\partial \bar{v}}{\partial z} - \frac{\partial \bar{w}}{\partial y} \qquad (10.35)$$

where ξ represents the x component of the rotation. Equation (10.34) may be considered as a stability criterion for straight flow, and the following may be deduced:

1. In absence of turbulent fluctuation components, circulation (secondary flow) does not exist; this applies to laminar-flow problems.
2. The terms on the right-hand side in Eq. (10.34) add up to zero if the turbulence is isotropic.
3. In channel flow where the turbulence is not isotropic, circulation (secondary flow) will develop spontaneously.

Equation (10.34) and its interpretation were given by EINSTEIN et al. (1958) and later by SHEN (1961). In a subsequent experimental investigation by SHEN (1961), EINSTEIN et al. (1964), and SHEN et al. (1968), the foregoing theoretical reasoning was tested. As a most interesting result it was found that two types of meander patterns could be distinguished. One type, occurring in a two-dimensional channel with a movable bed near critical flow condition ($N_F \approx 1$), is a special case of diagonal dune pattern, and seems to be the meander EXNER (1919) had referred to. The other type, occurring in a two-dimensional channel with movable bed and coarse gravel banks, formed a series of alternating holes. It was felt that these pronounced scour holes are induced by secondary currents which, in turn, are caused by the changing shear values (Reynolds stresses) between the smooth movable bed and the coarse gravel bed. The physical argument centers around the vorticity concept, but cannot be considered conclusive. YEN (1970) measured bed shear distribution in a meandering channel and also observed the resulting bed topography, but the results are not conclusive. Further research along this line may very well render a useful, but probably rather involved, explanation of the meandering.

It is worthwhile to note that the contributions by ADACHI (1967), ANANIAN (1967), and MURAMOTO (1967) consider the equation of motion in form of the Navier-Stokes equation, and the vorticity concept as the beginning for their calculations.

10.3.4 LONGITUDINAL RIVER PROFILES

When a whole river or a stretch of a river is in equilibrium—neither aggrading nor degrading—the slope decreases from source to mouth. Apparently SCHOKLITSCH (1930) was the first to formulate this problem. Analogous to the size decrease of particles along a watercourse, the slope of the watercourse S was made proportional to the particle size, such as

$$S = \alpha_s W_0 e^{-\varphi l} \tag{10.36}$$

where l = length of the river stretch under consideration
W_0, W = weight of the particle at the beginning and end of the river stretch, respectively
α_s = coefficient
φ = specific abrasion

With $l = 0$ and $S = S_0$, the following is obtained:

$$\frac{dz}{dl} = S = S_0 e^{-\varphi l} \tag{10.37}$$

After integration and introduction of the boundary condition $x = 0$, $z = z_0$,

$$(z_0 - z) = \frac{S_0}{\varphi}(1 - e^{-\varphi l}) \tag{10.38}$$

Relations similar to Eqs. (10.36) through (10.38) have been suggested by STERNBERG (1875), PUTZINGER (1919), FORCHHEIMER (1930), SHULITS (1936), and others. SCHOKLITSCH (1930) reports that profiles of the middle Rhine, Maas, Enns, and Mur agree with Eq. (10.38). SHULITS (1941) finds that the lower Mississippi, the lower Ohio, and a portion of the Colorado River fit this equation.

According to Eq. (10.37), the slope S at any point l along the watercourse can be determined if an initial slope S_0 and the specific abrasion φ are known. To explain the latter, we must consider the relationship established for the particle reduction along the stream. The reduction of the particle weight over a distance is made proportional to the particle weight, such that

$$\frac{dW}{dl} = -\varphi W \tag{10.39}$$

Integrating and assuming at the beginning that $l = 0$, $W = W_0$, the following is obtained:

$$W = W_0 e^{-\varphi l} \tag{10.40}$$

Equation (10.40) is generally attributed to STERNBERG (1875). Simple as it is, it agrees fairly well with many field observations and laboratory tests.

Table 10.5 Measured specific abrasion [*after* SCHOKLITSCH (*1930*)]

Rock	φ, m^{-1}
Marly limestone	0.0000167
Limestone	0.0000100
Dolomite	0.0000083
Quartz	0.0000033
Gneiss and granite	0.0000050–0.0000033
Amphibolite	0.0000035–0.0000020

The quantity φ is called the *specific abrasion* or *coefficient of abrasion*, and represents the loss of particle weight per unit weight and unit distance. It depends primarily on the material and the roundness of the particle. Table 10.5 shows measured specific abrasion for various rocks; a list with more than 80 different materials was published by SCHOKLITSCH (1933). In general, it may be expected that the rounder the particle, the lower the coefficient of abrasion. In a detailed investigation, SCHOKLITSCH (1933) has shown that the coefficient of abrasion φ is not a constant for a given material. It changes with the one-fourth power of the velocity, and is dependent on characteristics of the underlying sediments. SCHOKLITSCH (1930, 1933) has compiled the specific abrasion φ for European rivers, such as the Rhine, Iller, Mur, Danube, Gail, Traun, Lech, and Seine. Rather typical values are listed in Table 10.6.

Thus we may reiterate that, whenever the coefficient of abrasion is known, the slope of the watercourse may be determined with Eq. (10.37);[1] the same is true for the determination of particle weight when Eq. (10.40) is used.

[1] Simons (1969) has informed the author that recent work at Colorado State University has shown that hydraulic sorting may be of importance in the development of the longitudinal profile of sand bed streams.

Table 10.6 Coefficients of abrasion φ from river measurements

River	*Abrasion coefficients*	*Reporter*
Mur (at Graz, Austria)	0.0181 km^{-1}	FORCHHEIMER (1914)
Gail (Carinthia, Austria)	0.1054 km^{-1}	SCHOKLITSCH (1933)
Danube (Austria)	0.0230 km^{-1}	SCHOKLITSCH (1933)
Traun	0.0269 km^{-1}	SCHOKLITSCH (1933)
Lech	0.0937 km^{-1}	SCHOKLITSCH (1933)
Seine (Paris-Rouen)	0.0095 km^{-1}	SCHOKLITSCH (1933)
Mississippi (lower)	0.00162 mi^{-1}	SHULITS (1941)
Ohio River (lower)	0.00293 mi^{-1}	SHULITS (1941)

HACK (1957) and BRUSH (1961) have established a relationship between channel slope and channel length, and have tested it for a limited number of streams of the eastern United States. Using a probabilistic argument, LEOPOLD et al. (1962) and LANGBEIN (1964) constructed random-walk models and, thus, predicted the longitudinal river profile.

10.4 CONCLUDING REMARKS

There exists considerable similarity between equations determining the geometry of canals in regime and those determining the geometry of rivers in regime. This may be immediately recognized by inspecting Table 10.3. However, further research in this field is very necessary. The geometry of meanders has been expressed with a simple set of equations. Again, much will have to be learned before conclusive statements on the meandering problem may be given. Finally, there are included remarks on equilibrium river profiles.

The application of the knowledge gained by Chap. 10 is difficult and requires many an estimate, thus much experience. However, a great deal may be gained by reading BLENCH (1957, 1961, 1966) and MATTHES et al. (1956).

11
Bedform Mechanics

Consider a flat and even, stationary bed consisting of loose, cohesionless, and mobile solid particles of uniform size. If flow is established, bedforms are molded from the bed material. The geometry of these bedforms depends on various flow parameters. In turn, the various flow parameters are influenced by the bedforms.

The existence of bedforms and their influence on the flow can be observed, and is recognized and studied by hydraulicians and geomorphologists. Bedforms will be discussed first, in Sec. 11.1, and in Sec. 11.2 the effects of the bedforms on the flow resistance will be considered.

11.1 BEDFORMS

In this section we shall discuss pertinent experimental and theoretical investigations dealing with the creation, geometry, and stability of bedforms.

It should be kept in mind that there exist certain similarities between bedforms produced by water action on either the stream bed or on ocean floors and beaches, and bedforms produced by wind action on either sand surfaces or on snow. For a discussion on the foregoing, the interested reader is referred to SCHEIDEGGER (1961); a detailed treatment of aeolian features was given by BAGNOLD (1941). There exists a limited similarity between bedforms created at loose sand-fluid interfaces and waves created at fluid-fluid interfaces. BAGNOLD (1941, p. 144) pointed to the fact that in the latter a propagation of energy is encountered, while in the former, it "is merely a crumpling or heaping up of the surface."

Furthermore, since the present may help with the mystery of the past, bedforms may be one key to the understanding of those geologic formations which are a result of the fluid-sediment interface. Interesting contributions along these lines were prepared by HARMS et al. (1965) and ALLEN (1968). The latter publication gives also a very detailed account on bedforms and related features from a geologist-sedimentologist's point of view. Stratigraphy and sedimentation from the geologist's point of view were summarized by KRUMBEIN et al. (1963).

11.1.1 EARLIER DEVELOPMENTS

A most fascinating description of the creation, geometry, and movement of bedforms is found in DUBUAT (1786, p. 100).[1] It is apparent from this treatment that the triangular geometry of a bedform is recognized. The sand grain moves up a gentle slope, then arrives at a summit, and falls down the steep slope where it is sheltered from the fluid's action. The whole bedform advances little by little.

Using a mathematical approach, HELMHOLTZ (1888)[2] showed that a boundary between two fluids of different density moving with different velocities is subject to waves. BASCHIN (1899) contested that bedforms in streams can be explained with Helmholtz's model, and as such considered the loose sand to act almost like a fluid.

Other contributions are reviewed briefly by FORCHHEIMER (1914). Included is the work by Sainjon (1871), who took measurements on the Loire River and obtained a relation between water velocity and the advancement of the bedforms. Some studies on the Rhine River concerning the behavior of bedforms are reported by Jasmund (1911). Experimental studies have been carried out in a glass-walled flume by Deacon (1894). It was found that an increase in the flow velocity is responsible for an increase in the bedform velocity.

An important contribution to the further understanding of bedforms was made by BLASIUS (1910). The importance of the Froude number $N_F = \bar{u}/\sqrt{gD}$ was recognized, and depending upon the Froude number,

[1] The important lines are contained in ¶72, which were noted earlier in Sec. 5.3.3.
[2] This mathematical approach may be followed readily in LAMB (1945, p. 373).

different bedforms were observed. For $N_F > 1$, diagonally placed bars (*Bank-Kolk* or *Schrägbänke*) were reported; for $N_F < 1$, riffles (*Parallel-riffeln*) and bars were observed.

Very special attention deserves the research of GILBERT (1914). Not only did GILBERT (1914) provide the profession with a storehouse of good data but also his description of the "modes of transportation" reflects vividly the good eye of the observer. It is worthwhile to use GILBERT's (1914, p. 11) own words:

> When the conditions are such that the bed load is small, the bed is molded into hills, called dunes, which travel downstream. Their mode of advance is like that of eolian dunes, the current eroding their up-stream faces and depositing the eroded material on the downstream faces. With any progressive change of conditions tending to increase the load, the dunes eventually disappear, and the debris surface becomes smooth. The smooth phase is in turn succeeded by a second rhythmic phase in which a system of hills travel upstream. These are called antidunes, and their movement is accomplished by erosion on the downstream face and deposition on the upstream face. Both rhythms of debris move-ment are initiated by rhythms of water movement.

This description is graphically represented by Fig. 11.1. FORCHHEIMER (1930) pointed out that the Froude number N_F serves as a useful criterion; in streams with $N_F < 1$ dunes are expected, but in torrents with $N_F > 1$ antidunes are observed. GILBERT (1914) also remarked that antidunes travel much faster than dunes; their profiles are more symmetric than those of dunes, and are followed readily by the water surface. As far as the

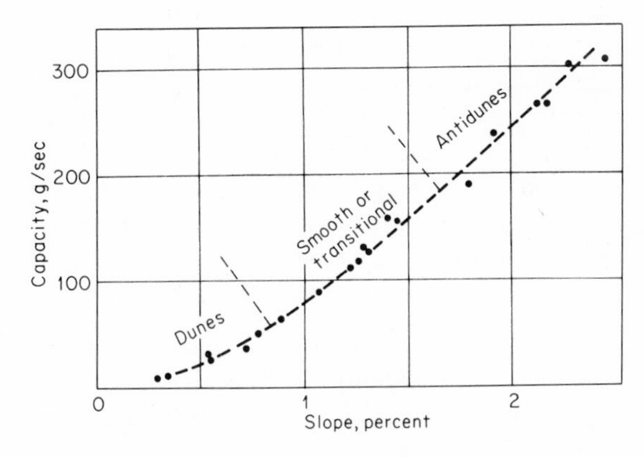

Fig. 11.1 A series of observations on sediment load and slope. [*After* GILBERT (*1914*).]

description of the bedform phenomenon goes, little has been added to GILBERT'S (1914) careful observations.

With the advent of good measuring equipment, bedforms received more and more attention in sediment-transport experiments. KRAMER (1935) remarked that bedforms (riffles) accompany the bedload movement, starting with the lowest stages. Apparently for the first time KRAMER (1935) correlated the appearance of bedforms with the shear stress on the bed τ_0, a concept which was further extended by SHIELDS (1936). KRAMER (1935) also stated that bedforms have an influence on the roughness value.

Early Soviet research carried out by Goncharov, Lapshin, and Roborovskaya was discussed by KONDRAT'EV (1962, p. 13). Experiments were used to establish the flow velocities at which bedforms are initiated and grow in size.

11.1.2 EXPERIMENTAL INVESTIGATIONS

11.1.2.1 C.S.U. contributions. The most extensive experimental studies on bedforms were made by the U.S. Geological Survey at the Colorado State University (C.S.U.). A summary of the data of all the flume experiments was presented by GUY et al. (1966). A total of 339 equilibrium runs were made to evaluate "the effects of the size of bed material, temperature of the flow, and fine sediment in the flow on the hydraulic and transport variables." Two recirculatory flumes were used. One of them, 8 ft wide, 2 ft deep, and 150 ft long, handled flow rates up to 22 cfs; slope adjustments from horizontal to 1.5 percent were possible. The other one, 2 ft wide, $2\frac{1}{2}$ ft deep, and 60 ft long, handled flow rates up to 8 cfs; slope adjustments from horizontal to 10 percent were possible. The sand used in the experiments was natural and, thus, nonuniform material. Ten different mixtures were used, the smallest one having a median diameter of $d = 0.19$ mm and the largest being $d = 0.54$ mm. The report by GUY et al. (1966) is most complete; it includes the description of the equipment, the operation procedure, the experimental variables and parameters, and the individual runs. The basic data are tabulated at the end of the report. Almost 100 photographs are included, illustrating bedforms and accompanying water surfaces.

As a result of this investigation, SIMONS et al. (1961) presented a classification of bedforms which has been widely accepted.[1] This classification is presented in Table 11.1; the typical bedforms are shown in Fig. 11.2.

Flow regime. In open-channel flow the Froude number N_F is often used as a flow criterion. The flow is tranquil (streaming) if the normal depth is greater than the critical depth; in this case the Froude number is $N_F < 1$.

[1] See, for example, LEOPOLD et al. (1964), HENDERSON (1966), and RAUDKIVI (1967), etc. Yet a different classification is suggested by ALLEN (1968).

Table 11.1 Classification of bedforms and other information [*after* SIMONS *et al.* (*1965*) *and* SIMONS *et al.* (*1966*)]

Flow regime	*Bedform*	*Bed material concentrations, ppm*	*Mode of sediment transport*	*Type of roughness*	*Roughness, $C/\sqrt{g}$*
Lower regime	Ripples Ripples on dunes Dunes	10–200 100–1,200 200–2,000	Discrete steps	Form roughness predominates	7.8–12.4 — 7.0–13.2
Transition	Washed-out dunes	1,000–3,000		Variable	7.0–20.0
Upper regime	Plane beds Antidunes Chutes and pools	2,000–6,000 2,000 → 2,000 →	Continuous	Grain roughness predominates	16.3–20 10.8–20 9.4–10.7

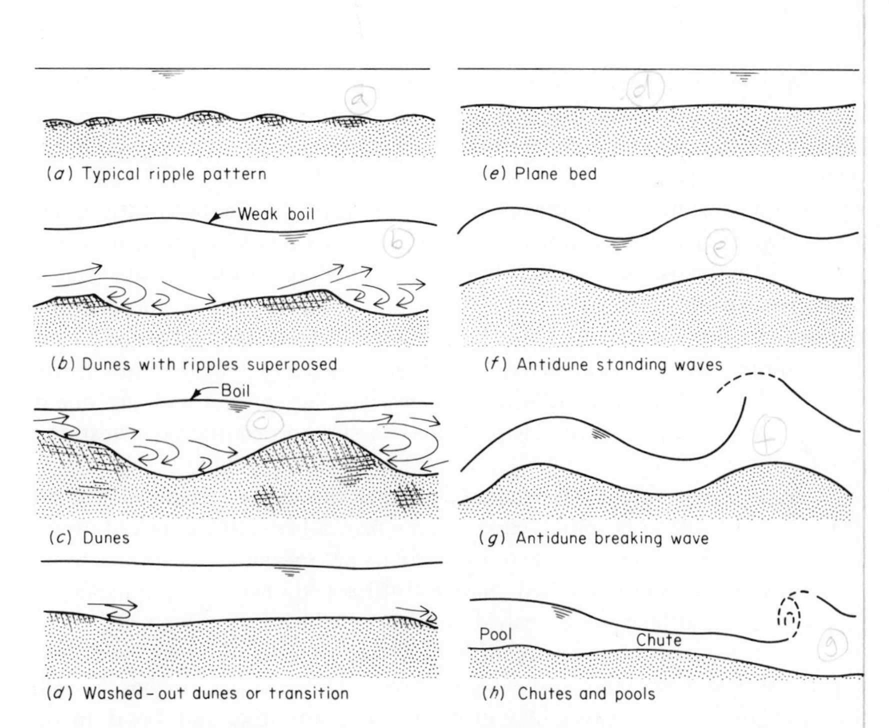

Fig. 11.2 Idealized bedforms in alluvial channels. [*After* SIMONS *et al.* (*1961*).]

The flow is rapid (shooting) if the normal depth is smaller than the critical one; in this case the Froude number is $N_F > 1$. If normal and critical depth are identical, the flow is critical, and the Froude number is $N_F = 1$. SIMONS et al. (1961) suggested to classify bedforms according to the Froude number of the flow, which is given by

$$N_F = \frac{\bar{u}}{\sqrt{g\bar{D}}} \tag{11.1}$$

where $\bar{u}$ is the flow velocity and $\bar{D}$ is the depth, both values averaged over the cross section. However, it must be pointed out that any *local* Froude number might differ considerably from the *averaged* Froude number; a large variation of Froude numbers may thus appear in any given cross section. Furthermore, the absolute magnitude of the Froude number is only qualitatively significant. For example, SIMONS et al. (1961) remarked that a certain bedform occurs in the laboratory flume at $N_F < 0.6$ and in a large, deep river at $N_F \ll 0.3$.

For the purpose of classification of bedforms, three flow regimes may be distinguished: (1) a lower flow regime with $N_F < 1$; (2) an upper flow regime with $N_F > 1$; and (3) a transition in between with $N_F \approx 1$.

Bedforms. Idealized sketches of the various bedforms are shown in Fig. 11.2. At extremely low velocities the critical shear stress $(\tau_0)_{cr}$ of the bed is not exceeded and sediment motion does not occur. An increase in the velocity may result in sediment movement and formation of bedforms. The common bedforms occurring in the lower regime have dune pattern and are triangular-shaped elements with a steep downstream and a gentle upstream slope. The downstream slope is inducive to flow separation and migrates in the downstream direction. The spacing and the geometry of the elements are random for an individual one, but uniform in the statistical sense. Individual dune patterns are referred to as *ripples* or *dunes*. *Ripples* are understood to be small bedforms, whereas *dunes* are larger ones and are out of phase with the water surface. Ripples may be superposed upon the upstream side of dunes.

As the flow velocity is increased further the upper regime is reached. The first bedform to be observed is a *plane bed*, a bed surface devoid of any "bedform." A further increase in the velocity causes the water surface to become unstable. The plane bed changes into a bedform—similar to and in phase with the surface wave—called *antidunes*,[1] which may remain stationary or move upstream or downstream. At lower Froude numbers, antidunes appear as standing sand and water waves. However, at higher Froude numbers, the surface waves may grow, become unstable, and break in the

[1] Note that antidunes as defined here do not necessarily move upstream.

upstream direction. If the latter occurs, the antidunes are destroyed, the bed becomes flat, and formation of antidunes starts all over again. Extremely strong antidune activity leads to *chutes and pools flow*. Shooting flow on the mounds of sediment runs into a pool where the flow is generally a streaming one.

Between the lower and the upper regime there is a *transition zone*. For a given flow condition, the bed configuration is erratic; it may range from developed dunes to a flat bed, or it may consist of a heterogeneous array of bedforms.

One bedform not necessarily linked with a flow regime, and probably reflecting more than any other bedform the three-dimensionality of natural streams, is the *bar*. Bars are large depositional features and have a length of the same order as the channel width or longer. Several different types of bars have been observed. In general it may be said that they have the geometry of a dune, but are much larger. Bars are created at high flow and may appear as little islands or peninsulas during low flow. Some worthwhile remarks on bars are contained in LEOPOLD et al. (1964) and ALLEN (1968).

The previously described bedforms have been observed in nature as well as in laboratory flumes. If there is an apparent discrepancy between such observations, we must realize the limitation of the experimental data. Experimental facilities rarely allow the modeling of three-dimensional effects. In the laboratory the shear stress is almost uniformly distributed over the bed and the sides are not erodible; such a hydraulic condition provides usually one dominant bedform. In nature the shear stress distribution may be such that a variety of bedforms appears at a given cross section. Therefore the formation of bars (meandering) is seldom achieved in the laboratory; it requires three-dimensional conditions. The effect of the width of laboratory channels on the bedforms was studied by CRICKMORE (1970). Height, length, and width of bedforms are reported to increase with increasing channel width. Under closely similar flow conditions reductions up to 30 percent are observed when 1.5-ft width and 5.0-ft width experiments are compared. An extensive laboratory investigation on three-dimensional bedforms was performed by ALLEN (1969).

SIMONS et al. (1965) contributed a few remarks on the stratification of bedforms. A kind of segregation was apparently absent in the upper regime, but the lower regime always showed some sorting or segregation. With ripples, the finer material was found to stay at the top, but the coarser material accumulated at the troughs. With dunes, the moving material formed a bedding plane parallel to the downstream side of the dune. Further remarks on stratification have been discussed by ALLEN (1968, chap. 5).

Sediment transport. Throughout the lower regime the solid particles make discrete steps; in the upper regime the bed material is nearly continuously in motion. In the lower regime the bed material moves mainly in contact

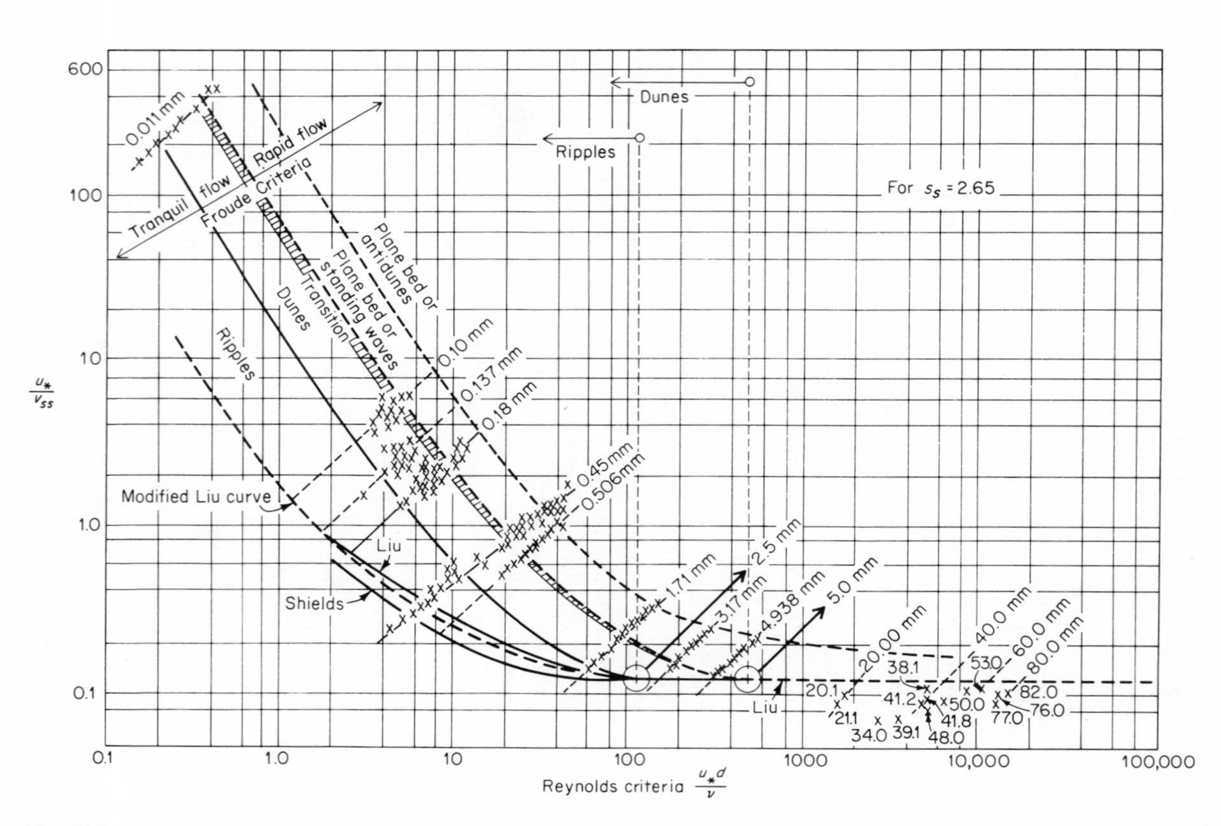

Fig. 11.3 Criteria for bedforms. [*After* Simons *et al.* (*1961a*).]

with the bed, as bedload, whereas in the upper regime the main mode of transport is suspension.

Representation of data. Since bedforms are the results of shear applied to a loose boundary, it seems logical to use a critical shear stress criterion for representation of the data. In fact, this had been successfully suggested by SHIELDS (1936), but the critical shear stress was not a popular parameter. In a similar way to SHIELDS (1936)—see Sec. 6.3.2.2—LIU (1957) derived a relation used at present:

$$\frac{u_*}{v_{ss}} = \text{fct}\left(\frac{du_*}{\nu}\right) \tag{11.2}$$

where u_* is the friction velocity and v_{ss} is the particle settling velocity. It can easily be shown that Eqs. (11.2) and (6.24) are related. LIU (1957) was interested in an equation for the ripple formation. SIMONS et al. (1961*a*) extended it to all bedforms. The resulting graph is given by Fig. 11.3. According to the data, both the particle Reynolds number and the Froude number of the flow play a major role. The functional relation for any bedform is thus given by

$$\frac{u_*}{v_{ss}} = \text{fct}\left(\frac{du_*}{\nu}, \frac{\bar{u}}{\sqrt{g\bar{D}}}\right) \tag{11.3}$$

According to Fig. 11.3, ripples and dunes are both limited to a lower Reynolds number range. Since field data have not been used, Fig. 11.3 has to be applied with extreme caution. Some data from the Rio Grande do not verify this plot; this was reported by CULBERTSON et al. (1964).

Finally, it is worthwhile to mention a study by GARDE et al. (1963) and one by BOGARDI (1965), wherein both suggested a relation similar to Eq. (11.3). A new relationship was suggested by SIMONS et al. (1963*a*); the stream power $\tau_0\bar{u}$ is plotted against the grain diameter, as obtained from a settling experiment. This relation, as given by Fig. 11.4, was obtained by utilizing not only laboratory data but also an extensive set of field data from the United States and the Punjab. ACAROGLU et al. (1968) established a criterion similar to Eq. (11.2). Thus they were able to correlate bedforms in open channels, streams, and closed conduits.

11.1.2.2 Other contributions. The geometry and movement of bedforms were investigated by TSUBAKI et al. (1953), who show from a dimensional consideration that both the length λ and the height ΔH of a bedform depend on

$$\frac{\lambda}{D} = \text{t}_1\left(\frac{\rho u_*^2}{\rho_s g d}\right) \tag{11.4}$$

$$\frac{\Delta H}{D} = \text{fct}_2\left(\frac{\rho u_*^2}{\rho_s g d}\right) \tag{11.5}$$

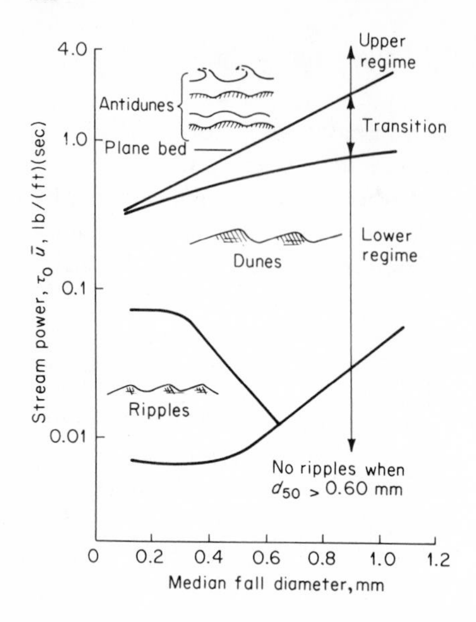

Fig. 11.4 Relation of bedforms to stream power and grain diameter. [*After* SIMONS *et al.* (*1963a*).]

The functional relation of these equations was obtained with experiments performed in an artificial irrigation canal of rectangular cross section. The length of the bedforms was observed as $80 < \lambda < 150$ cm and the height as $1 < \Delta H < 7$ cm. The data indicate that for a given sand a shear stress increase is responsible for a gradual growth of height ΔH and for a decrease of length λ. However, the suggested relation remains limited because the data used represent a small range of Froude numbers in the lower regime. With additional data available SHINOHARA et al. (1959) have pointed out that there exists no unique relation as given by Eq. (11.5); for Eq. (11.4) a function may be obtained, but some spread of the data is evident. The advance velocity of the bedform, c_B, was given by the relation

$$\frac{c_B R_h}{\sqrt{(\rho_s/\rho)gd^3}} = \text{fct}_3\left(\frac{\rho u_*^2}{\rho_s g d}\right) \tag{11.6}$$

With an increase of the right-hand side of Eq. (11.6), the left-hand side increases as well. Limited data by SHINOHARA et al. (1959) established this trend.

KONDRAT'EV (1962) undertook the worthwhile task of presenting and discussing some Soviet experimental research. Without much reservation or explanation KONDRAT'EV (1962) represents the data in terms of a dimensionless number, $\bar{u}/\sqrt{gD}$, which is well known as the Froude number. First, it was shown that a Froude number may be used as a bedform criterion. The Froude number was plotted against the grain diameter; data obtained by Goncharov and Lapshin were used. From such a graph, shown in Fig. 11.5, the initiation of grain motion (I), of bedforms (ridges) (II), and the washaway of bedforms (III) can be determined. Secondly, a relationship between the Froude number and a Reynolds number, in terms of the bedform velocity, is represented as shown in Fig. 11.6.

Geometrical properties of bedforms have been studied by YALIN (1964). By making some experimentally unsupported assumption, dimensionless variables for height ΔH and length λ of bedforms were derived, and the functional relations were obtained from experimental data. The height ΔH was shown to be related to the shear stress, the latter depending on the flow depth. A dimensionless expression for the bedform height was given by

$$\frac{\Delta H}{D} = \text{fct}_1\left(\frac{\tau_0}{(\tau_0)_{cr}}\right) = \text{fct}_2\left(\frac{D}{D_{cr}}\right) \tag{11.7}$$

where D_{cr} is the depth at which incipient motion takes place. A good many experiments were used to establish the function in Eq. (11.7). YALIN (1964) obtained

$$\frac{\Delta H}{D} = \frac{1}{6}\left(1 - \frac{D_{cr}}{D}\right) \tag{11.8}$$

Since $D_{cr} < D$, it was concluded that the height ΔH cannot exceed one-sixth of the flow depth, or

$$\frac{\Delta H}{D} < \frac{1}{6} \tag{11.8a}$$

NORDIN et al. (1965) suggested that a value of $\frac{1}{3}$ instead $\frac{1}{6}$ is more in place.

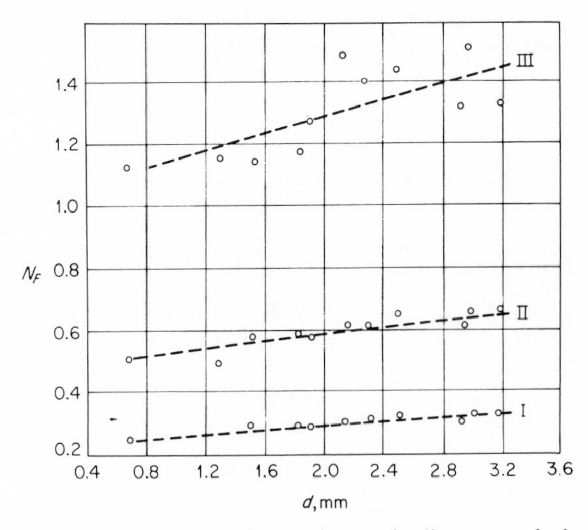

Fig. 11.5 Froude number against grain diameter; a bedform criterion. [*After* KONDRAT'EV (*1962*).]

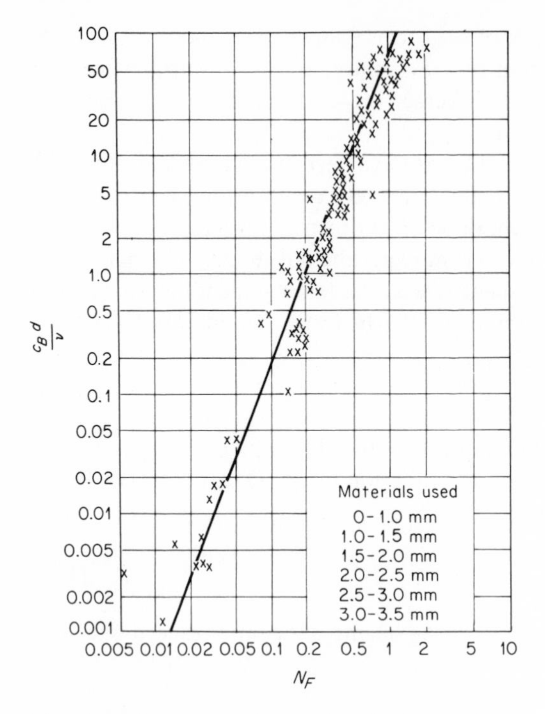

Fig. 11.6 Froude number against *bedform velocity* Reynolds number. [*After* KONDRAT'EV (*1962*).]

Again applying dimensional reasoning, YALIN (1964) showed that the length λ of bedforms is given by

$$\frac{\lambda}{d} = \text{fct}\left(\frac{\bar{D}}{d}\right) \tag{11.9}$$

for large-particle Reynolds numbers $u_* d/\nu$, and

$$\frac{\lambda}{d} = \text{const} \tag{11.10}$$

for small-particle Reynolds numbers. Plotting the experimental data, previous equations could be rewritten as

$$\frac{\lambda}{D} = 5 \tag{11.9a}$$

for $d \geq 0.38$ mm, and

$$\frac{\lambda}{d} = 1{,}000 \tag{11.10a}$$

for $d \leq 0.18$ mm. Neither of the two equations is restricted to subcritical flow. YALIN (1964) has reported that the limiting particle Reynolds number separating Eqs. (11.9) and (11.10)—is on the order of 20. Furthermore, it was suggested that the bedform given by Eq. (11.9*a*) is defined as a *dune*, and the one given by Eq. (11.10*a*) as a *ripple*.

A generalized empirical relation for bedforms that incorporated about 20 Soviet and other investigations was presented by ZNAMENSKAYA (1962) and considerably extended by ZNAMENSKAYA (1965, 1969). The latter contribution includes a valuable graph duplicated in Fig. 11.7. For a given hydraulic condition N_F and $\bar{u}$ and a given sediment material, the graph permits the determination of the kind of resulting bedform, the bedform velocity c_B, and of the bedform steepness $\Delta H/\lambda$. It is understood that a verification of the graph for a natural stream (Polomet River) was successful.

An elaborate experimental program was reported by ALLEN (1968) who tried to show a relationship between a fluid flow field and the resulting bedform. The bedforms studied are the (three-dimensional) "asymmetrical ripple marks" which correspond to ripples and dunes, according to the classification discussed in Sec. 11.1.2.1.

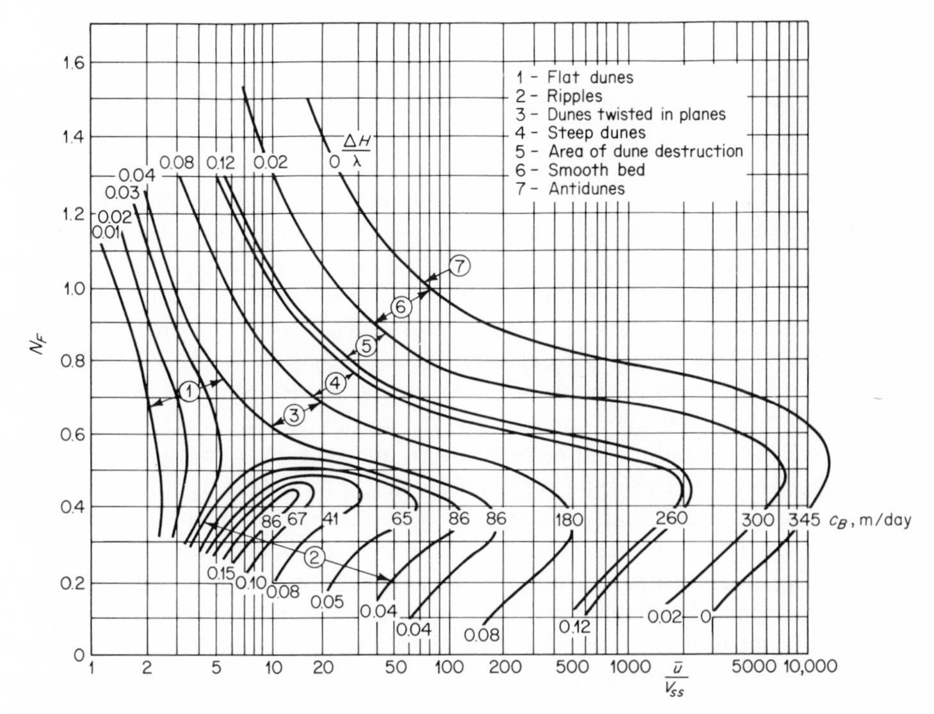

Fig. 11.7 Relationship of N_F vs. $\bar{u}/v_{ss}$, with c_B and $\Delta H/\lambda$ as parameters. [*After* ZNAMENSKAYA (*1969*).]

In a further study, ZNAMENSKAYA (1962) attempted to relate bedforms to hydrologic characteristics of the flow-duration curve. It was suggested that, at low water stages, bedforms are obtained, called *microforms*,[1] which may be regarded as bottom roughness. At flood stages, *macroforms*[1] are observed, which determine an additional local energy loss. The macroforms apparently include those bedforms which were previously referred to as *bars*.

In reality, such simple geometric forms as are usually described do not occur frequently. Much rather, the geometry of bedforms is highly irregular and usually presents an entire spectrum of wavelength and height. NORDIN et al. (1966) have applied techniques of autocovariance and spectral density analysis to describe bedforms; CRICKMORE et al. (1962) considered the probability of the profile ordinates. This is a new approach to the problem, and we may await interesting results.

Little information is available on the formation of bedforms from an initial plane bed. BAGNOLD (1941, p. 146) contested that:

> . . . the surface cannot be perfectly even, and we can imagine the existence of a number of tiny chance unevennesses. These may be caused, for instance, by more grains happening temporarily to be moved out of a certain small area than are moved into.

KONDRAT'EV (1962) and RAUDKIVI (1967) are led by visual and qualitative experiments as well as by a fluid-mechanics analogy in their descriptions of the formation of bedforms. KONDRAT'EV (1962) suggested that at certain flow velocities the movable boundary shows simultaneously furrows lying at right angles to the flow. The furrows are the result of an active lift mechanism. The sand which is carried away from these furrows is deposited downstream. The slopes of the furrows are steeper, the ones of the deposits are flatter; furrows and deposits together resemble a bedform (ridges). RAUDKIVI (1967) explained the formation of bedforms with a local *pile-up* of particles. The tendency of the pile-up is by chance; the cause may be the turbulent nature of the flow close to the boundary. As soon as a pile-up has formed, such a configuration gradually spreads out and begins to move. Owing to this pile-up the streamline will become curved and, in turn, a pressure change is experienced. Both KONDRAT'EV (1962) and RAUDKIVI (1967) point to the useful analogy of the fluid mechanics of bedforms and of streamlined bodies. RAUDKIVI (1967, p. 201) investigated experimentally a bedform (dune) and a negative step (abrupt expansion), and obtained interesting results. On the downstream side of the bedform, a number of short-lived eddies were observed and backflow was apparent. Exceeding about six bedform heights downstream from the crests, mainflow was observed. Between the regions of

[1] Both are classifications, according to the "Channel Processes Division of the State (Soviet) Hydrologic Institute (GGJ)."

backflow and mainflow, there was a zone of hesitancy where the grains moved in either direction. It was said that the first pile-up is responsible for a surface of flow discontinuity and high turbulence intensity. Therefore sand particles become easily entrained, and are carried along until the turbulence intensity decreases. At the location where the sand particles settle out, the second pile-up is formed.

An experimental investigation on the time response of bedforms to flow conditions was done by RAICHLEN et al. (1965). Steady flow was passed over an initial plane sand bed until eventually the formed bedforms were observed to be in equilibrium. It was found that length and height of the bedforms (dunes) as well as the friction factor reach 80 to 90 percent of their equilibrium value in half the time of development.

11.1.3 THEORETICAL STUDIES

Of all the studies that are concerned with a theoretical prediction of geometry and movement of bedforms, the one by EXNER (1925) is the most original and by now a classic treatment. More recently, various researchers have proposed analytical potential flow models. Models like these coupled with the equation of erosion, suggested by EXNER (1925), lead to interesting relationships between the Froude number N_F and geometric bedform dimensions. A most encouraging study considering the real fluid-sediment effects has been reported by ENGELUND et al. (1966). Each of the investigations has its merit under the given set of assumptions, and it is presently impossible to prefer one study to the other.

11.1.3.1 Exner's models. EXNER (1925) built a number of mathematical models of the water-loose bed interface, and compared the results qualitatively with laboratory experiments and limited field observations. None of the models gives an answer to the cause of the bedforms, but the models describe the geometry and movement of the bedforms. EXNER (1925) realized the difficulties in solving the entire problem, which becomes mathematically rather complex, and therefore progressed slowly by building more and more complicated models. First, the problem is handled as if frictional effects could be ignored and, later, the frictional effects are included. EXNER'S (1925) models will be discussed in the following.

11.1.3.1.1 MODELS WITHOUT FRICTION. Consider Fig. 11.8. The water depth is given as $h - \eta$, and the cross-sectional average velocity as $\bar{u}$. According to the law of continuity, we may write

$$(h - \eta)B\bar{u} = Q = \text{const} \tag{11.11}$$

where B is the channel width and Q is the flow rate. EXNER (1925) reasoned that erosion occurs if the flow velocity increases, and deposition occurs if it

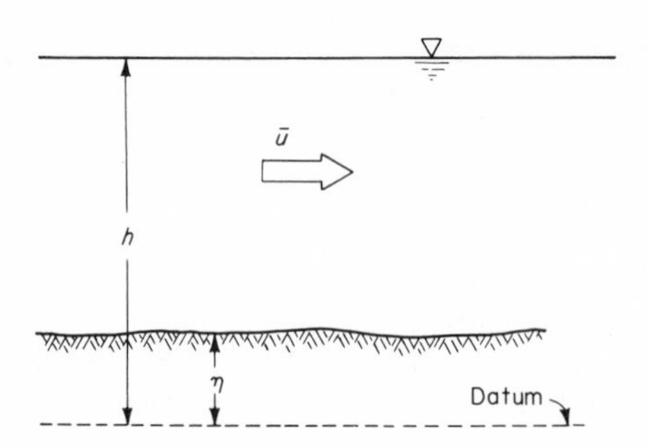

Fig. 11.8 Definition sketch for EXNER's (1925) models.

decreases in the downstream direction. As a result of this the bed level will change. This is expressed with Exner's erosion equation, or

$$\frac{\partial \eta}{\partial t} = -a_E \frac{\partial \bar{u}}{\partial x} \tag{11.12}$$

where a_E is the Exner erosion coefficient.

Constant width.[1] Combining Eq. (11.11) with Eq. (11.12) results in

$$\frac{\partial \eta}{\partial t} = \frac{a_E Q}{B} \frac{1}{(h - \eta)^2} \frac{\partial (h - \eta)}{\partial x} \tag{11.13}$$

With the approximation of h as constant, Eq. (11.13) becomes

$$\frac{\partial \eta}{\partial t} = -\frac{a_E Q}{B} \frac{1}{(h - \eta)^2} \frac{\partial \eta}{\partial x} \tag{11.14}$$

The solution of this differential equation is facilitated if an initial condition is imposed. Suppose, at the beginning, $t = 0$ and the bedform is given by a cosine function such as

$$\eta = A_0 + A_1 \cos \frac{2\pi x}{\lambda} \tag{11.15}$$

where λ is the (wave)length of the bedform and A_1 is its amplitude. Then the equation of the bedform for any given time is

$$\eta = A_0 + A_1 \cos \frac{2\pi}{\lambda}\left[x - \frac{a_E Q}{B(h - \eta)^2} t\right] \tag{11.16}$$

For a special case EXNER (1925) gives a graphical solution shown in Fig. 11.9. With Fig. 11.9 on hand and Eq. (11.16) in mind, the following conclusions may be drawn. The amplitude A_1 of the bedform is independent

[1] "Constant width and variable channel depth," one may call this model. As an analogous one, EXNER (1925) showed that "constant depth and variable width" gives an interesting model to study bank scour.

of time. The bedforms migrate downstream with a velocity of

$$c_B = \frac{a_E Q}{B(h - \eta)^2} \tag{11.17}$$

From Eq. (11.17) it is obvious that the crest of the bedform moves faster than the trough. Theoretically, an overhanging portion of the crest is predicted, but, practically, the angle of repose of the sand material makes this impossible. In general, the resulting bedform with a gentle slope on the upstream and a steeper one on the downstream side resembles rather strikingly certain bedforms in nature. Furthermore, from Eq. (11.17), we may reason that the bedform velocity will increase if the flow rate Q or the Exner erosion coefficient a_E increases.

Variable width. The foregoing analysis was for a canal with constant width; now the case of a variable width, $B = f(x)$, will be considered. Again assume a horizontal water surface, and the resulting erosion equation becomes

$$\frac{\partial \eta}{\partial t} = -\frac{a_E Q}{B} \frac{1}{(h - \eta)^2} \frac{\partial \eta}{\partial x} - \frac{a_E Q}{B^2} \frac{1}{(h - \eta)} \frac{\partial B}{\partial x} \tag{11.18}$$

Note that Eq. (11.18) becomes Eq. (11.14) for a constant-width channel. If the width is given as $B = f(x)$, then the general integral reads

$$B(h - \eta) = \text{fct}\left[\int \frac{dx}{f(x)} - \frac{a_E Q}{B^2(h - \eta)^2} t\right] \tag{11.19}$$

For a special set of conditions EXNER (1925) has given the graphical solution shown in Fig. 11.10. From Eq. (11.19) and Fig. 11.10 the following is

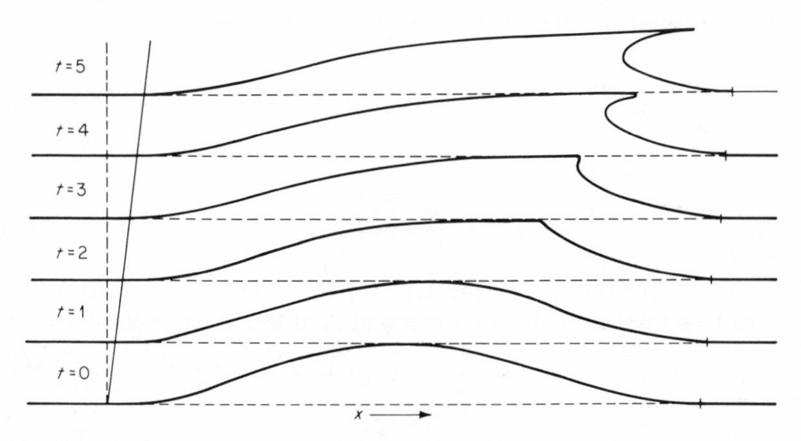

Fig. 11.9 Bedforms in a constant-width channel, according to Eq. (11.16); $A_0 = 1$, $A_1 = 1$, $\lambda = 20$, $h = 3$, and $a_E Q/B = 1$. [*After* EXNER (*1925*).]

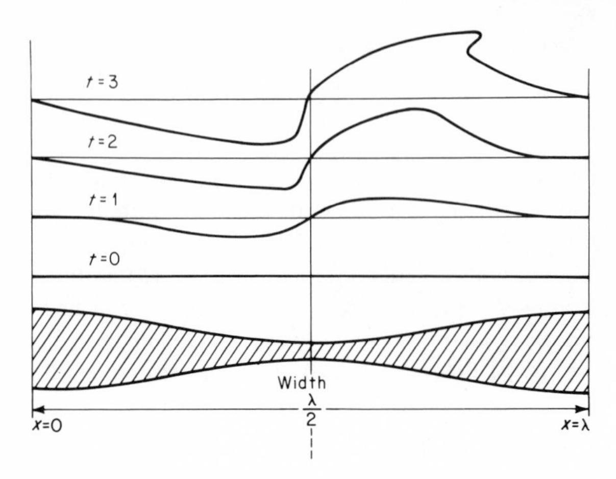

Fig. 11.10 Bedforms in a variable-width channel according to Eq. (11.19); $B = f(x) = A[\frac{5}{3} + \cos(2\pi/\lambda)x]$ at $t = 0$, $\eta = \eta_E = 1.5$, and $h = 3$. [*After* EXNER (*1925*).]

concluded. A migration of bedforms does not exist; the sand itself, however, travels downstream. In convergent parts of the channel erosion is evident and in divergent parts, its deposition. The deposition pattern resembles Fig. 11.9.

11.1.3.1.2 MODELS WITH FRICTION. Suppose the flow velocity $\bar{u}$ is given by the equation of motion or

$$\frac{\partial \bar{u}}{\partial t} = -\bar{u}\frac{\partial \bar{u}}{\partial x} - k\bar{u} + g_x - \frac{1}{\rho}\frac{\partial p}{\partial x} \tag{11.20}$$

with $k\bar{u}$ approximating frictional effects. If the pressure gradient is expressed by

$$\frac{\partial p}{\partial x} = \rho g \frac{\partial h}{\partial x} \tag{11.21}$$

then Eq. (11.20) becomes

$$\frac{\partial \bar{u}}{\partial t} = -\bar{u}\frac{\partial \bar{u}}{\partial x} - k\bar{u} + g_x - g\frac{\partial h}{\partial x} \tag{11.22}$$

Equation (11.22) together with Eqs. (11.11) and (11.12) represents the set of equations that approximates water and sand bed movements.

Constant width. Rearranging and differentiating these equations results in

$$\frac{\partial^2 \eta}{\partial t^2} - n\frac{\partial^2 \eta}{\partial x\,\partial t} + k\frac{\partial \eta}{\partial t} - a_E g\frac{\partial^2 \eta}{\partial x^2} = 0 \tag{11.23}$$

in which

$$n = \frac{gQ}{BU^2} - U$$

where U is a space-averaged velocity. The solution of Eq. (11.23) is obtained with the initial condition of a cosine function, given earlier with Eq. (11.15). The resulting integral reads

$$\eta = A_0 + A_1 e^{-(k/2-p)t} \cos \frac{2\pi}{\lambda} \left[x - \frac{n}{2p} \left(\frac{k}{2} - p \right) t \right] \tag{11.24}$$

where $p = \text{fct}\,(k,n,\lambda,a_E)$. We shall briefly discuss Eq. (11.24). EXNER (1925) has shown that the nonfriction solution to Eq. (11.23) is similar in its appearance to Eq. (11.16). If $k/2 - p$ is positive, the amplitude of the bedform decreases with time. The friction is responsible for this decrease; furthermore, the movement of the bedform is in the downstream direction. The longer bedforms migrate with a smaller velocity than the shorter ones. However, shorter bedforms have a faster decreasing amplitude than longer ones.

Variable width. The resulting equation for a variable-width channel becomes extremely involved and has not been analyzed. In an approximate way, EXNER (1925) argued the problem by making use of the energy principle. In general, friction has a tendency to level out the stream bed. The short bedforms decay rapidly, and the remaining long ones move rather slowly downstream. EXNER'S (1925) mathematical models and observations represent an important contribution to the bedform mechanics. The mathematical models focus around the physical observation:

> Erosion increases the area (space) and diminishes, therefore, the flow velocity, which in turn causes a reduction or an end to the erosion. Erosion has thus the tendency to destroy itself, and eventually steady state conditions are reached . . . .

EXNER (1925) wanted to check his model with his own experiments. The value of these experiments, however, must be considered rather limited. An extension of the foregoing study was performed by Velikanov (1936, 1955) and is discussed in some detail by RAUDKIVI (1967) and KONDRAT'EV (1962).

11.1.3.2 Potential flow models

11.1.3.2.1 GENERAL CONSIDERATIONS. For irrotational flow of an ideal incompressible fluid subject to influences of gravity, the following set of equations is applicable. If ϕ is the velocity potential, then the velocities in a two-dimensional problem are

$$u = \frac{\partial \phi}{\partial x} \qquad \text{and} \qquad v = \frac{\partial \phi}{\partial y} \tag{11.25}$$

Owing to continuity, we obtain

$$\frac{\partial^2 \phi}{\partial x^2} + \frac{\partial^2 \phi}{\partial y^2} = 0 \tag{11.26}$$

This is Laplace's equation for the determination of the function ϕ. The following boundary conditions are applicable. For solid boundary,

$$\frac{\partial \phi}{\partial n} = 0 \tag{11.27}$$

and for a free surface of $y = 0$,

$$\frac{\partial \phi}{\partial y} = -\frac{1}{g}\frac{\partial^2 \phi}{\partial t^2} \tag{11.28}$$

In addition, the pressure condition at the free surface is written as

$$\frac{p}{\rho} = -\frac{\partial \phi}{\partial t} - gy + \frac{p_0}{\rho} \tag{11.29}$$

The special problem of flow over a sinuous bottom was solved by MILNE-THOMSON (1960, pp. 388 and 406) and shall now be outlined; a sketch is shown in Fig. 11.11. A stream of a depth $\bar{D}$ flows with a speed $\bar{u}$ over a sand bed which is taken as a sinusoidal wave, itself moving with a velocity c_B, or,

$$y_B = a_B \sin \frac{2\pi}{\lambda}(x - c_B t) \tag{11.30}$$

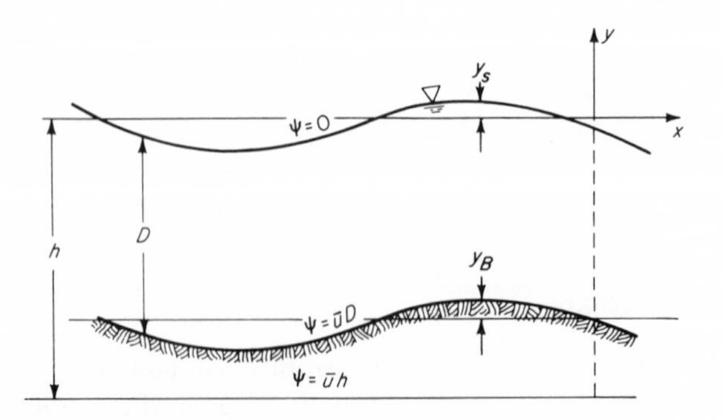

Fig. 11.11 Sketch of steady flow over a sinuous sand bed.

If the origin is taken in the free surface, the complex potential is given by

$$\omega = \bar{u}\left[z + \frac{a_S}{\sinh 2\pi h/\lambda} \cos \frac{2\pi}{\lambda} (z + ih - c_B t)\right] \tag{11.31}$$

where $z = (x + iy)$, and h is determined by

$$\bar{u}^2 = \frac{g\lambda}{2\pi} \tanh \frac{2\pi h}{\lambda} \tag{11.32}$$

and the profile of the free surface is given by

$$y_S = a_S \sin \frac{2\pi}{\lambda} (x - c_B t) \tag{11.33}$$

For sake of completeness, the complex potential, given by Eq. (11.30), can be separated into a real and an imaginary part. Thus the velocity potential ϕ and the stream function ψ are obtained, or

$$\begin{aligned} \phi &= \bar{u}x + a_S\bar{u}\,\frac{\cosh (2\pi/\lambda)(y + h)}{\sinh (2\pi/\lambda)h} \cos \frac{2\pi}{\lambda} (x - c_B t) \\ \psi &= \bar{u}y + a_S\bar{u}\,\frac{\sinh (2\pi/\lambda)(y + h)}{\sinh (2\pi/\lambda)h} \sin \frac{2\pi}{\lambda} (x - c_B t) \end{aligned} \tag{11.31a}$$

The free surface is given by the streamline $\psi = 0$, the reference streamline by $\psi = \bar{u}h$, and the streamline defining the sand bed by $\psi = \bar{u}D$. Putting this knowledge into the part of the complex potential that describes the stream function, the following is obtained:

$$y_B = \frac{a_S}{\sinh 2\pi h/\lambda} \sin \frac{2\pi}{\lambda} (x - c_B t) \sinh \frac{2\pi}{\lambda} (h - D) \tag{11.34}$$

which compares with Eq. (11.30) if

$$a_B = \frac{a_S}{\sinh 2\pi h/\lambda} \sinh \frac{2\pi}{\lambda} (h - D) \tag{11.35}$$

With a substitution of Eq. (11.32) into Eq. (11.35), we obtain

$$\frac{a_S}{a_B} = \frac{1}{\cosh (2\pi/\lambda)D - (g\lambda/2\pi\bar{u}^2) \sinh (2\pi/\lambda)D} \tag{11.36}$$

a relation relating the amplitude of the surface waves to the one of the bedforms. If

$$\bar{u}^2 > \frac{g\lambda}{2\pi} \tanh \frac{2\pi}{\lambda} D \tag{11.37}$$

the two waves (surface wave and bedform) are in phase, whereas if

$$\bar{u}^2 < \frac{g\lambda}{2\pi} \tanh \frac{2\pi}{\lambda} D \tag{11.37a}$$

the two waves are out of phase. Note that the quantity on the right-hand side of Eq. (11.37) is the speed of propagation of a wave with length λ, and in a water of depth D, or

$$c^2 = \frac{g\lambda}{2\pi} \tanh \frac{2\pi}{\lambda} D \tag{11.38}$$

Furthermore, note that for $\bar{u} = c$ the ratio a_S/a_B becomes infinite, and the assumption on which the solution was obtained breaks down.

11.1.3.2.2 THE SEDIMENT TRANSPORT PROBLEM. At the sand bed and water interface, another condition has to be satisfied. This is the sediment transport, and may be expressed with Exner's erosion equation, or

$$\frac{\partial y_B}{\partial t} = -a_E \frac{\partial \bar{u}}{\partial x} \tag{11.12}$$

Anderson (1953) had used Eq. (11.12), and suggested that it may define a bedform profile at a particular time for any given variation in the mean velocity. The complex potential provides information and calculation of the stream function and the velocity potential, as given with Eqs. (11.31a), which, in turn, allows calculation of the velocities according to Eqs. (11.25). For the special case of flow over a fixed sinusoidal sand bed, Anderson (1953) obtained an amplitude ratio of

$$\frac{a_S}{a_B} = \frac{1}{2} \cosh \frac{2\pi}{\lambda} D \tag{11.39}$$

Combining Eqs. (11.39) and (11.36) and writing it "in a more convenient form," Anderson (1953) obtained:

$$\frac{1}{N_F{}^2} = \frac{gD}{\bar{u}^2} = \frac{2\pi D}{\lambda}\left(\tanh \frac{2\pi D}{\lambda} - \frac{2}{\sinh 4\pi D/\lambda}\right) \tag{11.40}$$

Equation (11.40) suggests a relationship between the Froude number N_F of the flow and the relative wavelength. It was found in reasonably good agreement with the few data available; this may be seen in Fig. 11.12. Anderson (1953) argued that the value of

$$\cosh \frac{2\pi D}{\lambda} = 2 \tag{11.39a}$$

divides asymmetrical and symmetrical bedforms. If the left-hand side in Eq. (11.39a) is greater then 2, then the amplitude of the sand wave is less than

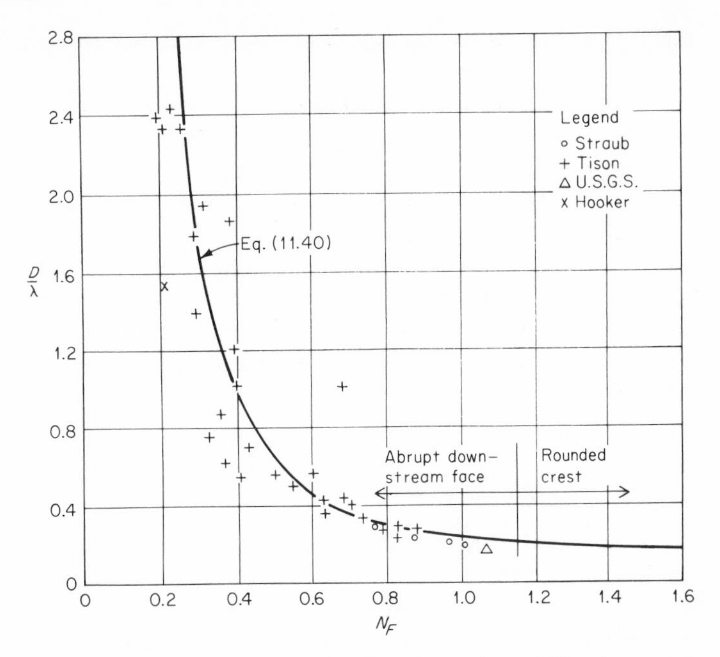

Fig. 11.12 Relative wavelength vs. Froude number. [*After* ANDERSON (*1953*).]

the one of the surface wave. Owing to continuity, the stream velocity at the crest is smaller than the one at the trough; therefore sediment tends to deposit on the crest and is carried to the downstream face. This is, according to ANDERSON (1953), indicative for asymmetrical bedforms with an abrupt downstream and a gentle upstream slope. If the left-hand side of Eq. (11.39*a*) is smaller than 2, the opposite is true. Relative larger velocities at the crest are causing erosion, and the resulting bedforms are symmetrical ones. The Froude number that corresponds to Eq. (11.39*a*) was given as $N_F = 1.15$ and is indicated in Fig. 11.12.

For the case where the bedforms move downstream, the bedform profile was given as

$$y_B = \frac{2a_B}{\cosh(2\pi D/\lambda)} \sin \beta t \cos\left(\frac{2\pi}{\lambda} x - \beta t\right) \tag{11.41}$$

and the speed c_B was given as

$$c_B = \frac{\beta\lambda}{2\pi} = \frac{(a_E \bar{u} 2\pi/\lambda) \cosh 2\pi D/\lambda}{2 \sinh 2\pi D/\lambda} \tag{11.42}$$

ANDERSON (1953) was unable to test the latter equation because of a lack of adequate data. Similar to ANDERSON (1953), the bedform phenomenon was described with the potential-flow concept by KONDRAT'EV (1962, p. 49).

KENNEDY (1963) used an Exner-type erosion equation to obtain the amplitude a_B of the bedform wave and the velocity c_B with which bedforms migrate. The sediment transport was given by

$$\frac{\partial(g_{st})}{\partial x} + B\frac{\partial \eta}{\partial t} = 0 \tag{11.43}$$

where g_{st} is the local rate of sediment transport per unit width in weight and B is the bulk specific weight of the sediment in the bed. Since Eq. (11.43) introduces another unknown quantity, namely g_{st}, it was suggested to relate it to a known parameter. There seems to be sufficient evidence that the sediment transport rate is related to the stream velocity, as KENNEDY (1963) suggested:

$$g_{st} = m\left[\frac{\partial \phi}{\partial x}(x - \delta, D, t)\right]^n \tag{11.44}$$

where m, n, and δ are constants and said to depend on depth and velocity of the flow and on the properties of fluid and sediment. The quantity δ represents the lag of the fluid and solid-particle velocity.

After substitution of Eqs. (11.31a) into Eq. (11.44) and expansion into a binomial series, the following is obtained:

$$g_{st} = m\bar{u}^n - \frac{2\pi}{\lambda} mn\bar{u}^n a_S \frac{\cosh 2\pi(h - D)/\lambda}{\sinh 2\pi h/\lambda} \times \sin\frac{2\pi}{\lambda}(x - \delta - c_B t) + O(u^2) \tag{11.45}$$

If higher-order terms are neglected, the net forward sediment transport rate is given as

$$\bar{g}_{st} = m_1\bar{u}^n \tag{11.46}$$

Expressions of the sediment transport, as given with Eqs. (11.45) and (11.46) together with Eqs. (11.30) and (11.35), provide for a differential equation for a_S, which, in turn, allowed KENNEDY (1963)[1] to give the following relation for the bedform velocity:

$$c_B = -\frac{n\bar{g}_{st}\, 2\pi}{\lambda B} \coth\frac{2\pi}{\lambda}(h - D)\cos\frac{2\pi}{\lambda}\delta \tag{11.47}$$

and for the bed amplitude,

$$a_B = a_S \frac{\sinh 2\pi(h - D)/\lambda}{\sinh 2\pi h/\lambda} \exp\left[t\,\frac{n\bar{g}_{st} 4\pi^2}{\lambda^2 B}\coth\frac{2\pi}{\lambda}(h - D)\sin\frac{2\pi}{\lambda}\delta\right] \tag{11.48}$$

[1] Relations similar to Eqs. (11.47) and (11.48) are given by KENNEDY (1964) for bedforms in closed conduits and in the desert.

Equation (11.48) may be interpreted such that small disturbances of the sand bed will increase exponentially if $2\pi/\lambda$ and δ make the exponential term in Eq. (11.48) positive. The amplitude cannot grow indefinitely and, indeed, bedforms are observed to attain certain maximum heights. Due to linearization in the development of Eq. (11.48), the maximum height of the bedform cannot be predicted.

Conditions for the occurrence of different bed configurations may be gotten by considering Eqs. (11.35), (11.47), and (11.48). Five different configurations are distinguished and summarized in Table 11.2. A detailed discussion of the tabulated arguments is given by KENNEDY (1963). Note, however, the similarity of this classification of bedforms when compared with the one of SIMONS et al. (1965), as was given in Table 11.1.

KENNEDY (1963) proceeded using various rather involved concepts of fluid-stability analysis, and obtained

$$N_F{}^2 = \frac{\bar{u}^2}{gD} = \frac{1 + (2\pi D/\lambda) \tanh 2\pi d/\lambda + (2\pi\delta/\lambda) \cot 2\pi\delta/\lambda}{(2\pi D/\lambda)^2 + [2 + (2\pi\delta/\lambda) \cot 2\pi\delta/\lambda](2\pi D/\lambda) \tanh 2\pi d/\lambda} \tag{11.49}$$

It was suggested that δ can be given as

$$\delta = jD \tag{11.50}$$

For the limiting case of $\delta \ll D$ or $j \to 0$, Eq. (11.49) reads

$$\lim_{j\to 0} N_F{}^2 = \frac{2 + (2\pi D/\lambda) \tanh 2\pi D/\lambda}{(2\pi D/\lambda)^2 + (6\pi D/\lambda) \tanh 2\pi D/\lambda} \tag{11.51}$$

a relation quite similar to Eq. (11.40) by ANDERSON (1953).

Table 11.2 List of bed configuration and conditions of their occurrence [*after* KENNEDY (1963)]

	Bed and surface profiles	$h - D$	$k\delta$	$\sin k\delta$	$\cos k\delta$	*Movement of bed features*	*Bed configuration*
			With $k = 2\pi/\lambda$				
1	In phase	Pos.	$0 < k\delta < \frac{\pi}{2}$	Pos.	Pos.	Upstream	Antidunes
2	In phase	Pos.	$\frac{\pi}{2}$	Pos.	Zero	None	Antidunes
3	In phase	Pos.	$\frac{\pi}{2} < k\delta < \pi$	Pos.	Neg.	Downstream	Antidunes
4a	No bed waves	Neg.	$\pi < k\delta \leqq \frac{3\pi}{2}$	Neg.	Pos.	—	Flat bed
4b	No bed waves	Neg.	$0 < k\delta < \pi$	Pos.	—	—	Flat bed
4c	No bed waves	Pos.	$\pi < k\delta < 2\pi$	Neg.	—	—	Flat bed
5	Out of phase	Neg.	$\frac{3\pi}{2} < k\delta < 2\pi$	Neg.	Pos.	Downstream	Dunes

Now it remains only an academic endeavor to show relations of N_F vs. $2\pi D/\lambda$ for various j values. According to Eq. (11.49), it cannot be denied that certain interesting arguments may be derived. A more readily useful relation, however, will be obtained when Eq. (11.51) is plotted as shown in Fig. 11.13 and compared with experimental data. In addition to the curve given by Eq. (11.51), the following relations are indicated in Fig. 11.13.

1. Maximum possible Froude number for long crested features $(N_F)_m$: According to Eq. (11.32) for $h \to \infty$, the minimum wavelength is given by

$$\lambda_{min} = \frac{2\pi}{g} \bar{u}^2 \tag{11.52}$$

If Eq. (11.52) is divided by $2\pi D$, or

$$(N_F)_m{}^2 = \frac{\lambda_{min}}{2\pi D} = \frac{\bar{u}^2}{gD} \tag{11.53}$$

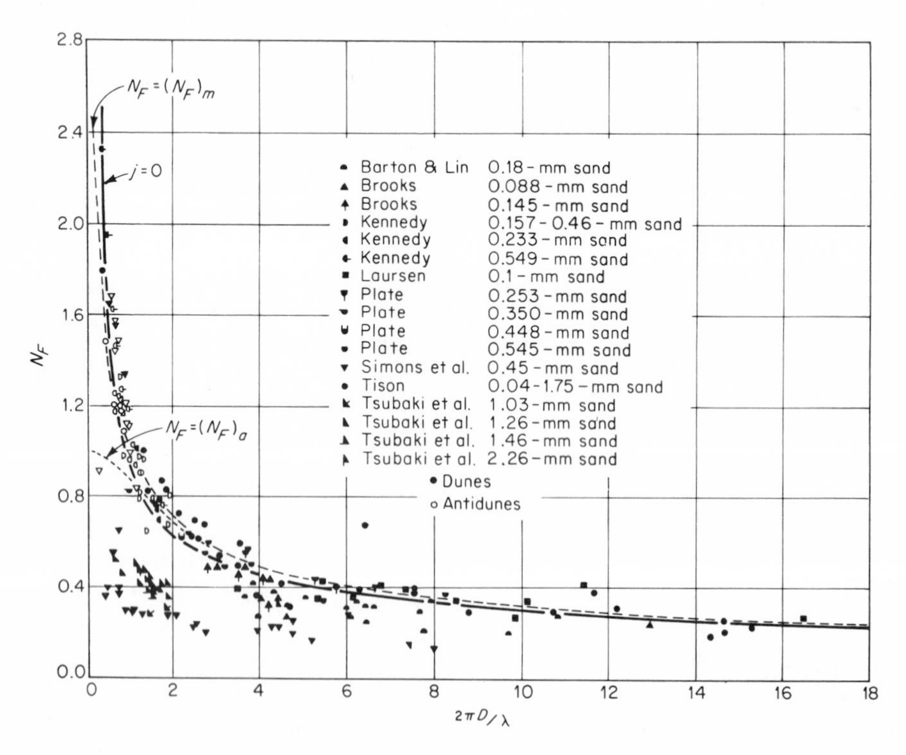

Fig. 11.13 Comparison of predicted and observed bedform regions. [*After* KENNEDY (*1963*).]

the maximum possible Froude number for two-dimensional waves is obtained. Values exceeding $(N_F)_m$ form three-dimensional, short-crested waves.

2. Minimum Froude number for antidunes and maximum Froude number for dunes: For the very special case[1] of $h = D$, Eq. (11.32) becomes, after dividing each side by gD,

$$(N_F)_a{}^2 = \frac{\bar{u}^2}{gD} = \frac{\tanh 2\pi D/\lambda}{2\pi D/\lambda} \tag{11.54}$$

It can be readily noticed that antidune data are in best agreement, while practically all dune data fall below the curve. Nevertheless, the overall results may be considered rather encouraging. KENNEDY (1963) remarked that experimental research should be oriented toward determination of the δ value and its relation to other parameters. Only then could Eqs. (11.49), (11.47), and (11.48) be checked properly. Until this is achieved, statements like the one by RAUDKIVI (1967) are in place. RAUDKIVI (1967, p. 193) writes:

> Generally, the analytical treatment is neat and informative, but it does not add much to the knowledge of the physics of the problem. It could be looked upon as an extension of Exner's and Anderson's work. Much of the results depends on this parameter, δ or j, about which nothing is known. With a different combination of parameters, the same potential flow problem could yield answers of different forms.

KENNEDY'S (1963) study was extended by REYNOLDS (1965) to the three-dimensional flow. A refinement of the sediment transport rate relation given with Eq. (11.44) was presented by HAYASHI (1970). TSUCHIYA et al. (1967) suggested to compound two waves which are generated simultaneously and have a different wavelength and velocity. The resulting relation does not compare with KENNEDY'S (1963) and ANDERSON'S (1953) relations.

11.1.3.3 Real fluid–sediment models. ENGELUND et al. (1966) considered the problem of bedforms in alluvial channels, and offered interesting ideas and conclusions. Their problem is one of stability, where, first, they try to account for the hydraulics of a two-dimensional, sinusoidal sand bed and, then, for the interaction of fluid flow and sand erosion.

The equations of motion for real fluid are set up together with the continuity equation for an incompressible fluid. Certain assumptions become necessary. After linearization, a differential equation for the unknown

[1] This statement requires agreement with KENNEDY'S (1963) postulate that "antidunes are any features for which $h > D$, while for dunes, $h < D$."

variation of the bottom shear stress τ_0 is obtained, or

$$\frac{\tau_0}{\gamma y} = I_0 - \frac{dh}{dx_1} - (1 - N_F{}^2)\frac{dy}{dx_1} - \frac{1}{6} N_F{}^2 D^2 \left(2\frac{d^3y}{dx_1{}^3} + 3\frac{d^3h}{dx_1{}^3}\right) \tag{11.55}$$

where the symbols used are identical with those of the original paper, and are described in Fig. 11.14. Another expression for the shear stress τ_0, which is independent of Eq. (11.55), was obtained by applying the knowledge of the boundary-layer theory:

$$\frac{\tau_0}{\gamma y} = I_0 - 3\frac{\eta}{D} I_0 + 0.072 \frac{N_F{}^2 D^2}{C'} \frac{d^3y}{dx_1{}^3} \tag{11.56}$$

Between the previous equations a relation for the variation of the depth may be worked out, and

$$\frac{dh}{dx_1} + (1 - N_F{}^2)\frac{dy}{dx_1} + N_F{}^2 D^2 \left[\left(\frac{1}{3} + \frac{0.072}{C'}\right)\frac{d^3y}{dx_1{}^3} + \frac{1}{2}\frac{d^3h}{dx_1{}^3}\right] - 3\frac{\eta}{D} I_0 = 0 \tag{11.57}$$

where $C' = C/\sqrt{g} = \bar{u}/u_*$.

Suppose we study a sinusoidal bed of

$$h = h_0 \sin kx_1 \tag{11.58}$$

A solution to Eq. (11.57) may be given as

$$\eta = \eta_0 \sin (kx_1 + \varphi) \tag{11.59}$$

Under this condition, the amplitude ratio is given by

$$\frac{\eta_0}{h_0} = \frac{1 - (kDN_F)^2/2}{[N_F{}^2 - 1 + A(kDN_F)^2]^2 + [3N_F{}^2/kDC'^2]^2} \tag{11.60}$$

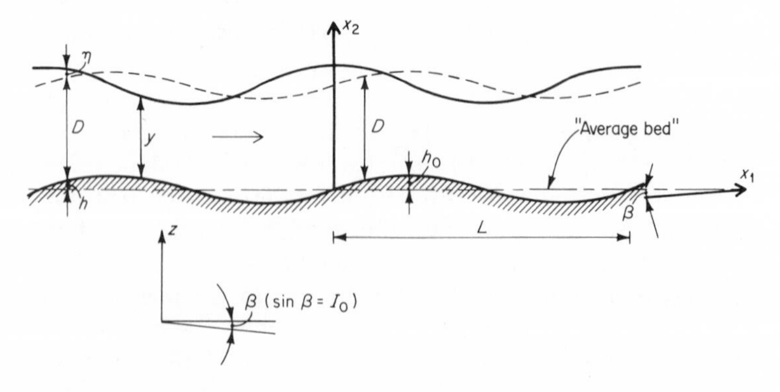

Fig. 11.14 Definition sketch for Sec. 11.1.3.3. [*After* ENGELUND *et al.* (1966).]

while the phase difference between the bed and depth is

$$\tan \varphi = \frac{3N_F^2/kDC'^2}{N_F^2 - 1 + A(kDN_F)^2} \tag{11.61}$$

with $A = \frac{1}{3} + 0.072/C'$. Concerning the shear stress, we obtain

$$\frac{\tau_0}{\gamma D I_0} = 1 - 2\frac{\eta_0}{D} \sin (kx_1 + \varphi) - 0.072C'(kD)^2 k\eta_0 \cos (kx_1 + \varphi) \tag{11.62}$$

Interesting observations can be made by comparison of Eq. (11.62) with Eq. (11.59), or

$$\frac{\tau_0}{\gamma D I_0} = 1 + a_\tau \sin (kx_1 + \varphi + \Psi_E) \tag{11.63}$$

With $\Psi_E = 0.036C'(kD)^3$ being the phase difference between the bed shear stress and the depth. The amplitude a_τ is given as

$$a_\tau = 2\frac{\eta_0}{D}\sqrt{1 + \tan^2 \Psi_E}$$

ENGELUND et al. (1966) emphasized that the maximum of the bed shear stress is located upstream of the minimum depth.

The theory, as given with the foregoing equations, was checked against experiments and found to be in reasonably good agreement.

Another expression for the shear stress at the bed is derived if the characteristics of the loose sand bed and the sediment transport are considered. Again, it is beyond the present scope to discuss details which the interested reader will find in the original exposition by ENGELUND et al. (1966). For sediment transport in nonuniform flow, ENGELUND et al. (1966) develop a relation between the bed shear and the geometry of the bedform, or

$$\frac{\tau_0(x_1)}{\gamma} = \frac{\tau_{0m}}{\gamma} + \zeta h(x_1 + l) \tag{11.64}$$

where τ_{0m} is the mean value of the bed shear, ζ is a dimensionless bedform velocity, and l is an assumed general *delay distance* between the sediment discharge and the bed shear stress, which depends on the turbulence structure. It should be pointed out that l is probably similar to the average step A_L, introduced by EINSTEIN (1950), or to the δ value used by KENNEDY (1963). ENGELUND et al. (1966) established experimentally a useful empirical relation of

$$kl = f(N_F) \tag{11.65}$$

which facilitates the determination of the delay distance; some theoretical discussions were included.

The three independent equations, Eqs. (11.55), (11.56), and (11.64), describe the dynamic interaction between the fluid flow and the movable bed. Any kind of perturbation could be applied, but ENGELUND et al. (1966) studied a sinusoidal one, given by Eq. (11.58). The stability of an original sand bed was investigated next by determining whether or not the amplitude of the disturbances will increase, decrease, or remain unchanged.

Stability conditions were investigated for different N_F vs. C' combinations and have been plotted in the *stability* diagram, which is reproduced in Fig. 11.15. Regions indicated as *plane bed* are stable against two-dimensional disturbances. *Antidune* regions are those where the bedforms travel upstream; in *dune* regions, the opposite is true. For a given set of hydraulic conditions, it is possible, according to Fig. 11.15, to predict the resulting bedform and its stability.

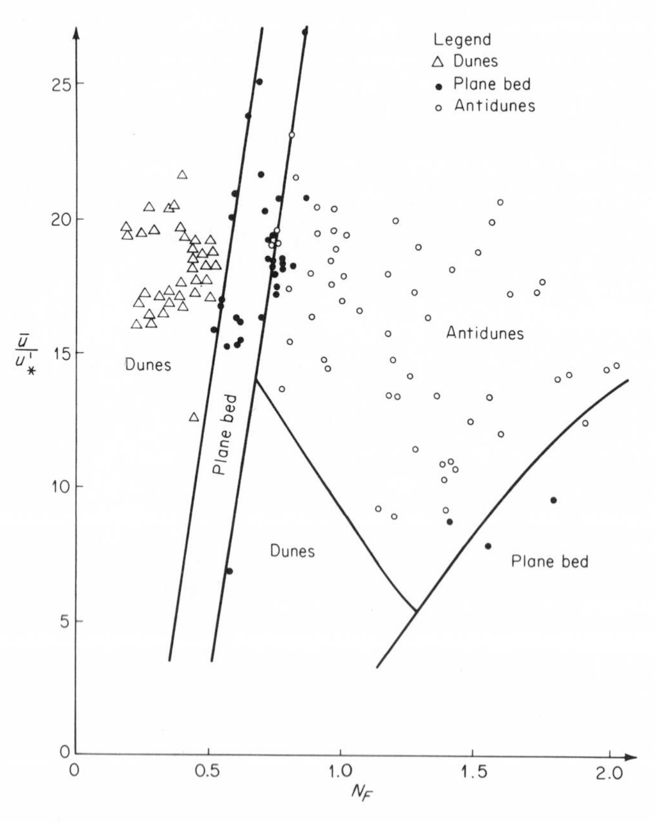

Fig. 11.15 Stability diagram; the bedforms are indicated. [*After* ENGELUND *et al.* (1966).]

As yet, only the two-dimensional problem had been considered. ENGELUND et al. (1966) generalized the stability calculations to three-dimensional disturbances and obtained preliminary but encouraging results.

In a similar way, but not as exhaustive as ENGELUND et al. (1966), the problem of the bedforms was handled by KONDRAT'EV (1962), RAUDKIVI (1967), and MATSUNASHI (1967).

11.1.3.4 Other investigations. The kinematics of bedforms was considered by ERTEL (1966) and FÜHRBÖTER (1967). This rather simple approach to downstream-progressing bedforms leads to relations for the height and velocity of bedforms. However, the coefficients and exponents involved have not been checked out with experimental data.

By applying the concept of the specific energy and continuity, KONDRAT'EV (1964) showed a rather straightforward relationship between the profile of the bottom and the one of the free surface when bedform movement took place. The results and conclusions are not new and may be found in previously discussed theories. This study established beyond doubt the importance of the Froude number in bedform mechanics. An interesting feature of KONDRAT'EV'S (1964) study is the kinematic scheme of the bottom roller, i.e., the vortex which develops behind bedforms.

HILL (1966) put forward some ideas concerning two kinds of instabilities, namely, dunes replacing flat beds and ripples replacing flat beds. These two distinct phenomena are governed by different processes and are discussed by HILL (1966). Subsequently, HILL et al. (1969) proposed a functional relationship which governs the problem, such as

$$\frac{u_* d}{\nu} = f\left(\frac{gd^3}{\nu^2}\right)$$

Limited data were used to establish this relation.

11.2 BEDFORMS AND FLOW RESISTANCE

11.2.1 GENERAL REMARKS

Consider a plane bed. The resistance to the flow is given by

$$\tau_0 = \tau_0' \tag{11.66}$$

The shear stress τ_0' is due to the grain roughness and is often referred to as the *surface drag*.

Consider a bed with bedforms superimposed. The resistance to the flow is given by

$$\tau_0 = \tau_0' + \tau_0'' \tag{11.67}$$

The additional shear stress τ_0'' is due to the bedforms and is frequently referred to as the *form drag*. That the total resistance τ_0 can be expressed as the sum of the individual components has been discussed by EINSTEIN et al. (1950). Of course, this holds as long as the separate components do not exert a mutual influence on one another. Equation (11.67) could be rewritten, and we obtain[1]

$$\tau_0 = \gamma S(R_h' + R_h'') \tag{11.68}$$

where R_h' and R_h'' are the hydraulic radii due to grain roughness and bedforms, respectively. Furthermore, we can obtain from Eq. (11.67) the following shear velocities relationship:

$$u_*^2 = u_*'^2 + u_*''^2 \tag{11.69}$$

11.2.2 FLOW RESISTANCE IN THE ABSENCE OF BEDFORMS

The resistance to steady uniform flow is commonly expressed with the Weisbach-Darcy relation, or

$$S = \frac{h_L}{L} = f\frac{1}{4R_h}\frac{\bar{u}^2}{2g} \tag{11.70}$$

where f is the Weisbach-Darcy friction factor. Between Eq. (11.70) and other frequently used flow formulas, such as the Chezy and the Manning equation, there exist the following well-known relations. For the Chezy C value,

$$C = \sqrt{\frac{8g}{f}} \tag{11.71}$$

and for Manning's n value (in metric units),

$$n = R_h^{1/6}\sqrt{\frac{f}{8g}} \tag{11.72}$$

There exists also a relationship between the logarithmic velocity distribution and the friction factor, such as

$$\frac{\bar{u}}{u_*} = \sqrt{\frac{8}{f}} \tag{11.73}$$

The Manning flow formula, given by

$$\bar{u} = \frac{1}{n} R_h^{2/3} S^{1/2} \tag{11.74}$$

[1] This equation can be written as $\tau_0 = \gamma R_h(S' + S'')$. [See Sec. 7.3.2.3.]

is possibly the most popular equation presently in use. It is therefore not surprising that much information on the n value is available. CHOW (1959) has tabulated n values in what seems to be a most comprehensive list; it is given in Table 11.3. An interesting approach to the evaluation of the roughness coefficient for natural channels is presented by BARNES (1967). With the aid of descriptive data and color photographs from 50 different stream channels of the United States "typical" n-values are suggested. The range of $0.024 < n < 0.075$ has been covered; the surveyed channels are considered to be stable.

It should be mentioned that n values, as given in Table 11.3 and in other tabulations, do sometimes include both surface and form drag effects. Knowledge of the n value permits calculation of the flow resistance according to Eq. (11.74).

Another possibility for the determination of flow resistance exists by solving Eq. (11.70). However, its solution depends on knowledge of the f value, which can be expressed with Eq. (11.73). The f value depends on the flow Reynolds number and the roughness of the channel. Whereas very good charts are available for the f value in pipe hydraulics, channel hydraulics still lacks this valuable information. It has been shown by GRAF (1964) and other investigators that starting with the logarithmic velocity distribution, equations for the f value may be derived:

$$f' = \left[2.0 \log \left(\frac{\bar{u}R_h'}{\nu} \sqrt{f}\right) + 0.32\right]^{-2} \tag{11.75}$$

for a smooth channel and

$$f' = \left[2.0 \log\left(\frac{R_h'}{k_s}\right) + 2.2\right]^{-2} \tag{11.76}$$

for a rough channel; these equations apply for plane channels only and, thus, R_h' replaces R_h and f' replaces f. Unfortunately, very little information is available on the equivalent (*Nikuradse*) sand roughness k_s, and these equations must remain, therefore, limited in their use. An extensive discussion of the f value may be found in SILBERMAN et al. (1963).

A rather useful relation was given by EINSTEIN (1950) as

$$\frac{\bar{u}}{u_*} = 5.75 \log\left(12.27 \frac{R_h'X}{k_s}\right) \tag{11.77}$$

The apparent roughness k_s/X must be obtained from Fig. 7.10, where X is a correction factor incorporating the smooth-transition–rough-wall effects. It was assumed that $k_s \approx d_{65}$. Note the similarity between Eqs. (11.77) and (7.62). Due to the presence of suspended matter, Eq. (11.77) may have to be modified; for a discussion on this aspect, the reader is referred to Sec. 8.3.2.3.2.

Table 11.3 Values for the Manning *n*-value [*after* CHOW (*1959*)]

Type of channel and description	*Minimum*	*Normal*	*Maximum*
A. Closed conduits flowing partly full			
A-1. Metal			
a. Brass, smooth	0.009	0.010	0.013
b. Steel			
1. Lockbar and welded	0.010	0.012	0.014
2. Riveted and spiral	0.013	0.016	0.017
c. Cast iron			
1. Coated	0.010	0.013	0.014
2. Uncoated	0.011	0.014	0.016
d. Wrought iron			
1. Black	0.012	0.014	0.015
2. Galvanized	0.013	0.016	0.017
e. Corrugated metal			
1. Subdrain	0.017	0.019	0.021
2. Storm drain	0.021	0.024	0.030
A-2. Nonmetal			
a. Lucite	0.008	0.009	0.010
b. Glass	0.009	0.010	0.013
c. Cement			
1. Neat, surface	0.010	0.011	0.013
2. Mortar	0.011	0.013	0.015
d. Concrete			
1. Culvert, straight and free of debris	0.010	0.011	0.013
2. Culvert with bends, connections, and some debris	0.011	0.013	0.014
3. Finished	0.011	0.012	0.014
4. Sewer with manholes, inlet, etc., straight	0.013	0.015	0.017
5. Unfinished, steel form	0.012	0.013	0.014
6. Unfinished, smooth wood form	0.012	0.014	0.016
7. Unfinished, rough wood form	0.015	0.017	0.020
e. Wood			
1. Stave	0.010	0.012	0.014
2. Laminated, treated	0.015	0.017	0.020
f. Clay			
1. Common drainage tile	0.011	0.013	0.017
2. Vitrified sewer	0.011	0.014	0.017
3. Vitrified sewer with manholes, inlet, etc.	0.013	0.015	0.017
4. Vitrified subdrain with open joint	0.014	0.016	0.018
g. Brickwork			
1. Glazed	0.011	0.013	0.015
2. Lined with cement mortar	0.012	0.015	0.017
h. Sanitary sewers coated with sewage slimes, with bends and connections	0.012	0.013	0.016
i. Paved invert, sewer, smooth bottom	0.016	0.019	0.020
j. Rubble masonry, cemented	0.018	0.025	0.030

Table 11.3 (*Continued*)

Type of channel and description	*Minimum*	*Normal*	*Maximum*
B. Lined or built-up channels			
B-1. Metal			
a. Smooth steel surface			
1. Unpainted	0.011	0.012	0.014
2. Painted	0.012	0.013	0.017
b. Corrugated	0.021	0.025	0.030
B-2. Nonmetal			
a. Cement			
1. Neat, surface	0.010	0.011	0.013
2. Mortar	0.011	0.013	0.015
b. Wood			
1. Planed, untreated	0.010	0.012	0.014
2. Planed, creosoted	0.011	0.012	0.015
3. Unplaned	0.011	0.013	0.015
4. Plank with battens	0.012	0.015	0.018
5. Lined with roofing paper	0.010	0.014	0.017
c. Concrete			
1. Trowel finish	0.011	0.013	0.015
2. Float finish	0.013	0.015	0.016
3. Finished, with gravel on bottom	0.015	0.017	0.020
4. Unfinished	0.014	0.017	0.020
5. Gunite, good section	0.016	0.019	0.023
6. Gunite, wavy section	0.018	0.022	0.025
7. On good excavated rock	0.017	0.020	—
8. On irregular excavated rock	0.022	0.027	—
d. Concrete bottom float finished with sides of			
1. Dressed stone in mortar	0.015	0.017	0.020
2. Random stone in mortar	0.017	0.020	0.024
3. Cement rubble masonry, plastered	0.016	0.020	0.024
4. Cement rubber masonry	0.020	0.025	0.030
5. Dry rubble or riprap	0.020	0.030	0.035
e. Gravel bottom with sides of			
1. Formed concrete	0.017	0.020	0.025
2. Random stone in mortar	0.020	0.023	0.026
3. Dry rubble or riprap	0.023	0.033	0.036
f. Brick			
1. Glazed	0.011	0.013	0.015
2. In cement mortar	0.012	0.015	0.018
g. Masonry			
1. Cemented rubble	0.017	0.025	0.030
2. Dry rubble	0.023	0.032	0.035
h. Dressed ashlar	0.013	0.015	0.017
i. Asphalt			
1. Smooth	0.013	0.013	—
2. Rough	0.016	0.016	—
j. Vegetal lining	0.030	—	0.500
C. Excavated or dredged			
a. Earth, straight and uniform			
1. Clean, recently completed	0.016	0.018	0.020

Table 11.3 (*Continued*)

Type of channel and description	Minimum	Normal	Maximum
C. Excavated or dredged (*Continued*)			
2. Clean, after weathering	0.018	0.022	0.025
3. Gravel, uniform section, clean	0.022	0.025	0.030
4. With short grass, few weeds	0.022	0.027	0.033
b. Earth, winding and sluggish			
1. No vegetation	0.023	0.025	0.030
2. Grass, some weeds	0.025	0.030	0.033
3. Dense weeds or aquatic plants in deep channels	0.030	0.035	0.040
4. Earth bottom and rubble sides	0.028	0.030	0.035
5. Stony bottom and weedy banks	0.025	0.035	0.040
6. Cobble bottom and clean sides	0.030	0.040	0.050
c. Dragline-excavated or dredged			
1. No vegetation	0.025	0.028	0.033
2. Light brush on banks	0.035	0.050	0.060
d. Rock cuts			
1. Smooth and uniform	0.025	0.035	0.040
2. Jagged and irregular	0.035	0.040	0.050
e. Channels not maintained, weeds and brush uncut			
1. Dense weeds, high as flow depth	0.050	0.080	0.120
2. Clean bottom, brush on sides	0.040	0.050	0.080
3. Same, highest stage of flow	0.045	0.070	0.110
4. Dense brush, high stage	0.080	0.100	0.140
D. Natural streams			
D-1. Minor streams (top width at flood stage <100 ft)			
a. Streams on plain			
1. Clean, straight, full stage, no rifts or deep pools	0.025	0.030	0.033
2. Same as above, but more stones and weeds	0.030	0.035	0.040
3. Clean, winding, some pools and shoals	0.033	0.040	0.045
4. Same as above, but some weeds and stones	0.035	0.045	0.050
5. Same as above; lower stages; more ineffective slopes, sections	0.040	0.048	0.055
6. Same as 4, but more stones	0.045	0.050	0.060
7. Sluggish reaches, weedy, deep pools	0.050	0.070	0.080
8. Very weedy reaches, deep pools, or floodways with heavy stand of timber and underbrush	0.075	0.100	0.150
b. Mountain streams, no vegetation in channel, banks usually steep, trees and brush along banks submerged at high stages			
1. Bottom: gravels, cobbles, and few boulders	0.030	0.040	0.050
2. Bottom: cobbles, large boulders	0.040	0.050	0.070

Table 11.3 (Continued)

Type of channel and description	*Minimum*	*Normal*	*Maximum*
D. Natural streams (*Continued*)			
D-2. Flood plains			
a. Pasture, no brush			
1. Short grass	0.025	0.030	0.035
2. High grass	0.030	0.035	0.050
b. Cultivated areas			
1. No crop	0.020	0.030	0.040
2. Mature row crops	0.025	0.035	0.045
3. Mature field crops	0.030	0.040	0.050
c. Brush			
1. Scattered brush, heavy weeds	0.035	0.050	0.070
2. Light brush and trees, in winter	0.035	0.050	0.060
3. Light brush and trees, in summer	0.040	0.060	0.080
4. Medium to dense brush, in winter	0.045	0.070	0.110
5. Medium to dense brush, in summer	0.070	0.100	0.160
d. Trees			
1. Dense willows, summer, straight	0.110	0.150	0.200
2. Cleared land with tree stumps, no sprouts	0.030	0.040	0.050
3. Same as above, but with heavy growth of sprouts	0.050	0.060	0.080
4. Heavy stand of timber, a few down trees, little undergrowth, flood stage below branches	0.080	0.100	0.120
5. Same as above, but with flood stage reaching branches	0.100	0.120	0.160
D-3. Major streams (top width at flood stage >100 ft). The n value is less than that for minor streams of similar description, because banks offer less effective resistance.			
a. Regular section with no boulders or brush	0.025	—	0.060
b. Irregular and rough section	0.035	—	0.100

Thus either Eq. (11.75) or Eq. (11.76) or Eq. (11.77) together with Eq. (11.73), may be used to calculate the surface drag.

Before discussing the bedforms and their effect on the resistance, it may be worthwhile to quote important conclusions of an ASCE Task Committee on Friction Factors in Open Channels under the chairmanship of SILBERMAN et al. (1963):

> There is great utility in expressing resistance to steady, fully developed flow in uniform channels in terms of the Darcy-Weisbach f. Manning's n is also useful for this purpose, but has certain limitations, which are explained; it is especially useful for fully rough flow. For smooth flow

and partly rough flow, pipe resistance diagrams of the Moody type may be used for estimating f (and obtaining n therefrom). Definite values of f or n for movable bed channels are difficult to specify, as is the criterion for determining when fixed-bed channels become movable.

and

. . . At the present (1961) stage of knowledge, if applied with judgement, both n and f are probably equally effective in the solution of practical problems.

11.2.3 FLOW RESISTANCE IN THE PRESENCE OF BEDFORMS

It appears that bedforms and their flow resistances have been studied by two different groups. One of these considers the shear stress and shear velocity as criteria; the other takes the friction factor f. Despite the fact that the two approaches are related by Eq. (11.73), it will facilitate the discussion to treat them in separate sections.

11.2.3.1 Flow resistance expressed with the friction velocity. In Sec. 11.1.2 it was remarked that bedforms change if the rates of sediment

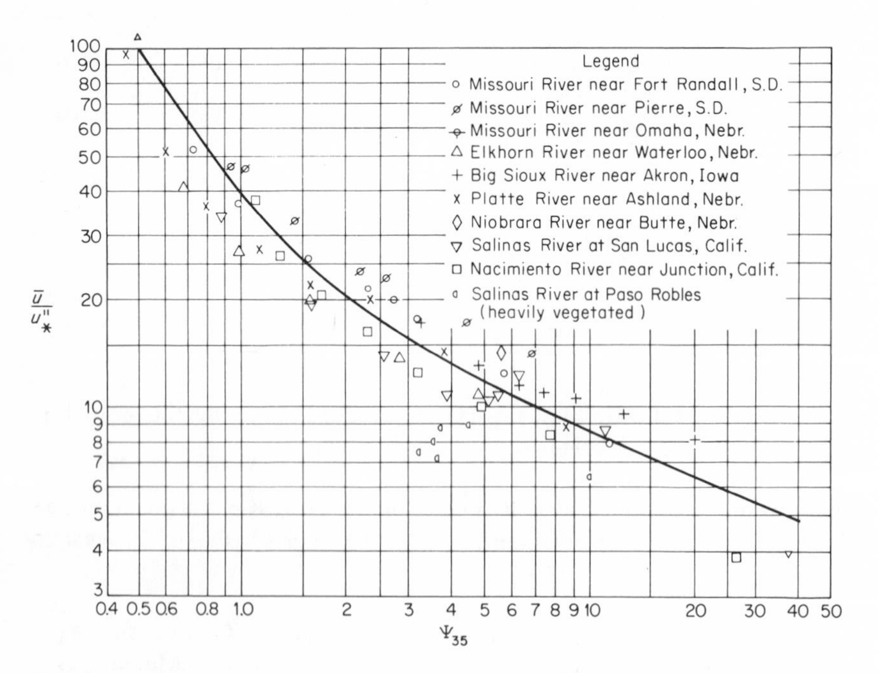

Fig. 11.16 Flow resistance due to bedforms. [*After* EINSTEIN *et al.* (*1952*).]

transport change. Inspection of Table 11.1 confirms this statement. It is therefore likely that a certain relation exists between the flow resistance due to bedforms and the total sediment transport. This was pointed out by EINSTEIN (1950) who suggested a functional relation such as

$$\frac{\bar{u}}{u''_*} = \text{fct}\,(\Psi'_{35}) \tag{11.78}$$

where Ψ'_{35} is the intensity of shear on representative particles, and is given by

$$\Psi'_{35} = \frac{\rho_s - \rho}{\rho} \frac{d_{35}}{R'_h S} \tag{11.79}$$

where d_{35} is the sieve size in the bed material of which 35 percent is finer.

At the present there exists no analytical function for Eq. (11.78), but an empirical relation has been established. Actual river measurements from the northwestern United States were used by EINSTEIN et al. (1952) to determine the foregoing relation. The relation obtained from these data is shown in Fig. 11.16. Realizing the difficulties involved in working with field data, we must agree that the degree of scatter is not excessive. EINSTEIN et al. (1952) interpret Fig. 11.16 as follows:

> The effect of irregularities (bedforms) is to distort the flow pattern. When the discharge is least, the distortion of the flow pattern is greatest; as witness the meandering of natural streams at low flows. As the discharge increases and hence the sediment transport along the bed also increases, the distortion of the flow pattern becomes less and less because the alinement of flow becomes progressively straighter. Consequently, one may expect that the additional friction loss, u''_*, diminishes as the discharge increases.

The relation of Fig. 11.16 was confirmed by NORDIN (1964) with data from the Rio Grande, and by HARRISON et al. (1967) with field data from the Missouri and Mississippi. Laboratory data were found to be in disagreement with the proposed curve. VANONI et al. (1957) reported substantial deviation especially for smaller Ψ'_{35} values. EINSTEIN et al. (1959) derived a separate curve from the analysis of flume data. These flume data and river data curves are shown in Fig. 11.17. Figure 11.17 also presents laboratory data from a 8-ft-wide flume by SIMONS et al. (1966). It is evident that for small and large Ψ'_{35} values the data depart systematically from either curve.[1] The apparent difference between flume and field data is little understood. VANONI et al. (1957) speculate that the wall effect in flume experiments or the scale effect may be a possible explanation.

[1] The low $\bar{u}/u''_*$ values at the small Ψ'_{35} values are due to formation of antidunes.

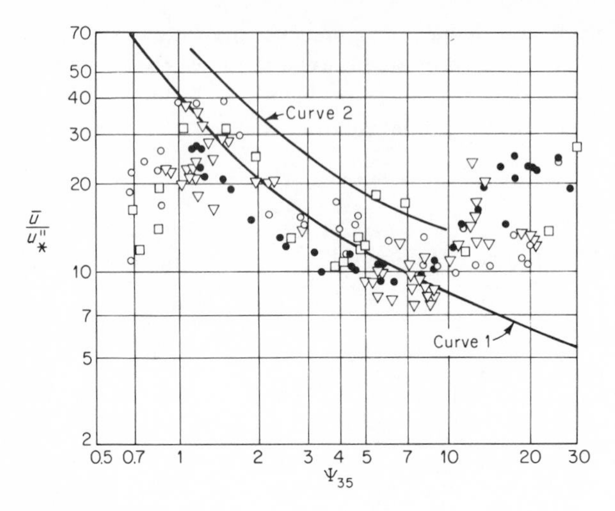

Fig. 11.17 Flow resistance due to bedforms; curve 1—river data, curve 2—flume data. [*After* SIMONS *et al.* (*1966*).]

The relation given by Eq. (11.78) by EINSTEIN (1950) was extended by SHEN (1962). Based on physical reasoning and experimental evidences, SHEN (1962) suggested a relation, such as

$$\frac{\bar{u}}{u''_*} = \text{fct}\left(\Psi_{35}, \frac{v_{ss}d}{\nu}\right) \tag{11.80}$$

which includes a Reynolds number in terms of the settling velocity. Both flume and field data were utilized to establish Eq. (11.80). The resulting graph for flume data only is given by Fig. 11.18; field data follow the same trend. For values of $\Psi_{35}/\lambda' < 1.5$, a plane bed is reported. Data exceeding this value can have two branches. One for fine sediments ($v_{ss}d/\nu < 100$), with dunes and ripples; this branch is essentially given by the relation of EINSTEIN et al. (1952). The other one is for coarse sediments ($v_{ss}d/\nu > 100$) and, interestingly enough, dunes develop. The latter statement is in agreement with the data given in Fig. 11.4. Analysis of data from the Mondego River by CUNHA (1967) are in general agreement with the findings by SHEN (1962).

Considering the theory of dimensions, YALIN (1964) developed a relation for the flow resistance of a stream, such as

$$\frac{\bar{u}}{u_*} = \text{fct}\left[\frac{du_*}{\nu}, \frac{\rho u_*^2}{(\gamma_s - \gamma)d}, \frac{D}{d}, \frac{\rho}{\rho_s}\right] \tag{11.81}$$

Equation (11.81) includes essentially only those parameters used in other relations, like the one by EINSTEIN (1950), and thus does not add much to current knowledge.

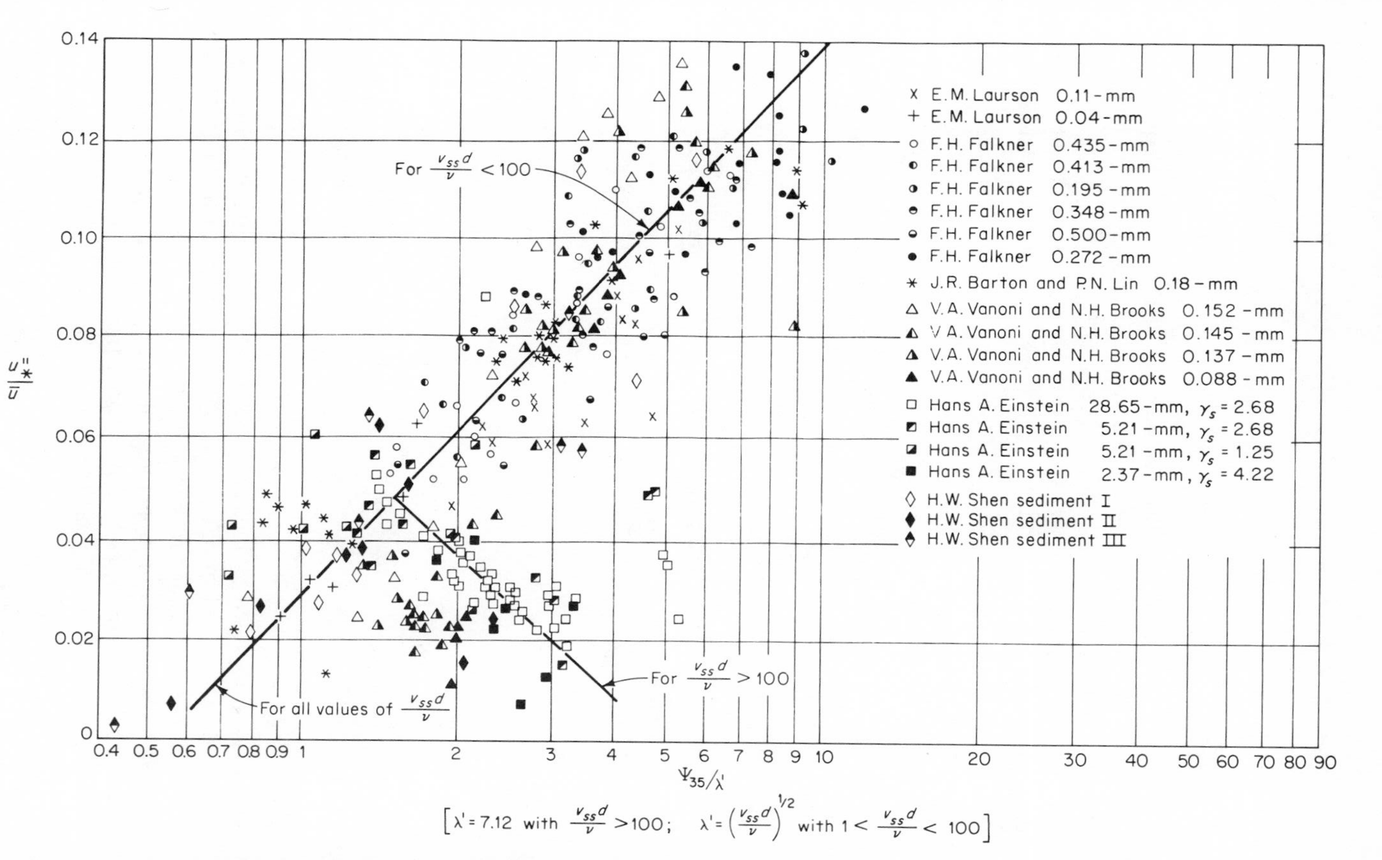

Fig. 11.18 Flow resistance due to bedforms; flume data. [*After* SHEN (*1962*).

With the aid of flume and field data, GARDE et al. (1966) suggested the following relation:

$$\frac{\bar{u}}{\sqrt{[(\gamma_s - \gamma)/\gamma]gd}} = K_G\left(\frac{R_h}{d}\right)^{2/3}\left(\frac{\gamma S}{\gamma_s - \gamma}\right)^{1/2} \tag{11.82}$$

It was pointed out that Eq. (11.82) has the form of the Manning equation. The coefficient K_G is given by $K_G = 3.2$ for ripples and dunes, and by $K_G = 6.0$ for the transition and antidunes(?).

An interesting approach was suggested by ENGELUND (1966) who expressed the flow resistance due to bedforms with the expansion-loss equation

$$\Delta h = \alpha \frac{\bar{u}^2}{2g}\left(\frac{\Delta H}{D}\right)^2$$

If this knowledge is introduced into Eq. (11.67), and the resulting equation is divided by γD, the following is obtained:

$$\frac{\tau_0}{\gamma D} = \frac{\tau_0'}{\gamma D} + \frac{\bar{u}^2}{2g\lambda}\alpha\left(\frac{\Delta H}{D}\right)^2 \tag{11.83}$$

where λ and ΔH are the bedform length and height, respectively. Let

$$\Theta = \frac{DS}{(\rho_s/\rho - 1)d}$$

$$\Theta' = \frac{D'S}{(\rho_s/\rho - 1)d} \tag{11.84}$$

and

$$\Theta'' = \frac{1}{2}N_F{}^2\frac{\alpha(\Delta H)^2}{(\rho_s/\rho - 1)\,d\lambda}$$

Then Eq. (11.83) reads

$$\Theta = \Theta' + \Theta'' \tag{11.85}$$

This relation was suggested by ENGELUND et al. (1967) and plotted with help of the flume data reported by GUY et al. (1966); it is given in Fig. 11.19. Note that a lower and upper regime and a transition zone in between are distinguished. In part of the upper regime—the plane bed and standing-wave zone—no expansion losses take place and, therefore, $\Theta = \Theta'$.

Recently RAUDKIVI (1967) has suggested a relation of

$$\frac{\bar{u}}{\sqrt{u_*^2 - (u_*^2)_{cr}}} = \text{fct}\left[\frac{\rho u_*^2}{(\gamma_s - \gamma)d}\right] \tag{11.86}$$

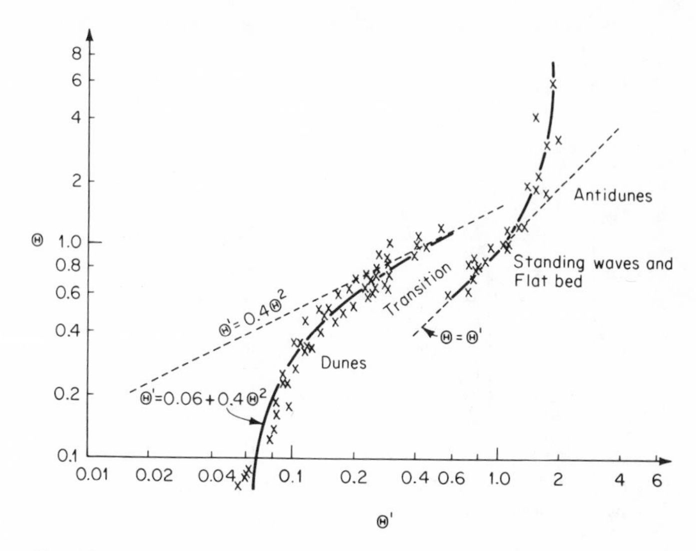

Fig. 11.19 Flow resistance given by Eq. (11.83). [*After* ENGELUND *et al.* (*1967*).]

where $(u_*)_{cr}$ is the threshold value of the shear velocity. But owing to appreciable scatter—at the best a trend is exhibited—this relation can as yet not be recommended for use. Another interesting beginning is given by ZNAMENSKAYA (1967) who analyzed experimentally the microstructure of the turbulence caused by the bedforms, and obtained an empirical relation between the geometry of the bedforms and the flow resistance.

At this point it is worth remarking that BAJORUNAS (1952) investigated the effect of bedforms on Manning's n value. Analogous to Eq. (11.67), we obtain

$$n = n' + n'' \tag{11.87}$$

Similar to EINSTEIN's (1950) relation, which is given by Eq. (11.78),

$$n'' = \text{fct}(\Psi'_{35}) \tag{11.88}$$

was established with the use of actual river data.

Finally, we should not miss pointing to a convincing demonstration of Eq. (11.67), or

$$\tau_0 = \tau_0' + \tau_0'' \tag{11.67}$$

In Fig. 11.20 RAUDKIVI (1967) presents a demonstration as to how the shear stress τ_0 varies owing to an increase in the average velocity $\bar{u}$. For sake of comparison, the surface drag τ_0' corresponding to the plane bed resistance is shown with a dashed line; the difference $\tau_0 - \tau_0'$ represents the form drag.

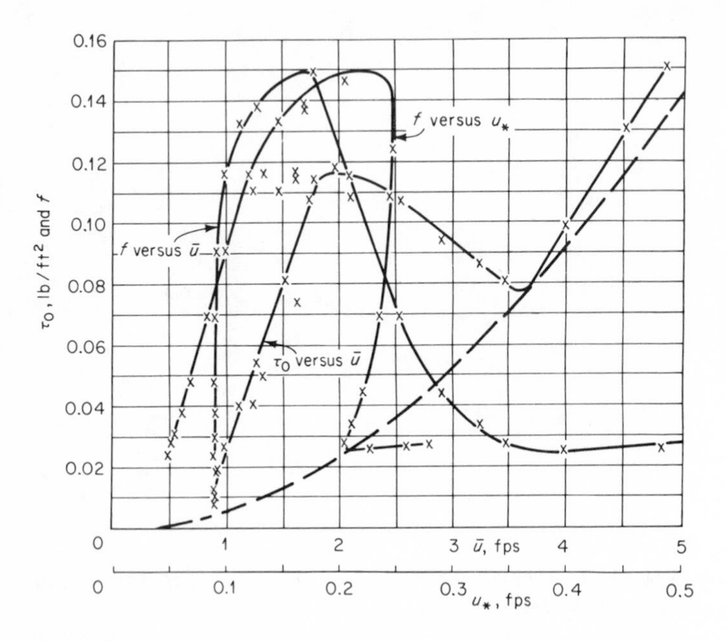

Fig. 11.20 Flow resistance due to bedforms. [*After* RAUDKIVI (*1967*).]

In this experiment, it appears that, at $\bar{u} \approx 3.7$ fps, a plane bed exists and, thus, the upper regime begins. Immediately below $\bar{u} \approx 3.7$ fps, there should be the transition zone, while the maximum shear stress τ_0 corresponds to the steepest bedforms. Also, note that certain shear stress values may occur at several different velocities. (Take, for example, $\tau_0 = 0.10$ lb/ft^2; it may be obtained at three different values of average velocities.) This points to the fact that the shear stress alone does not describe the flow adequately if bedforms are present.

11.2.3.2 Flow resistance expressed with the friction factor. Consider Fig. 11.20.[1] Note that the f value increases rapidly as soon as bedforms are created, reaches a certain maximum value, and then decreases as the bedforms are washed out; in the transition region the friction factor f is as low as it was at the initiation of the bed material motion, and increases slightly when the bedforms of the upper regime are formed. The described trend is found in both the f vs. $\bar{u}$ and the f vs. u_* relationships. A certain friction factor f may result from one or more average velocities and from one or more shear velocities. Thus neither the average velocity nor the shear velocity is always sufficient to describe the flow completely.

[1] Similar relations are presented by SIMONS et al. (1961), SIMONS et al. (1966), and KENNEDY et al. (1963).

The foregoing problem has been considered by BROOKS (1958) and VANONI et al. (1957). It was expressed that such variation of the roughness is due to the sediment transport, and is caused by two processes. The principal cause is the existence or lack of bedforms; a secondary cause is the damping effect of the suspended sediment material. That the latter effect is much smaller than the former was shown by VANONI et al. (1960). Since both of these effects often act simultaneously, it becomes difficult to separate them. BROOKS (1958) showed experimentally in the laboratory that neither the flow velocity nor the bed material transport can be expressed as single-valued functions of the shear stress or of any combination of slope and flow depth. The cause for the lack of uniqueness was believed to be the variation of bedforms. Subsequently, EINSTEIN et al. (1958*a*), KENNEDY et al. (1963), and NORDIN (1964) have shown that BROOKS' (1958) experiments covered a rather limited range of flow and sediment conditions, namely, the transition region between the upper and lower regimes.

Considerable research toward an analytical or empirical relation for prediction of the flow resistance was done by SIMONS et al. (1961*a*), SIMONS et al. (1966), and RICHARDSON et al. (1967). Dimensional analysis may be employed such that the number of variables is reduced to dimensionless numbers. A relation that correlated the available data rather well was given by SIMONS et al. (1961*a*) as

$$f = \text{fct}\left[a\left(\frac{u_* d}{\nu} + b\right)^c N_F^e \frac{\gamma_s - \gamma}{\gamma} \frac{d}{SD} \right] \tag{11.89}$$

where a, b, c, and e are constants to be determined experimentally.

RICHARDSON et al. (1967) used the velocity distribution law and determined the empirical values with laboratory and field data. Since

$$\sqrt{\frac{f}{8}} = \frac{C}{\sqrt{g}} = \frac{\bar{u}}{u_*} \tag{11.90}$$

where C is the Chezy coefficient, the following resistance equations are suggested:

1. For a plane bed with little or no sediment transport,

$$\frac{C}{\sqrt{g}} = 5.9 \log \frac{D}{d_{85}} + 5.44 \tag{11.91}$$

2. For a plane bed with appreciable sediment transport,

$$\frac{C}{\sqrt{g}} = 7.4 \log \frac{D}{d_{85}} \tag{11.92}$$

3. For ripples (in English units),

$$\frac{C}{\sqrt{g}} = \left(7.66 - \frac{0.3}{u_*}\right) \log D + \frac{0.13}{u_*} + 11 \tag{11.93}$$

4. For dunes and antidunes,

$$\frac{C}{\sqrt{g}} = 7.4 \log \frac{D}{d_{85}} \sqrt{1 - \frac{\Delta R_h S}{R_h S}} \tag{11.94}$$

The term $\Delta R_h S$ is the increase of $R_h S$ due to form roughness, and is given in Fig. 11.21.

Resistance values in terms of $C/\sqrt{g}$ for the different flow regimes as they occur in the flume experiments are reported by RICHARDSON et al. (1967) and are included into Table 11.1. If there exists a lack of a pronounced trend of $C/\sqrt{g}$, it is probably due to the fact that bedforms are not mutually exclusive occurrences in space or time.

With the combined knowledge of Eqs. (11.69) and (11.73), we may write

$$f = f' + f'' \tag{11.95}$$

This equation was briefly discussed by SILBERMAN et al. (1963) and investigated by VANONI et al. (1967). Both research groups suggested that the f' value may be obtained from a pipe friction diagram using $D = 4R_h$, such that

$$f' = \text{fct}(N_R, k_s/4R_h) \tag{11.96}$$

VANONI et al. (1967) used laboratory data and found that a relation for f'' may be obtained which reasonably explains the phenomenon. The relation applicable for hydrodynamically rough beds was given as

$$\frac{1}{\sqrt{f''}} = 3.5 \log \frac{R_h}{e\,\Delta\bar{H}} - 2.3 \tag{11.97}$$

where $\Delta\bar{H}$ is the mean height of the bedforms and e is the exposure parameter, $e = A/A_s$ (A is the total area and A_s is the horizontal projection of the lee face of the bedforms). VANONI et al. (1967) concluded that the roughness length for bedform channel flow is given by $e\,\Delta\bar{H}$. For the case where e is not determined, an equation in terms of the bedform steepness is given as

$$\frac{1}{\sqrt{f''}} = 3.3 \log \frac{\lambda R_h}{(\Delta\bar{H})^2} - 2.3 \tag{11.98}$$

Note that Eqs. (11.97) and (11.98) are apparently restricted to the lower regime. For flat-bed flow, LOVERA et al. (1969) proposed to represent the friction factor in a format similar to the Moody diagram for pipe friction.

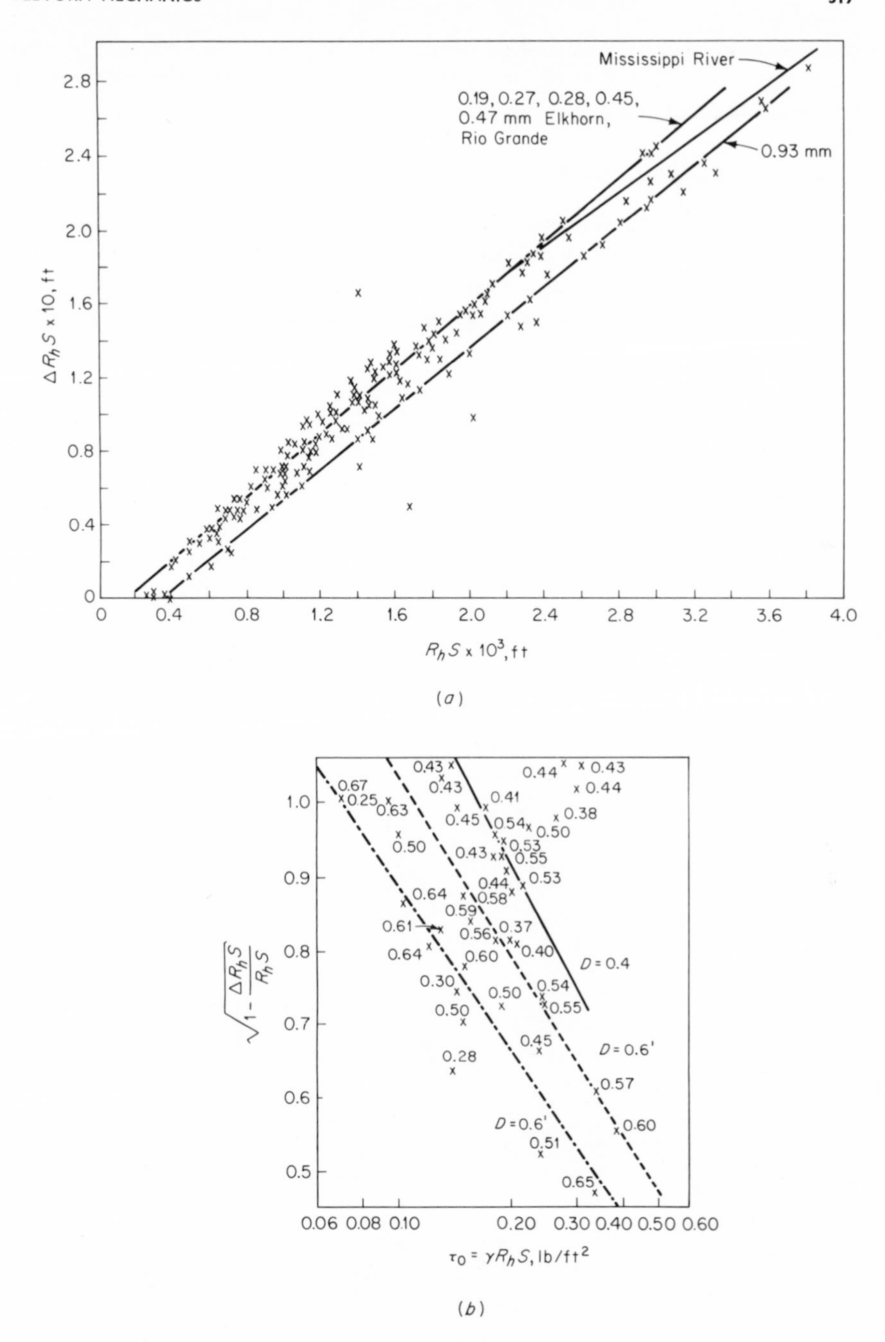

Fig. 11.21 Correction term $\Delta R_h S$ as used in Eq. (11.94). (*a*) For dune bed configuration; (*b*) for antidune bed configuration. [*After* RICHARDSON *et al.* (*1967*).]

The friction factor f was plotted against the Reynolds number, $\bar{u}R_h/\nu$, with the relative roughness R_h/d_{50} as a parameter. Many laboratory and natural data yielded a promising result.

11.2.4 CONCLUDING REMARKS

From the preceding discussion it must be learned that recognition of the formation of bedforms as well as their effect on the flow resistance is paramount for river engineering. Many of the bed material load equations take either explicitly or implicitly care of this effect. In no other relation,

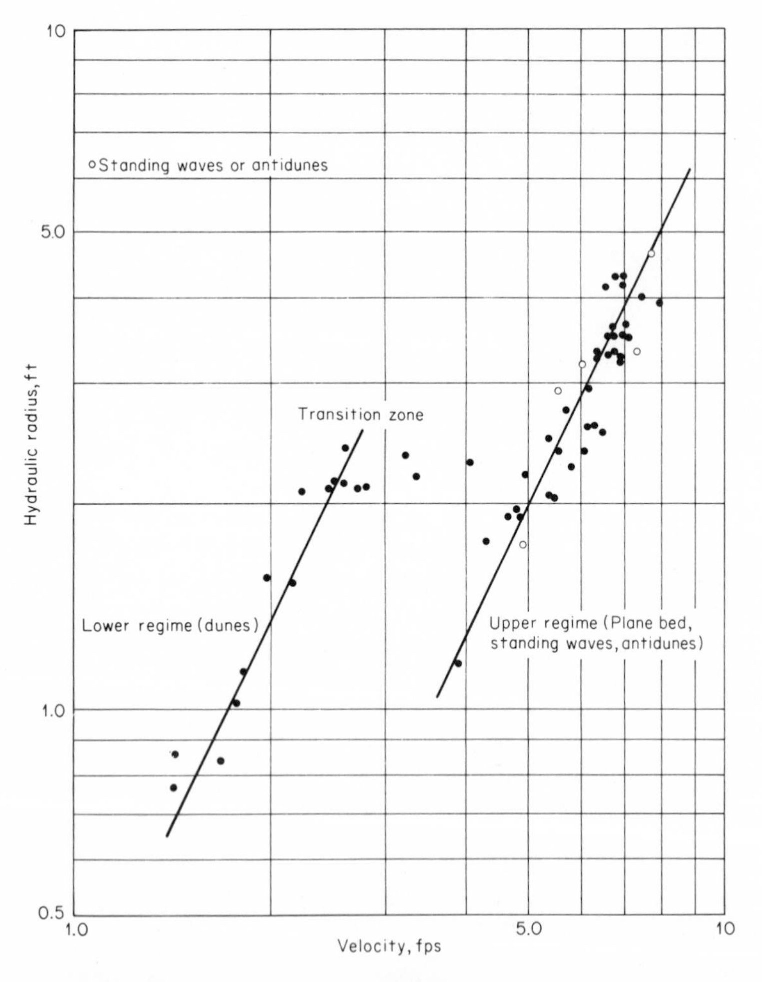

Fig. 11.22 Relation of hydraulic radius to velocity for Rio Grande near Bernalillo. [*After* NORDIN (*1964*).]

however, do bedforms play a more important role than in the one by EINSTEIN (1950). As a matter of fact, EINSTEIN (1950) proposed the relation given by Eq. (11.68) in connection with Fig. 11.16 in order to facilitate the solution of the bed material load equation, which had been discussed in Sec. 9.2.1.

Knowledge of existence and behavior of bedforms is also helpful in explaining certain anomalies in stage-discharge relations. Discontinuous rating curves (see Fig. 11.22) or rating curves with loops (see Fig. 9.21) may be interpreted in this way. However, we should be careful to attribute any kind of anomaly to this phenomenon, because more often than not the presence of the washload may be responsible for it. As was discussed earlier in Sec. 9.1 and shall be repeated here, under identical flow conditions the washload may alter drastically, depending on the availability of the material in the watershed. If adequate hydraulic data are lacking, EINSTEIN et al. (1952) proposed to use Fig. 11.16 for a determination of the stage-discharge relation. How this may be done is illustrated in a sample calculation given in Sec. 9.4.2.

Special attention was paid to the stage-discharge relations by SIMONS et al. (1962). It was found that in the upper regime the depth-discharge relation is reasonably stable; a certain stage corresponds to a certain discharge regardless of rising or falling stages. This cannot be observed in the lower regime where loops and/or crosses in the depth-discharge relation are readily noticeable. At a given stage, the discharge may vary, depending on rising or falling stages. The magnitude of the variation between the two branches depends strongly on the bedforms, but also on flow rate and sediment load. EINSTEIN et al. (1958) suggested that strong anomalies in stage-discharge relations do not occur frequently in natural streams. Nevertheless, NORDIN (1964) presented a very striking example for the Rio Grande given in Fig. 11.22. It was found that flow in the upper and lower regimes is reasonably stable, with an unstable transition zone between. Further evidence to the discontinuity in stage-discharge relation was given by CULBERTSON et al. (1964), who analyzed many sandbed cross sections in the middle Rio Grande basin. Again, the discontinuity occurred between the lower and upper regime. A rating curve with a pronounced loop was discussed by CAREY et al. (1957). This was explained with the existence and the formation of bedforms. Further evidence of discontinuities in rating curves for some streams of the western United States was reported by DAWDY (1961). Figure 11.22 exhibits clearly, and other field data often agree, that velocities frequently double in the transition zone. The same trend is exhibited if the hydraulic radius is plotted vs. the suspended sand; for a discussion on this issue, the reader is referred to CULBERTSON et al. (1964).

12
Cohesive-material Channels

12.1 INTRODUCTORY REMARKS

Consider a plane stationary bed in a channel or stream with liquid flowing over it. The flowing liquid is responsible for hydrodynamic forces being exerted on the individual solid particle of the bed.

If attention is focused on an assembly of cohesionless, loose and solid particles,[1] then the forces acting on the particles are the hydrodynamic drag and lift forces, given as F_D and F_L, respectively, and the submerged weight W of the particle. This is shown in Fig. 12.1*a* for a horizontal bed. The angle of repose is given by φ, and a force relation between the sum of the forces acting on the particle in the tangential and normal direction may be written as

$$\tan \varphi = \frac{F_D}{W - F_L} \tag{12.1}$$

[1] This particular case was treated extensively in Secs. 6.1 through 6.3.

Equation (12.1) is a simplified form of Eq. (6.2). The drag and lift forces as well as the submerged weight of the particle were expressed earlier in Eqs. (6.3) through (6.5). By introducing these relations into Eq. (12.1) and expressing the shear stress with Eq. (6.16), or

$$\tau_0 = \gamma R_h S \tag{6.16}$$

the following is obtained:

$$\frac{\tau_0}{(\gamma_2 - \gamma)d} = A_1'' \tag{12.2}$$

where, A_1'' is a sediment coefficient. Of course, Eqs. (12.2) and (6.17) are almost identical.

The better part of Chap. 6 was devoted to the determination of a particular shear stress, called the critical shear $(\tau_0)_{cr}$, at which scour or erosion would start. Chaps. 7 through 9 are concerned with the case of sediment transport where the available shear stress exceeds the critical shear stress.

In this chapter, however, we are not interested in the sediment-transport problem as modeled for cohesionless particles, but rather in the one where the wetted perimeter of the conveyance system is composed of cohesive material. The forces acting on a particle which is part of a cohesive bed are the hydrodynamic drag and lift forces F_D and F_L, respectively, the cohesive or physicochemical force F_C, and the submerged weight W of the particle. This is shown in Fig. 12.1*b* for a horizontal bed. A relationship for the angle of repose may be obtained analogous to Eq. (12.1), or[1]

$$\tan \varphi = \frac{F_D \pm F_C}{W - F_L} \tag{12.3}$$

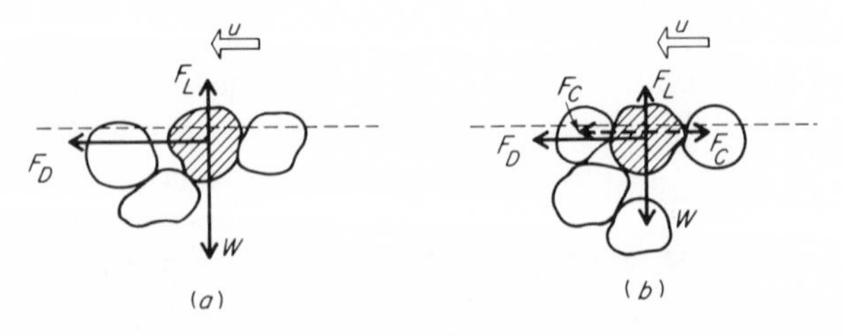

Fig. 12.1 Forces diagram on particles in (*a*) cohesionless and (*b*) cohesive beds.

[1] Actually, F_C could also appear in the denominator because the cohesion forces act in all directions.

and, analogous to Eq. (12.2), a relation for the shear stress may be written as

$$\frac{\tau_0}{(\gamma_s - \gamma)d} = A_1'' + C_0 \tag{12.4}$$

where C_0 is a coefficient of cohesion of the material. Special attention must be given to the fact that neither the cohesive force F_C nor the coefficient of cohesion C_0 can be uniquely expressed as a function of particle size or some other variable. Thus it is obvious that the importance of Eq. (12.4) rests with the experimental findings.

In the practice of soil mechanics, the shear stress of a soil corresponding to failure may be approximated by[1]

$$\tau = \sigma \tan \varphi + C \tag{12.5}$$

where τ is the shearing strength or shearing resistance, σ is the effective pressure, and C is the *cohesion*. This equation, discussed by TERZAGHI et al. (1968, p. 103), is known as *Coulomb's* equation, and is rather similar to Eq. (12.4).

An equation such as Eq. (12.4) should be valid if all of the forces F_D, F_L, F_C, and W are active. In the absence of the cohesive force, Eq. (12.4) becomes Eq. (12.2), whereas for purely cohesive material, i.e., for material where the cohesive force is much larger than the other forces, the relation becomes

$$\frac{\tau_0}{(\gamma_s - \gamma)d} = C_0 \tag{12.6}$$

The word *cohesive* appears rather frequently in this chapter, and it may be worthwhile to discuss elements of cohesive materials.

12.2 COHESIVE MATERIALS

12.2.1 GENERAL REMARKS

Cohesive (soil) material may be defined as soil material in which cohesive forces play an important role. In engineering practice, cohesive material consists of a mixture of clay-sized (colloidal) particles, of silt-sized particles, and sometimes of sand-sized particles. The upper limit of clay-sized particles is, according to the U.S. Department of Agriculture scale and the International (Atterberg) scale, equal to $d = 2\mu$. However, it is not sufficient for the foregoing mixture to include only clay-sized particles, as among these clay-sized particles must also be clay minerals, in addition to nonclay

[1] This equation is known to the student of soil mechanics as the equation of the *rupture line*.

minerals and possible organic material. It is the clay mineral that is responsible for the existence of physicochemical forces.

12.2.2 CLAY MINERALS AND THEIR PROPERTIES

Most clay minerals are crystalline and have a sheeted structure. There are three important groups of clay minerals, known as the *kaolinites*, the *montmorillonites*, and the *illites*. There exist other clay minerals, such as halloysites, chlorites, vermiculites, etc., but, generally, in minor amounts. Two different structural units are distinguished. They are, according to GRIM (1962, p. 8):[1]

> One unit consists of two sheets of closely packed oxygens or hydroxyls in which aluminum, iron, or magnesium atoms are embedded in octahedral coordination, so that they are equidistant from six oxygens or hydroxyls . . . .
> The second unit is built of silica tetrahedrons. In each tetrahedron, a silicon atom is equidistant from four oxygens or hydroxyls, if needed to balance the structure, arranged in the form of a tetrahedron with the silicon atom at the center. The silica tetrahedral groups are arranged to form a hexagonal network, which is repeated indefinitely to form a sheet . . . .

Most of the clay minerals are composed of these two units, but there are some fibrous ones that are composed of different units. For example, the structure of kaolinite is composed of a single tetrahedral sheet and a single alumina octahedral sheet, and the structure of montmorillonite is made up of two silica tetrahedral sheets with a central alumina octahedral sheet.

Common to all clay minerals is their flake-shaped appearance, with their major dimension often considerably smaller than 2 μ. Their color varies from white to yellow to green. As far as the origin and occurrence of clay minerals are concerned, the reader is referred to a discussion by GRIM (1962). Almost all types of clay minerals have been identified in the common soil types where they are formed by the weathering process. What kind of clay mineral will develop depends on the parent material (geology) and the climate. Clay minerals exist in both recent and ancient sediments. They have also been synthesized from mixtures of oxides.

The nature of the physicochemical forces experienced by a clay particle has been subject of much scientific endeavor. The present discussion must be brief and can only be qualitative. The interested reader, however, is referred to a more applied treatment of the topic by GRIM (1962) and to a scholarly, scientific one by KRUYT (1952).

[1] The same reference brings brief descriptions and diagrammatic sketches of the structure of clay minerals.

Table 12.1 Cation-exchange capacity of clay minerals, in milliequivalents per 100 g [*after* GRIM (*1962*)]

Mineral	*Cation-exchange capacity*
Kaolinite	3– 15
Illite	10– 40
Montmorillonite	80–150

An oversimplified picture of the physicochemical forces in action distinguishes between repulsive and attractive forces; both of these appear at the surface of the particle. It is not uncommon to find an excess charge on the lattice structure of the clay mineral. For example, substitution of trivalent aluminum for quadrivalent silicon (in the tetrahedral sheet) results in an unbalanced charge in the structural lattice. There are at least two causes for residual electric charges: (1) Flat surfaces of clay minerals carry negative charges as a result of an unbalanced substitution within the lattice; (2) broken edges of the clay particles may carry both positive and negative charges. As a result of this, a negatively charged clay mineral has a certain capacity to absorb cations. The cation-exchange capacity of some clay minerals is given in Table 12.1. However, due to the existing positively charged clay minerals, there is also a limited capacity to absorb anions. Usually, the anion-exchange capacity is of minor importance when compared with the cation-exchange capacity, and it is also much less understood.

From the foregoing discussion we may conclude that clay minerals having residual electric charges have thus the ability to hold certain cations and anions. These held ions, however, are retained in an exchangeable state. Common exchangeable cations are calcium, magnesium, potassium, hydrogen, etc.; common exchangeable anions are sulfate, chlorine, phosphate, etc. The rate of ion exchange depends on the concentration and valency of ions present in the dispersion media; in other words, it depends on the chemical composition of the flowing water. Each clay mineral is surrounded by an *electrical double layer*. There is an inner layer firmly attached to the surface by action of the most negative residual ions, and there is an outer layer of electrostatically attracted movable positive ions. The potential difference between the inner and outer layer is commonly called the ζ (*zeta*) *potential*. The ζ potential is higher for a monovalent than for a polyvalent series of ions; for a given valency it increases with diminishing ionic radius of the absorbed ions, a fact explained by the greater ionic hydration associated with the smaller ionic radius. Thus it follows that the ζ potential decreases the following order, for monovalent ions:

$$Li^+ > Na^+ > K^+ > Rb^+ > Cs^+$$

and for bivalent ones,

$$Mg^{++} > Ca^{++} > Sr^{++} > Ba^{++}$$

Furthermore, the larger the ζ potential, the looser an ion is held, and the easier it might be replaced. Thus the replacing power of ions increases with a decrease in the ζ potential; for example, Ca^{++} will more easily replace Na^+ than Na^+ will replace Ca^{++}.[1]

Coming back to the relation of the repulsive and attractive forces, the following may be said: As long as the residual electric charges of the clay particles are not satisfied, repulsive forces dominate; the more the residual charges can be satisfied, the more the attractive forces will become of importance. When in colloidal dispersions in a solution the repulsive forces dominate, the clay minerals will remain in suspension for a long time; when attractive forces are of importance, collisions of clay minerals may occur and an active settling takes place. This is the *sedimentation-flocculation* problem, discussed in Sec. 12.4. Once the flocs have settled out, they are deposited on the bottom where they may remain as loose flocs or become compacted. A loose or compacted floc bed subjected to shearing forces of flowing water may be scoured. This is the *initial scour problem*, discussed in Sec. 12.3.

12.2.3 COHESIVE MATERIALS AND THEIR PROPERTIES

An agglomerate of clay minerals, nonclay minerals, and organic material make up the clay-sized fraction composition of cohesive material. As far as the properties of the cohesive material are concerned, none of the fractions influences it more than the clay mineral fraction.

In problems of interest to the hydraulic engineer, cohesive materials are mostly in contact with water. The interaction of water with the various fractions of cohesive material, but with clay minerals in particular, represents an extremely complicated and an apparently little understood problem. Most, if not all, of the properties of cohesive materials depend on this interaction. GRIM (1962, p. 34) distinguished between two different categories of water, namely, *liquid* water and *nonliquid* water. Although most of the water in a pore is liquid water, there is also a thin film of nonliquid water on the surface of the clay minerals. It was suggested that there exists a certain organization of the water molecules, which is influenced by attractive forces and by the location and nature of the absorbed anions which tend to hydrate. For example, a sodium montmorillonite, which has a large hydration radius, would build up a thicker film than a calcium montmorillonite. Thus the amount of absorbed water varies, depending on the cohesive

[1] Use is made of this exchange mechanism in softening the drinking water, which is one of the oldest applications of ionic exchange.

material. Frequently it is implied that the water film acts as a lubricant between flakes.

In the following, we shall discuss some of the properties of cohesive materials expected to be useful for a limited classification of cohesive-material channels.

12.2.3.1 Consistency (plasticity). The consistency of cohesive materials depends on the water content. It is found convenient to describe the consistency in terms of certain limiting values. Frequently the limiting values by *Atterberg* are used. They are the liquid limit, the plastic limit, and the plasticity index. GRIM (1962, p. 205) quotes the following definitions:

> *Liquid limit*, L_w, is the moisture content expressed as a percentage by weight of the oven-dried soil at which the soil will just begin to flow when jarred slightly.
>
> *Plastic limit*, P_w, is the lowest moisture content expressed as a percentage by weight of the oven-dried soil at which the soil can be rolled into threads $\frac{1}{8}$ inch in diameter without breaking into pieces. Soils which cannot be rolled into threads at any moisture content are considered non-plastic.
>
> *Plasticity index*, I_w, is the difference between the liquid limit and the plastic limit. It is the range of moisture content in which a soil is plastic. When the plastic limit is equal to or greater than the liquid limit, the plastic index is recorded as zero.

Casagrande (1932) observed that many properties are well correlated by means of the plasticity chart, given in Fig. 12.2. Six regions are labeled on this chart, with the ordinate as the plasticity index I_w and the abscissa as the liquid limit L_w. From numerous data, which are tabulated by GRIM (1962, p. 207), it is quite evident that no single plastic-limit or liquid-limit value is characteristic of a particular clay mineral. Nevertheless, some preliminary conclusions can be drawn. Montmorillonites, especially the ones saturated with sodium or lithium, show extremely high liquid and plastic limits. GRIM (1962) observed that the soil-water relationship that determines the consistency is explained in terms of existing bonding forces, which are largely a result of the charge on the particles. For a sodium montmorillonite, the high limit values must be considered as a result of their ability to disperse into extremely small particles with a large effective surface area. On the other hand, the limit values depend not only on the chemistry of the mineral phase but also on the chemistry of the liquid phase. In general, the plasticity index for montmorillonites may exceed 600; the lowest plasticity index is reported for sodium kaolinite, which is about 1. Nonclay

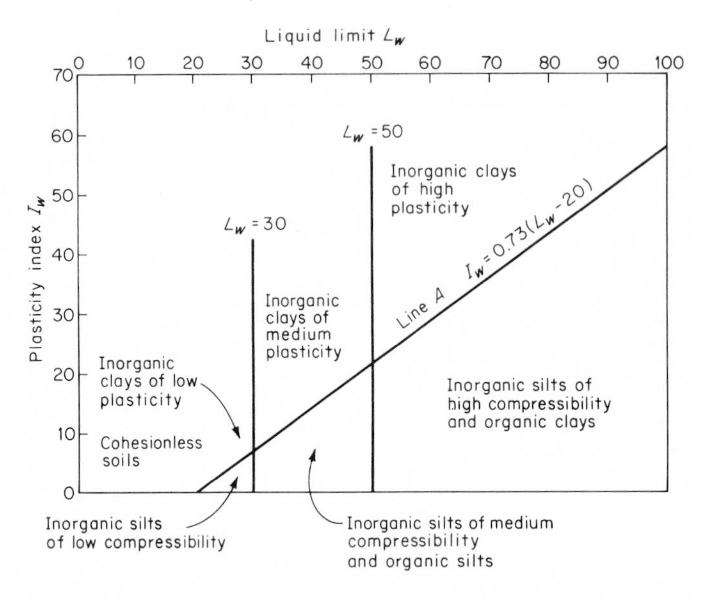

Fig. 12.2 Plasticity chart. [*After* TERZAGHI *et al.* (*1968*).]

components, such as quartz or calcite, have a tendency to reduce the plasticity. How organic matter, when part of the cohesive material, affects the plasticity is little understood.

An additional limiting value, the *liquidity index* I_l, is defined as the ratio of the natural water content of the soil minus the plastic limit to the plasticity index. The Atterberg limits and the plasticity chart serve as useful criteria for the description of the cohesive material to the hydraulic engineer.

12.2.3.2 Soil aggregates. The solid phase of the soil is divided into particles of different sizes. Generally, these particles do not exist in a dispersed state, but rather in an aggregated one. The particle-size distribution is an essential characteristic of the cohesive material. Its range might include everything from sand, with $d \approx 2$ mm (and possibly larger material), to clay, with $d \leqq 2\ \mu$. Determination of the sizes may be achieved with various methods, which are discussed in textbooks on *soils*; for example, in BODMAN (1960) or TERZAGHI et al. (1968).

Aggregates, according to TERZAGHI et al. (1968), may be classified qualitatively by visual inspection in the field, and quantitatively by laboratory and field tests. Qualitative criteria are texture, structure, and consistency of the aggregate. The *texture* is the degree of fineness and uniformity. The schematized arrangement of soil particles—the packing—is determining the *structure* of a soil. The degree of adhesion between soil particles is expressed by the *consistency*. The quantitative properties are important in defining the overall properties of a cohesive material.

The proportion of nonsolid volume, or pore space, present in an assemblage of soil particles may be expressed as the porosity or as the void ratio. With v as the total volume and v_v as the total volume of the voids, the *porosity* p is given as

$$p = \frac{v_v}{v} \tag{12.7}$$

and the *void ratio* ϵ is

$$\epsilon = \frac{v_v}{v - v_v} \tag{12.8}$$

At this point it should be remarked that the void ratio ϵ is a good indicator of the compressibility (pressure) of the soil.

Table 12.2 Properties of soil aggregates of typical soils in a natural state [*after* TERZAGHI *et al.* (*1968*)]

Description	Porosity, p, %	Void ratio, ϵ	Water content, w, %	Unit weight g/cm³ γ_d	Unit weight g/cm³ γ	Unit weight lb/ft³ γ_d	Unit weight lb/ft³ γ
1. Uniform sand, loose	46	0.85	32	1.43	1.89	90	118
2. Uniform sand, dense	34	0.51	19	1.75	2.09	109	130
3. Mixed-grained sand, loose	40	0.67	25	1.59	1.99	99	124
4. Mixed-grained sand, dense	30	0.43	16	1.86	2.16	116	135
5. Glacial till, very mixed-grained	20	0.25	9	2.12	2.32	132	145
6. Soft glacial clay	55	1.2	45	—	1.77	—	110
7. Stiff glacial clay	37	0.6	22	—	2.07	—	129
8. Soft slightly organic clay	66	1.9	70	—	1.58	—	98
9. Soft very organic clay	75	3.0	110	—	1.43	—	89
10. Soft bentonite	84	5.2	194	—	1.27	—	80

w = water content when saturated, in percent of dry weight.
γ_d = unit weight in dry state.
γ = unit weight in saturated state.

The *water (moisture) content* w, expressed in percent, is given with the ratio of weight of water to the dry weight of the aggregate. The natural water content w of a cohesive material (normally loaded) is usually close to the liquid limit L_w. The *degree of saturation* is

$$S_w(\%) = \frac{\epsilon_w}{\epsilon} 100 \tag{12.9}$$

where ϵ_w is the volume occupied by water per unit volume of solid matter.

Another important property of a soil aggregate is its *unit weight*, defined by the weight of the aggregate, as the sum of the soil and water fraction per unit volume.

Some properties for various soil aggregates are tabulated in Table 12.2.

12.3 SCOUR CRITERIA

In Sec. 12.1 shear stress relations for a cohesionless soil were given by

$$(\tau_0)_{cr} = \text{fct}(d) \tag{12.10}$$

and for a cohesive soil by

$$(\tau_0)_{cr} = \text{fct}(d,C_0) \tag{12.11}$$

where d is the sediment-particle diameter and C_0 is the coefficient of cohesion. Once such a relationship for the critical shear stress $(\tau_0)_{cr}$ at which incipient sediment movement begins is known, any channel can be designed according to Eq. (6.16). Equation (12.10) was subject of an extensive discussion in Chap. 6. Now we have the task to establish proper relations between the shear stress and the coefficient of cohesion, as given with Eq. (12.11). The shear stress of cohesive material depends on the composition of the different fractions (clay and nonclay minerals and organic matter), the particle-size distribution and the particles' shape, the packing, and probably on other items. Also geologic events, such as compression or stratification, may be of importance.

In the following, laboratory, flume, and field investigations will be examined for useful relationships for the critical tractive (shear) force. To the hydraulic engineer, the critical tractive force is the criterion for initial motion or scour, and is called the scour criterion. To the soils engineer, it is the failure criterion, defined by the *shear strength*, expressed as the peak (maximum) shear stress.

12.3.1 LABORATORY INVESTIGATIONS

Studies with specially designed apparatus have been of help to define Eq. (12.11). Specimens of soils to be tested were taken from the field—hopefully

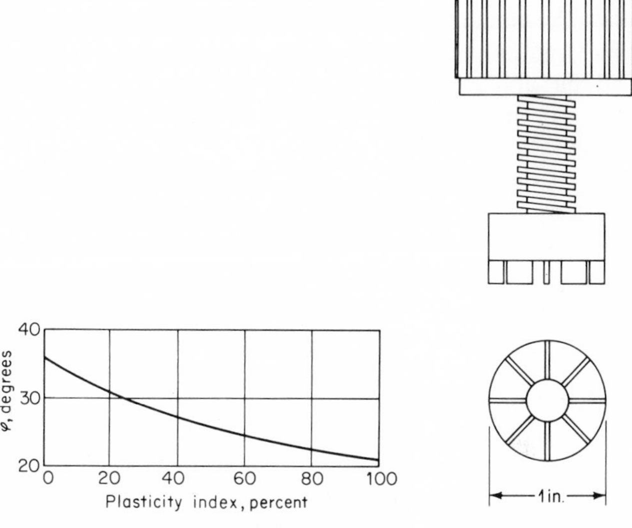

Fig. 12.3 Relationship between φ and plasticity index I_w. [*After* TERZAGHI *et al.* (*1968*).]

Fig. 12.4 Torvane, a vane-shear apparatus, for determination of the shear strength of cohesive materials. [*After* TERZAGHI *et al.* (*1968*).]

undisturbed—into the laboratory and the necessary tests were performed. Engineers in the area of soil mechanics have gained considerable experience and, therefore, for any kind of soil exploration and investigation soil mechanics books should be consulted.

Among the various types of equipment, a frequently used one is the *triaxial* apparatus, adequately described by TERZAGHI et al. (1968). It is shown that the results of triaxial tests on normally loaded cohesive soils can be expressed satisfactorily by approximating Eq. (12.5) with

$$\tau = \sigma \tan \varphi$$

Under such a condition, the values of φ are found to be in functional relation with the plasticity index I_w, as shown in Fig. 12.3. This and some further experimental evidence, discussed by TERZAGHI et al. (1968), point out that the plasticity index I_w is an acceptable parameter to express the coefficient of cohesion in Eq. (12.11). GRIM (1962) presents a graph on which the shear strength is related to the liquidity index I_l; but plasticity index and liquidity index are somewhat related.

Another most versatile equipment is the *vane-shear apparatus*, which can be used in both laboratory and in situ investigations. The torvane, a portable unit, is shown in Fig. 12.4. The eight vanes that are attached to a spindle are pressed into the soil specimen. The assembly is then rotated, and thus applies a torque through a spring until, eventually, the soil specimen fails. Proper calibration of the indicator on the spring gives a direct reading of the shear stress.

In the following, we shall focus on investigations done by hydraulic engineers. Each of these studies took a soil specimen obtained in the field and tested it in an apparatus. Whether the obtained test results are fully indicative of the cohesive material in the field remains an open question.

An early attempt to obtain critical-shear-stress values for cohesive material was made by DUNN (1959). A *jet test apparatus* was developed. The shear stress was obtained by means of a submerged jet of water which discharges perpendicularly on the soil surface, and then spreads and flows essentially parallel to the soil surface. The magnitude of the shear stress on the test sample was directly measured by replacing part of the soil surface with a shear plate, similar to the ones discussed in Sec. 6.5.2.1. The critical value was obtained, according to DUNN (1959), the following way:

> The jet was then positioned above the soil sample and the head of water on the nozzle was slowly increased . . . With each additional increase in H (head), a small additional amount of soil was carried off the surface, followed by clearing of the water in the container. When H reached the critical value, the rate of erosion decreased, the water became cloudy, and no subsequent clearing occurred. The critical point was definite and reproducible in the clay samples.

The critical shear stress of the sample was then related to the shear strength obtained with a vane-shear apparatus. The data from these experiments give reason to conclude that the best way of estimating the critical shear stress $(\tau_0)_{cr}$ proved to be the plasticity index I_w; the applicable range is $2 < I_w < 16$. It was also pointed out that methods based on the characteristics of the particle-size-distribution curves are useful in the correlation.

An instrument similar to the jet test apparatus was used by MOORE et al. (1962) to determine the relative scour resistance of different kinds of cohesive sediments. The results of this study have shown the depth of scour to be proportional to the logarithm of the time during which erosion occurred. A device, called the *rotating-cylinder test apparatus*, was suggested by MOORE et al. (1962) and is shown in Fig. 12.5. This instrument was developed with the idea that special care must be taken to minimize the effect of variation in the shear stress with respect to time and space. The

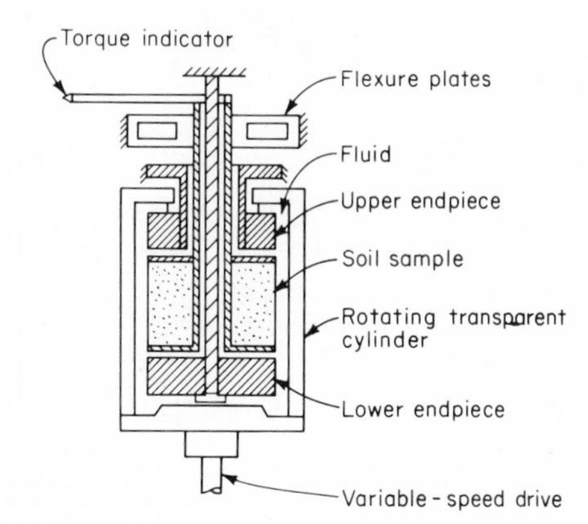

Fig. 12.5 The rotating-cylinder test apparatus. [*After* MOORE *et al.* (*1962*).]

principle of this instrument is also found in various viscometers. The inner cylinder containing the soil sample is stationary, the outer cylinder is rotating, and a shear-producing fluid exerts shear stress at the soil-sample surface. The critical shear stress was obtained by observing the soil sample through the transparent outer cylinder until cracking, chipping, or sloughing off developed. The beginning of scour was accompanied by a sudden jerking motion at the torque indicator. It is of special interest how MASCH et al. (1963) describe scour as it develops:

> At lower speeds, it was possible to observe what might be called a washing of the surface, i.e. the flaking of small soil particles from the surface of the sample. As the speed and the shear were increased, the sample was observed through the transparent cylinder until the critical shear was reached at which time appreciable quantities of sediment came loose from the sample, and the water in the annulus became cloudy.

Above observation was made with Taylor Marl soil; its properties are listed in Table 12.3.

REKTORIK et al. (1964), using the rotating-cylinder test apparatus, find it a convenient and relatively simple means of indexing soils in relation to critical shear stress. Natural and artificial soils from Texas, the physical properties of which are listed in Table 12.3, were studied. Most of the soil tested for scour showed a relationship between the critical shearing stress, moisture content, and void ratio; a relationship to the plasticity index was not apparent. MASCH et al. (1968) show such a relationship with Fig. 12.6 for all soils listed in Table 12.3. Most of the tested specimens indicate clearly that an increase in the moisture content corresponds to a decrease in the critical shear stress. No explanation could be given for San Saba soil, which exhibits an opposite trend.

Table 12.3 Physical properties of some clay-type sediment from Texas [*after* MASCH *et al.* (*1968*)]

Properties	*Lake Charles K319*	*Lufkin K116*	*Houston K177*	*Houston K177A*	*Houston K177B*	*San Saba*	*Taylor Marl*
	Soil name and number						
Texture	*Clay or silty clay*	*Clay or clay loam*	*Clay or silty clay*	*Clay or silty clay*	*Clay or silty clay*	*Silty clay*	—
Liquid limit	56.4	49.4	43.7	44.7	48.7	47.7	47
Plastic limit	22.0	15.9	20.5	17.7	18.0	22.0	21
Plasticity index	34.4	33.5	23.2	27.0	30.7	25.7	26
Clay, %	46.2	40.3	55.5	55.5	55.5	44.7	50
Mean particle size, mm	0.0019	0.0084	0.0033	0.0033	0.0033	0.0020	0.0048

All the tests performed with the rotating-cylinder test apparatus were done with disturbed (remolded) soil specimens, being either formed by hand or with an extruder. This, however, leaves open some questions: Is there a relationship and, if so, what relationship exists between a disturbed and an undisturbed soil sample, and between the soil as it is on the site (in situ)?

A *tractive-force test tank* was developed, built, and used by THOMAS et al. (1961). The test tank is a 35-in.-diameter cylindrical tank; inside is a

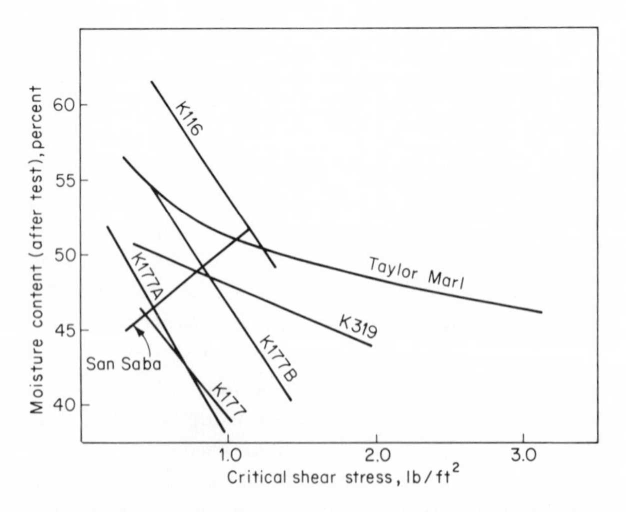

Fig. 12.6 Critical shear stress vs. moisture content for Texas soils. [*After* MASCH *et al.* (*1968*).]

variable-speed motor driving the impeller blades. The undisturbed soil sample is inserted into the bottom of the cylindrical tank, being flush with the bottom. The velocity of the blades is changed gradually until the soil sample begins to erode. In a further study, CARLSON et al. (1962) did extensive studies on 46 soils, measuring their critical shear stress, liquid limit, plasticity index, soil density, percent maximum proctor density, shrinkage limit, gradation, and vane-shear strength. Using multiple linear correlation, formulas were derived for the critical tractive force as a function of several of the seven soil variables. Out of a total of 38 formulas, the 7 best combinations of 7 or fewer variables always included the plasticity index and the liquid limit, and gave correlation coefficients for all data from 0.71 to 0.79. This paper also includes an example of determining the critical tractive forces for a given soil material by using the multiple-linear-correlation method.

KARASEV (1964) contests that the erosion of cohesive material does not proceed particle by particle, but rather aggregate by aggregate. Based on

Table 12.4 Comparison of experimental data with formulas [*after* KARASEV (*1964*)]

No. of points	*Material*	*Diameter of aggregate, mm*	*Adhesion, kg/cm² σ_M*	H_M	*Scouring velocities of the stream, m/sec*		
					Experiments	*Mirtskhulava's formula*	*Formula*
	A. Aggregated materials						
1	Medium loam	2.2	0.0023	60	1.74	1.64	1.83
2	Rammed clay	3.0	0.0056	60	2.16	2.38	2.82
3	Rammed clay	4.0	0.0049	60	2.06	2.16	2.36
4	Heavy loam	2.7	0.0038	60	2.03	2.04	2.22
5	Heavy loam	4.0	0.0023	60	1.20	1.53	1.70
6	Clay	4.0	0.0025	60	2.20	1.58	1.75
7	Clay	4.0	0.0029	60	1.30	1.69	1.86
8	Clay	6.0	0.0020	60	1.43	1.42	1.45
9	Clay	4.0	0.0029	60	1.54	1.69	1.86
10	Compacted clay	0.8	0.0052	60	2.23	2.83	3.20
11	Heavy loam	1.0	0.0031	60	2.14	2.13	2.35
	B. Dispersed materials						
12	Medium loam	4.0	0.0076	60	1.91	2.61	2.88
13	Clay	1.5	0.0072	20	3.06	2.46	2.68
14	Clay	3.2	0.0080	20	2.35	2.21	2.40
15	Clay	4.0	0.0044	20	2.87	1.59	1.93
16	Clay	0.8	0.0080	60	2.40	3.52	2.42
17	Clay	4.0	0.0056	60	1.54	2.28	2.50

this a model of erosion was developed, and the scouring velocity was given by[1]

$$(\bar{u})_{cr} = 0.142C\left[\frac{2d(\gamma_s - \gamma) + 3\sigma_M\left(1.2 + 8\dfrac{D}{D_a}\right)}{\gamma}\right]^{1/2}$$

where C = Chezy coefficient
σ_M = information on the cohesion
D_a = atmospheric pressure

A comparison between the preceding formula and experimental data, as well as with another formula, is given in Table 12.4; agreement may be considered reasonable. The experiments performed by Mirtskhulava (1960) were such that erosion was achieved by placing the specimen under water and subjecting it to "a descending flooded jet from a nozzle."

12.3.2 FLUME INVESTIGATIONS

Early investigations are reported by DuBuat (1786) who determined the velocity at which "pottery clay" will start to be eroded. However, it is only recently that researchers have paid attention to the systematic investigation of the erosion problem in cohesive channels.

Smerdon et al. (1959) carried out a flume study on cohesive soils. The section of the flume where the thoroughly mixed soil was placed was 18 ft long, and was located approximately in the middle section of a 60-ft-long flume. The thickness of the soil sample was $2\frac{1}{2}$ in. Eleven soils from Missouri were selected for the tests. The shear stress τ_0 was computed according to Eq. (6.15); the critical shear stress $(\tau_0)_{cr}$ was defined as the shear stress corresponding to the general movement of the bed material. The properties of the soils under consideration were plasticity index, mean particle size, dispersion ratio, percent clay, and phi-mean particle size. The investigators concluded that the critical shear stress correlated well with the plasticity index I_w; this can be seen in Fig. 12.7. The regression line for the data is also given in Fig. 12.7. This study was essentially extended by Lyle et al. (1965), but Texas soils, similar in their physical properties to the ones tabulated in Table 12.3, were used. Although the earlier study by Smerdon et al. (1959) concluded that the plasticity index I_w correlates well with the critical shear stress, the research by Lyle et al. (1965) suggests that both the plasticity index I_w and the void ratio ϵ are of importance. This is of considerable significance, because the void ratio ϵ is an indication of the soil compaction. A typical result is shown with Fig. 12.8. Figures 12.7 and 12.8 show clearly an increase of the critical shear stress with an increase in the plasticity index. The latter figure shows also an

[1] In the translation of the paper, it is not entirely clear whether this is really the average velocity $\bar{u}$.

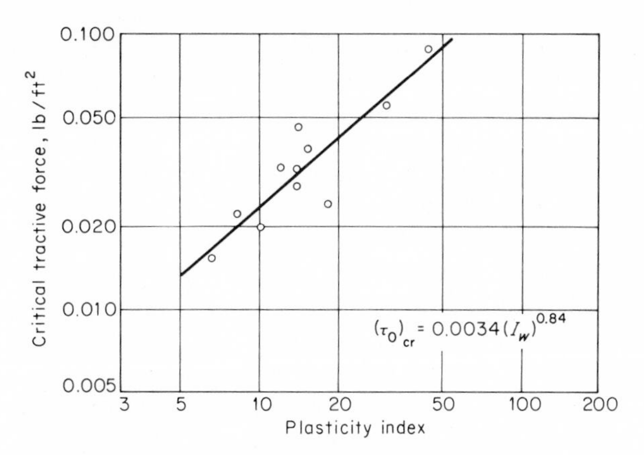

Fig. 12.7 Critical shear stress vs. plasticity index for Missouri soils. [*After* SMERDON *et al.* (*1959*).]

increase of the critical shear stress with a decrease in void ratio or an increase in compaction.

An investigation was described by ENGER (1963), where a relatively short, 8-ft-long recirculatory flume was used. Disturbed soil samples of 8 in. in diameter were inserted in the middle of the flume. The only useful relations for the soils tested could be given for the critical shear stress $(\tau_0)_{cr}$ and for the moisture content at which the soil specimen was compacted.

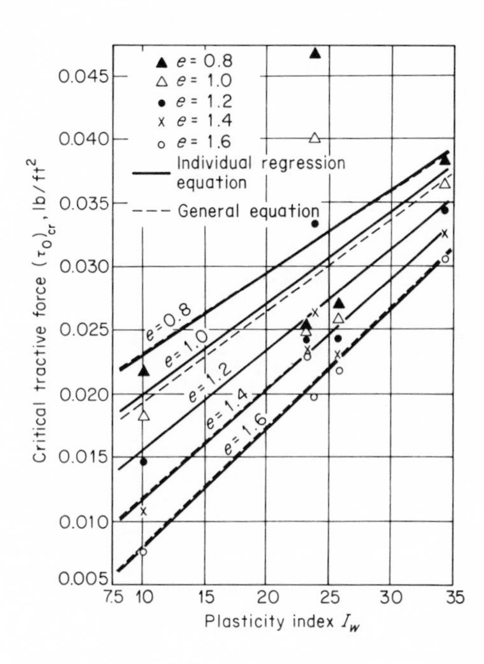

Fig. 12.8 Critical shear stress vs. plasticity index for Texas soils. [*After* LYLE *et al.* (*1965*).]

ABDEL-RAHMAN (1963) reports of a study in a flume of 10½ m long and having a 6-m measuring reach. The tests were done for "pure clayey soil (Opalinuston)" and mixtures of this clay material with sand. The major objective of this investigation was to establish "a relation for the mean erosion depth as a function of the tractive stress of water ($\gamma R_h S$) and the vane-shear strength of the bed soil." A typical experimental result is shown in Fig. 12.9. At the beginning of the experiment, strong erosion is evident; the eroded material is found in form of moving bedload and/or suspended load. After a certain time, which is almost independent of the shear stress, the rate of erosion decreases, and eventually no erosion is noticeable. Curves of bedload rate vs. time exhibit a rather similar trend as that shown by Fig. 12.9. Such a trend was also reported by PARTHENIADES (1965) and KRONE (1962) when plotting suspended-sediment concentration vs. time. ABDEL-RAHMAN (1963) made the interesting observation that the bed surface during the erosion process becomes covered with grooves of various dimensions. (This is very well demonstrated in the report by a time series of photographs, up to 243 hours after the start of the test.) Furthermore, it was noticed that a gluey layer developed on the surface of the bed when

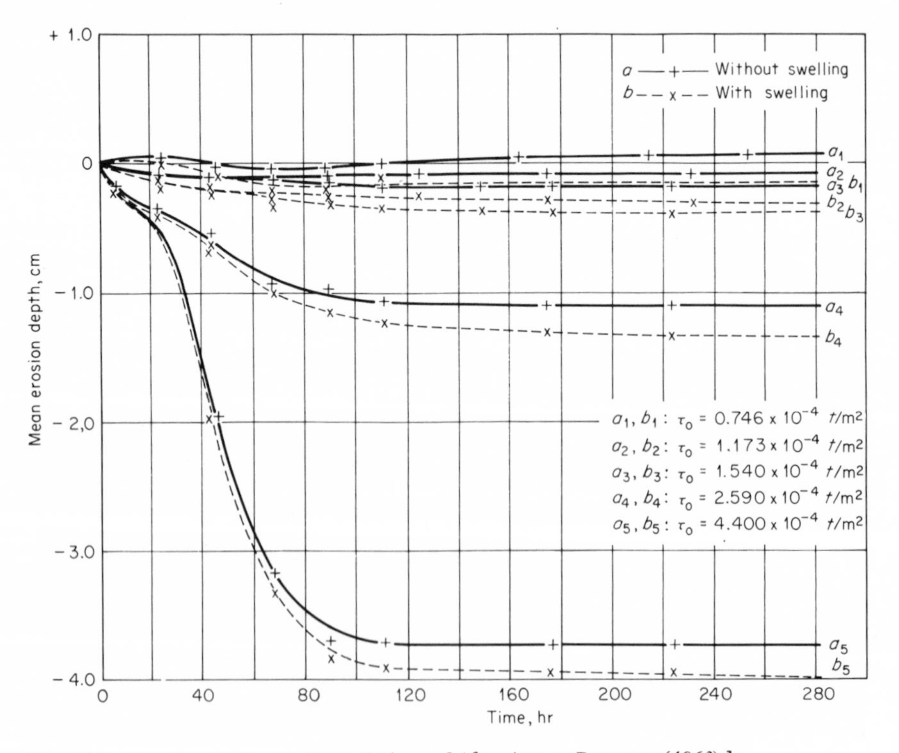

Fig. 12.9 Erosion depth vs. time relation. [*After* ABDEL-RAHMAN (*1963*).]

steady state was reached. For the limited available data, an empirical relation between the depth of erosion at steady-state condition and the mean shear stress and vane-shear strength of the soil after the flow was suggested. KRONE (1962) obtained data in a 3-ft-wide and 100-ft-long flume, and observed that scour is a resuspension process caused by bed failure and erosion. An extensive laboratory program was conducted by ZELLER (1969) who presently still analyzes these data. Directly or indirectly the velocity, concentration, and bed shear distributions were measured. The tests were performed with clay suspension over clay beds in 0.6-m-wide and 12.0-m-long channels.

Further, it is worthwhile to mention some research dealing with erosion[1] of a loose, cohesive bed made up of previously flocculent suspended clay material which had settled to the bottom. Two investigations, one by INGERSOLL et al. (1960) and another one by PARTHENIADES (1965), present limited experimental research, but conclusive results have not been obtained so far.

Very little is known about bedforms of cohesive beds. Interesting but very limited experiments by ALLEN (1969*a*) have given the following results. With an increase in mean flow velocity—but not necessarily in the flow Reynolds number—the bed showed (1) longitudinal rectilinear grooves; (2) longitudinal meandering grooves; (3) flute marks; and (4) transverse erosional markings.

12.3.3 FIELD INVESTIGATIONS

For the same types of soils—taken from cohesive channels in the western part of the United States—that were tested with the jet test apparatus described in Sec. 12.3.1, direct field tests are reported. DUNN (1959) performed in situ shear tests with a vane-shear apparatus mounted on a tripod base. The test, which took place with water in the canals, furnished values of shear strengths. Proper calibration of the vane-shear apparatus allowed for the calculation of tractive force of the particular soil in the field. DUNN's (1959) results are encouraging.

FLAXMAN (1963) advanced the concept that "the erosion resistance of cohesive soils can be determined by unconfined compressive strength tests of saturated undisturbed soil samples." This conclusion was reached after an investigation of 13 streams in 6 western states of the United States. The entire study consisted of a laboratory and a field investigation. The unconfined compressive strength[2] was obtained in the laboratory. It was

[1] INGERSOLL et al. (1960) call it *resuspension* and define it as "the entrainment into the flow of particles that have once settled to the floor."

[2] According to TERZAGHI et al. (1968), the unconfined compressive strength is a measure of the soil consistency; it is the load per unit area at which unconfined prismatic or cylindrical soil specimens fail in a simple compression test. The consistency is usually described as soft, medium, stiff, and hard.

shown by a multiple regression analysis that the unconfined compressive strength is dependent on the plasticity index, on the dry density of the soil, and on the percentage finer than 5 μ. In the field, data were obtained to calculate the "tractive power" given by the product of slope, hydraulic radius, specific weight of water, and average velocity, or $SR_h\gamma\bar{u}$. By observations in the field—agreeable highly subjective determinations—the boundary between "stable" and "eroding" was obtained. The resulting plot of Fig. 12.10, showing the tractive power vs. the unconfined compressive strength, may serve, if further pursued, as a useful tool for design engineers.

12.3.4 DESIGN CRITERIA

The important study of FORTIER et al. (1926) on *permissible canal velocities* has paid attention to what was called *colloid matter*. The result of this study has been presented in Table 6.1. All the information accumulated for the preparation of this table is based on personal deductions and long-range experience with individual canals. FORTIER et al. (1926, p. 951) summarize their task by stating:

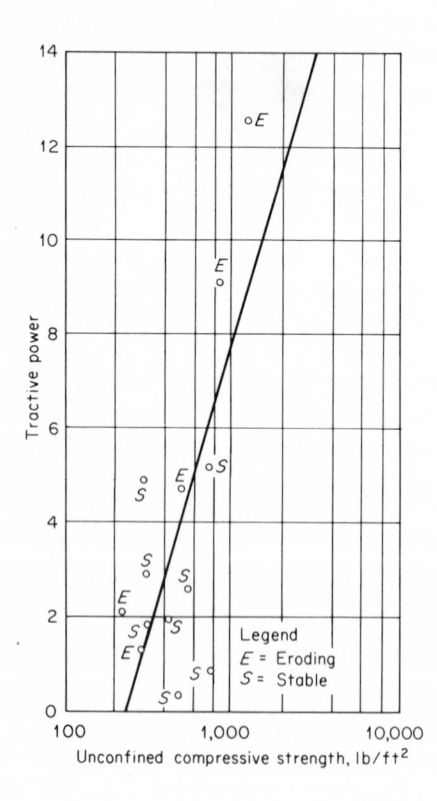

Fig. 12.10 Relationship between tractive power and the unconfined compressive strength. [*After* FLAXMAN (*1963*).]

> After a careful study of all the data presented and in the absence of experimental data bearing on the subject, . . . [they] . . . recommend the values given in Table 6.1 for maximum permissible velocities . . . .

Notice that the figures given in Table 6.1 are for depths of 3 ft or less. Much of the important conclusions drawn by FORTIER et al. (1926) are quoted in Sec. 6.2.2 and should be read. Of the 14 different soil materials used in the classification, at least 4 of them may be useful for the design of cohesive bed channels. It was correctly recognized that cohesive material, whether found in the channel bed itself or suspended in the water, has a tendency to resist erosive effects. FORTIER's et al. (1926) paper—in particular their table, i.e., Table 6.1—has been extensively quoted and recommended for use. DAVIS' (1942, p. 461) "Handbook of Applied Hydraulics" points to the fact that, although a final solution is not in sight, a safe design criterion is suggested by the values of Table 6.1. Reporting on studies on the design of stable channels by the U.S. Bureau of Reclamation, LANE (1953) could not report new developments, but recommended as "the only data on safe tractive forces available" the ones presented by FORTIER et al. (1926) and given in Table 6.1. Only recently, in a comprehensive survey on erosion of cohesive sediments, the task committee of the American Society of Civil Engineers [MASCH et al. (1968)] reprinted the table (Table 6.1) of FORTIER et al. (1926). Also CHOW (1959, p. 165) lists in a table the values recommended by FORTIER et al. (1926).

According to CHOW (1959), some Russian research was reported in 1936 which contained valuable information on permissible velocities for cohesive soils. It was suggested that void ratio and clay content adequately describe the cohesive material. CHOW (1959) presented the data in terms of the critical tractive force; the relation is given in Fig. 12.11. Further information on limiting tractive-force values may be found in WITTMANN's (1955, p. 717) River-Engineering in the "Taschenbuch für Bauingenieure" (Handbook for Civil Engineers). Many of the data accumulated by European engineers have been summarized by GARBRECHT (1961). In a plot of permissible velocity vs. pore volume, given in Fig. 12.12, the data show some spread, but a certain trend is obvious.

An extensive field investigation was done at the U.S. Bureau of Reclamation, and is reported by ENGER et al. (1960) and THOMAS et al. (1961). Forty-six reaches, located throughout the western part of the United States, were selected. The maximum operating discharges varied from 2 to 3,000 cfs. The soil specimens were tested in the laboratory for various properties. Using all of these data, an electronic computer was employed to develop multiple linear correlations; this was briefly discussed in Sec. 12.3.1. Equations of the following type were obtained:

$$(\tau_0)_{cr} = a + b(I_w) + c(L_w) + d(D\%) + e(M_\Phi)$$

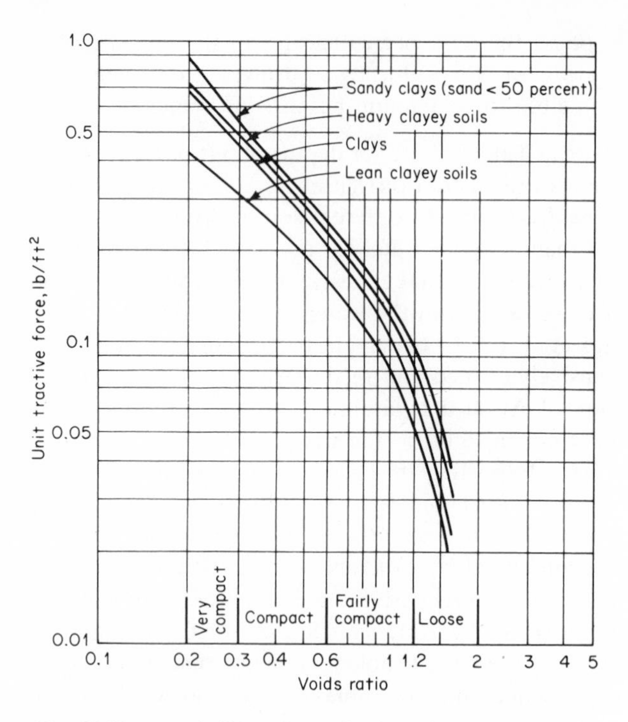

Fig. 12.11 Permissible unit tractive force for canals vs. void ratio converted from USSR data on permissible velocity. [*After* Chow (*1959*).]

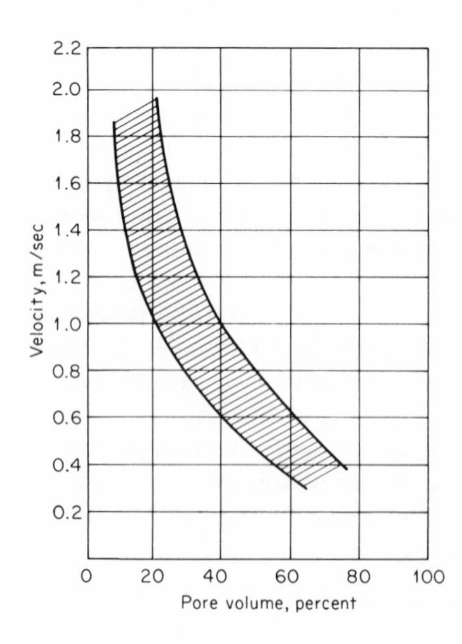

Fig. 12.12 Permissible velocity vs. pore volume. [*After* Garbrecht (*1961*).]

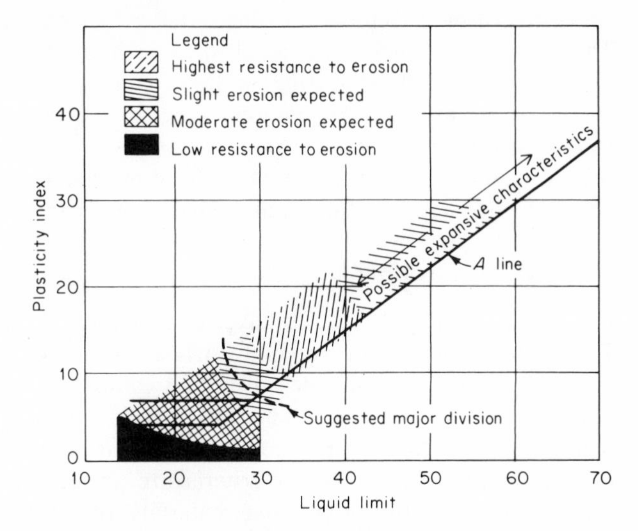

Fig. 12.13 Suggested trend of erosion characteristics for fine graded cohesive soils with respect to plasticity. [*After* GIBBS (*1962*).]

where a, b, c, d, e = constants [to be obtained from THOMAS et al. (1961) table]

I_w = plasticity index

L_w = liquid limit

$D\%$ = the in-place percent maximum soil density

M_Φ = mathematical description of soil gradation

The same set of data was also used by GIBBS (1962). The important feature of this study is that field data have been incorporated as well. By field inspection of the test reaches, the reaches were classified as stable, slight erosion, moderate, and appreciable erosion. The laboratory tests aided in establishing meaningful correlations. GIBBS (1962) believes that the plasticity chart, in particular the region of the A line, "is a logical form for evaluating the characteristics of erosion because it is well known in the unified classification procedure." The result of this research is given in Fig. 12.13.

In Chap. 10 the regime concept was discussed. The various contributions to the present knowledge of the regime concept include scattered remarks on cohesive channels. SIMONS et al. (1963), in suggesting a design method for alluvial channels, have paid considerable attention to channels with beds or banks of cohesive materials. The data used to establish regime equations included data from canal studies in India and the United States. The regime equations in a general form were given by Eqs. (10.18) through (10.20); coefficients appearing in these equations are tabulated in Table 10.2, and are different for different types of material. A typical plot—the hydraulic radius vs. the discharge relation—is shown in Fig. 12.14. The regime equations permit, provided the flow rate Q is given, expedient calculation of average width, depth, velocity, and channel slope.

12.3.5 CLOSING REMARKS

Few research and field studies have been performed with cohesive materials. The research performed on undisturbed—but more often disturbed—soil specimens, with specially designed test equipment or in the flume itself, offers as yet no general conclusions. Field tests (in situ) are extremely scarce, but would allow a better and more accurate evaluation for establishing further design criteria. Scour criteria are presently empirically related to the various properties of cohesive materials, but seldom with the properties of clay minerals which play an important role in the clay material. A more direct relation between scour and properties of the clay mineral fraction of the clay material would be of great help. A beginning of such a trend may be observed in the contributions by KARASEV (1964), PARTHENIADES (1965), and GRISSINGER (1966). Some interesting observations may be found in GRIM (1962, p. 272). Most of the research points to the fact that erosion of cohesive soils takes place in aggregates rather than particle by particle; the latter is the case for noncohesive soil material. Small amounts of cohesive material have the tendency to stabilize the channel. Although it is beyond

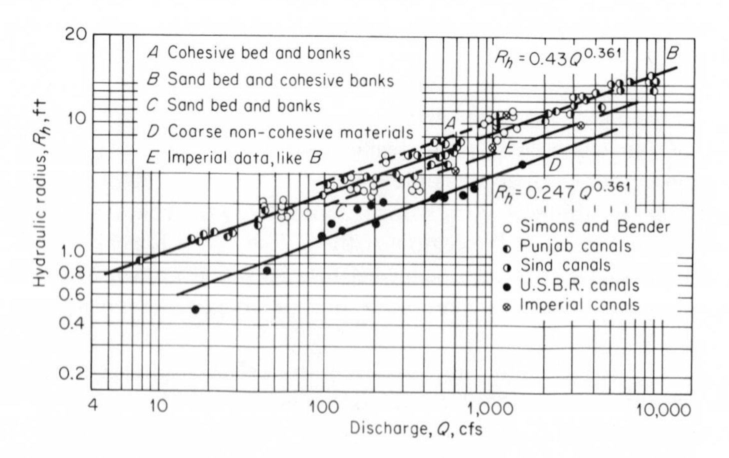

Fig. 12.14 Regime relation; hydraulic radius vs. flow rate. [*After* SIMONS *et al.* (*1963*).]

the present task, it should be remarked that *grassed* channels exhibit a similar stabilizing effect. Experiments have shown that an increase in vegetation inhibits erosion, but also affects the velocity distribution. For a further discussion of grassed channels, the reader is referred to CHOW (1959). Finally, we should like to express hope that forthcoming research may shed further light on this rather complex problem.

12.4 SEDIMENTATION-FLOCCULATION PROBLEM

12.4.1 GENERAL REMARKS

Suppose a suspension of solid particles experiences mainly gravity forces; thus there are large particles which have a tendency to settle to the bottom. An explanation of this problem was given in Chap. 8. On the other hand, imagine a suspension of solid particles where physicochemical forces are dominant; it is a dispersed suspension of very small particles which may or may not settle out, even after a considerable time has lapsed. These very small particles are made up of cohesive material. If many of the very small particles come together and form flocs, the effective weight of such an agglomerate would increase and sedimentation will occur. This entire process is frequently referred to as *flocculation*. The flocs are loose, and irregular clusters and the original particles can still be recognized. Edge-to-edge arrangements of particles and parallel arrangements may be found. Studies of flocculation of dispersed fine particles have been carried out extensively; much of the knowledge is presented by KRUYT (1952).

According to EINSTEIN et al. (1961), who follow KRUYT (1952), the formation of flocs may be achieved by a reduction of the repulsive forces acting between the particles and/or by the motion and collision of the particles. The reduction of the repulsive forces is obtained by a chemical change in the environment. Attractive forces become of growing importance if the residual charges on the clay mineral are satisfied. For example, suspended cohesive matter will stay in suspension in the river flow if the chemical environment does not change. However, if the river meets a saline environment of the estuary or if discharge of certain waste material alters the environment, formation of flocs may occur.

Collision of particles may be the result of Brownian motion, or of internal-shear motion, or of differential settling. Yet it must be realized that not every collision is such that colliding particles will stick together and form a larger aggregate. In general, we may distinguish three kinds of collisions: (1) An elastic rebound is the outcome of a collision; the result is a *stable suspension*. (2) A nonelastic rebound is the result of a collision; cohesive matter will form flocs rather fast. This is referred to as a *rapid flocculation.* (3) Elastic and nonelastic rebounds are the result of a collision; some particles will stick together and flocculate, whereas others will repel each other. This is referred to as a *slow flocculation.*

It has been pointed out that the ζ potential often serves as an indication for flocculation. For a rapid flocculation to take place, the ζ potential should be low, and many collisions result in adhesion of the particles. At higher ζ potentials there are fewer adhesion-causing collisions. The ζ potential was discussed in Sec. 12.2.2.

For rapid perikinetic[1] flocculation of a dilute, monodisperse system, the following relation is given by KRUYT (1952, p. 281):

$$n_t = \frac{n_0}{1 + t/T} \tag{12.12}$$

where n is the number of all sizes of particles per cubic centimeter (cm³) at zero time (n_0) and at time $t(n_t)$, and T is the time of flocculation given by

$$T = \frac{1}{4\pi D_f R n_0} \tag{12.13}$$

where R is the particle diameter in specific cases, and, in general, the distance between the centers of two particles, and D_f is the diffusion constant. For water as the medium of dispersion, at about 20°C, Eq. (12.13) becomes

$$T \approx \frac{2 \cdot 10^{11}}{n_0} \tag{12.13a}$$

For slow flocculation of a dilute, monodisperse system, Eq. (12.12) holds true, but Eq. (12.13) becomes

$$T = \frac{1}{4\pi D_f R n_0 \alpha} \tag{12.14}$$

where α expresses the fraction of successful collisions.

The probability of a normal Brownian collision was given by KRUYT (1952, p. 290) as

$$I = 4\pi D_f R n_0 \tag{12.15}$$

Orthokinetic flocculations, which are caused by a velocity gradient in the flow field, are also investigated by KRUYT (1952, p. 289). The probability of collision is given by

$$J = \frac{4}{3} n_0 R^3 \frac{du}{dy} \tag{12.16}$$

[1] Perikinetic—due to Brownian movement alone.

where du/dy is the velocity gradient of a laminar current. The ratio of the two collision probabilities is given by the ratio J/I. KRUYT (1952, p. 291) reports of the following calculations:

> For, $du/dy = 1\ \text{sec}^{-1}$, and a particle diameter of 10^{-5} cm., the ratio, J/I, is of the order of 10^{-3} and completely negligible, but when the diameter of at least one of the particles is larger than 1 μ, the ratio becomes larger than unity, and for particles of 10 μ, the orthokinetic flocculation is far more important than the normal perikinetic flocculation.

Once the flocs have formed and settle out—different floc sizes have different settling velocities—the smaller flocs apparently disappear by joining with the larger ones, and the whole flocculation process is speeded up.

12.4.2 ENGINEERING INVESTIGATIONS

Rather extensive investigations by EINSTEIN et al. (1961) and KRONE (1962) have been performed on cohesive sediments in the estuarial region of the San Francisco Bay in California. In most estuarial regions, the salinity exceeds 1 g/liter, and thus nearly every collision results in an adhesion of the particles. EINSTEIN et al. (1961) apply, therefore, the model of rapid flocculation in order to calculate the probability of collision from Brownian motion, given by Eq. (12.15). Assuming the suspended-sediment concentration to be about 0.1 g/liter and a primary particle of 0.3 μ, Eq. (12.15) gives $I = 0.014\ \text{sec}^{-1}$, or a collision on a primary particle at start of flocculation at every 69 sec. Due to collisions of primary particles with other primary particles and/or flocs, the settling velocity increases, but also the total number of particles is reduced; in turn, the distance between primary particles and/or flocs increases, and thus the frequency of collision decreases. With an additional calculation, using Eq. (12.12), KRONE (1962) demonstrated that after a lapse of 12 hr of flocculation, the flocs that resulted from the collision of the primary particles still did not have a sufficiently high settling velocity, and may be considered to remain in suspension for all practical purposes. The contribution of the internal shear of the flowing water to the flocculation should not be overlooked. EINSTEIN et al. (1961) have shown that the ratio J/I is unity for $du/dy = 1\ \text{sec}^{-1}$ and $d = 2\ \mu$. However, with an increase in floc size, the ratio J/I increases very rapidly. For example, for $du/dy = 1\ \text{sec}^{-1}$ and $d = 10\ \mu$, the ratio J/I becomes 120. From this, EINSTEIN et al. (1961) concluded that after a reasonably slow flocculation due to Brownian motion, internal shear accelerates flocculation considerably. It was suggested that under strong shearing rates, which may be encountered under wave conditions in unstable flow, in nonsteady and nonuniform flow, or with strong turbulence, large flocs are caused rather rapidly, and the flocculation process is considerably increased. However, high shearing rates might also become responsible for breaking up established flocs and thus slow down the rate of flocculation, a fact which is pointed out in TRASK (1939, p. 274) and discussed by KRONE (1962). The latter contest that the maximum floc size depends on the local shearing rate; but whether or not the flocs attain the maximum possible size depends, in turn, on the opportunities for a collision of the suspended clay particles.

The flocculation process is completed when the flocs arrive on the bottom. EINSTEIN et al. (1961) describe the deposition process as follows:

> At a density of 10 grams per liter, the flocs make bodily contact and form a continuous cohesive mass of the same density and strength as the individual large flocs. This mass can support a measurable shear stress without permanent deformation, at least for restricted duration. Upon prolonged stress, it will very slowly creep. Such creep may be either a shear motion or a compaction, or both, depending on the types of stress applied. Compaction is particularly important because it causes the strength of the deposit to increase as more and more bonds are made.

A rather well-defined interface exists between the flocculated cohesive material—frequently referred to as *fluid mud*—and the overlying water. According to studies by KRONE (1962) with cohesive sediments from San Francisco Bay, fluid mud should not exist below sediment concentrations of about 10 g/liter; an upper limit is given at about 170 g/liter. Measurements of the viscosity of fluid mud indicate a *Bingham*-fluid[1] behavior. Furthermore, it was shown that direct proportionality exists between the yield stress values (shear strength) of the flocculent suspension and the sediment concentration. KRONE (1962) conducted research on the deposition of clay sediments from flowing salt water in two different laboratory flumes. The results offered some interesting observations; however, it appears premature to draw definite conclusions from these limited experimental investigations. One of the conclusions may be worth mentioning. KRONE (1962) said: "Suspended sediment can be deposited only at bed shear less than a critical value (0.8 dynes/cm^2); deposition occurs more rapidly at lower shears."

PARTHENIADES (1965) finds that the critical velocity—the velocity above which all clay stays in suspension and below which all clay deposits—for cohesive sediment of San Francisco Bay is about 0.5 fps. The value was obtained experimentally in a 1-ft-wide and 60-ft-long steel flume. Recent

[1] A Bingham fluid follows the two-parameter model, $\tau = \tau_i + \mu(du/dy)$; it remains rigid when $\tau < \tau_i$, but flows similar to a newtonian fluid when $\tau > \tau_i$.

experiments in an annular rotating channel are reported by ETTER et al. (1968). When studying the depositional behavior of kaolinite, it was found that an initial period of relatively rapid sediment deposition is followed by an equilibrium suspended-sediment concentration. This observation is in general agreement with the experimental results of KRONE (1962).

Flocculation, as described in this section, is of considerable concern and importance in saline sediment transport. In fresh-water regions, flocculation may occur if the chemical composition of the water changes, which, in turn, affects the ζ potential. The collision-causing mechanisms are mainly the velocity gradient [see Eq. (12.16)] and the relative settling velocity. There exists a lack of research on flocculation of cohesive matter in fresh-water uniform-channel flow.

12.5 TRANSPORTATION PROBLEM

There exists little information on the transport of cohesive sediment at the present; this is especially true from the engineer's point of view. EINSTEIN et al. (1961) have stated that the transport of cohesive sediment—at least in estuaries—is determined by flocculation characteristics of the suspended cohesive particles. As has been discussed earlier, we may anticipate two modes of transport:

1. Particles which have an extremely low settling velocity will stay in suspension and thus will be transported as suspension.
2. Particles which get a chance to settle to the bottom and form the fluid mud may either move as connected fluid mud or become resuspended and then move as suspension.

As far as the gravity flow of the connected fluid mud is concerned, this type of flow is limited to bottom slopes which are sufficiently steep to initiate and maintain its flow. KRONE (1962) remarked that rather high concentrations of sediment may be transported in this way, but over limited distances only. Fluid mud may also be moved by a shear-induced motion. If shear is applied to the top layer, the motion of flocs or particles is such that they are entrained by the turbulent eddies, become thus resuspended, and move as suspension. KRONE (1962) suggests that cohesive-matter transport by estuarial water is largely in form of suspension. Visual observations in a laboratory flume affirmed this. There exists presently no reason to believe that the same mode of transport prevails in channels and streams. The fine cohesive particles are nothing but a fraction of the washload, which was discussed in Sec. 9.1.

What was earlier referred to as moving *fluid mud* seems to be similar to what is known in the literature as *density current*, turbidity current, underflow, and occasionally as stratified flow. Owing to the flocculated material, turbidity currents have a specific gravity of their own. Thus they are able to maintain their individuality as a separate body of liquid and move along the bottom of a watercourse or on the ocean floor. KUENEN (1952) has succeeded in obtaining a rather simple relation for the average velocity of the density current $\bar{u}_D$ as

$$\bar{u}_D = C_D \sqrt{\frac{\rho_{susp} - \rho}{\rho} D_D S} \tag{12.17}$$

where D_D = thickness of the flow
ρ_{susp} = density of the density current
C_D = a constant

KUENEN (1952) has given the following values for the constant, obtained from experiments as

$$C_D \approx 140 \qquad \text{cm}^{1/2}/\text{sec}$$

and has suggested a value of

$$C_D \approx 280 \qquad \text{cm}^{1/2}/\text{sec}$$

for underflow in reservoirs. It should be pointed out that Eq. (12.17) is a Chezy-type formula and was first suggested by O'BRIEN (1937). Experiments by KUENEN (1952) and others indicated that Eq. (12.17) is valid for a bottom steepness of $\alpha > 20°$. It is apparent that, for extremely small bed slopes, the driving force will be the density difference. Although Eq. (12.17) has an empirical character, Hinze (1960) and HARLEMAN (1963) have shown that it is essentially a fluid dynamically sound equation. The interface between the density current and the overlying fluid material depends on the relative velocity between the two media. As long as the relative velocities are low, the interface is smooth and rather distinct; at reasonable relative velocities it becomes possible for mixing to occur at the interfaces. As a result of the mixing process, sediment particles or flocs become entrained in the overlying fluid material.

Density currents in streams have their origin quite frequently in flood discharges of tributaries which drain an area composed of easily erodible and clayey soils. The occurrence of extremely high suspended-sediment concentrations of streams in the semiarid southwestern United States has been presented by BEVERAGE et al. (1964). It was found that some tributaries of the Rio Grande and the Colorado River are notorious for their high sediment loads. Studying the data from the files of the U.S. Geological Survey, BEVERAGE et al. (1964) reported 44 stream flow samples which exceed 40 percent sediment by weight, or 400,000 ppm. For example, it was reported that "the Rio Puerco contributes about 3% of the stream flow of

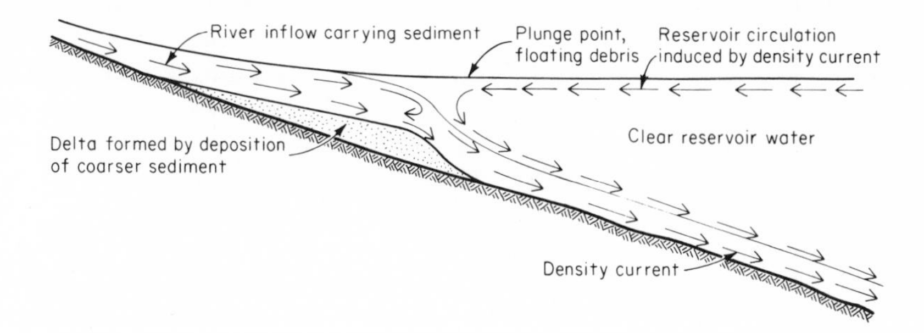

Fig. 12.15 Sketch of a flow pattern in a reservoir and the occurrence of a density current. [*After* HARLEMAN (*1963*).]

the Rio Grande below the juncture and about 45% of the sediment." It appears that the water-sediment mixture of the Rio Puerco when joining the Rio Grande continues its journey as an underflow in the same concentrated form down the river. Eventually, it either mixes with the overlying water or it deposits the solid material. Density currents, as described, are usually of local origin and limited in duration and quantity. Good observations or measurements of such phenomena are scarce. Density currents occur during high stages, high velocities, and strong sediment transport, all conditions not conducive to making detailed field investigations. There exists limited evidence that density currents may be occurring in the Colorado River, the Rio Grande, the lower Mississippi, the lower Arkansas River, the Nile, and some other rivers draining alluvial plains.

Sediment-laden rivers entering a lake or a reservoir of relatively clear water deposit their sediment load as follows: The coarse material, due to the weight, will settle out first and is deposited as a delta at the head of the reservoir. The fine material will be carried for longer distances, depending on the sizes of the particles. Some of the fine matter may not be deposited at all, and passes through the reservoir either after a general mixing has occurred as an overflow or as a density current moving along the bottom and passing through properly located outlets. The flow pattern of a reservoir with a pronounced density current is schematically sketched in Fig. 12.15. Good photographic evidence of density currents occurring in nature and in laboratories may be found in BELL (1942). The deposition pattern of the intensively studied Lake Mead is given in Fig. 12.16, which shows the profiles of the sediment accumulation over a period of 13 years.

As a result of field studies done at various reservoirs and reported by BELL (1942), we may conclude that density currents contain particles usually smaller than $d < 20\ \mu$. Calculation of the settling rate of such particles shows that a current of about 3 cm/sec is sufficiently strong to keep these

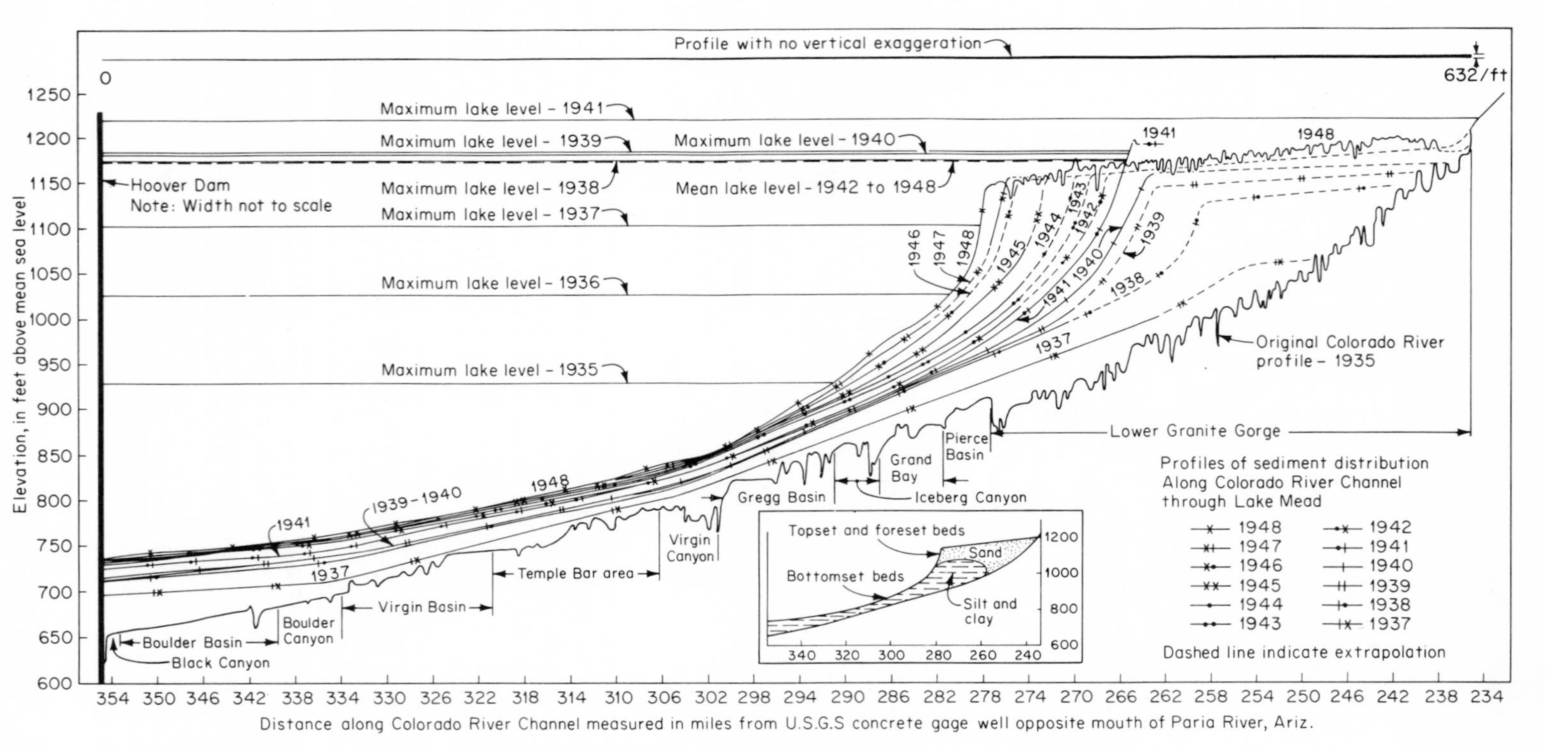

Fig. 12.16 Deposition pattern along the Colorado River through Lake Mead. [*After* SMITH *et al.* (*1954*).]

particles in suspension. This seems to be in agreement with studies reported by GROVER et al. (1937), who described that at three different periods during 1935 an underflow of very turbid water passed through Lake Mead, an artificial lake of a length of about 80 miles established on the Colorado River above the Boulder (now Hoover) Dam. It was suggested that the underflow moved through the reservoir essentially unmixed.

Density currents also occur in oceans. They may be initiated by the sediment load that a river discharges into the oceans or by any other tectonic or weather stimuli. SCHEIDEGGER (1961) discussed their morphological importance, and suggested that turbidity currents may be powerful agents in shaping submarine features. The role of density currents in the formation of sedimentary rock is more and more recognized by geologists.

13
Sediment Measuring Devices

13.1 INTRODUCTORY REMARKS

Various instruments have been developed to measure the solid-material discharge. Such measurements are necessary to determine directly the amount of sediment load and/or to establish or check analytical or empirical relations which permit direct calculation of the sediment load. Very few of the developed instruments are universally accepted. It thus becomes necessary to use the instruments with extreme care, and within the range of the hydraulic and sediment parameters as specified by the manufacturers.

Sediment transport occurs in two ways. One is referred to as *bedload*, the other one is *suspended load*; together they make up the *total load*. An extensive discussion on various kinds of transports and their predictions is given earlier. To facilitate the treatment in this chapter, Sec. 13.2 will be devoted to bedload measuring devices, Sec. 13.3 to suspended-load measuring devices, and Sec. 13.4 to total-load measuring devices.

At this point it shall be pointed out that, in order to distinguish between bedload and suspended load, we would need a clear line of demarcation between the two. At the present there exists no generally agreed upon definition for this demarcation; this has been discussed in Chaps. 7 and 8. This makes the problem of measuring correct quantities difficult and at times dubious.

Measurement and analysis of the particle size and related data will not be discussed in Chap. 13. The reader is referred to a review paper by GUY et al. (1969).

13.2 BEDLOAD MEASURING DEVICES

13.2.1 GENERAL REMARKS

More often than not, the bedload is measured as the bedload rate in weight per unit time and unit width, and is given the symbol g_s. To obtain g_s, we measure the weight of the bedload material which passes a certain location at a given cross section in a given time. Two different methods, one of direct and one of indirect measurement, may provide information on the bedload g_s. The direct methods will be described in Sec. 13.2.2, and the indirect methods will be discussed in Sec. 13.2.3.

However, before beginning these discussions, it is worthwhile to point out certain remarks which are more or less common to all bedload measuring devices. In order to measure the bedload g_s, it is very often necessary to introduce a piece of equipment—referred to as the *measuring device*—into the very layer where the bedload moves. The mere presence of this device is frequently disturbing the flow pattern sufficiently so that behavior and intensity of the bedload are altered. Furthermore, it is often difficult to give the equipment correct vertical and horizontal alignment with the bedload flow. In addition, it is very difficult to have a device which collects all the size fractions of the bedload, from the coarsest to the finest grains.

If a correct measurement of the bedload is desired, the following should be kept in mind:

1. Bedload measuring devices should be calibrated and their efficiency, the ratio of sampled to actual bedload, should be determined. It is not only very difficult to determine the efficiency of a given device but also the efficiencies are variable and, thus, uncertain. Efficiencies of bedload samplers have been determined in laboratory flumes with fixed and movable beds by way of testing scale models. A brief discussion on calibration of bedload samplers is given by HUBBELL (1964).
2. At any cross section in a stream, the bedload g_s is subject to fluctuations with respect to space and time. The first is explained by the shear stress distribution over the cross section, as given in Fig. 6.11; the latter is due to the fact that the bedload transport—as can be seen from field data in

Fig. 7.17—represents an unsteady phenomenon. It is thus desirable to obtain long-term measurements at various points throughout a cross section.

3. Wrong measurements may be obtained by improper operation of the entire measuring equipment. Under such circumstances, the sampler may scoop up the bed material and/or may collect suspended-load material. Proper care must be taken in the selection of a reliable timing device.
4. Bedforms apparently influence the sampling procedure. Little information on this point is available, but HUBBELL (1964) offered some interesting remarks. The relations of the geometry and size of bedforms and the measuring equipment have a strong influence on the efficiency of the equipment.
5. The inevitable question as to what is the true bedload (its thickness) is not answered to everybody's satisfaction. Especially, if equations are to be established, extreme care should be taken in reporting the data. EINSTEIN (1948) suggested that the thickness of layer within which bedload moves is different for each watercourse; it should be at least equal to the diameter of the largest grain.

Any discussion of bedload samplers must remain incomplete, because many agencies all over the world have developed their own devices. However, these measuring devices are, at least in principle, often quite similar and, thus, may readily be discussed.

13.2.2 DIRECT MEASUREMENTS

To facilitate the discussion, the samplers are treated in different sections: box- and basket-type samplers, pan-type samplers, pit-type samplers, and other ones.

13.2.2.1 Box- and basket-type samplers. Bedload samplers of this type consist of a pervious container where bedload accumulates, of a supporting frame and cables to make the sampler portable, and of a vane(s) to give the sampler the appropriate direction. The sampling operation consists of lowering the sampler to the bed and, on contact with the bed, the front gate of the sampler opens and a timer is released. The water and the bedload enter the box or basket, experience a velocity reduction which is often aided by a screen, and thus the bedload is deposited in the trap. At the end of the measurement, the gate is closed, the measuring time is recorded, and the sampler now containing the bedload is lifted. The bedload can be removed and carefully measured.

There are various sources of error in using such devices. In addition to those already mentioned in Sec. 13.2.1, the following may be of concern.

Due to the presence of a box or basket sampler, aided at times with additional screens or obstructions, a noticeable flow resistance is established. This is often responsible for a pressure increase at the entrance to the sampler, accompanied by a velocity reduction. Such an effect is felt especially by the water-sediment layer close to the bed, and, as a result, some of the bedload material accumulates in front of the sampler, causing the efficiency of the measuring device to vary and the sample to be unrepresentative. Certain modifications in the design of the equipment have been suggested to remedy this serious disadvantage. Such samplers are commonly known as *pressure-difference* samplers. To ensure that intake velocity and stream velocity are identical, the sampler walls diverge toward the rear. A representative portion of the bedload may enter the sampler, and the tendency of accumulation of bedload within the sampler is increased. Efficiencies of such samplers are reported to be very good.

A further source of error may be a result of the samplers not sitting immediately on the bed, thus permitting a sort of local scour action underneath the sampler. The bottom wall of the sampler should be made of a material which will readily adjust to the available bedform. Material such as loose sheets of chain link or rubber has been suggested, and is used in some types of samplers. Another precaution to be taken is that a sampler should not be filled to the total sampler volume. Usually, the calibration of the sampler will determine its proper range, but, in general, filling in excess of 30 percent of the total sampler volume should be avoided.

Bedload samplers of the box- and basket-type have been developed and built over the years in various designs. A typical box-basket sampler is the so-called *Mühlhofer sampler*, designed and used by MÜHLHOFER (1933) to measure bedload in rivers of the Tyrol. This sampler is shaped like a box and made out of screens; it is reproduced with its metric dimensions in Fig. 13.1. The original design of the sampler had a rigid bottom plate, but this was displaced in favor of loosely woven iron rings which conformed more readily with the bedforms.

Earlier samplers include those of Schaffernak (1922) and Born (1928). The *Born sampler*, described by JAROCKI (1963), was designed for measuring bedload consisting primarily of fine materials. This catcher was improved over the years and is in use in Poland.

The sampler of MÜHLHOFER (1933), who started to use it as early as 1928, experienced further modifications and minor alterations. These were done by EHRENBERGER (1931) and NESPER (1937) in Austria, by the *Eidgenössische Amt für Wasserwirtschaft* (1939) in Switzerland and by the *Laboratoire de Beauvert* (1942) in France. In a comparative study by NOVAK (1959), it was recommended that MÜHLHOFER'S (1933) samplers have an average efficiency of about 45 percent, EHRENBERGER'S (1931) about 60 percent, and NESPER'S (1037) about 40 percent.

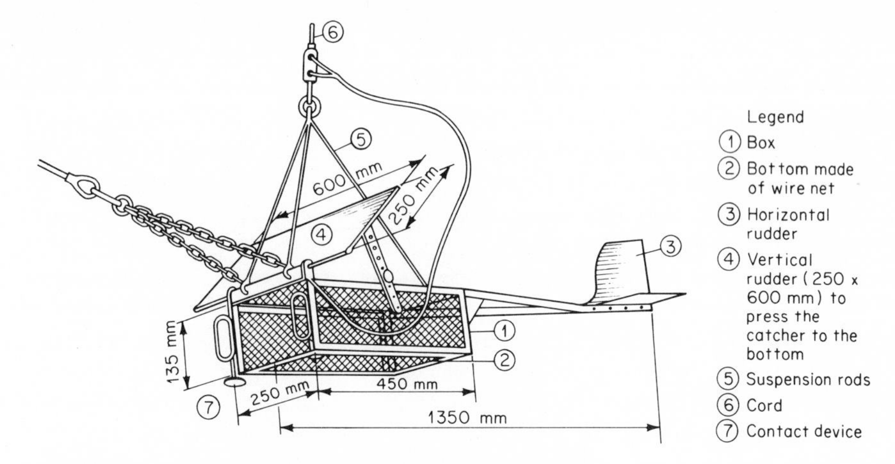

Fig. 13.1 Mühlhofer sampler. [*After* JAROCKI (*1963*).]

The more recent designs of bedload samplers are of the pressure-difference type, as discussed previously. Probably one of the better and wider known of this type is the so-called *Arnhem sampler* or bedload transport meter (*BTMA*), shown in Fig. 13.2 and first suggested by SCHAANK (1937). According to the *Waterloopkundig Laboratorium* in Delft, Netherlands (1968), the following information is available. Its purpose is to measure the bedload of coarse sand and fine gravel. The sampler is mounted on a frame, and a leaf spring presses it to the bottom. The mouth of the sampler has an opening of $0.085 \times 0.05\ m^2$; the basket located behind is of wire meshing, and its size is $0.53 \times 0.15\ m^2$. Bed material too coarse to pass the wire meshing remains in the sampler. The instrument is robustly built, weighs about 57 kg, and must be handled by a davit or derrick. The overall maximum dimensions are 1.83 m long and 0.89 m wide. Another bedload sampler was developed in the Netherlands by VINCKERS et al. (1953), which came to be known under the name *Sphinx*.[1] Flow enters through a rather small opening and is gradually spiraled around, and, thus, bedload accumulates in the sampler and eventually leaves the sampler at a larger opening. Both the Arnhem and the Sphinx samplers have, according to NOVAK (1959), a very good sampler efficiency.

A box-type sampler, the height of which is increased toward the rear, was designed by KAROLYI (1947). Its overall layout and metric dimensions are given in Fig. 13.3. The entire instrument weighs about 90 kg and the box is made of steel plates. The sediment-water mixture enters the sampler. The rear section of the box has a perforated dividing wall. Beneath this wall,

[1] Diephuis (1969) of the Delft Hydraulics Laboratory informs the author that the Sphinx sampler is not in use anymore.

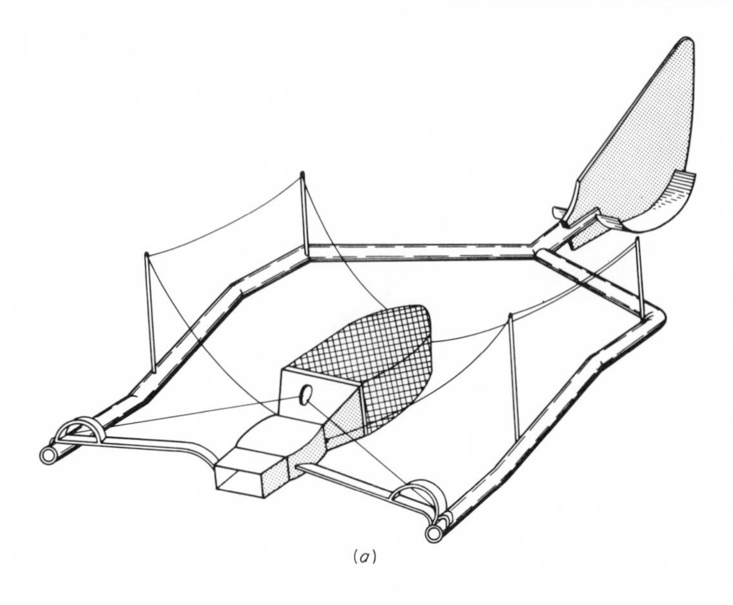

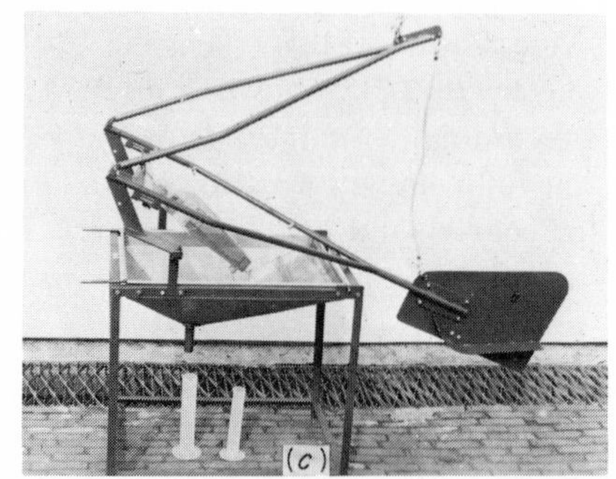

Fig. 13.2 Arnhem sampler (BTMA). (*a*) The Arnhem sampler [*after* HUBBELL (*1964*)]. (*b*) The new Arnhem sampler with an improved frame construction. (*c*) The emptying of the instrument; the catch is measured volumetrically. [The photographs are provided by Diephuis (1969) from the Delft Hydraulics Laboratory and are made available by the Van Essen N.V. company in Delft. This modified version is not yet in use in the Netherlands; however, application is expected.]

the bedload gets trapped, and the clear water rises and leaves at the exit of the sampler. The *Karolyi sampler* was tested and improved by NOVAK (1959). The flow pattern within the sampler was examined carefully and it was this study which led NOVAK (1959) to redesign the geometry of the Karolyi sampler. The *VUV sampler*, as it is often referred to, is given in Fig. 13.4. Of all the samplers tested by NOVAK (1959)—and there are quite a few—the sampling efficiency of the VUV-type sampler was the best, being 70 percent.

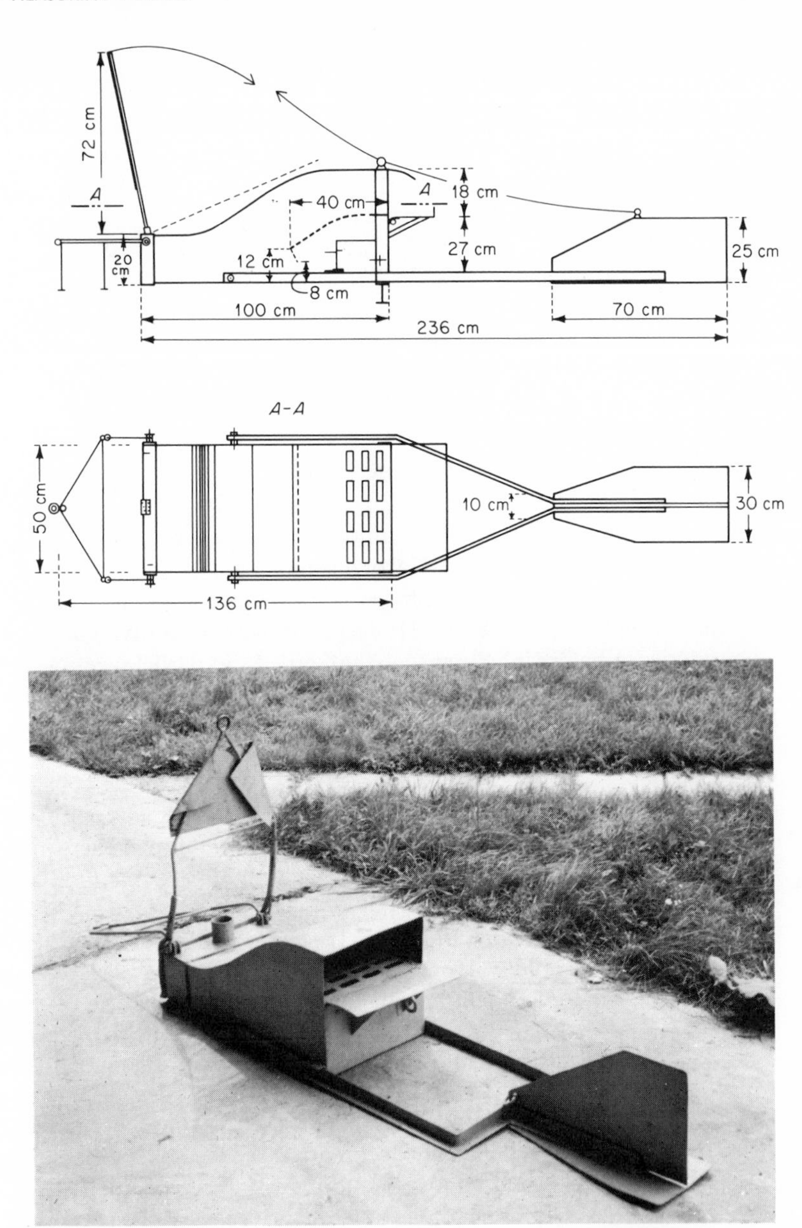

Fig. 13.3 Karolyi sampler [*after* Novak (*1959*)]. [The photograph is of a sampler presently at use at the *River Training Experimental Station* and was provided by Stelczer (1969).]

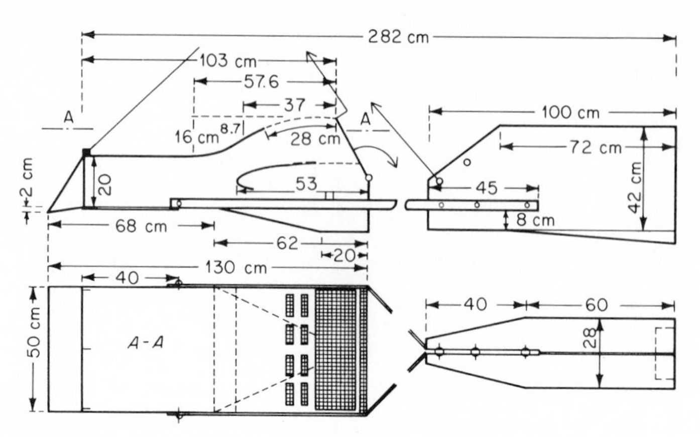

Fig. 13.4 VUV sampler. [*After* NOVAK (*1959*).]

HUBBELL (1964) reported of a sampler developed by Uppal et al. (1958), which is rather similar to the Karolyi samplers.

Another box-type sampler was developed by the U.S. Corps of Engineers and is discussed by HUBBELL (1964). In a diverging rectangular box a series of baffles is housed such that settled-out particles get a good chance to be retained. The principle of this sampler is a combination of a box-type and a pan-type sampler. Samplers of similar design, suggested by Soviet engineers, are reviewed by NOVAK (1959).

13.2.2.2 Pan-type samplers. Bedload samplers of the pan-type consist of a pan with a bottom and two sidewalls. Within this pan there may or may not be a baffle system to retard the water-sediment mixture and, thus, trap the sediment. Samplers of this type have been advanced by Soviet researchers, such as Shamov, Poljakov, Losiebsky, and others. A discussion is given by JAROCKI (1963), HUBBELL (1964), and NOVAK (1959). Outside the Soviet Union there exists little experience with such samplers. However, it was recommended that this kind of sampler be limited to streams with low velocities and small bedload rates.

13.2.2.3 Pit-type samplers. Depressions (pits) in the channel bottom may be installed to catch and accumulate the bedload. If a mechanical device is installed which removes continuously the accumulated sediment, a continuous record of the bedload rate at the measuring section is obtained. The efficiency of such a sampler is reported as rather high. HUBBELL (1964) reported that Mühlhofer in Austria used a pit sampler without a pumping device; a similar system is used in Poland according to DEBSKI (1965). An elaborate sediment measuring station of the pit-type with a pump system was installed and studied at the Enoree River in the United States, and is reported in DOBSON et al. (1940). A semiportable pit-type sampler was used by EINSTEIN (1948). A contemplated design of a portable pit sampler was suggested by HUBBELL (1964) and is given in Fig. 13.5.

13.2.2.4 Other types of samplers. A device consisting of a nozzle and a pump was suggested by Hiranandani (1943) and reviewed by HUBBELL (1964). The nozzle is located within the bedload layer, and is used to withdraw a point-integrated sample.

13.2.3 INDIRECT MEASUREMENTS

13.2.3.1 Calculation by measuring the bed material. Since most of the available methods to measure bedload are rather involved—frequently not too accurate but always costly—analytical methods may be used to obtain the bedload. Analytical methods were discussed in Chap. 7 and require knowledge of the bed composition. Thus the bed composition has to be determined experimentally. EINSTEIN (1950) remarked that, in fact, the bed material composition at each stage should be known. However, such information is seldom available, and the bed sampling at a representative river stage is usually performed.

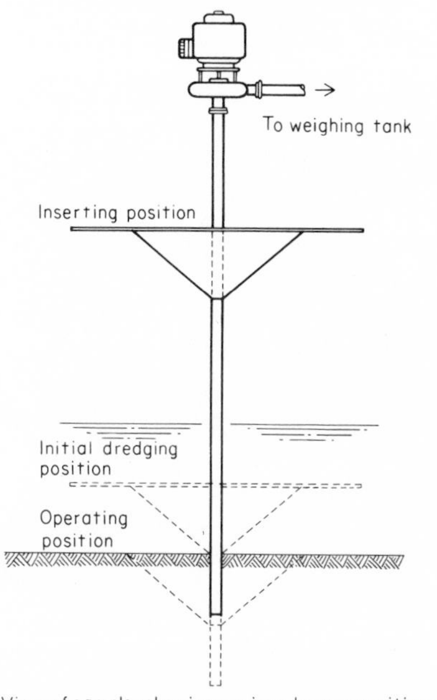

Fig. 13.5 Portable pit sampler. [*After* HUBBELL (*1964*).]

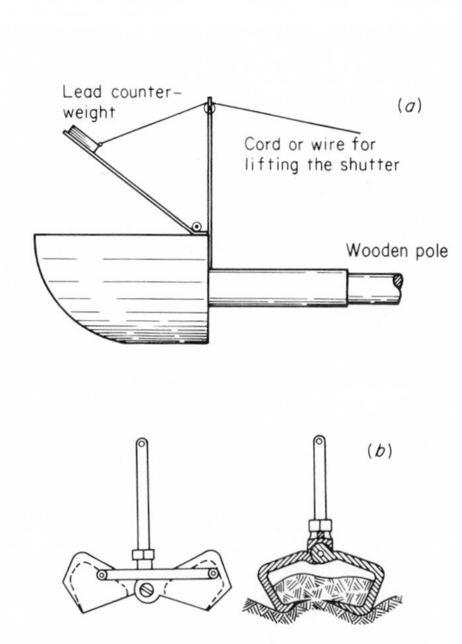

Fig. 13.6 Grabbing devices for sampling bed material. (*a*) Scoop. (*b*) Dredges. [*After* JAROCKI (*1963*).]

Most measuring devices used for sampling the bed under flowing water have one disadvantage in common, namely, that fine particles are often lost while the equipment is recovered. Bed material samplers could be either grabbing devices or boring pipes. Two rather simple but common grabbing devices are the scoop and the dredge, both shown in Fig. 13.6. A rather interesting sampler was developed by the INTERAGENCY COMMITTEE ON WATER RESOURCES[1] (1963) in the United States. A steel-made streamlined sampler, shown in Fig. 13.7, houses a scoop bucket. Upon contact with the bed, a heavy coil spring swings the bucket out, which takes a sample of the top 2 in. of the stream bed. Two samplers of this kind were developed and

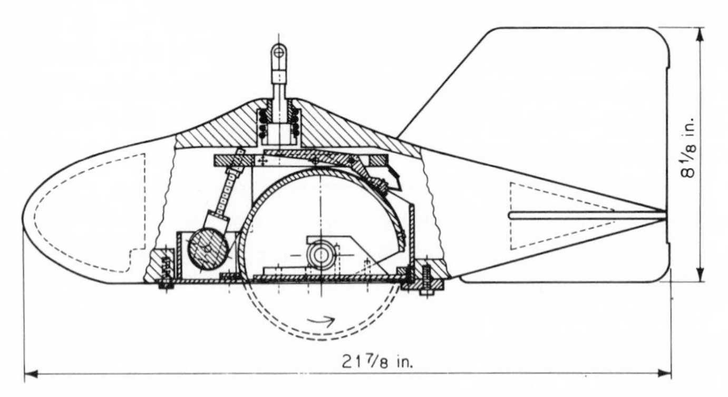

Fig. 13.7 Bed material sampler US BM-60. [*After* INTERAGENCY COMMITTEE ON WATER RESOURCES (*1963*).]

[1] This committee, essentially initiated in 1939, develops improved sediment sampling instruments and techniques for measurement and analyses of sediment load in streams.

recommended for field use: a heavy one ($\approx$100 lb), US BM-54, and a lighter one ($\approx$30 lb), US BM-60. Both the BM-54 and BM-60 are sealed with a rubber gasket when the bucket is closed. As a result, these samplers retain all of the fines in the bed mixture except those that reside on the surface and are scoured away when the sampler approaches the stream bed.

Boring samplers consist of a hollow cylinder or pipe which is forced into the bed material. Discussions on some different boring samplers are given by JAROCKI (1963) and the INTERAGENCY COMMITTEE ON WATER RESOURCES (1963). The latter designed an instrument and recommended it for use.

13.2.3.2 Sound samplers. The bedload scraping along the bottom creates audible sound waves. Acoustic instruments are designed to pick up these waves. The equipment is simple, consisting of an underwater microphone located at a certain distance from the bottom, an amplifier, and a recorder. Most of these instruments have underwater microphones of various designs. Some of the samplers are dircussed by HUBBELL (1964). One design, the Beauvert Laboratory hydrophonic detector, discussed by HUBBELL (1964), has the microphone mountea on a triangular base plate which, in turn, rests on the bed. The sound produced by particle-to-particle and particle-to-instrument collisions is recorded. This sampler was used by BRAUDEAU (1951). IVICSICS (1956) described a sampler where the microphone is housed in a streamlined body, similar in its design to the one given by Fig. 13.7. This instrument does not influence the bedload movement because the sampler is suspended a few meters above the bottom of the bed. None of the sound samplers discussed so far give but qualitative information on the bedload movement.

Also based on sound waves, but entirely different from the foregoing method, is a device based on ultrasonic sound waves. A transmitter and a receiver of sound waves are located so that the sediment-water mixtures pass between the two. Different amounts of energy are absorbed at different concentrations. A sampler of this type was developed by SMOLTCZYK (1955) for the measurement of bedload rates consisting of fine sand. Measurements obtained with such samplers produce interesting and reliable results, as was shown by DEBSKI (1965).

13.2.3.3 Tracking of bedforms. The determination of bedload in the lower flow regime ($N_F \leq 1$) is possible with Eq. (13.1), or

$$q_s = (1 - m)c_B \frac{\overline{\Delta H}}{2} + C \tag{13.1}$$

where m = porosity of the bed

c_B = average velocity of the bedforms

$\overline{\Delta H}$ = average height of bedforms

C = constant of integration and may be taken as equal to zero

Equation (13.1) was derived in Sec. 7.5.1 and was given as Eq. (7.78).

In clear and/or shallow water, the velocity of the bedforms, c_B, may be directly observed; the bedform height $\overline{\Delta H}$ can be measured too. Equation (13.1) allows calculation of the bedload rate q_s, provided the porosity m is known. In deep and less clear water, continuous depth measurements have to be performed. Maps of bottom (depth) contours for different times may be evaluated such that bedform velocity and height are available and the bedload rate may be calculated. The problem, however, is one of obtaining accurate, meaningful, and possibly continuous depth records. HUBBELL (1964) discussed the method of ultrasonic sounding in detail. KARAKI et al. (1961) developed an ultrasonic instrument to measure surface and stream bed profiles simultaneously under dynamic conditions. An interesting device was presented by ARNBORG (1953). A box with both ends open is lowered until it rests on the bed and the bedload can move through. On one side of the two vertical walls there are vertical rows of photocells, each one facing a lamp at the corresponding position on the other wall. Since the light intensity is altered by passage of sediment material, the thickness of the bedforms could be determined continuously.

13.2.3.4 Other methods. EINSTEIN (1948) suggested that a survey of the delta growth might be a method for determining the bedload. This method is restricted to rivers which discharge into lakes or reservoirs and deposit the bedload in the form of a delta, as shown in Fig. 12.15. In this way, a reasonable estimate of the bedload rate may be obtained.

Another method suggested by EINSTEIN (1948) is to measure the bedload at locations where it is suspended by means of suspended-load samplers. Sediment that usually moves in contact with the bed often becomes suspended below rapids, river constrictions, or other places where the turbulence is increased locally. BENEDICT et al. (1953) have applied this knowledge and developed the *turbulence flume method*, which is discussed in Sec. 13.4.2.2.

13.3 SUSPENDED-LOAD MEASURING DEVICES

13.3.1 GENERAL REMARKS

The suspended load q_{ss} is usually given as the suspended-load rate in volume per unit time and unit width. To obtain q_{ss}, the sediment concentration C is determined from samples of a water-sediment mixture, and the corresponding water discharge q is measured. The suspended load is then obtained as

$$q_{ss} = Cq \tag{13.2}$$

Since the previous depends on the assumption that fluid and solid particles move with the same velocity, Eq. (13.2) is only good in the upper layers of the flow, and fails in the bed layer. Furthermore, the concentration of sediment C is not uniformly distributed over the cross section—there is a pronounced vertical and horizontal distribution—and, therefore, samples at various locations throughout the cross section have to be measured.

Although a number of samplers have been developed, many of those can be treated within three separate sections: Instantaneous samplers, to be discussed in Sec. 13.3.2, enclose instantaneously a certain volume of a water-sediment mixture, integrating samplers, to be discussed in Sec. 13.3.3, collect, over a certain distance or at a given point, a certain volume of a water-sediment mixture; and continuously recording devices will be discussed in Sec. 13.3.4.

Prior to discussing the kinds of samplers and their operations, it is necessary to point out facts common to most all suspended-load measuring devices. Some of these remarks are similar to those on bedload samplers, which were discussed in Sec. 13.2.1. To obtain a truly representative sample, the presence of the sampler should cause little disturbance on the flow pattern of the water-sediment mixture. This is best achieved if a small nozzle projects from the collector unit into the flow. Such a design should also ensure that the velocity in the nozzle is identical to the ambient stream velocity, a condition which is ordinarily not met with vertical samplers. Furthermore, it is important that the sampler has the proper vertical and horizontal orientation toward the flow.

In order to obtain good measurements of the suspended load, the following points should be kept in mind:

1. The suspended matter is very much subject to timely fluctuations; it is the turbulent-velocity component that keeps the solid matter in suspension. Thus it is extremely desirable to obtain samples collected over a reasonable time period. Instantaneously sampled specimens should be treated with caution.
2. Improper operation of the measuring device may cause the sampler to collect bedload material. Since the velocity of the bedload is different from the ambient liquid velocity, this violates Eq. (13.2), and the whole sample will be in error.
3. Depending on the desired degree of accuracy of the measurements, the number and location of sampling vertical(s) should be selected. The

common methods in use are given and briefly discussed by the INTERAGENCY COMMITTEE ON WATER RESOURCES (1963, p. 39), and are given here as follows:

a. single vertical at midstream;
b. single vertical at thalweg or point of greatest depth;
c. verticals at $\frac{1}{4}$, $\frac{1}{2}$, and $\frac{3}{4}$ width;
d. verticals at $\frac{1}{6}$, $\frac{1}{2}$, and $\frac{5}{6}$ width;
e. four or more verticals at mid-points of equal-width sections across the stream;
f. verticals at the centroids of sections of equal water discharge.

4. At a given vertical, the suspended load can be sampled by a depth-integration method or by a point metered or point-integration method at one or more points along the vertical. The depth integration is useful for determination of the suspended-load rate. The point method is used if the vertical suspended-load distribution itself is desired. Proper selection of sampling points in a vertical is of importance, and was discussed by the INTERAGENCY COMMITTEE ON WATER RESOURCES (1963, p. 46).
5. To facilitate the calculation of the suspended-load rate according to Eq. (13.2), suitable data of velocities (or discharges) should be obtained simultaneously.
6. Suspended-load samplers measure everything that is suspended, and thus measure both the suspended fraction of the bed material load and the wash-load. This must be taken into consideration when the measurements are analyzed.

13.3.2 INSTANTANEOUS SAMPLERS

The principle of operation of an instantaneous sampler is simple. An open container, either in a vertical or horizontal position, is lowered to the desired location. Then some kind of a device releases a mechanism which traps the sample of the water-sediment mixture. As can be imagined, numerous samplers based on such a principle have been designed all over the world; only a few of those will be discussed here. Generally, they may have most any capacity, but usually they vary from trapping about 1 to 5 liters of mixture. They are introduced into the water by rods (for shallow depth) or by cables or ropes.

A typical horizontal sampler, discussed by JAKUSCHOFF (1932), is shown in Fig. 13.8. It consists of a pipe, two cover plates, metal weight, and a rope. Upon release of the metal weight, the cover plates enclose a sample of the sediment-water mixture into the pipe. This design of a sampler was frequently used in Sweden. Similar in design are the Joukowsky samplers, used in

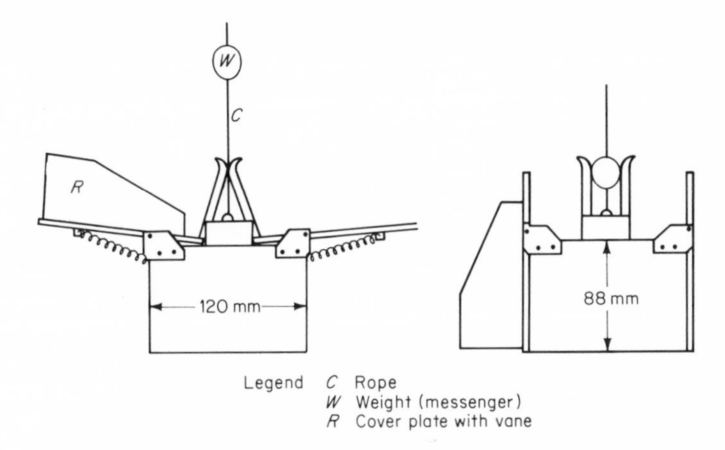

Fig. 13.8 Instantaneous horizontal sampler. [*After* JAKUSCHOFF (*1932*).]

Russia, and the *sonde de collet*, used in France; both are discussed by JAROCKI (1963). Again similar, but with a more refined mechanical design, are the *Tennessee Valley Authority samplers* and the *Tait-Binckley samplers*, which are described by the INTERAGENCY COMMITTEE ON WATER RESOURCES (1963, p. 52).

A typical vertical sampler is shown in Fig. 13.9 and described by JAKUSCHOFF (1932). It consists of a cylinder and a cover plate, a metal weight, and a cable. If the device is in the sampling position, the metal weight, called messenger, is sent down the cable, releases the cover plate, which drops and encloses the sample into the cylinder. Similar in design is the *Eakin sampler*, described by JOHNSON et al. (1940), and the *Polish Hydrographical Institute sampler*, described by JAROCKI (1963). The instantaneous vertical sampler is used extensively in oceanography, where it is well known under the name *Nanson bottle.*

13.3.3 INTEGRATING SAMPLERS

There exist two different types of integrating samplers. The first obtains a sample at a given point, where it remains for a certain time; this is the point-integration method. The second obtains a sample while it is moved vertically through a given distance; this is the depth-integration method. Each of these samplers consists of a container, an intake nozzle, and an air escape vent, and, at times, of a closing and opening mechanism.

13.3.3.1 Point-integrating samplers. The principle of this sampler is well demonstrated by the sampling bottles shown in Fig. 13.10. Bottles are

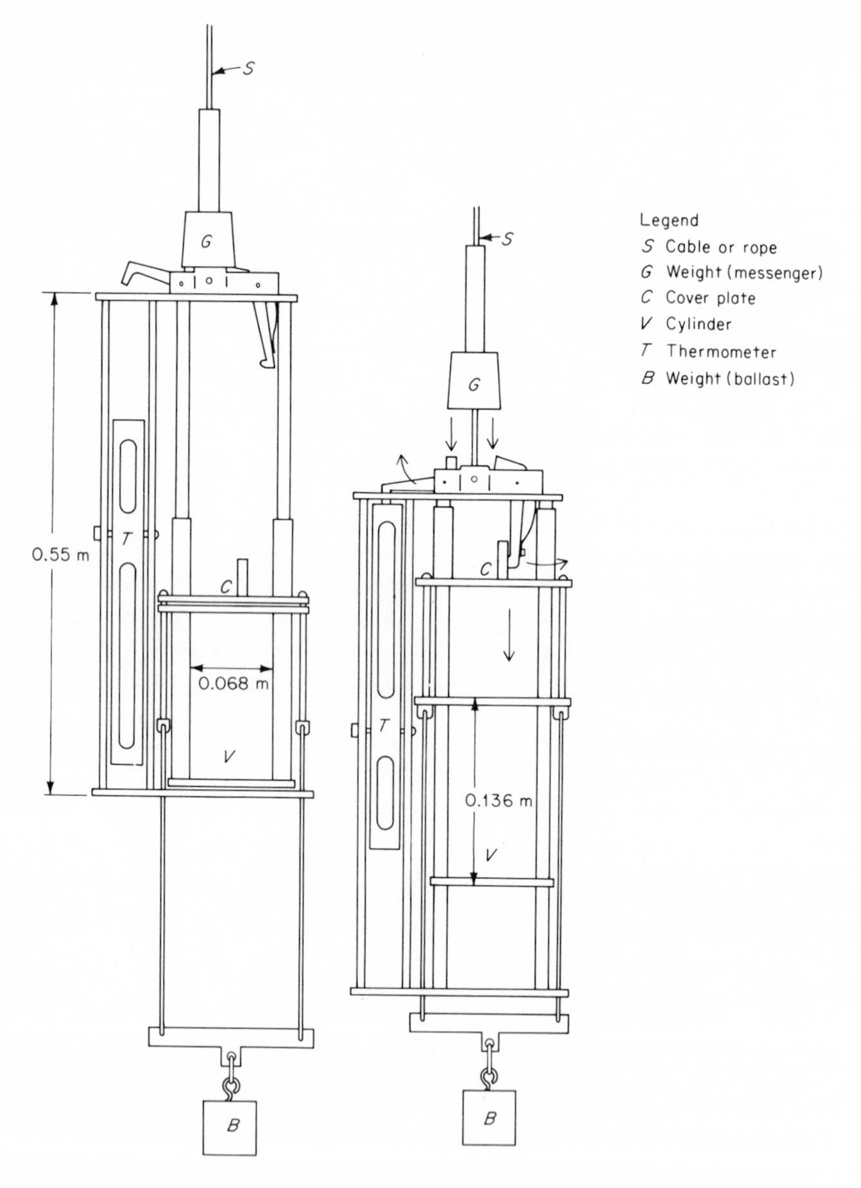

Fig. 13.9 Instantaneous vertical sampler. [*After* JAKUSCHOFF (*1932*).]

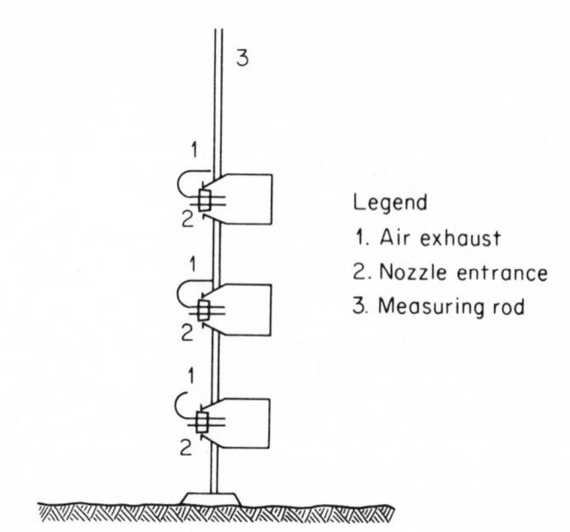

Fig. 13.10 Sampling bottles at different depths.

placed in positions where samples are to be obtained. Such a sampler fills gradually and causes little disturbance to the flow. If the sampler can be placed quickly into the sampling position, no valves are necessary; otherwise control valves to open and close the tubes are essential.

The *Gluschkoff bellows*, described by JAKUSCHOFF (1932) and shown in Fig. 13.11, operate on this principle. A deflated rubber bag is brought into the measuring position. The draw tube at the rubber bag is now held into the flow, and the rubber bag begins to fill. Just before the rubber bag is filled, the draw tube is turned and the filling time recorded. Both the flow rate and the suspended load can be determined. Ordinary bottles, such as fruit jars or milk bottles, have been used extensively. JOHNSON et al. (1940) discussed and used such a simple device; so does JAROCKI (1963). Over the years the principle has not changed, but certain modifications in the design have been made. A most extensive development program was undertaken by the INTERAGENCY COMMITTEE ON WATER RESOURCES (1963),[1] which shall be

[1] It should be noted that many of the samplers discussed in this report were developed, used, and described in various earlier reports by the same agency.

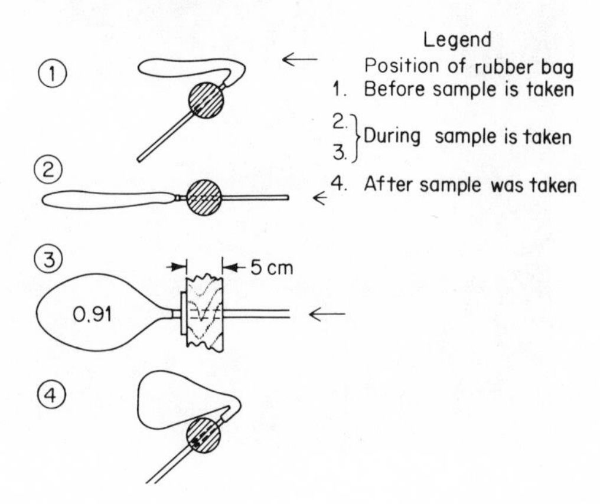

Fig. 13.11 Gluschkoff's bellows. *After* DEBSKI (*1965*).]

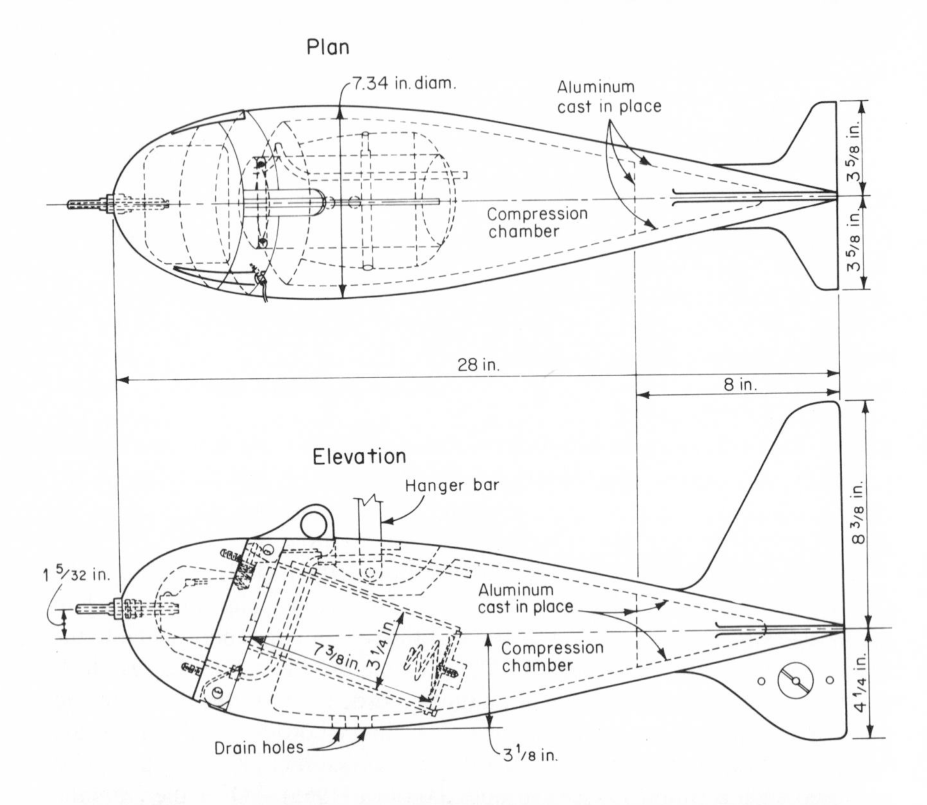

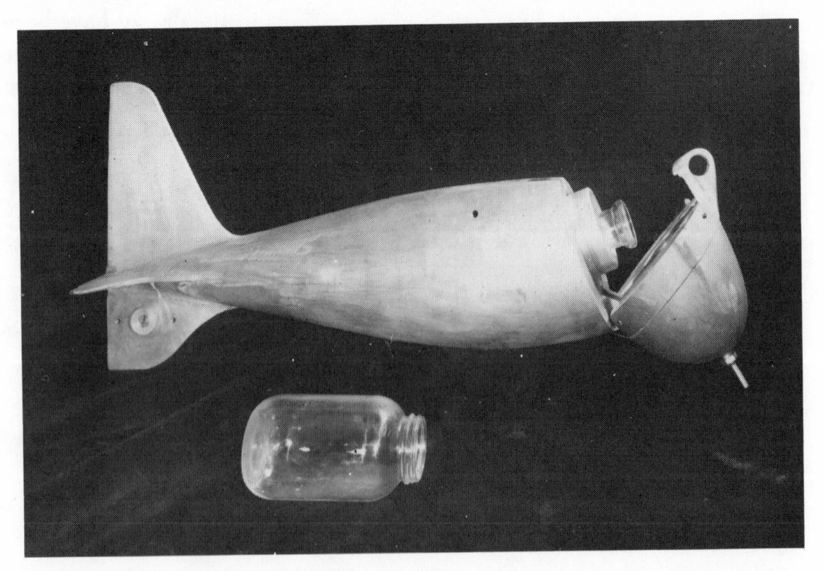

Fig. 13.12 Point-integrating sampler US P-61. [*After* Interagency Committee on Water Resources (*1963*).]

discussed briefly. The agency developed and recommended for use two of their samplers, known as US P-46R, described in detail by NELSON et al. (1950), and US P-61; the latter is shown in Fig. 13.12. Both samplers weigh about 100 lb and have a streamlined appearance. They are made of cast bronze and equipped with tail vanes. A removable bottle is located in a cavity of the sampler. An intake nozzle of $\frac{3}{16}$ in. points into the flow and the air exhaust leaves the body of the sampler on the side. Operations of intake and exhaust are controlled by a valve. There is also a pressure-equalizing chamber where the air pressure in the container is equalized with the external hydrostatic head at the nozzle at any depth. In addition to these two samplers, the same agency developed heavier samplers, up to 300 lb; those are used in extremely deep streams with high velocities. A sampler developed in France, the *Neypric silt sampler*, is rather similar to the previously discussed instruments.

Some point-integrating samplers are designed such that they collect the samples automatically and the suspended-load concentration is subsequently determined in the laboratory. One such sampler, called the *single-stage sampler*, was developed by the INTERAGENCY COMMITTEE ON WATER RESOURCES (1963) for automatic collection of samples without immediate attention from field personnel. Several of these units may be mounted at one station. The sampler consists of a bottle mounted on a rack. A stopper with two holes, for a siphon-shaped air exhaust and intake tube, seals the

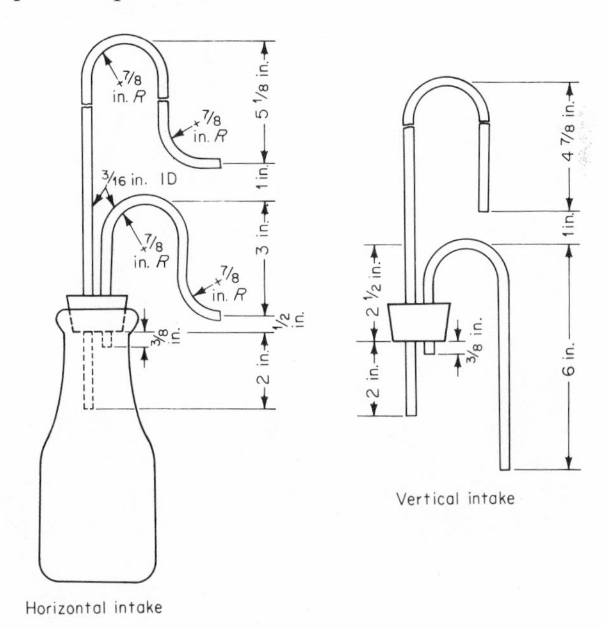

Fig. 13.13 Single-stage suspended-load sampler. [*After* INTERAGENCY COMMITTEE ON WATER RESOURCES (*1963*).]

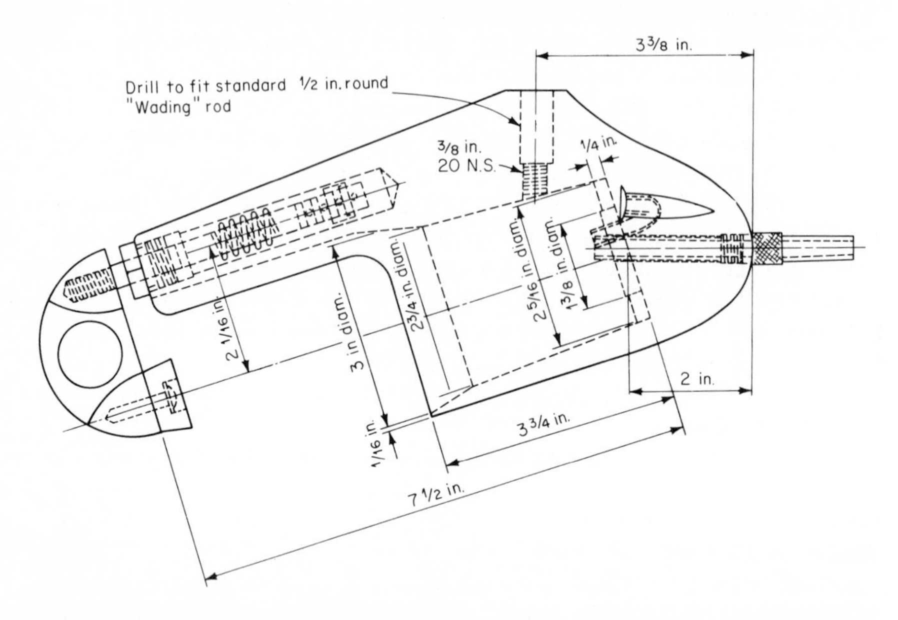

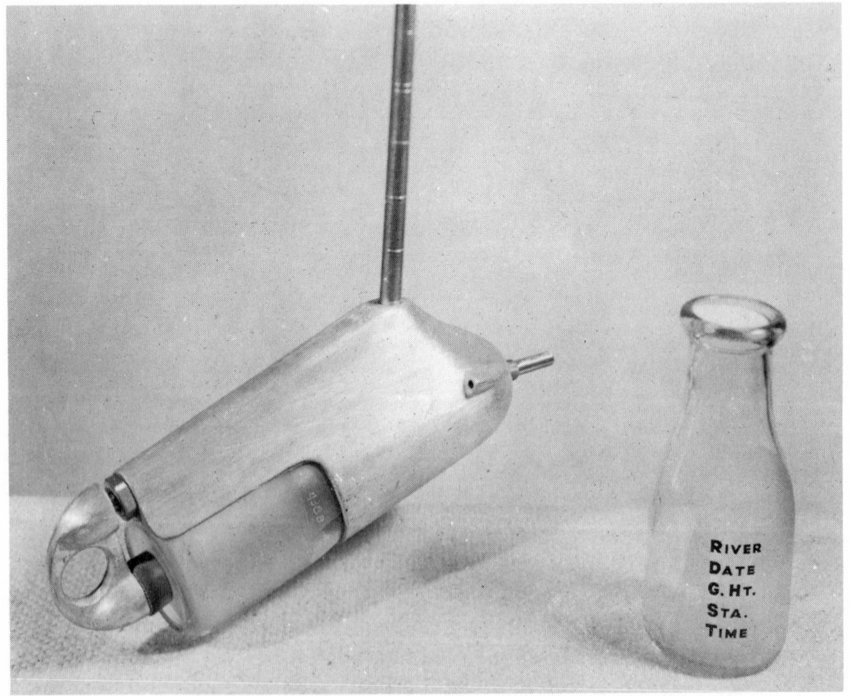

Fig. 13.14 Depth-integrating sampler US DH-*48*. [*After* INTERAGENCY COMMITTEE ON WATER RESOURCES (*1963*).]

bottle. The water-sediment mixture enters whenever the water level rises to the elevation at which the intake tube is placed. A single-stage sampler with both a vertical and a horizontal intake is shown in Fig. 13.13. The big advantage of this equipment is that it can sample at flood stage when manual sampling would be difficult. Extremely valuable data on the suspended-load concentrations may thus be obtained.

Another point-integrating sampler is the *pumping sampler*, which was also developed by the INTERAGENCY COMMITTEE ON WATER RESOURCES (1963). The sampler consists of a pumping system with the intake located in the stream and the outlet going to a bottling and/or flushing system. At a preset location in the stream and at a preset time interval, a sample is taken and collected in a sediment-recording system. Three systems have been suggested: the accumulative-weight recorder, the volume recorder, and the individual-sample bottling. Although the pumping sampler is still being field-tested, it may soon be used as a routine recording system. According to HUBBELL (1969), a number of different pumping samplers have recently been constructed and are used routinely. Each of these samplers has been designed to fill a particular need. As a result, a variety of pumping samplers are in use in the United States, but no "standard" sampler exists. Pumping samplers have also been suggested by Polakov (1948) and by the Soviet Hydrological Institute; they are reviewed by JAROCKI (1963). A principle similar to the one discussed previously is used in many hydraulic laboratory investigations. Samples are siphoned or pumped from the flume through pitot tubes like samplers. The experimental procedure was described by VANONI (1941).

13.3.3.2 Depth-integrating samplers. A depth-integrating sampler is essentially similar to a point-integrating sampler without a control valve; thus, it is simpler. The latter one can, of course, be used without modification as a depth-integrating sampler. The sampler begins to collect the mixture when it enters the water and stops on leaving the water. During this time and over the distance traveled, it collects the average concentration of suspended solids. The sampler should cover the distance with a constant speed. For a depth of 15 ft or less, depth integration is performed on a round-trip basis, i.e., the sampler is lowered and raised in an open position. For a depth of 30 ft or less, depth integration is performed with a single trip, i.e., the sampler is lowered in an open position but, upon contact with the bed, a foot trigger closes the intake and exhaust line, and is raised in a closed position. For a depth in excess of 30 ft, more than one sampling trip is necessary; a sampler with a valve system may be desirable.

The INTERAGENCY COMMITTEE ON WATER RESOURCES (1963) has studied many different depth-integrating samplers and recommended the following: US DH-48, US DH-59, and US D-49. The US D-49 sampler is the heaviest,

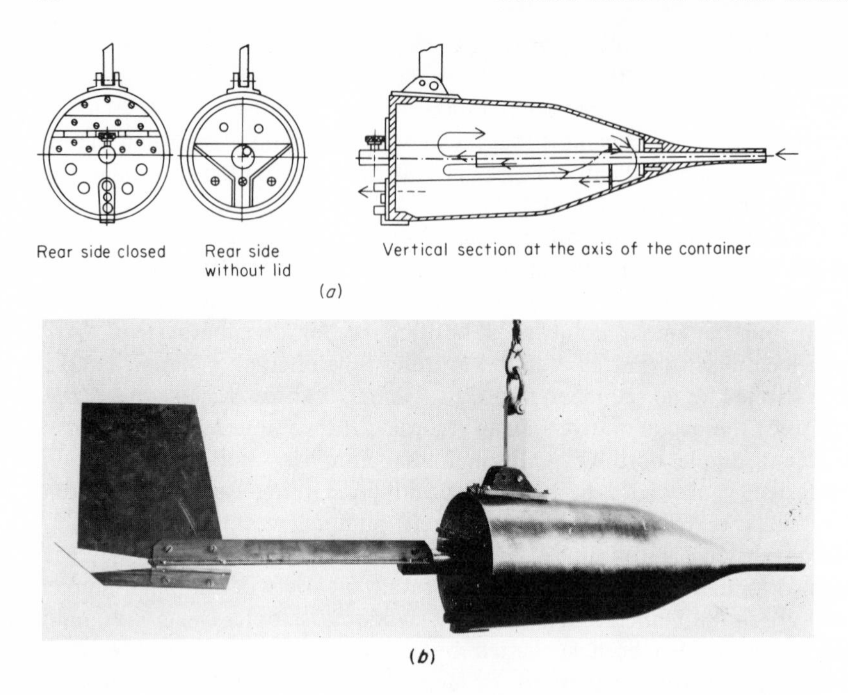

Fig. 13.15 The Delft bottle. (*a*) Schematic sketch, (*b*) suspended on a cable. [*The information is provided by Diephuis (1969) of the Delft Hydraulics Laboratory and is made available by the company Van Essen N.V. at Delft.*]

weighing around 60 lb, and is in its appearance similar to the one shown in Fig. 13.12. The US DH-48, shown in Fig. 13.14, is of cast aluminum and weighs only $4\frac{1}{2}$ lb. With a wading rod attached, it can be used as a hand sampler.

Similar to the United States samplers is a sampler discussed by JAROCKI (1963). The *Delft bottle* is of different design. It is made of a pipe of 0.31 m in diameter. In front is a nozzle of 0.022 m in diameter pointing into the flow; in the rear is a cover plate with holes. The water-sediment mixture enters the nozzle. Owing to a sharp velocity decrease in the wide chamber of the sampler, the solid particles settle out and the water leaves the sampler. It is claimed that, due to the flow-through principle, a large volume of mixture can be sampled. The sampler collects the suspended load rate g_{ss} directly. The Delft bottle is illustrated in Fig. 13.15.

13.3.4 CONTINUOUSLY RECORDING SAMPLERS

Several automatically and continuously recording sediment measuring devices have been suggested. The presence of the suspended solid particles makes the liquid less transparent; this effect may be used to measure concentration. Any device with a light-emitting source and a photocell located opposite will do this. The reading is taken with an ammeter which has to be calibrated for every situation because of the variability of the sediment and of the daylight.

The use of this principle was recognized by Kalitin (1923), who himself suggested different designs of the instrument. The work of Kalitin was presented by JAKUSCHOFF (1932). A simple photometer, as it was called, consists of a pipe with two holes opposite to each other. On one side is the light source, and on the other one is the photometer. The pipe, which is aligned with a vane, is then submerged, and the water-sediment mixture passes through it. Another device was suggested, which has the form of a fork, with a light source on one end and the photocell on the other. A fork-type meter, the so-called *Davall siltmeter*, was recently developed in England by the Research Department of the *British Transport Docks Board.* It is understood that this instrument is used in both the field and the laboratory. A similar device was described by BHATTACHARYA et al. (1969), and was developed primarily for laboratory use. The INTERAGENCY COMMITTEE ON WATER RESOURCES (1963) has studied the possibility of using the turbidity method, but feels that several instrumentation difficulties have to be overcome before such a device can be used in the field.

Two other devices that measure continuously the concentration are under investigation at the INTERAGENCY COMMITTEE ON WATER RESOURCES (1963). One is an electronic sensing equipment which measures and counts individual particles when passed through an aperture. Its principle is that the resistance between two electrodes is altered if solid particles displace part of the liquid in the aperture. The other one is an ultrasonic device. The attenuation of sound waves passing through a solid-liquid mixture varies with the concentration of the particles. A gage which measures the concentration in flowing water, based on the attenuation of x-rays, was described by MURPHREE et al. (1968). The sensing unit is a radioactive (Cadmium-109) source which emits x-rays. The rays that pass through the turbid water are recorded by a scintillation detector and converted into electric pulses which are recorded. Limited field studies have shown promise.

13.4 TOTAL-LOAD MEASURING DEVICES

13.4.1 GENERAL REMARKS

In general, the total load rate g_{st} is obtained by addition of the bedload rate g_s and the suspended-load rate g_{ss}. Thus one way to obtain a measurement of the total load rate g_{st} is to measure its fractions. However, it must be kept in mind that, in the measurements of its fractions, certain limitations are encountered, which were discussed in Secs. 13.2.1 and 13.3.1. Because of these limitations the determination of the total (or bed material) load frequently involves more than a simple addition of these fractions. In the following discussion indirect and direct (tracing methods) measurements will be considered.

13.4.2 INDIRECT MEASUREMENTS

13.4.2.1 With a calculation and measurement procedure. Under certain conditions it might be necessary to obtain a sample of the bed material and calculate the bedload and suspended load for each size fraction. A summation of the bedload and suspended load is the total load. On the other hand, it might be possible to measure the bedload. Analytical determination of the suspended load is required to get the total load. If the suspended-load rate can be measured, the bedload rate has to be calculated. The summation is the total load that does not include the washload.

13.4.2.2 With a measurement procedure. A method to measure the total load at such locations where it moves entirely in suspension was suggested by EINSTEIN (1948). Suspended-load samplers can thus be used to measure the total load that includes the washload. Due to a locally increased turbulence, the bedload is forced into suspension. This may happen below rapids, at stream constrictions, and other locations, or the required conditions are obtained artificially by constructing a *turbulence flume*. The latter consists of a series of turbulence-inducing obstructions, such as baffles or sills, which are distributed over a short section of the channel. The turbulence flume was developed and tested in both the laboratory and field by BENEDICT et al. (1953). One purpose of the model study was to determine the geometry, size, and spacing of the baffles. The recommended design for the prototype structure was as follows. The turbulence flume was to be built underneath a bridge crossing. It consisted of a reinforced concrete floor being 82 ft wide and 38 ft long, with an end sill at the end of the floor. Removable baffle plates, 1 ft high and 2 ft wide, were placed normal to the flow direction; their longitudinal spacing was 2 ft, and their lateral spacing was 6 ft from center to center.

Total-load measurements may also be made at locations where the entire total load settles out. Proper evaluation of surveys of sediment deposits can lead to good estimates of the total load.

13.4.3 TRACING METHODS

Tracing of labeled particles in a sediment mixture helps to obtain information on the movement of sediments. The motion of labeled particles is studied and measured, and from this knowledge a qualitative and/or quantitative determination of the entire sediment transport is deduced. Tracer techniques offer a unique approach for studying the motion of discrete particles. By these methods we may eventually obtain a very good picture of the sediment-transport process. Tracing techniques have helped already to understand such involved and complicated sedimentation problems as exist in estuaries, along coastlines, and in oceans. By now there is a good deal of literature available, much of which was summarized by COURTOIS (1966). Theoretical models of tracer movement have been discussed by CRICKMORE et al. (1962), HUBBELL et al. (1964), TODOROVIC et al. (1966), and YANO et al. (1969).

The operation of obtaining data with any tracing method usually includes the following steps. An appropriate tracer is selected. This tracer is then supplied to the sediment to be studied. The labeled sediment is introduced into the hydraulic environment. After a certain time period the tracer has spread and data collection may begin. The collected data are now subject to qualitative and/or quantitative interpretation.

13.4.3.1 Types of tracers. The tracer material and the labeled sediment particle(s) should fulfill the following requirements:

1. A labeled and an unlabeled solid particle must react to the forces responsible for sediment motion in the same way.
2. The physical and/or chemical properties of the traced particle(s) must be distinguishable and/or detectable with appropriate equipment. Depending on the desired accuracy of the data and the sensitivity of the measuring device, the amount of tracing material must be selected.
3. The tracer on or in the solid particle should be durable, at least for the time over which the experiment or study extends.
4. The tracer should not be hazardous to the biological environment.
5. The cost of producing traced particles should be reasonable.

At present there are three different types of tracers in use: radioactive tracers, paint and fluorescent tracers, and density tracers.

13.4.3.1.1 RADIOACTIVE TRACERS. Labeling of solid particles with a radioactive tracer may be obtained in different ways. One method is the *irradiation* method. Suppose an inactive isotope exists in a natural or artificial solid particle. By neutron irradiation in a nuclear reactor this isotope can be activated and then emits detectable radiation. Artificial sediment has been made out of glass, having the same density as quartz. Incorporated in the glass is an inactive isotope. KRONE (1960) recommended the use of natural gold or scandium for a sand transport study. The glass is ground and sorted, and a size distribution is selected which matches the sediment to be investigated. Immediately prior to the test the ground glass is irradiated. On the other hand, there are some natural sediments which contain already inactive isotopes. CRICKMORE (1961) discussed the use of such a sand and some of the difficulties in using it as a quantitative tracer in a laboratory flume.

Another method is by *sorption or coating* of radioactive elements on the natural sand. Various labels have been used, and the technique of sorption varies from one problem to the other. KRONE (1960) felt that gold-198 met the requirements when studying clayey sediment transport in an estuary. The tracer is prepared by adding a small amount of radioactive gold chloride to the wet sediment. HUBBELL et al. (1963) used sand particles labeled with iridium-192. The sand was mixed with an iridium-192 solution, and then was baked at 700°F for several hours. Another sorption technique was employed by MCDOWELL (1963). The sand was to sorb scandium-46 chloride, and was heated to 840°C to convert the chloride to the more stable oxide. Further sorption techniques are found in some of the references in COURTOIS (1966).

Investigators concerned with the movement of pebbles have yet explored another method. Tracers were inserted into holes which were drilled into the natural or artificial pebbles and then sealed in resin. Various techniques employed in the process were reviewed by KIDSON et al. (1962).

At the beginning of this section some requirements were cited; now some further observations are offered. It is recommended that isotopes be used which emit hard gamma radiation rather than soft beta radiation. Gamma radiation is only partly absorbed by sand and water, and thus it facilitates an in situ radioactivity measurement; beta particles are easily absorbed. The half-life of the tracer should be long enough to conduct the experiment, but short enough to minimize interference between successive experiments. Elements known to be required in the metabolic processes of organisms should be used with utmost care, if at all. There exists some indication [see KRONE (1960)] that sorbed labels can be prepared—at reasonable cost—in large amounts.

13.4.3.1.2 PAINT AND FLUORESCENT TRACERS. Paint and synthetic resin finishes have been used as tracers. However, as pointed out by KIDSON et al. (1962), there are certain disadvantages involved. Although these finishes are rather cheap, they are limited in their use to the larger fractions of the sediment. Silt and finer matter will begin to stick together. Also, since these finishes are subject to abrasion, heat processing may become essential.

A *fluorescent tracer* is material attached to the solid grain made up of organic dyes which fluoresce, catalysts and resins. Descriptions of fluorescent tracers are given by KIDSON et al. (1962) and TELEKI (1963); the latter reference describes also the actual production procedure. A disadvantage of this tracing method lies in the difficulty of obtaining the samples, which often have to be collected in the dark with fluorescent lamps.

13.4.3.1.3 DENSITY TRACERS. The study of source and long-term movement of sediments has been frequently approached by considering some natural characteristics of the sediment. Texture, density, angularity, radioactivity, or other properties of a sediment have invariably been used as tracers.

Radioactivity, wherever sufficiently strong to be detected, can be used and measured with instruments discussed previously. Among geologists, it is popular to trace heavy and light minerals. According to KRUMBEIN et al. (1963, p. 139) there are about 30 heavy minerals which are reasonably diagnostic of geologic source material. Some of the heavy minerals used for this purpose are hornblende, biotite, muscovite, augite, zircon, etc. The study of the distribution of heavy minerals is helpful for the determination of sand movement along coastlines, as was done by CHERRY (1965), and in estuaries, as done by BYRNE et al. (1967).

13.4.3.2 Measurement of tracers. After proper selection of a tracer, it has to be applied to the sediment material whose motion is being investigated. If the bedload motion is of major importance, the tracer should be released as close to the bed as practicable. Investigating the dispersion of bed material, HUBBELL et al. (1963) decided to release radioactive tracers as a line source and to record the longitudinal distribution as time passed. Frequently the tracer material is shipped or flown in special containers to the test site where it is dumped. When tracing suspended-load material, special care should be taken to add the tracer at a concentration similar to the one of its environment. A direct (in situ) and an indirect method to measure sediment tracer distribution have been used.

13.4.3.2.1 DIRECT METHOD. The direct or in situ method can, at the present, only be used if a radioactive tracer emits gamma rays which are not entirely absorbed by sand or water. Measuring devices have been designed such that they can be towed over the measuring section and provide a continuous record of the tracer abundance. KRONE (1960) has undertaken the task to design and build an underwater detector. A detector in a watertight housing is mounted on a sled which can be dragged by a boat along the bottom of the stream. The detector is connected by a cable to the electronic equipment on the boat. This general design was successfully used in field investigations by KRONE (1960), HUBBELL et al. (1963), CUMMINGS et al. (1963), and KLINGEMAN et al. (1965). The detector may be either a Geiger counter, as used by CRICKMORE (1961), or a scintillation counter, as used by KRONE (1960) and many others. A brief discussion on these devices was given by KRONE (1960) and MCDOWELL (1963). If a quantitative evaluation of the data is desired, certain precautions are necessary. The detector is supposed to record only the radioactivity at a given point. Such a reading may be masked due to the fact that radioactive material emits radiation in all directions and that a background radiation may exist.

13.4.3.2.2 INDIRECT METHOD. Another method of measuring the tracer distribution is to take samples and, subsequently, analyze them in the laboratory. Where quantitative data are desired, it is essential that samples be uniform in thickness. Radioactive material with beta emission is analyzed by autoradiographic means with x-ray films. Whenever the material can be distinguished by size and/or color, a scanning procedure is used. This may be done by simple visual counting of the colored fluorescent particle with the aid of a luminoscope. An electronic photometric counter for fluorescent particles was suggested by DeVries (1967). If heavy minerals are used as tracers, they have to be separated from the rest of the mixture. A common method is to use bromoform as a separating liquid, the density of which should be such that heavy minerals settle at the bottom while the light one remains suspended. The heavy minerals are subsequently identified with the aid of a petrographic microscope.

After adding up all the advantages of the direct measuring method, it may be suggested that radioactive tracers are most suitable for short-term (days or months) sedimentation studies. For long-term (year or centuries) investigations, natural indicators are recommended.

13.4.3.3 EVALUATION OF TRACER STUDIES. Tracer studies have led to interesting and important conclusions on qualitative and quantitative sediment motion. Release of a tracer at locations to be studied, and subsequent areal and temporal measurements of tracer abundance, determine source, deposition, and erosion of sediment material. Direction of sediment motion as well as its distributions, durations, and rates of movement may also be investigated. Most studies have as their scope the providing of qualitative answers; this is especially so if investigations are done under complicated hydraulic conditions, such as in estuaries and along coastlines. There exist a few studies which deal with the quantitative question, namely, to determine the bed material discharge.

The transport rate of bed material per unit width is given by the product of the average transport velocity and the average thickness of the movement. Experiments were carried out to evaluate these quantities. Flume studies were done by CRICKMORE (1961), McDOWELL (1963), and HUBBELL et al. (1964). A field investigation on a 1,800-ft-long and 50-ft-wide stretch of the North Loup River in Nebraska was done by HUBBELL et al. (1964). In either case, radioactive tracers were used and were placed fairly uniformly across the entire width at a given section. After certain time intervals, the distribution of the tracer particles along the channel was measured. From these data the average transport velocity was determined, which is given by the rate of movement of the centers of gravity of the distributions. To obtain the average thickness, the vertical distribution of labeled particles was measured. The results were encouraging.

14
Model Laws

Many problems in hydraulic engineering are sufficiently complicated so that the equations of fluid mechanics cannot be applied or do not supply answers. Then the use of a model, which is a miniature prototype, might lead to a much desired answer.

The essentials of the model laws, as applicable to channel and/or river models, will be discussed in this chapter. Further useful information on the theory, construction, and operation of models will be found in FREEMAN (1929), VOGEL (1935), STEVENS et al. (1942), ALLEN (1947), and RZHANITSYN (1960). These references also include many illustrative examples. A special laboratory study on the meandering of alluvial rivers is reported by FRIEDKIN (1945).

14.1 SIMILITUDE AND DIMENSIONLESS NUMBERS

A model is a simulated prototype. The similitude is the indication of the given relationship between the two phenomena. Complete dynamic similarity between model and prototype exists if identical types of forces are parallel and have the same model prototype ratio at all points of the corresponding flow fields. In short, dynamic similarity is the similarity of forces. The types of forces which may affect a flow field in a channel investigation are inertia F_I, pressure F_P, gravity F_G, and viscosity F_V. Thus the following equations may be formed, or

$$\frac{F_I{}^m}{F_I{}^p} = \frac{F_P{}^m}{F_P{}^p} = \frac{F_G{}^m}{F_G{}^p} = \frac{F_V{}^m}{F_V{}^p} = \text{const} \tag{14.1}$$

where the superscripts m and p denote model and prototype, respectively. From Eq. (14.1), the following relations may be derived:

$$\frac{F_I{}^m}{F_P{}^m} = \frac{F_I{}^p}{F_P{}^p} \tag{14.2}$$

$$\frac{F_I{}^m}{F_G{}^m} = \frac{F_I{}^p}{F_G{}^p} \tag{14.3}$$

$$\frac{F_I{}^m}{F_V{}^m} = \frac{F_I{}^p}{F_V{}^p} \tag{14.4}$$

It will be desirable to express the types of forces by following fundamental relationships:

$$\begin{aligned} F_I &= \rho L^3\left(\frac{V^2}{L}\right) = \rho V^2 L^2 \\ F_P &= (\Delta p)L^2 \\ F_G &= \rho L^3 g \\ F_V &= \mu\left(\frac{dV}{dL}\right)L^2 = \mu VL \end{aligned} \tag{14.5}$$

where L = length dimension

Δp = pressure dimension

V = velocity dimension

g = gravitational constant

μ = coefficient of viscosity

ρ = density

Introducing the knowledge of Eqs. (14.5) into Eqs. (14.2) through (14.4) the following is obtained:

$$\left(\frac{\rho V^2}{\Delta p}\right)^p = \left(\frac{\rho V^2}{\Delta p}\right)^m \tag{14.6}$$

$$\left(\frac{V^2}{gL}\right)^p = \left(\frac{V^2}{gL}\right)^m \tag{14.7}$$

$$\left(\frac{VL\rho}{\mu}\right)^p = \left(\frac{VL\rho}{\mu}\right)^m \tag{14.8}$$

Equations (14.6) through (14.8) are dimensionless numbers of dynamic similarity and can be derived by dimensional analysis. Equation (14.6) is known as the *Euler number*, expressing the ratio of inertia force to pressure force, or

$$\mathrm{Eu} = V\sqrt{\frac{\rho}{\Delta p}} \tag{14.9}$$

Equation (14.7) is known as the *Froude number*, expressing the ratio of inertia force to gravity force, or

$$\mathrm{Fr} = \frac{V}{\sqrt{gL}} \tag{14.10}$$

and Eq. (14.8) is known as the *Reynolds number*, expressing the ratio of inertia force to friction force, or

$$\mathrm{Re} = \frac{VL\rho}{\mu} \tag{14.11}$$

Furthermore, it must be remarked that only two equations of Eqs. (14.6) through (14.8) are independent, while the third one is a dependent one. (It is usually the inertia-pressure relation which is automatically equal for prototype and model if the other relations are equal.) Thus dynamic similarity is ensured only if two equations are simultaneously satisfied. However, owing to certain limitations, such as available space, fluids, and others, it becomes at times really difficult to satisfy these equations. For some problems it is not necessary to fulfill the two simultaneous equations, because one or the other force may be negligible or altogether absent. Take, for example, the problem of a submerged airfoil. Gravity forces do not affect the flow field and, thus, a Reynolds analogy ensures complete dynamic similitude. Also take, for example, the motion of an unstreamlined ship. Viscosity forces can be ignored, and a Froude analogy establishes complete dynamic similitude and, thus, all the modeling parameters.

A rather promising approach was recently suggested by BARR (1969) who provides scaling procedures in terms of similitude theory. HERBERTSON (1969) has applied it for mobile bed models. This method will not be discussed, and the reader is referred to the original contributions.

14.2 RIVER AND CHANNEL MODELS

It was shown previously that in problems where either gravity or viscosity forces predominate, no trouble is encountered in establishing the model parameters. Important modeling ratios for gravity and viscosity models are summarized in Table 14.1. In situations in which both forces play an important role, certain difficulties may occur in fulfilling dynamic similitude.

Under these circumstances, both a Froude and a Reynolds analogy would have to be satisfied simultaneously. The Froude law was given with Eq. (14.7), and is written now as

$$\left(\frac{V^2}{L}\right)^p = \left(\frac{V^2}{L}\right)^m \tag{14.7a}$$

with the assumption that the gravitational constant cannot be varied. The Reynolds law was given with Eq. (14.8), or

$$\left(\frac{VL\rho}{\mu}\right)^p = \left(\frac{VL\rho}{\mu}\right)^m \tag{14.8}$$

Eliminating the velocity from the foregoing two equations, we obtain

$$\frac{\nu^p}{\nu^m} = \left(\frac{L^p}{L^m}\right)^{3/2} \tag{14.12}$$

When choosing a reasonable scale ratio of L^p/L^m, it will be almost impossible to select a fluid which can satisfy the kinematic viscosity ratio, since the range of viscosity among common fluids (especially water) is limited. Unfortunately, in many problems of river and open-channel flow, both the gravity and the frictional forces are active in a larger or smaller degree.

14.2.1 FIXED-BED MODELS

By way of introduction to the movable-bed models, the fixed-bed models will be discussed first, and it is hoped that much can be learned from them.

Table 14.1 Modeling ratios

	Model parameters		
	Reynolds number	*Froude number*	*Froude number (distorted model)*
Velocity, $\frac{V^m}{V^p}$	$\frac{L^p}{L^m}\frac{\rho^p}{\rho^m}\frac{\mu^m}{\mu^p}$	$\left(\frac{L^m}{L^p}\right)^{1/2}$	$\left(\frac{L_V^m}{L_V^p}\right)^{1/2}$
Flow rate, $\frac{Q^m}{Q^p}$	$\frac{L^m}{L^p}\frac{\rho^p}{\rho^m}\frac{\mu^m}{\mu^p}$	$\left(\frac{L^m}{L^p}\right)^{5/2}$	$\left(\frac{L_V^m}{L_V^p}\right)^{3/2}\frac{L_H^m}{L_H^p}$
Force, $\frac{F^m}{F^p}$	$\left(\frac{\mu^m}{\mu^p}\right)^2\frac{\rho^p}{\rho^m}$	$\left(\frac{L^m}{L^p}\right)^3\frac{\rho^m}{\rho^p}$	$\frac{\rho^m}{\rho^p}\frac{L_H^m}{L_H^p}\left(\frac{L_V^m}{L_V^p}\right)^2$
Time, $\frac{t^m}{t^p}$	$\left(\frac{L^m}{L^p}\right)^2\frac{\rho^m}{\rho^p}\frac{\mu^p}{\mu^m}$	$\left(\frac{L^m}{L^p}\right)^{1/2}$	$\frac{L_H^m}{L_H^p}\left(\frac{L_V^m}{L_V^p}\right)^{-1/2}$

14.2.1.1 Undistorted models. In fixed-bed models it has become common practice to use as the frictional force criterion not the Reynolds number but an empirical frictional relationship, for example, the Manning formula. The application of the Manning equation as a friction criterion requires not only turbulent flow in the prototype but also in the model. To ensure turbulent flow in the model, however, is not an easy task.[1] The Manning equation for prototype and model may be given as

$$\left(\frac{R_h^{2/3}S^{1/2}}{\bar{u}\cdot n}\right)^p = \left(\frac{R_h^{2/3}S^{1/2}}{\bar{u}\cdot n}\right)^m \tag{14.13}$$

With the ratio of vertical distances equaling the ratio of horizontal distances, and with the hydraulic radius replaced by the length dimension, we obtain

$$\left(\frac{L^{2/3}}{V\cdot n}\right)^p = \left(\frac{L^{2/3}}{V\cdot n}\right)^m \tag{14.14}$$

The existing gravity forces are expressed with Froude's law, given by Eq. (14.7*a*), or

$$\left(\frac{V^2}{L}\right)^p = \left(\frac{V^2}{L}\right)^m \tag{14.7a}$$

Putting Eq. (14.7*a*) into Eq. (14.14), we get

$$\frac{n^p}{n^m} = \left(\frac{L^p}{L^m}\right)^{1/6} \tag{14.15}$$

In order to obtain realistic n values in Eq. (14.15), it is necessary to keep the length ratio relatively small, thus making the value of the model, at times, questionable. To illustrate the point, take the n value for a natural stream with stony sections as $n = 0.05$, and the model with gunite linings with an n value of $n = 0.02$. According to Eq. (14.15), the scale ratio is

$$\frac{L^p}{L^m} = \left(\frac{n^p}{n^m}\right)^6 = \left(\frac{0.05}{0.02}\right)^6 \approx 240 \tag{14.16}$$

In river hydraulics, however, we may encounter problems where such a scale ratio is too small for a desired convenience or for the available physical facility.

On the other hand, it is still necessary to check whether the flow in the model is turbulent, so that Manning's friction formula truly applies. It is customary to conduct river models with water, and thus with the same (or almost the same) kinematic viscosity of the prototype. The Reynolds number

[1] Einstein (1969) pointed out to the author that the Manning equation gives very bad results when the relative roughness becomes small.

ratio then reads

$$\frac{\text{Re}^p}{\text{Re}^m} = \frac{(VL)^p}{(VL)^m} \tag{14.8a}$$

Expressing the velocity with the Froude law, given with Eq. (14.7*a*), we obtain

$$\frac{\text{Re}^p}{\text{Re}^m} = \left(\frac{L^p}{L^m}\right)^{3/2} \tag{14.17}$$

or, after introducing the scale ratio of this example into Eq. (14.16),

$$\left(\frac{L^p}{L^m}\right)^{3/2} = (240)^{3/2} = 3{,}700 \tag{14.18}$$

The scale ratio as chosen might very well give a model Reynolds number[1] in the laminar flow range. This is not permissible because the Manning formula applies to turbulent conditions only.

Thus it must be pointed out that not every scale ratio may be a desirable one for at least two reasons. First, the scale ratio—given with Eq. (14.16)—may have a magnitude which for the sake of convenience, accuracy, or economy becomes unacceptable. Secondly, not every scale ratio, whether or not satisfying these conditions, produces turbulent flow in the model, which, however, is a necessary requirement to be fulfilled. To overcome these difficulties, if and when they arise, it is necessary to distort the model.

14.2.1.2 Distorted models. If the model has a larger vertical scale than a horizontal one, then we speak of a distorted model; such models are geometrically distorted.

Since gravity forces in the vertical scale govern the flow velocity, the Froude law may be written as

$$\left(\frac{V^2}{L_V}\right)^p = \left(\frac{V^2}{L_V}\right)^m \tag{14.19}$$

where the subscript V stands for the vertical scale and subscript H will stand for the horizontal scale. Additional model parameters for a distorted model are given in Table 14.1. The frictional relationship will again be expressed with the Manning formula, given by Eq. (14.13). Introducing Eq. (14.19) and expressing the slope S as

$$S = \frac{L_V}{L_H} \tag{14.20}$$

[1] According to ALLEN (1947, p. 24), the Reynolds number is $\text{Re}^m = 1{,}400$ (where L^m is the hydraulic mean depth), below which flow is laminar.

we obtain the following:

$$\left(\frac{R_h^{2/3}L_V{}^{1/2}}{L_V^{1/2}nL_H^{1/2}}\right)^p = \left(\frac{R_h^{2/3}L_V^{1/2}}{L_V^{1/2}nL_H^{1/2}}\right)^m \tag{14.21}$$

Solving Eq. (14.21) for the n value gives

$$\frac{n^p}{n^m} = \left(\frac{R_h{}^p}{R_h{}^m}\right)^{2/3}\left(\frac{L_H{}^m}{L_H{}^p}\right)^{1/2} \tag{14.22}$$

Because the hydraulic radius R_h is more a function of the vertical scale L_V than of the horizontal scale L_H, and because the horizontal scale ratio is greater than the vertical scale ratio, the ratio of the n values could be less than unity. Thus the roughness of the model may exceed the one of the prototype which, in turn, ensures turbulent flow in the model. In fact, the large n values may no longer be achieved by making the surface very rough, but rather by putting small obstacles such as nails, metal strips, or pebbles on the model river bed. Quite frequently the model roughness must be obtained by a trial-and-error procedure; a known prototype flow rate is scaled according to the model law, and then the model roughness is adjusted until the calculated model flow rate is obtained.

It appears that distortion within limits is very helpful, and will be shown to be even more helpful when movable-bed models are discussed. STEVENS et al. (1942) advise that the horizontal scale ratio for models of rivers and harbors should be

$$100 < \frac{L_H{}^p}{L_H{}^m} < 2{,}000 \tag{14.23}$$

while the vertical scale ratio was given as

$$50 < \frac{L_V{}^p}{L_V{}^m} < 150 \tag{14.24}$$

However there are disadvantages due to the departure from complete dynamic similitude that should be kept in mind. The U.S. BUREAU OF RECLAMATION (1953) states them as follows:

(*a*) velocities are not necessarily correctly reproduced in magnitude and direction;
(*b*) some of the flow details are not correctly reproduced;
(*c*) slopes of cuts and fills are often too steep (larger than the angle of repose) to be molded in sand . . . ;
(*d*) there is an unfavorable psychological effect on the observer who views distorted models.

To sum it up, in a distorted fixed-bed model, Eq. (14.22) may be used for design purposes. Two of the three ratios, n^p/n^m, $R_h{}^p/R_h{}^m$, and $L_H{}^p/L_H{}^m$, may be chosen arbitrarily, while the remaining one will be found from Eq. (14.22). If both requirements, the one of the Froude and of the Manning equations, cannot be satisfied simultaneously to reproduce prototype events, then a reasonable model can be designed according to the Manning equation only. STEVENS et al. (1942, p. 40) remarked that this is possible without obtaining results with appreciable errors.

14.2.2 MOVABLE-BED MODELS

In studying movable-bed models we face the hydraulically complex problem of sediment transport, and thus the design and proper understanding of these models remain extremely intricate. In addition to a gravity criterion expressed with the Froude law and a friction criterion expressed presently with the Manning formula, other criteria introducing the mechanics of sediment transport have to be used. In the following we shall distinguish between an implicit or more empirical method and an explicit or more rational one.

14.2.2.1 Empirical approach. In fixed-bed models scale distortion is frequently of great help; the distorted scale ratio controls the bed roughness, which is expressed by Eq. (14.22). With movable-bed models it is the type of bed roughness, the bed configuration, and the bed-material motion which determine the roughness. When a model is distorted, the longitudinal slope is increased. This has a direct influence on the velocity profile which, in turn, has a direct bearing on the sediment movement. Since it is difficult to control the roughness, it is equally difficult to control the velocity profile; thus dynamic similarity may be destroyed. At the same time, a distortion will allow for an easier bed material movement, since the shear stress is proportional to the slope. It seems that relaxing the requirements on the fluid-flow similarity gives a better bed-material-movement similarity. However, STEVENS et al. (1942) recommended that the distortion ratio of movable-bed models be never greater than about 6, or

$$\left(\frac{L^p}{L^m}\right)_V : \left(\frac{L^p}{L^m}\right)_H \approx 1:6$$

as a compromise between the sediment-motion and water-motion similarity. This value is in general agreement with Soviet model investigations, as shown by RZHANITSYN (1960).

The essentials of the empirical approach are summarized by STEVENS et al. (1942): "If a model can be adjusted to reproduce events that have occurred in the prototype, it should indicate events that will occur in the prototype."

In such a model with its low velocities and shallow depths, a very light sediment material must be used. Light particles may be either small particles or particles with low specific gravity. Generally, it is unreasonable to apply the scale ratio to the prototype sand grains, since the resulting model grains would get much too small; it is, however, customary to alter the specific gravity. If coal dust ($\rho_s/\rho \approx 1.3$) or plastics ($\rho_s/\rho \approx 1.2$), both commonly used materials, are chosen, preliminary studies will determine a critical scour velocity at a certain depth. Also known is the Manning n-value for either material. Using the Manning formula, given with Eq. (14.13), the slope ratio that gives the critical velocity at the depth required can be determined. Therefore, with the slope ratio and vertical scale ratio being known quantities, the horizontal scale ratio can be determined, or

$$\left(\frac{L^p}{L^m}\right)_H = \left(\frac{L^p}{L^m}\right)_V \left(\frac{S^p}{S^m}\right)^{-1} \tag{14.25}$$

The basic design parameters have now been deduced, the model may be built accordingly, and then is ready for verification. The model is supplied with past known flow events, and it is then noted whether the model has experienced similar bed configurations as existed in the prototype. If not, an adjustment is required, which is often obtained by tilting the model and, as such, represents a further distortion. Much of the art of model testing is, indeed, involved in the verification process, and the interested reader is advised to consult FREEMAN (1926), STEVENS et al. (1942), ALLEN (1947), and numerous available reports concerned with model testing. For example, STEVENS et al. (1942) have suggested that lengthening the time of low flow periods and shortening the time of high flows give the model a bed configuration similar to the one of the prototype. The time it takes for a bed of a model to reproduce the changes in the prototype is called the *sedimentation time*, and is different from the time scale governing the fluid flow.

Such models as described in this section give qualitative and not quantitative results. The trial-and-error operation is tedious, costly, and time consuming, but has frequently been used with considerable success. Finally it should be mentioned that BLENCH (1955) suggested to apply the regime concept to model studies. BLENCH (1964) contests:

> A large portion of the time consuming trial-and-error inevitable in major quantitative models may be spent while the model adjusts to the scales required by nature. Most of this portion can be saved by using regime formulas to forecast the natural scales consequent on imposing the permissible three basic arbitrary ones. The regime formulas are usable as a first approximation provided the model is devised so that the coefficients required for river work are the same as in the prototype.

The underlying principle is that one river is considered as the model of another one. The distinct advantage lies with the fact that various model parameters can be varied and the resulting effect can be studied. Any consistent set of regime equations may be used; for example, Eqs. (10.13) through (10.15) become

$$\frac{L_H{}^p}{L_H{}^m} = \left(\frac{Q^p}{Q^m}\right)^{1/2} \tag{10.13a}$$

$$\frac{L_V{}^p}{L_V{}^m} = \left(\frac{Q^p}{Q^m}\right)^{1/3} \tag{10.14a}$$

$$\frac{S^p}{S^m} = \frac{(L_V/L_H)^p}{(L_V/L_H)^m} = \left(\frac{Q^p}{Q^m}\right)^{-1/6} \tag{10.15a}$$

Additionally, it might become necessary to apply also the relation for meandering streams. By simultaneous solution of Eqs. (10.13*a*) and (10.14*a*) we get

$$\frac{L_V{}^p}{L_V{}^m} = \left(\frac{L_H{}^p}{L_H{}^m}\right)^{2/3}$$

which implies that, if one of the scale ratios is fixed, the other one may be calculated accordingly. Furthermore, it is suggested that prototype bed sand can be used in models, a method which is quite popular among designers of models. For a detailed discussion of this method and its application to practical problems, the reader is referred to BLENCH (1957).

14.2.2.2 Rational approach. With the previously discussed empirical approach for designing movable-bed models, the qualitative similarity may be achieved; however, it is very difficult to produce quantitative similarity. By way of a rational approach the latter one may be obtained, at least in an approximate fashion. Similarity relations for movable-bed models have been advanced by EINSTEIN (1944*a*) and EINSTEIN et al. (1954*a*). In the following a simplified version of these relations will be presented.

Scour criteria. Applying dimensional analysis as, for example, RAUDKIVI (1967) has done, we may show that, for the initial-motion problem, a grouping results which is commonly known as the *Shields parameters*. SHIELDS (1936) expressed such a relationship as

$$\frac{(\tau_0)_{cr}}{(\gamma_s - \gamma)d} = \text{fct}\left(\frac{du_*}{\nu}\right) \tag{6.24}$$

This relation is determined experimentally and is given in Fig. 6.7; it has been discussed in Sec. 6.3.2.2. The left-hand side of Eq. (6.24) is the dimensionless critical shear stress, and the right-hand side is the particle Reynolds number or relative sublayer thickness.

When studying problems of initial scour, it becomes necessary to add the critical shear stress and sublayer criteria to the gravity and frictional criteria. The first two criteria, when applied to a distorted model, were given by Eq. (14.22). Introducing into this equation the empirical relation by Strickler between the bed particle diameter and the n value, the following is obtained:

$$\left(\frac{d^p}{d^m}\right)^{1/6} = \frac{n^p}{n^m} = \left(\frac{R_h{}^p}{R_h{}^m}\right)^{2/3}\left(\frac{L_H{}^m}{L_H{}^p}\right)^{1/2} \tag{14.26}$$

Since $\tau_0 = \gamma R_h S$,[1] the critical-shear-stress criterion becomes

$$\frac{\gamma^p}{\gamma^m}\frac{R_h{}^p}{R_h{}^m}\frac{L_V{}^p}{L_V{}^m}\left[\frac{L_H{}^p}{L_H{}^m}\frac{d^p}{d^m}\frac{(\gamma_s-\gamma)^p}{(\gamma_s-\gamma)^m}\right]^{-1} = 1 \tag{14.27}$$

and since $u_*^2 = \tau_0/\rho$, the sublayer criterion becomes

$$\left(\frac{d^p}{d^m}\right)^2 \frac{\gamma^p}{\gamma^m}\frac{R_h{}^p}{R_h{}^m}\frac{L_V{}^p}{L_V{}^m}\frac{L_H{}^m}{L_H{}^p}\left(\frac{\nu^m}{\nu^p}\right)^2\frac{\rho^m}{\rho^p} = 1 \tag{14.28}$$

For all practical purposes, model and prototype liquids are identical, i.e., $\gamma^p/\gamma^m = 1$ and $\nu^p/\nu^m = 1$. There remain four independent variables,[2] L_V, L_H, d, and $\gamma_s - \gamma$, and there exist three equations, Eqs. (14.26) through (14.28). The problem is completely determined, if one of the four parameters may be chosen, while the remaining three variables are found as solutions of the equations. At the outset of this example, a scale-distorted model was assumed. This is a convenient and practical assumption, because an undistorted model can be used only in a limited number of river models. Furthermore, from Eqs. (14.27) and (14.28) it can be seen that reducing the specific gravity differences $(\gamma_s - \gamma)^m$ in the model requires the model grain size d^m to exceed the prototype grain size.

Various researchers have argued that some of the model laws can be relaxed without much harm to the overall investigation. EINSTEIN et al. (1954*a*) suggested that the frictional criterion, the Froude criterion, or the sublayer criterion may absorb further distortions. Thus it is obvious that as long as both model and prototype produce turbulent flow around the particle, the sublayer criterion is unnecessary. When the Froude number is very small, which is the case in deep and/or slowly moving rivers, the gravity criterion loses its significance. The frictional criterion may have to deviate from the exact similarity if such is necessary for practical reasons. A further

[1] More correctly, the hydraulic radius R_h should be replaced by R_h', the hydraulic radius associated with the grain roughness. Thus it is assumed that bedforms do not exist and $R_h \approx R_h'$.

[2] The hydraulic radius R_h is a function of L_V and L_H, and is found by a trial procedure for any given river cross section.

distortion has been suggested, which exists when the model is not only vertically distorted but also tilted.

In summary, in order to establish scour criteria for a model-prototype investigation, four criteria—the frictional, the gravitational, the critical shear stress, and the sublayer criterion—have to be fulfilled to achieve exact similarity. The designer of the model has one degree of freedom; one variable is chosen, the remaining three variables are calculated. However, under certain circumstances, small deviations from the exact similarity may be allowed—and may even become necessary—and then it is possible to select more than a single variable arbitrarily.

Sediment transport. In problems where it is necessary to obtain quantitative information on the motion of sediments, it becomes essential to introduce at least one further criterion. Any sediment-transport equation, as, for example, discussed in Chaps. 7 through 9, which is dimensionally homogeneous should provide such a modeling criterion. By inspection it will become obvious that, for example, the bedload relations given by SHIELDS (1936) with Eq. (7.14), by KALINSKE (1947) with Eq. (7.18), by MEYER-PETER et al. (1948) with Eq. (7.42), or by EINSTEIN (1950) with Eq. (7.71) are dimensionally homogeneous. There are, however, bedload relations which are dimensionally not homogeneous, such as the one given by DUBOYS (1879) with Eq. (7.6) or by SCHOKLITSCH (1934) with Eq. (7.34). The choice of either dimensional homogeneous equation is not a strong restriction to the generality of the method, as was pointed out by EINSTEIN et al. (1954*a*).

In any case, a further unknown value, the bedload rate, and a further relationship, the bedload criterion, are added to the already existing relations, given by Eqs. (14.26) through (14.28). Using, for example, the bedload relation as suggested by EINSTEIN (1950), or

$$\Phi = \frac{g_s}{\gamma_s}\sqrt{\frac{\rho}{\rho_s - \rho}\frac{1}{gd^3}} \tag{7.51}$$

where Φ is the intensity of bedload transport and g_s is the bedload rate in weight per unit time and unit width, the values of Φ must be the same for corresponding sections in prototype and model. If model and prototype liquids are identical, the bedload criterion reads

$$\frac{\Phi^p}{\Phi^m} = \frac{g_s{}^p}{g_s{}^m}\frac{\gamma_s{}^m}{\gamma_s{}^p}\left[\frac{(\gamma_s - \gamma)^m}{(\gamma_s - \gamma)^p}\right]^{1/2}\left(\frac{d^m}{d^p}\right)^{3/2} = 1 \tag{14.29}$$

There are five independent variables, L_V, L_H, d, $\gamma_s - \gamma$, and g_s, and there exist four equations, Eqs. (14.26) through (14.29). Again, the problem is completely determined; one of the five parameters may be chosen, the remaining ones are found as solutions to the previous equations. Once again, just as was outlined previously, some of the model laws may be relaxed.

In addition to the bedload, a river or canal may also carry suspended load. The impossibility of obtaining suspended-load rates in models which reproduce bedload rates was bridged by EINSTEIN et al. (1954*a*) in the following way. A relationship betweeen bedload rate and total-load rate was introduced as

$$\frac{g_s^{\,p}}{g_s^{\,m}} \frac{g_{st}^{\,m}}{g_{st}^{\,p}} B_E = 1 \qquad (14.30)$$

where B_E represents some kind of distortion factor. Since $g_s^{\,p}$ and $g_s^{\,m}$ may be determined as previously outlined, and since $g_{st}^{\,p}$ and $g_{st}^{\,m}$ may be computed according to Eq. (9.4), the distortion factor B_E is known.

Another matter of concern is the knowledge of two time scales. There is the hydraulic time, i.e., the time it takes for a fluid particle to travel with a certain velocity through a certain distance, or

$$\frac{V^p}{V^m} \frac{t^p}{t^m} \frac{L_H^{\,m}}{L_H^{\,p}} = 1 \qquad (14.31)$$

There is the sedimentation time t_s indicating the duration of individual flows. This time scale indicates the time intervals required by a certain total-load rate to fill a certain volume, or

$$\frac{g_{st}^{\,p}}{g_{st}^{\,m}} \frac{\gamma_s^{\,m}}{\gamma_s^{\,p}} \frac{L_V^{\,m}}{L_V^{\,p}} \frac{t_s^{\,p}}{t_s^{\,m}} \frac{L_H^{\,m}}{L_H^{\,p}} = 1 \qquad (14.32)$$

The total-load rate may be replaced by the relation given by Eq. (14.30). The sedimentation time scale is directly dependent on the stages; as a result a sliding time scale appears to become necessary in many cases.

In summary, the rational approach to movable-bed models presents the variables involved and a set of similarity criteria. The designer has one degree of freedom; this means that one variable may be selected arbitrarily, the other variables must be computed from the equations. Under certain circumstances, one or two of the similarities may be relaxed; thus the designer has more than one degree of freedom. The major credit for putting forward the so-called *rational approach* must be given to EINSTEIN (1944*a*) and EINSTEIN et al. (1954*a*). Applying this method to a Rhine River model, EINSTEIN (1944*a*) reported very close agreement between calculated and measured ratios. A complete discussion of the similarity of distorted movable-bed models is given in EINSTEIN et al. (1954*a*), which also includes a sample calculation on a fictitious model. Because the presently discussed approach has still many complications, EINSTEIN et al. (1954*a*) suggest that:

> . . . it is absolutely necessary to verify any such model and its scales. Such a verification consists in the reproduction of a known prototype development in the model by similar flows. Only if such a verification

is possible and successful can the model be depended upon for the prediction of future developments . . . .

Finally, it is worthwhile pointing to an interesting investigation by RZHANITSYN (1960), where it is suggested to determine the model scales from characteristics of the streams of the river system, which, in turn, involves a hydrological and morphological knowledge of the structure of the river net. Model scales computed in this manner are in good agreement when compared with model scales as derived from a set of equations, as given previously.

REFERENCES FOR PART THREE

ABDEL-RAHMAN, N. M. (1963): The Effects of Flowing Water on Cohesive Beds, *Mitt. VAWE, Eidg. Techn. Hochschule, Zürich*, no. 56.

ACAROGLU, E. R., and W. H. GRAF (1968): The Modes of Sediment Transport and Their Related Bedforms in Conveyance Systems, *Bull. Intern. Assoc. Sci. Hydr.*, XIIIe année, no. 3.

ACKERS, P. (1964): Experiments on Small Streams in Alluvium, *Proc. Am. Soc. Civil Engrs.*, vol. 90, no. HY4.

ADACHI, S. (1967): A Theory of Stability of Streams, *Intern. Assoc. Hydr. Res., 12th Congr., Fort Collins.*

ALLEN, J. (1947): "Scale Models in Hydraulic Engineering," Longmans and Green, London.

ALLEN, J. R. L. (1968): "Current Ripples, Their Relation to Pattern of Water and Sediment Motion," North-Holland, Amsterdam.

——— (1969*a*): Erosional Current Marks of Weakly Cohesive Mud Bed, *J. Sediment. Petrol.*, vol. 39, no. 2.

——— (1969): On the Geometry of Current Ripples in Relation to Stability of Fluid Flow, *Geografiska Annaler*, vol. 51A.

ANANIAN, A. K. (1967): An Approximate Theory of Secondary Flows at the Bend of the River Course, *Intern. Assoc. Hydr. Res., 12th Congr., Fort Collins.*

———, and E. T. GERBASHIAN (1965): About the System of Equations of the Movement of Flow Carrying Suspended Matter, *J. Hydr. Res.*, vol. 3, no. 1.

ANDERSON, A. G. (1942): Distribution of Suspended Sediment in a Natural Stream, *Trans. Am. Geophys. Union*, vol. 23.

——— (1953): The Characteristics of Sediment Waves Formed by Flow in Open Channels, *3d Midwest. Conf. Fluid Mech., Univ. of Minnesota, Minneapolis.*

——— (1967): On the Development of Stream Meanders, *Intern. Assoc. Hydr. Res., 12th Congr., Fort Collins.*

ANSLEY, R. W. (1963): "Open Channel Transport of Fluidized Solids," Ph.D. thesis, Univ. of Alberta.

APMANN, R. P., and R. R. RUMER (1970): Diffusion of Sediment in Developing Flow, *Proc. Am. Soc. Civil Engrs.*, vol. 96, no. HY1.

Arnborg, L. (1953): The Sand-Trap, An Apparatus for Direct Measurement of Bed-Load Transportation in Rivers, *Geografiska Annaler*, Heft 2.

Bagnold, R. A. (1941): "The Physics of Blown Sand and Desert Dunes," W. Morrow, New York.

——— (1956): Flow of Cohesionless Grains in Fluids, *Proc. Roy. Soc.* (*London*), *Phil. Trans.*, ser. A, vol. 249, no. 964.

——— (1966): An Approach to the Sediment Transport Problem from General Physics, *U.S. Geol. Survey*, *Prof. Paper 422-J.*

Bajorunas, L. (1952): River Channel Roughness: A Discussion, *Trans. Am. Soc. Civil Engrs*, vol. 117.

Barekyan, A. S. (1962): Discharge of Channel Forming Sediments and Elements of Sand Waves, *Soviet Hydrol.* (*Am. Geophys. Union*), no. 2.

——— (1963): Method of Determining Permissible Velocities, *Soviet Hydrol.* (*Am. Geophys. Union*), no. 3.

Barnes, H. H. (1967): Roughness Characteristics of Natural Channels, *U.S. Geol. Survey*, *Water Supply Paper 1849.*

Barr, D. I. (1969): Method of Synthesis—Basic Procedures for the new Approach to Similitude, *Water Power* (*London*), vol. 21, April and May.

Baschin, O. (1899): Die Entstehung wellenähnlicher Oberflächenformen, *Z. Gesellschft. Erdkunde*, Bd. XXXIV.

Bates, R. E. (1939): Geomorphic History of the Kickapoo Region, Wisconsin, *Bull. Geol. Soc. Am.*, vol. 50.

Benedict, P. C., M. L. Albertson, and M. Q. Matejka (1953): Total Sediment Load Measured in Turbulence Flume, *Proc. Am. Soc. Civil Engrs.*, vol. 79, no. HY.

Bell, H. S. (1942): Density Currents as Agents for Transporting Sediments, *J. Geol.*, vol. L, no. 5.

Beverage, J. P., and J. K. Culbertson (1964): Hyper-concentrations of Suspended Sediment, *Proc. Am. Soc. Civil Engrs.*, vol. 90, no. HY6.

Bhattacharya, P. K., J. R. Glover, and J. F. Kennedy (1969): Field Tests of an X-ray Sediment-Concentration Gage: A Discussion, *Proc. Am. Soc. Civil Engrs.*, vol. 95, no. HY1.

Bisal, F., and K. Nielsen (1962): Movement of Soil Particles in Saltation, *Canad. J. Soil Sci.*, vol. 42.

Bishop, A. A., D. B. Simons, and E. V. Richardson (1965): Total Bed-Material Transport, *Proc. Am. Soc. Civil Engrs.*, vol. 91, no. HY2.

Blasius, H. (1910): Über die Abhängigkeit der Formen der Riffeln und Geschiebebänke vom Gefälle, *Z. Bauwesen*, Jgg. LX.

Blench, T. (1951): Regime Theory for Self-Formed Sediment-Bearing Channels, *Proc. Am. Soc. Civil Engrs.*, vol. 77.

——— (1955): Scale Relations among Sand-Bed Rivers including Models, *Proc. Am. Soc. Civil Engrs.*, vol. 81, no. HY.

——— (1957): "Regime Behavior of Canals and Rivers," Butterworth, London.

——— (1961): Hydraulics of Canals and Rivers of Mobile Boundary, in "Butterworth's Civil Engineering Reference Book," 2d ed., Butterworth, London.

——— (1964): "River engineering," Illustrated Notes, Dept. of Civ. Eng., Univ. of Alberta, Edmonton, Canada.

——— (1966): "Mobile-Bed Fluviology," T. Blench, Alberta, Canada.

———, and B. Erb (1957): La Théorie du Régime appliquée à l'Analyse des Résultats experimentaux concernant les Transports de Fond, *La Houille Blanche*, no. 2.

———, and M. A. Qureshi (1964): Practical Regime Analysis of River Slopes, *Proc. Am. Soc. Civil Engrs.*, vol. 90, no. HY2.

Bodman, G. B. (1960): "Notes in Soil Physics," Dept. of Soil Phys., Univ. Calif., Berkeley, Calif.

Bogardi, J. L. (1958): The Total Sediment Load of Streams: A Discussion, *Proc. Am. Soc. Civil Engrs.*, vol. 84, no. HY6.

——— (1965): European Concepts of Sediment Transportation, *Proc. Am. Soc. Civil Engrs.*, vol. 91, no. HY1.

Bondurant, D. C. (1958): The Total Sediment Load of Streams: A Discussion, *Proc. Am. Soc. Civil Engrs.*, vol. 84, no. HY6.

Braudeau, G. (1951): Quelques Techniques pour l'Étude et la Mesure du Débit solide, *La Houille Blanche no. Special A*.

Bretting, A. E. (1958): Stable Channels, *Acta Polytech. Scand.*, Ci. 1.

Brooks, N. H. (1958): Mechanics of Streams with Movable Bed of Fine Sand, *Trans. Am. Soc. Civil Engrs.*, vol. 123.

——— (1963): Calculation of Suspended-Load Discharge from Velocity and Concentration Parameters, *Proc. Fed. Interagency Sedim. Conf., U.S. Dept. Agr., Misc. Publ. no. 970.*

——— (1963*a*): Boundary Shear Stress in Curved Channels: A Discussion, *Proc. Am. Soc. Civil Engrs.*, vol. 89, no. HY3.

Brown, C. B. (1950): Sediment Transportation, in "Engineering Hydraulics" (H. Rouse, ed.), Wiley, New York.

Brush, L. M. (1961): Drainage Basins, Channels, and Flow Characteristics of Selected Streams in Central Pennsylvania, *U.S. Geol. Survey, Prof. Paper 282-F.*

———, H. W. Ho, and S. R. Singamsetti (1962): A Study of Sediment in Suspension, *Intern. Assoc. Sci. Hydr., Commiss. Land Erosion (Bari), Pub. no. 59.*

Bugliarello, G., and E. D. Jackson (1964): Random Walk Study of Convective Diffusion, *Proc. Am. Soc. Civil Engrs.*, vol. 90, no. EM4.

Byrne, J. V., and L. D. Kulm (1967): Natural Indicators of Estuarine Sediment Movement, *Proc. Am. Soc. Civil Engrs.*, vol. 93, no. WW2.

Callander, R. A. (1969): Instability and River Channels, *J. Fluid Mech.*, vol. 36/3.

Camp, T. R. (1945): Sedimentation and the Design of Settling Tanks, *Trans. Am. Soc. Civil Engrs.*, vol. 111.

Carey, W. C., and M. D. Keller (1957): System Changes in the Beds of Alluvial Rivers, *Proc. Am. Soc. Civil Engrs.*, vol. 83, no. HY4.

Carlson, E. J., and P. F. Enger (1962): Studies of Tractive Forces of Cohesive Soils in Earth Canals, *U.S. Dept. Inter., Bureau of Recl., Rept. no. Hyd.-504.*

Carstens, M. R. (1952): Accelerated Motion of Spherical Particles, *Trans. Am. Geophys. Union*, vol. 33.

——— (1966): An Analytical and Experimental Study of Bed Ripples Under Water Waves, *Quart. Repts.* 8 and 9, *Georgia Inst. of Tech., School of Civil Engineering, Atlanta.*

Casey, H. J. (1935): Über Geschiebebewegung, *Mitt. Preuss. Versuchsanst. Wasser, Erd, Schiffsbau*, Berlin.

Chang, F. M., D. B. Simons, and E. V. Richardson (1967): Total Bed-Material Discharge in Alluvial Channels, *Intern. Assoc. Hydr. Res., 12th Congress, Fort Collins.*

Chang, Y. (1939): Laboratory Investigations of Flume Traction and Transportation, *Trans. Am. Soc. Civil Engrs.*, vol. 104.

Chepil, W. S. (1958): The Use of Evenly Spaced Hemispheres to Evaluate Aerodynamic Forces on Soil Surfaces, *Trans. Am. Geophys. Union*, vol. 39, no. 3.

——— (1961): The Use of Spheres to Measure Lift and Drag on Wind-Eroded Soil Grains, *Proc. Soil. Sci. Soc. Am.*, vol. 25, no. 5.

Cherry, J. (1965): Sand Movement along a Portion of the Northern California Coast, *U.S. Army Coastal Engr. Res. Center, Tech. Mem. no. 14.*

Chien, N. (1954): The Present Status of Research on Sediment Transport, *Proc. Am. Soc. Civil Engrs.*, vol. 80.

——— (1954*a*): Meyer-Peter Formula for Bed-Load Transport and Einstein Bed-Load Function, *Univ. Calif. Inst. of Eng. Res., no. 7.*

——— (1955): A Concept of Lacey's Regime Theory, *Proc. Am. Soc. Civil Engrs.*, vol. 81.

Chitale, S. V. (1970): River Channel Patterns, *Proc. Am. Soc. Civil Engrs.*, vol. 96, no. HY1.

Chiu, C. L., and J. E. McSparran (1966): Effect of Secondary Flow on Sediment Transport, *Proc. Am. Soc. Civil Engrs.*, vol. 92, no. HY5.

Chow, V. T. (1959): "Open-Channel Hydraulics," McGraw-Hill, New York.

Christiansen, J. E. (1935): Distribution of Silt in Open Channels, *Trans. Am. Geophys. Union*, 16th Annual Meeting, pt. II.

Colby, B. R. (1963): Fluvial Sediments, *U.S. Geol. Survey, Bull. 1181-A.*

——— (1964): Practical Computations of Bed-Material Discharge, *Proc. Am. Soc. Civil Engrs.*, vol. 90, no. HY2.

———, and C. H. Hembree (1955): Computations of Total Sediment Discharge, Niobrara River near Cody, Nebraska, *U.S. Geol. Survey, Water Supply Paper 1357.*

———, and D. W. Hubbell (1961): Simplified Method for Computing Total Sediment Discharge with the Modified Einstein Procedure, *U.S. Geol. Survey, Water Supply Paper 1593.*

———, and C. H. Scott (1965): Effects of Water Temperature on the Discharge of Bed-Material, *U.S. Geol. Survey, Prof. Paper 462-G.*

COLEMAN, N. L. (1967): A Theoretical and Experimental Study of Drag and Lift Forces, *Intern. Assoc. Hydr. Res., 12th Congress, Fort Collins.*

COURTOIS, G. (1967): Emploi des Radioéléments en Sédimentologie, *Proc. Symp., Isotopes in Hydrology, IAEA, Vienna.*

CRICKMORE, M. J. (1961): The Use of Irradiated Sand for Tracer Studies in Hydraulic Models, *La Houille Blanche*, no. 6.

——— (1970): Effect of Flume Width on Bedform Characteristics, *Proc. Am. Soc. Civil Engrs.*, vol. 96, no. HY2.

———, and G. H. LEAN (1962): The Measurement of Sand Transport by Means of Radioactive Tracers, *Proc. Roy. Soc.*, pt. A, vol. 266.

CULBERTSON, J. K., and D. R. DAWDY (1964): A Study of Fluvial Characteristics and Hydraulic Variables, Middle Rio Grande, New Mexico, *U.S. Geol. Survey, Water Supply Paper 1498-F.*

CUMMINGS, R. S., and L. J. INGRAM (1963): Use of Radioisotopes in Sediment Transport Studies, *Proc. Fed. Interagency Sedimentation Conf., U.S. Dept. of Agric., Misc. Publ. no. 970.*

CUNHA, L. V. (1967): About the Roughness in Alluvial Channels with Comparatively Coarse Bed Material, *Intern. Assoc. Hydr. Res., 12th Congress, Fort Collins.*

DAVIS, C. V. (editor) (1942): "Handbook of Applied Hydraulics," McGraw-Hill, New York.

DAWDY, D. R. (1961): Depth-Discharge Relations of Alluvial Streams—Discontinuous Rating Curves, *U.S. Geol. Survey, Water Supply Paper 1498-C.*

DEBSKI, K. (1955): "Continential Hydrology," vol. I, "Hydrometry," translated (1965) from Polish, U.S. Dept. Interior.

DE VRIES, M. (1962): Schwankungen im Geschiebe in natürlichen Wasserläufen, *Conf. Sci., 4/3, Budapest.*

——— (1965): Considerations about Nonsteady Bed-Load Transport in Open Channels, *Delft Hyd. Lab., Pub. no. 36.*

DOBBINS, W. E. (1943): Effect of Turbulence on Sedimentation, *Trans. Am. Soc. Civil Engrs.*, vol. 109.

DOBSON, G. C., and J. W. JOHNSON (1940): Studying Sediment Loads in Natural Streams, *Civil Engr. (Am. Soc. Civil Engrs.)*, vol. 10, no. 2.

DONAT, J. (1929): Über Sohlangriff und Geschiebetrieb, *Wasserwirtschaft*, Heft 26, 27.

DUBOYS, M. P. (1879): Le Rhone et les Rivieres a Lit affouillable, *Mem. Doc., Ann. Pont et Chaussees*, ser. 5, vol. XVIII.

DUBUAT, P. (1786): "Principes d'Hydraulique," 2d ed. (1st ed., 1779), 2 Books, De L'Imprimerie de Monsieur, Paris.

DUNN, I. S. (1959): Tractive Resistance of Cohesive Channels, *Proc. Am. Soc. Civil Engrs.*, vol. 85, no. SM3.

DUPUIT, H. P. (1865): "Traite de la Conduite et de la Distribution des Eaux," Paris.

Egiazaroff, J. V. (1965): Calculation of Nonuniform Sediment Concentrations, *Proc. Am. Soc. Civil Engrs.*, vol. 91, no. HY4.

Ehrenberger, R. (1931): Direkte Geschiebemessungen an der Donau bei Wien und deren bisherige Ergebnisse, *Wasserwirtschaft*, Heft 34.

Einstein, A. (1926): Die Ursache der Mäanderbildung der Flussläufe und des sogenannten Baerschen Gesetzes, *Naturwissenschaften*, Heft 11.

Einstein, H. A. (1942): Formulas for the Transportation of Bed-Load, *Trans. Am. Soc. Civil Engrs.*, vol. 107.

——— (1944): Bed-Load Transportation in Mountain Creeks, *U.S. Dept. Agric., Soil Conserv. Serv., TP-55.*

——— (1944*a*): Conformity between Model and Prototype: A Symposium: A Discussion, *Trans. Am. Soc. Civil Engrs.*, vol. 109.

——— (1948): Determination of Rates of Bed-Load Measurement, *Proc. Fed. Interagency Sedimentation Conf., U.S. Dept. Interior.*

——— (1950): The Bed-Load Function for Sediment Transportation in Open Channel Flows, *U.S. Dept. Agric., Soil Conserv. Serv., T.B. no. 1026.*

——— (1964): River Sedimentation, in "Handbook of Hydrology" (V. T. Chow, editor), McGraw-Hill, New York.

———, A. G. Anderson, and J. W. Johnson (1940): A Distinction Between Bed-Load and Suspended-Load in Natural Streams, *Trans. Am. Geophys. Union*, vol. 21, pt. 2.

———, and R. B. Banks (1950): Fluvial Resistance of Composite Roughness, *Trans. Am. Geophys. Union*, vol. 31, no. 4.

———, and N. L. Barbarossa (1952): River Channel Roughness, *Trans. Am. Soc. Civil Engrs.*, vol. 117.

———, and N. Chien (1954): Second Approximation to the Solution of the Suspended-Load Theory, *Univ. Calif. Inst. Eng. Res.*, no. 3.

———, and ——— (1954*a*): Similarity of Distorted River Models with Movable Bed, *Proc. Am. Soc. Civil Engrs.*, vol. 80, no. HY.

———, and ——— (1955): Effects of Heavy Sediment Concentration Near the Bed on Velocity and Sediment Distribution, *Univ. Calif. Inst. Eng. Res.*, no. 8.

———, and ——— (1958): Mechanics of Streams with Movable Beds of Fine Sand: A Discussion, *Trans. Am. Soc. Civil Engrs.*, vol. 123.

———, and E. S. El-Samni (1949): Hydrodynamic Forces on a Rough Wall, *Rev. Mod. Phys.*, vol. 21, no. 3.

———, and G. Kalkanis (1959): Sand Deposits in Canals, *Univ. Calif. Inst. Engrs. Res.*, series no. 93/6.

———, and R. B. Krone (1961): Estuarial Sediment Transport Patterns, *Proc. Am. Soc. Civil Engrs.*, vol. 87, no. HY2.

———, and H. Li (1958): Secondary Currents in Straight Channels, *Trans. Am. Geophys. Union*, vol. 39, no. 6.

———, and H. W. Shen (1964): A Study on Meandering in Straight Alluvial Channels, *J. Am. Geophys. Union*, vol. 69, December.

EISNER, F. (1932): Offene Gerinne, in "Handbuch der Experimentalphysik," vol. 4, pt. 4, Akademische Verlagsgesellschaft, Leipzig.

ELATA, C., and A. T. IPPEN (1961): The Dynamics of Open Channel Flow with Suspensions of Neutrally Buoyant Particles, *MIT Hydrodyn Lab.*, *T.R.* no. 45.

ELDER, J. W. (1959): The Dispersion of Marked Fluid in Turbulent Shear Flow, *J. Fluid Mech.*, vol. 5, no. 4.

ENGELS, H. (1905): Untersuchungen über die Bettausbildung . . . , *Z. Bauwesen*, Jgg. LV.

ENGELUND, F. (1966): Hydraulic Resistance in Alluvial Streams, *Proc. Am. Soc. Civil Engrs.*, vol. 92, no. HY2.

——— (1969): Dispersion of Floating Particles in Uniform Channel Flow, *Proc. Am. Soc. Civil Engrs.*, vol. 94, no. HY4.

———, and E. HANSEN (1966): Investigations of Flow in Alluvial Streams, *Acta Polytech. Scand.*, Ci-35.

———, and ——— (1967): "A Monograph on Sediment Transport in Alluvial Streams," Teknisk Forlag, Copenhagen.

ENGER, P. F. (1961): Tractive Force Fluctuations Around an Open Channel Perimeter, presented at Amer. Soc. Civil Engrs. Convention, Phoenix, Arizona.

——— (1963): Canal Erosion and Tractive Force Study, *U.S. Recl.*, *Rept. no. Hyd.-532.*

———, et al. (1960): Progress Report No. 3, Canal Erosion and Tractive Force Study, *U.S. Bureau Recl.*, *Rept. no. Gen.-26.*

ERTEL, H. (1966): Kinematik und Dynamik formbeständig wandernder Transversaldünen, *Monatsber.*, *Deut. Akad. Wiss.*, vol. 8/10.

ETTER, R. J., et al. (1968): Depositional Behavior of Kaolinite in Turbulent Flow, *Proc. Am. Soc. Civil Engrs.*, vol. 94, no. HY6.

EXNER, F. M. (1919): Zur Theorie der Flussmäander, *Sitzber. Akad. Wiss. Wien*, pt. II*a*, vol. 128, Heft 10.

——— (1925): Über die Wechselwirkung zwischen Wasser und Geschiebe in Flüssen, *Sitzber. Akad. Wiss. Wien*, pt. II*a*, Bd. 134.

FISCHER, H. B. (1966): A Note on the One-Dimensional Dispersion Model, *J. Air, Water Pollution*, vol. 10.

——— (1967*a*): Analytic Prediction of Longitudinal Dispersion Coefficients in Natural Streams, *Intern. Assoc. Hydr. Res.*, *12th Congress*, *Fort Collins.*

——— (1967*b*): The Mechanics of Dispersion in Natural Streams, *Proc. Am. Soc. Civil Engrs.*, vol. 93, no. HY6.

——— (1968): Methods for Predicting Dispersion Coefficients in Natural Streams, *U.S. Geol. Survey*, *Prof. Paper 582-A.*

FLAXMAN, E. M. (1963): Channel Stability in Undisturbed Cohesive Soils, *Proc. Am. Soc. Civil Engrs.*, vol. 89, no. HY2.

FORCHHEIMER, P. (1914): "Hydraulik," 1st ed. (2d ed., 1924, and 3d ed., 1930), Teubner, Leipzig/Berlin.

FORTIER, S., and F. C. SCOBEY (1926): Permissible Canal Velocities, *Trans. Am. Soc. Civil Engrs.*, vol. 89.

FRANCO, J. J. (1968): Effects of Water Temperature on Bed-Load Movement, *Proc. Am. Soc. Civil Engrs.*, vol. 94, no. WW3.

FRANK, P., and R. MISES (1927): "Differential- und Integralgleichungen," pt. 2/XIV, Dover-Vieweg, New York, Braunschweig.

FREEMAN, J. R. (1929): "Hydraulic Laboratory Practice," Am. Soc. Mech. Engrs., New York.

FRIEDKIN, J. F. (1945): Meandering of Alluvial Rivers, *U.S. Waterways Exper. Stat.*, Vicksburg, Mississippi.

FÜHRBÖTER, A. (1967): Zur Mechanik der Strömungsriffel, *Mitteil. Franzius-Inst.*, no. 29, Hannover.

GARBRECHT, G. (1961): Erfahrungswerte über die zulässigen Strömungsgeschwindigkeiten in Flüssen und Kanälen, *Wasser und Boden*, vol. 5.

GARDE, R. J., and M. L. ALBERTSON (1958): The Total Sediment Load of Streams, *Proc. Am. Soc. Civil Engrs.*, vol. 84, no. HY6.

———, and J. DATTATRI (1963): Investigations of the Total Sediment Discharge of Alluvial Streams, *Res. J., Univ. of Roorkee.*

———, and K. R. RAJU (1963): Regime Criteria for Alluvial Streams, *Proc. Am. Soc. Civil Engrs.*, vol. 89, no. HY6.

———, and ——— (1966): Resistance Relationships in Alluvial Channel Flow, *Proc. Am. Soc. Civil Engrs.*, vol. 94, no. HY4.

GESSLER, J. (1965): Der Geschiebetriebbeginn bei Mischungen untersucht an natürlichen Abpflästerungserscheinungen in Kanälen, *Mitteil. VAWE, Eidgen. Techn. Hochschule, Zürich*, no. 69.

GESSNER, J. B., and J. B. JONES (1965): On Some Aspects of Fully-Developed Turbulent Flow in Rectangular Channels, *J. Fluid Mech.*, vol. 23, pt. 4.

GHETTI, A. (1952): Sulla forma dei profili transversali di equilibrio degli alvei mobili, *Giorn. Genio Civile*, Fas. 6.

GIBBS, H. J. (1962): A Study of Erosion and Tractive Force Characteristics in Relation to Soil Mechanics Properties, *U.S. Bureau Recl.*, no. Em-643.

GILBERT, K. G. (1914): The Transportation of Debris by Running Water, *U.S. Geol. Survey, Prof. Paper 86.*

GLOVER, R. E. (1964): Dispersion of Dissolved or Suspended Materials in Flowing Streams, *U.S. Geol. Survey, Prof. Paper 433-B.*

———, and Q. L. FLOREY (1951): Stable Channel Profiles, *U.S. Bureau of Recl., Hydr.*, 325.

GODA, T. (1953): A Study on the Mechanism of Transportation of Suspended Sediment, *Mem. Faculty Eng., Kyoto Univ.*, vol. XV, no. IV.

GOSH, S., and N. ROY (1970): Boundary Shear Distribution in Open Channel Flow, *Proc. Am. Soc. Civil Engrs.*, vol. 96, no. HY4.

GRAF, W. H. (1964): Gesetze der turbulenten Geschwindigkeitsverteilung in geschlossenen Rohren und offenen Gerinnen, *Schweizer Bauz.*, Jgg. 82, no. 53.

———, and E. R. Acaroglu (1968): Sediment Transport in Conveyance Systems (Part 1), *Bull. Intern. Assoc. Sci. Hydr.*, XIIIe année, no. 2.

Grass, A. J. (1970): Initial Instability of Fine Bed Sand, *Proc. Am. Soc. Civil Engrs.*, vol. 96, no. HY3.

Grim, R. E. (1962): "Applied Clay Mineralogy," McGraw-Hill, New York.

Grissinger, E. H. (1966): Resistance of Selected Clay Systems to Erosion by Water, *Water Resour. Res.*, vol. 2, no. 1.

Grover, N. C., and C. L. Howard (1938): The Passage of Turbid Water through Lake Mead, *Trans. Am. Soc. Civil Engrs.*, vol. 103.

Guy, H. P. (1964): An Analysis of Some Storm Variables Affecting Stream Sediment Transport, *U.S. Geol. Survey, Prof. Paper 462-E.*

———, D. B. Simons, and E. V. Richardson (1966): Summary of Alluvial Channel Data from Flume Experiments, 1956–1961, *U.S. Geol. Survey, Prof. Paper 462-J.*

——— et al. (1969): Sediment Measuring Techniques: Laboratory Procedures (Task Committee Report), *Proc. Am. Soc. Civil Engrs.*, vol. 95, no. HY5.

Hack, J. T. (1957): Studies of Longitudinal Stream Profiles in Virginia and Maryland, *U.S. Geol. Survey, Prof. Paper 294-B.*

Halbronn, G. (1949): Remarque sur la Theorie de l' "Austausch" appliquée au Transport des Materiaux en Suspension, *Intern. Assoc. Hydr. Res., 3d Congr., Grenoble.*

Hanson, E. (1966): Bed-Load Investigation in Skive-Karup River, *Trans. Ingenioeren*, no. 3.

Harleman, D. R. (1963): Sediment Transportation Mechanics: Density Currents (*Progress Report of Task Comm*), *Proc. Am. Soc. Civil Engrs.*, vol. 89, no. HY5.

Harms, J. C., and R. K. Fahnestock (1965): Stratification, Bedforms, and Flow Phenomena (with an Example from the Rio Grande), *Am. Assoc. Petrol. Geologists*, Special Publication no. 12.

Harrison, A. S., and H. C. Lidicker (1963): Computing Suspended Sand Loads from Field Measurement, *Proc. Fed. Interagency Sedim. Conf., U.S. Dept. Agric., Misc. Publ. no. 970.*

———, and W. J. Mellema (1967): Movable Bed Model for Alluvial Channel Studies, *Intern. Assoc. Hydr. Res., 12th Congr., Fort Collins.*

Hayashi, T. (1970): Formation of Dunes and Antidunes in Open Channels, *Proc. Am. Soc. Civil Engrs.*, vol. 96, no. HY2.

Helley, E. J. (1969): Field Measurement of the Initiation of Large Bed Particle Motion in Blue Creek, *U.S. Geol. Survey, Prof. Paper 562-G.*

Helmholtz, H. (1888): Über atmosphaerische Bewegungen, *Sitzber. Königl. Preuss. Akad. Wiss., Berlin.*

Henderson, F. M. (1966): "Open Channel Flow," Macmillan, New York.

Herbertson, J. G. (1969): Scaling Procedures for Mobile Bed Hydraulics Models in Terms of Similitude Theory, *J. Hydr. Res.*, vol. 7, no. 3.

HEYNDRICKX, G. A. (1948): "Het transport van bodemmateriaal door stromend water," Drukkerij G.I.G., Brussels.

HILL, H. M. (1966): Bedforms Due to a Fluid Stream, *Proc. Am. Soc. Civil Engrs.*, vol. 92, no. HY2.

———, V. S. SRINIVASAN, and T. UNNY (1969): Instability of Flat Bed in Alluvial Channels, *Proc. Am. Soc. Civil Engrs.*, vol. 95, no. HY5.

HINO, M. (1963): Turbulent Flow with Suspended Particles, *Proc. Am. Soc. Civil Engrs.*, vol. 89, no. HY4.

HINZE, J. O. (1958): "Turbulence," McGraw-Hill, New York.

HJULSTRÖM, F. (1935): The Morphological Activity of Rivers as Illustrated by Rivers Fyris, *Bull. Geol. Inst. Uppsala*, vol. 25 (chap. III).

HOUSEHOLDER, M. K., and V. W. GOLDSCHMIDT (1969): Turbulent Diffusion and Schmidt Number of Particles, *Proc. Am. Soc. Civil Engrs.*, vol. 95, no. EM6.

HUBBELL, D. W. (1964): Apparatus and Techniques for Measuring Bed-Load. *U.S. Geol. Survey, Water Supply Paper 1748.*

——— (1969): Written communication to the author.

———, and D. Q. MATEJKA (1959): Investigations of Sediment Transportation, *U.S. Geol. Survey, Water Supply Paper 1476.*

———, and W. W. SAYRE (1963): Application of Radioactive Tracers in the Study of Sediment Motion, *Proc. Fed. Interagency Sedim. Conf., U.S. Dept. Agric., Misc. Publ. 970.*

———, and ——— (1964): Sand Transport Studies with Radioactive Tracers, *Proc. Am. Soc. Civil Engrs.*, vol. 90, no. HY3.

HUNT, J. N. (1954): The Turbulent Transport of Suspended Sediment in Open Channels, *Proc. Roy. Soc. London*, vol. 224A.

HURST, H. E. (1929): The Suspension of Sand in Water, *Roy. Soc. London*, series A, vol. 157.

HWANG, L., and E. M. LAURSEN (1963): Shear Measurement Technique for Rough Surfaces, *Proc. Am. Soc. Civil Engrs.*, vol. 89, no. HY2.

INGERSOLL, A. C., and R. MCLAUGHLIN (1960): The Suspension of Flocculent Solids in Sedimentation Basins, *Calif. Inst. Techn., Final Rept.*, May.

INGLIS, C. C. (1938): Relationship between Meander Belts . . . , *Central Irrig. Hydrodyn. Res. Stat. Poona (India), Tech. Note no. 12.*

——— (1947): Meanders and Their Bearing on River Training, *Instn. of Civil Engrs., Maritime and Waterways Eng. Div. Meeting.*

——— (1948): Historical Note on Empirical Equations, Developed by Engineers in India for Flow of Water and Sand in Alluvial Channels, *Intern. Assoc. Hydr. Res., 2d Meeting, Stockholm.*

——— (1949): The Effect of Variations in Charge and Grade on the Slopes and Shapes of Channels, *Intern. Assoc. Hydr. Res., 3d Meeting, Grenoble.*

INTERAGENCY COMMITTEE ON WATER RESOURCES (1963): A Study of Methods used in Measurement and Analysis of Sediment Loads in Streams, *Rept. No. 14*, Minneapolis, Minnesota.

IPPEN, A. T., and P. A. DRINKER (1962): Boundary Shear Stress in Curved Trapezoidal Channels, *Proc. Am. Soc. Civil Engrs.*, vol. 88, no. HY5.

———, and R. P. VERMA (1953): The Motion of Discrete Particles along the Bed of a Turbulent Stream, *Intern. Assoc. Hydr. Res., 5th Congress, Minneapolis.*

IVICSICS, L. (1956): A görgett hordalek zörejenek megfigyelese, *Különlenyomat a Hidrologiai Közlömy*, no. 4.

JAKUSCHOFF, P. (1932): Schwebestoffbewegung in Flüssen in Theorie und Praxis, *Wasserwirtsch.*, Jgg. 25.

JAROCKI, W. (1963): "A Study of Sediment," translated from Polish (1957), Nat. Sci. Found. and U.S. Dept. of Int., Washington, D.C.

JEFFERSON, M. (1902): The Limiting Widths of Meander Belts, *Natl. Mag.*, October.

JEFFREYS, H. (1929): On the Transport of Sediments by Streams, *Proc. Cambridge Phil. Soc.*, vol. XXV.

JOBSON, H. E., and W. W. SAYRE (1970): Vertical Transfer in Open Channel Flow, *Proc. Am. Soc. Civil Engrs.*, vol. 96, no. HY3.

JOHNSON, J. W. (1943): Laboratory Investigations on the Bed-Load Transportation and Bed Roughness, *U.S. Dept. Agric., Soil Conserv. Serv., TP-50.*

JORISSEN, A. (1938): Etude expérimentale du Transport solide des Cours d'Eau, *Rev. Univ. Mines*, 8e series, tome XIV, no. 3.

KABELAC, O. (1957): Rivers under Influence of Terrestrial Rotation, *Proc. Am. Soc. Civil Engrs.*, vol. 83, no. WW1.

KALINSKE, A. A. (1940): Suspended Material Transportation Under Non-Equilibrium Conditions, *Trans. Am. Geophys. Union*, vol. 21.

——— (1942): Criteria for Determining Sand-Transport by Surface-Creep and Saltation, *Trans. Am. Geophys. Union*, vol. 23, pt. 2.

——— (1947): Movement of Sediment as Bed-Load in Rivers, *Trans. Am. Geophys. Union*, vol. 28, no. 4.

———, and C. PIEN (1943): Experiments on Eddy-Diffusion and Suspended Material Transportation in Open Channels, *Trans. Am. Geophys. Union*, vol. 24.

KARAKI, S. S., E. E. GRAY, and J. COLLINS (1961): Dual Channel Stream Monitor, *Proc. Am. Soc. Civil Engrs.*, vol. 87, no. HY6.

KARASEV, I. F. (1964): The Regimes of Eroding Channels in Cohesive Materials, *Soviet Hydrol. (Am. Geophys. Union)*, vol. 6.

KAROLYI, Z. (1947): Kiserleter a Hordalekfogoval, *Vizügyi Közlemenyek*, C. 1–4.

KENNEDY, J. F. (1963): The Mechanics of Dunes and Antidunes in Erodible Bed Channels, *J. Fluid Mech.*, vol. 16/4.

———, and N. H. BROOKS (1963): Laboratory Study of an Alluvial Stream at Constant Discharge, *Proc. Fed. Interagency Sedim. Conf., U.S. Dept. Agric., Misc. Publ. no. 970.*

KENNEDY, R. G. (1895): The Prevention of Silting in Irrigation Canals, *Min. Proc., Inst. Civil Engrs.*, vol. CXIX.

KIDSON, C., and A. P. CARR (1962): Marking Bench Materials for Tracing Experiments, *Proc. Am. Soc. Civil Engrs.*, vol. 88, no. HY4.

Klingeman, P. C., and W. J. Kaufman (1965): Transport of Radionuclides with Suspended Sediment in Estuarine Systems, *Sanitary Eng. Res. Lab., Univ. Calif.*, no. 65–15.

Kondrat'ev, N. (1962): "River Flow and River Channel Formation," translation from Russian (1959), *Natl. Sci. Found. and U.S. Dept. Interior.*

——— (1964): Kinematic Pattern of a Stream with the Dune Structure of the Bottom, *Soviet Hydrol.* (*Am. Geophys. Union*), vol. 6.

Kramer, H. (1935): Sand Mixtures and Sand Movement in Fluvial Models, *Trans. Am. Soc. Civil Engrs.*, vol. 100.

Kresser, W. (1964): Gedanken zur Geschiebe- und Schwebstoffführung der Gewässer, *Österr. Wasserwirtsch.*, Jgg. 16, Heft 1/2.

Krey, H. (1925): Grenzen der Übertragbarkeit der Versuchsergebnisse, *Z. angew. Math. Mech.*, Bd. 5, Heft 6.

Krone, R. B. (1960): Methods of Tracing Estuarial Sediment Transport Processes, *Hydr. Eng. Lab., Univ. Calif., Berkeley*, October.

——— (1962): Flume Studies of the Transport of Sediment in Estuarial Shoaling Processes, *Hydr. Eng. Lab., Univ. Calif., Berkeley*, June.

Krumbein, W. C., and L. L. Sloss (1963): "Stratigraphy and Sedimentation," Freeman, San Francisco.

Kruyt, H. R. (editor) (1952): "Colloid Science," vol. 1, Elsevier, Amsterdam.

Kuenen, P. H. (1952): Estimated Size of the Grand Banks Turbidity Current, *Am. J. Sci.*

Lacey, G. (1929): Stable Channels in Alluvium, *Min. Proc. Inst. Civil Engrs.*, vol. 229.

——— (1935): Uniform Flow in Alluvial Rivers and Canals, *Min. Proc. Inst. Civil Engrs.*, vol. 237.

——— (1937): Stable Channels in Erodible Material: A Discussion, *Trans. Am. Soc. Civil Engrs.*, vol. 102.

——— (1946): A General Theory of Flow in Alluvium, *J. Inst. Civil Engrs.*, vol. 27, no. 1.

Lamb, H. (1945): "Hydrodynamics" (1st ed., 1879), Dover, New York.

Lane, E. W. (1937): Stable Channels in Erodible Material, *Trans. Am. Soc. Civil Engrs.*, vol. 102.

——— (1953): Progress Report on Studies on the Design of Stable Channels of the Bureau of Reclamation, *Proc. Am. Soc. Civil Engrs.*, vol. 79.

———, E. J. Carlson, and O. S. Hanson (1949): Low Temperature Increases Sediment Transportation in Colorado River, *Civil Eng.* (*Am. Soc. Civil Engrs.*), vol. 16.

———, and A. A. Kalinske (1939): The Relation of Suspended to Bed Material in Rivers, *Trans. Am. Geophys. Union*, vol. 20.

———, and ——— (1941): Engineering Calculations of Suspended Sediment, *Trans. Am. Geophys. Union*, vol. 20.

Langbein, W. B. (1964*a*): Geometry of River Channels, *Proc. Am. Soc. Civil Engrs.*, vol. 90, no. HY2.

——— (1964*b*): Profiles of Rivers of Uniform Discharge, *U.S. Geol. Survey, Prof. Paper 501-B.*

———, and L. B. LEOPOLD (1966): River Meanders—Theory of Minimum Variance, *U.S. Geol. Survey, Prof. Paper 422-H.*

LAURSEN, E. M. (1958): The Total Sediment Load of Streams, *Proc. Am. Soc. Civil Engrs.*, vol. 84, no. HY1.

LEIGHLY, J. B. (1932): Toward a Theory of the Morphologic Significance of Turbulence in the Flow of Water in Streams, *Univ. Calif. Pub. Geogr.*, vol. 6, no. 1.

——— (1934): Turbulence and the Transportations of Rock Debris by Streams, *Geogr. Rev.*, vol. XXIV, no. 3.

LELIAVSKY, S. (1955): "An Introduction to Fluvial Hydraulics," Constable, London.

LEOPOLD, L. B., and W. B. LANGBEIN (1962): The Concept of Entropy in Landscape Evolution, *U.S. Geol. Survey, Prof. Paper 500-A.*

———, and T. MADDOCK (1953): The Hydraulic Geometry of Stream Channels and Some Physiographic Implications, *U.S. Geol. Survey, Prof. Paper 252.*

———, and J. MILLER (1956): Ephemeral Streams—Hydraulic Factors and Their Relation to the Drainage Net, *U.S. Geol. Survey, Prof. Paper 282-A.*

———, and M. G. WOLMAN (1957): River Channel Patterns: Braided, Meandering and Straight, *U.S. Geol. Survey, Prof. Paper 282-B.*

———, and ——— (1960): River Meanders, *Bull. Geol. Soc. Am.*, vol. 71.

———, ———, and J. P. MILLER (1964): "Fluvial Processes in Geomorphology," Freeman, San Francisco.

LEUTHEUSSER, H. J. (1963): Turbulent Flow in Rectangular Ducts, *Proc. Am. Soc. Civil Engrs.*, vol. 89, no. HY3.

LIGGETT, J. A., C. CHIU, and L. S. MIAO (1965): Secondary Currents in a Corner, *Proc. Am. Soc. Civil Engrs.*, vol. 91, no. HY6.

LINDLEY, E. S. (1919): Regime Channels, *Proc. Punjab Eng. Congr.*, vol. VII.

LIU, H. K. (1957): Mechanics of Sediment-Ripple Formation, *Proc. Am. Soc. Civil Engrs.*, vol. 83, no. HY2.

LOVERA, F., and J. F. KENNEDY (1969): Friction-Factor for Flat-Bed Flows in Sand Channels, *Proc. Am. Soc. Civil Engrs.*, vol. 95, no. HY4.

LUNDGREN, H., and J. JONSSON (1964): Shear and Velocity Distribution in Shallow Channels, *Proc. Am. Soc. Civil Engrs.*, vol. 90, no. HY1.

LYLE, W. M., and E. T. SMERDON (1965): Relation of Compaction and Other Soil Properties to Erosion and Resistance of Soils, *Trans. Am. Soc. Agric. Eng.*, vol. 8, no. 3.

MACDOUGALL, C. H. (1933): Bed Sediment Transportation in Open Channels, *Trans. Am. Geophys. Union.*

MADDOCK, T., Jr. (1959): The Behavior of Straight Open Channels with Movable Beds, *U.S. Geol. Survey, Prof. Paper 622-A.*

MAJUMDAR, H., and M. R. CARSTENS (1967): Diffusion of Particles by Turbulence: Effect of Particle Size, *Water Res. Center, WRC-0967, Georgia Inst. Techn. Atlanta.*

Masch, F. D., W. Epsey, and W. L. Moore (1963): Measurement of the Shear Resistance of Cohesive Sediments, *Proc. Fed. Interagency Sedim. Conf., U.S. Dept. Agric., Misc. Publ. no. 970.*

——— et al. (1968): Erosion of Cohesive Sediment (Task Comm. Rept.), *Proc. Am. Soc. Civil Engrs.*, vol. 94, no. HY4.

Matsunashi, J. (1967): On a Solution of Bed Fluctuations in Open Channels with Movable Bed of Steep Slope, *Intern. Assoc. Hydr. Res., 12th Congr., Fort Collins.*

Matthes, G. H. (1941): Basic Aspects of Stream-Meanders, *Trans. Am. Geophys. Union.*

———, and J. H. Stratton (1956): River Engineering, in "American Civil Engineering Practice" (Edit. by R. Abbet), Wiley, New York.

Matyukhin, V. J., and O. N. Prokofyev (1966): Experimental Determination of the Coefficient of Vertical Turbulent Diffusion in Water for Settling Particles, *Soviet Hydrol. (Am. Geophys. Union)*, no. 3.

Mavis, F. T., and L. M. Laushey (1948): A Reappraisal of the Beginning of Bed Movement-Competent Velocity, *Intern. Assoc. Hydr. Res., 2d Meeting, Stockholm.*

———, T. Liu, and E. Soucek (1937): The Transportation of Detritus by Flowing Water—II, *Univ. of Iowa, Studies in Eng.*, no. 341.

McDowell, L. L. (1963): Sediment Movement as Defined by Radioactive Tracers, *Proc. Fed. Interagency Sedim. Conf., U.S. Dept. of Agric., Misc. Publ. no. 970.*

Mei, C. C. (1969): Nonuniform Diffusion of Suspended Sediment, *Proc. Am. Soc. Civil Engrs.*, vol. 65, no. HY1.

Meyer, L. D., and E. J. Monke: Mechanics of Soil Erosion by Rainfall and Overland Flow, *Trans. Am. Soc. Agric. Eng.*, vol. 8(4).

Meyer-Peter, E. (1949): Quelques Problèmes concernant le Charriage des Matières solides, *Soc. Hydrotechn. France*, no. 2.

——— (1951): Transport des Matières solides en general et Problèmes speciaux, *Bull. Genie Civil d'Hydraul. Fluviale*, Tome V.

———, H. Favre, and A. Einstein (1934): Neuere Versuchsresultate über den Geschiebetrieb, *Schweiz. Bauzeitung*, vol. 103, no. 13.

———, and R. Müller (1948): Formulas for Bed-Load Transport, *Intern. Assoc. Hydr. Res., 2d Meeting, Stockholm.*

———, and ——— (1949): Eine Formel zur Berechnung des Geschiebetriebes, *Schweiz. Bauzeitung*, Jgg. 67, no. 3.

Miller, R. L., and R. J. Byrne (1966): The Angle of Repose for a Single Grain on a Fixed Rough Bed, *J. Sedimentology*, vol. 6.

Milne-Thomson, L. (1960): "Theoretical Hydrodynamics," 4th ed., Macmillan, New York.

Moore, W. L., and F. D. Masch (1962): Experiments on the Scour Resistance of Cohesive Sediments, *Am. Geophys. Union, J. Geoph. Res.*, vol. 67, no. 4.

Morris, H. M. (1963): "Applied Hydraulics in Engineering," Ronald, New York.

MÜHLHOFER, L. (1933): Untersuchungen über Schwebstoff- und Geschiebeführung des Inns nächst Kirchbichl, *Wasserwirtschaft*, Heft 1-6.

MÜLLER, R. (1943): Theoretische Grundlagen der Fluss- und Wildbachverbauungen, *Mitteil. VAWE, Eidgen. Techn. Hochschule*, Zürich, no. 4.

MURAMOTO, Y. (1967): Secondary Flows in Curved Open Channels, *Intern. Assoc. Hydr. Res., 12th Congress, Fort Collins.*

MURPHREE, C. E., et al. (1968): Field Test on X-Ray Sediment-Concentration Gage, *Proc. Am. Soc. Civil Engrs.*, vol. 94, no. HY2.

NASH, J. E. (1959): A Study of Bank-Full Discharge of Rivers in England and Wales: A Discussion, *Proc. Inst. Civil Engrs.*, vol. 14.

NEILL, C. R. (1967): Mean Velocity Criterion for Scour of Coarse Uniform Bed-Material, *Intern. Assoc. Hydr. Res., 12th Congress, Fort Collins.*

NELSON, M. E., and P. C. BENEDICT (1950): Measurement and Analysis of Suspended Sediment Load in Streams, *Proc. Am. Soc. Civil Engrs.*, vol. 76, no. HY.

NEMENYI, P. (1933): "Wasserbauliche Strömungslehre," Barth, Leipzig.

——— (1946): The Transportation of Suspended Sediment by Water: A Discussion, *Trans. Am. Soc. Civil Engrs.*, vol. 111.

NESPER, F. (1937): Die Internationale Rheinregulierung: III, *Schweiz. Bauzeitung*, Bd. 110, no. 12.

NIXON, M. (1959): A Study of Bank-Full Discharge of Rivers in England and Wales, *Proc. Inst. Civil Engrs.*, vol. 12.

NORDIN, C. F. (1964): Aspects of Flow Resistance and Sediment Transport, Rio Grande near Bernalillo, New Mexico, *U.S. Geol. Survey, Water Supply Paper 1498-H.*

———, and J. A. ALGERT (1965): Geometrical Properties of Sand Waves: A Discussion, *Proc. Am. Soc. Civil Engrs.*, vol. 91, no. HY5.

———, and ——— (1966): Spectral Analyses of Sand Waves, *Proc. Am. Soc. Civil Engrs.*, vol. 92, no. HY5.

———, and G. R. DEMPSTER (1963): Vertical Distribution of Velocity and Suspended Sediment, Middle Rio Grande, New Mexico, *U.S. Geol. Survey, Prof. Paper 462-B.*

———, and J. P. BEVERAGE (1964): An Expression for Bed-Load Transportation: A Discussion, *Proc. Am. Soc. Civil Engrs.*, vol. 90, no. HY1.

NORGAARD, R. B. (1968): Streamflow Fluctuations, Bar Roughness, and Bed-Load Movement: A Hypothesis, *Water Resources Res.*, vol. 4, no. 3.

NOVAK, P. (1959): Vyzkum Funkce a Ucinnosti Pristroju na Mereni Splavenin, *Vyzkum Ustav Vodohospodarsky, Praha*, Sesit 99.

O'BRIEN, M. P. (1933): Review of the Theory of Turbulent Flow and Its Relation to Sediment Transport: *Trans. Am. Geophys. Union.*

——— (1936): Notes on the Transportation of Silt by Streams, *Trans. Am. Geophys. Union*, vol. 17.

——— (1937): The Passage of Turbid Water through Lake Mead: A Discussion, *Trans. Am. Soc. Civil Engrs.*, vol. 103.

———, and B. D. Rindlaub (1934): The Transportation of Bed-Load by Streams, *Trans. Am. Geophys. Union.*

Orlob, G. T. (1960): Eddy Diffusion in Homogeneous Turbulence, *Trans. Am. Soc. Civil Engrs.*, vol. 125.

Owen, P. R. (1964): Saltation of Uniform Grain in Air, *J. Fluid Mech.*, vol. 20, pt. 2.

Pang, Y. H. (1939): Abhängigkeit der Geschiebebewegung von der Kornform und der Temperatur, *Mitteil., Preuss. Versuchsanst. Wasser, Erd, Schiffsbau, Berlin*, Heft 37.

Partheniades, E. (1965): Erosion and Deposition of Cohesive Soils, *Proc. Am. Soc. Civil Engrs.*, vol. 91, no. HY1.

Pettis, C. R. (1937): Stable Channels in Erodible Material: A Discussion, *Trans. Am. Soc. Civil Engrs.*, vol. 102.

Pezzoli, G. (1963): Il moto di transporto solido e l'applicazionne ai modelli idraulici, *VIII Convegna de Idraulici, Pisa.*

Piest, R. F., et al. (1970): Sediment Sources and Sediment Yields (Progress Report of Task Committee), *Proc. Am. Soc. Civil Engrs.*, vol. 96, no. HY6.

Prandtl, L. (1926): Über die ausgebildete Turbulenz, *2d Internat. Congr. Applied Mech., Zürich.*

Preston, J. H. (1954): The Determination of Turbulent Skin Friction by Means of Pitot Tubes, *J. Roy. Aeron. Soc.*, vol. 58.

Putzinger, J. (1919): Das Ausgleichsgefälle geschiebeführender Wasserläufe und Flüsse, *Z. Österr. Ing.-u. Arch.-Verein*, no. 71.

Raichlen, F., and J. F. Kennedy (1965): The Growth of Sediment Bedforms from an Initially Flattened Bed, *Intern. Assoc. Hydr. Res., Proc. 11th Congr., Leningrad.*

Raudkivi, A. J. (1967): "Loose Boundary Hydraulics," Pergamon, Oxford.

Reitz, W. (1936): Über Geschiebebewegung, *Wasserwirtsch. und Tech.*, no. 28–30.

Rektorik, R. J., and E. T. Smerdon (1964): Critical Shearing Forces in Cohesive Soils from a Rotating Shear Apparatus, *Annual Meeting, Am. Soc. Agric. Engrs.*

Replogle, J. A., and V. T. Chow (1966): Tractive-Force Distribution in Open Channels, *Proc. Am. Soc. Civil Engrs.*, vol. 92, no. HY2.

Reynolds, A. J. (1965): Waves on the Erodible Bed of an Open Channel, *J. Fluid Mech.*, vol. 22/I.

Richardson, E. G. (1934): The Transport of Silt by a Stream, *Phil. Mag. and J. Sci.*, vol. 16, no. 114.

Richardson, E. V., and D. B. Simons (1967): Resistance to Flow in Sand Channels, *Intern. Assoc. Hydr. Res., 12th Congress, Fort Collins.*

Rouse, H. (1937): Modern Conceptions of the Mechanics of Turbulence, *Trans. Am. Soc. Civil Engrs.*, vol. 102.

——— (1938): Experiments on the Mechanics of Sediment Suspension, *Proc. 5th Intern. Congr. Appl. Mech., Cambridge, Mass.*

——— (1964): Sediment Transportation Mechanics: Suspension of Sediment (Progress Report of Task Committee): A Discussion, *Proc. Am. Soc. Civil Engrs.*, vol. 90, no. HY1.

ROZOVSKII, I. L. (1957): Flow of Water in Bends of Open Channels, *Acad. Sci. Ukrain. SSR*, translation from Russian (1961) by Natl. Sci. Found. and U.S. Dept. Interior.

RUBEY, W. W. (1933): Equilibrium-Conditions in Debris-Laden Streams, *Trans. Am. Geophys. Union.*

——— (1938): The Force Required to Move Particles on a Stream Bed, *U.S. Geol. Survey, Prof. Paper 189-E.*

RZHANITSYN, N. A. (1960): Morphologic and Hydrologic Regularities of Structures of the River Net, *Gydrometeozdit*, translated from Russian by U.S. Dept. Agric. and U.S. Dept. Interior.

SAYRE, W. W. (1967): "Dispersion of Mass in Open-Channel Flow," Ph.D. thesis, Colorado State Univ.

——— (1969): Dispersion of Silt Particles in Open Channel Flow, *Proc. Am. Soc. Civil Engrs.*, vol. 95, no. HY3.

———, and J. M. CHANG (1968): A Laboratory Investigation of Open-Channel Dispersion Processes for Dissolved, Suspended, and Floating Dispersants, *U.S. Geol. Survey, Prof. Paper 433-E.*

SCHAAK, E. M. (1937): Eine Vorrichtung zur Messung des Geschiebes an der Flussohle: Diskussion, *Intern. Assoc. Hydr. Struct. Res., 1st Congr., Berlin.*

SCHAFFERNAK, F. (1922): "Neue Grundlagen für die Berechnung der Geschiebeführung in Flussläufen," Deutike, Vienna.

SCHEIDEGGER, A. (1961): "Theoretical Geomorphology," Springer, Berlin.

SCHLICHTING, H. (1968): "Boundary Layer Theory," McGraw-Hill, New York.

SCHMIDT, W. (1925): Der Massenaustausch in freier Luft und verwandte Erscheinungen, in "Probleme der kosmischen Physik," Bd. 7, Hamburg.

SCHMITT, H. (1966): Zur Mechanik der Geschiebebildung, *Mitt. Planck-Institut Strömungsforsch.*, no. 37, Göttingen.

SCHOKLITSCH, A. (1914): "Über Schleppkraft und Geschiebebewegung," Engelmann, Leipzig.

——— (1926): "Die Geschiebebewegung an Flüssen und an Stauwerken," Springer, Vienna.

——— (1930): "Handbuch des Wasserbaues," Springer, Vienna (2d ed., 1950), English translation (1937) by S. Shulits.

——— (1933): Über die Verkleinerung der Geschiebe in Flussläufen, *Sitzber. Akad. Wiss. Wien*, Abt. II*a*, vol. 142, no. 8.

——— (1934): Geschiebetrieb und die Geschiebefracht, *Wasserkraft & Wasserwirtsch.*, Jgg. 39, Heft 4.

SCHROEDER, K. B., and C. H. HEMBREE (1956): Application of the Modified Einstein Procedure for Computation of Total Sediment Load, *Trans. Am. Geophys. Union*, vol. 37, no. 2.

SCHUMM, S. A. (1963): Sinuosity of Alluvial Rivers on the Great Plains, *Bull.*, *Geol. Soc. Am.*, vol. 74, Sept.

SHEN, H. W. (1961): A Study on Meandering and Other Bed Patterns in Straight Alluvial Channels, *Univ. of Calif.*, *Water Res. Center*, *Contr. No. 33.*

——— (1962): Development of Bed Roughness in Alluvial Channels, *Proc. Am. Soc. Civil Engrs.*, vol. 88, no. HY3.

———, and S. KOMURA (1968): Meandering Tendencies in Straight Alluvial Channels, *Proc. Am. Soc. Civil Engrs.*, vol. 94, no. HY4.

———, and S. VEDULA (1969): A Basic Cause of a Braided Channel, *Proc. 13th Congr. Intern. Assoc. Hydr. Res.*, *Kyoto*, vol. 5–1.

SHIELDS, A. (1936): Anwendung der Ähnlichkeitsmechanik und Turbulenzforschung auf die Geschiebebewegung, *Mitteil. Preuss. Versuchsanst. Wasser*, *Erd*, *Schiffsbau*, *Berlin*, no. 26.

SHINOHARA, K., and T. TSUBAKI (1959): On the Characteristics of Sand Waves Formed upon the Beds of the Open Channels and Rivers, *Rept.*, *Res. Inst. Appl. Mech.*, *Hyushu Univ.*, *Japan*, vol. VIII, no. 25.

SHUKRY, A. (1949): Flow around Bends in an Open Channel Flume, *Trans. Am. Soc. Civil Engrs.*, vol. 115.

SHULITS, S. (1935): The Schoklitsch Bed-Load Formula, *Engineering* (*Great Britain*), June.

——— (1936): Fluvial Morphology in Terms of Slope, Abrasion, and Bed-Load, *Trans. Am. Geophys. Union.*

——— (1941): Rational Equation of River-Bed Profile, *Trans. Am. Geophys. Union.*

SILBERMAN, E., et al. (1963): Friction Factors in Open Channels (Progress Report of the Task Committee), *Proc. Am. Soc. Civil Engrs.*, vol. 89, no. HY2.

SINELTSHIKOV, V. (1967): About Turbulent Diffusion of Suspended Particles, *Intern. Assoc. Hydr. Res.*, *12th Congr.*, *Fort Collins.*

SIMONS, D. B. (1967): River Hydraulics, *Intern. Assoc. Hydr. Res.*, *12th Congr.*, *Fort Collins.*

———, and M. L. ALBERTSON (1963): Uniform Water Conveyance in Alluvial Material, *Trans. Am. Soc. Civil Engrs.*, vol. 128/I.

———, and E. V. RICHARDSON (1961): Forms of Bed Roughness in Alluvial Channels, *Proc. Am. Soc. Civil Engrs.*, vol. 87, no. HY.

———, and ——— (1963*a*): A Study of Variables Affecting Flow Characteristics and Sediment Transport in Alluvial Channels, *Proc. Fed. Interagency Sedim. Conf.*, *U.S. Dept of Agric.*, *Misc. Publ. 970.*

———, and ——— (1966): Resistance to Flow in Alluvial Channels, *U.S. Geol. Survey*, *Prof. Paper 422-J.*

———, ———, and M. L. ALBERTSON (1961*a*): Flume Studies using Medium Sand (0.45 mm), *U.S. Geol. Survey*, *Water Supply Paper 1498-A.*

———, ———, and W. L. HAUSHILD (1962): Depth-Discharge Relations in Alluvial Channels, *Proc. Am. Soc. Civil Engrs.*, vol. 88, HY5.

———, ———, and C. F. Nordin (1965): Sedimentary Structures Generated by Flow in Alluvial Channels, *Am. Assoc. Petrol. Geologists*, Special Publ. no. 12.

———, ———, and ——— (1965*a*): Bed-load Equation for Ripples and Dunes, *U.S. Geol. Survey, Prof. Paper 462-H.*

Smerdon, E. T. (1964): Effect of Rainfall on the Critical Tractive Force in Channels with Shallow Flow, *Trans. Am. Soc. Agric. Engrs.*, vol. 7, no. 3.

———, and R. P. Beasley (1959): The Tractive Force Theory Applied to Stability of Open Channels in Cohesive Soils, *Univ. of Missouri, Agric. Exper. Stat. Res. Bull. 715.*

Smith, W. O., et al. (1954): Comprehensive Survey of Lake Mead: 1948–1949, *U.S. Geol. Survey.*

Smoltczyk, H. U. (1955): Beitrag zur Ermittlung der Feingeschiebe—Mengenganglinie, *Inst. Wasserbau, Tech. Univ. Berlin, Mittl. No. 43.*

Sooky, A. A. (1969): Longitudinal Dispersion in Open Channels, *Proc. Am. Soc. Civil Engrs.*, vol. 95, no. HY4.

Stall, J. B., N. L. Rupani, and P. K. Kandaswamy (1958): Sediment Transport in Money Creek, *Proc. Am. Soc. Civil Engrs.*, vol. 84, no. HY1.

Stanton, T. E., D. Marshal, and C. N. Bryant (1920): On the Conditions at the Boundary of a Fluid in Turbulent Motion, *Proc. Roy. Soc.*, vol. 97A.

Stebbings, J. (1963): The Shapes of Self-Formed Model Alluvial Channels, *Proc. Inst. Civil Engrs.*, vol. 25.

Sternberg, H. (1875): Untersuchungen über das Längen und Querprofil geschiebeführender Flüsse, *Z. Bauwesen*, vol. 25.

Stevens, J. C., et al. (1942): "Hydraulic Models," Committee of the Hydr. Div., Am. Soc. Civil Engrs., New York.

Straub, L. G. (1935): Missouri River Report, in *House Document 238*, p. 1135, 73d Cong., 2d Sess., *U.S. Government Printing Office*, Washington, D.C.

——— (1936): Transportation of Sediment in Suspension, *Civil Eng.* (*Am. Soc. Civil Engrs.*), vol. 6, no. 3.

——— (1939): Laboratory Investigation of Flume Traction and Transportation: A Discussion, *Trans. Am. Soc. Civil Engrs.*, vol. 65.

Strauss, V. (1962): K Problematike Priameho Merania tangencialnych, *Vodohospodarsky Casopis*, *Bratislava*.

——— (1964): Möglichkeiten der direkten Messung von Schubspannungen im hydraulischen Versuchswesen, *Sci. Hydr. Conf.*, *Bucharest*.

Sutherland, A. J. (1967): Proposed Mechanism for Sediment Entrainment by Turbulent Flow, *J. Geophys. Res.*, vol. 72, no. 24.

Taylor, G. (1954): Dispersion of Matter in Turbulent Flow Through a Pipe, *Proc. Roy. Soc. London*, vol. 223A.

Tchen, C. (1947): "Mean Value and Correlation Problems Connected with the Motion of Small Particles Suspended in a Turbulent Fluid," D.Sc. Dissertation, Technische Hogeschool, Delft.

TELEKI, P. G. (1963): A Summary of the Production and Scanning of Fluorescent Tracers, *Proc. Fed. Interagency Sedim. Conf., U.S. Dept. of Agric., Misc. Publ. 970.*

TERZAGHI, K., and R. B. PECK (1968): "Soil Mechanics in Engineering Practice," Wiley, New York.

THACKSTON, E. L., and P. A. KRENKEL (1967): Longitudinal Mixing in Natural Streams, *Proc. Am. Soc. Civil Engrs.*, vol. 93, no. SA5.

THOMAS, C. W., and P. F. ENGER (1961): Use of an Electronic Computer to Analyze Data from Studies of Critical Tractive Forces for Cohesive Soils, *Intern. Assoc. Hydr. Res., 9th Congr., Dubrovnik.*

TIFFANY, J. B., and C. B. BENTZEL (1935): Sand Mixtures and Sand Movement in Fluvial Models: A Discussion, *Trans. Am. Soc. Civil Engrs.*, vol. 100.

TISON, L. J. (1953): Recherches sur la Tension limité d'Entrainment des Materiaux constitutifs du Lit, *Intern. Assoc. Hydr. Res., 5th Congr. Minneapolis.*

TODOROVIC, P. et al. (1967): A Contribution to the Kinetic Theory of Bed-Material Discharge, *Proc. Symp. Isotopes in Hydrology, IAEA, Vienna.*

TOFFALETI, F. B. (1969): Definitive Computations of Sand Discharge in Rivers, *Proc. Am. Soc. Civil Engrs.*, vol. 95, no. HY1.

TRACY, H. J. (1965): Turbulent Flow in a Three-Dimensional Channel, *Proc. Am. Soc. Civil Engrs.*, vol. 91, no. HY6.

TRASK, P. D. (1939): "Recent Marine Sediments," Dover, New York.

TSUBAKI, T., T. KAWASUMI, and T. YASUTOMI (1953): On the Influence of Sand Ripples upon the Sediment Transport in Open Channels, *Rept., Res. Inst. Appl. Mech., Kyushu Univ., Japan*, vol. II, no. 8.

TSUCHIYA, A., and ISHIZAKI, K. (1967): The Mechanics of Dune Formation in Erodible-Bed Channels, *Intern. Assoc. Hydr. Res., 12th Congr., Fort Collins.*

TSUCHIYA, Y. (1969): On the Mechanics of Saltation of a Spherical Sand Particle in a Turbulent Stream, *Proc. 13th Congr. Intern. Assoc. Hydr. Res., Kyoto.*

U.S. BUREAU OF RECLAMATION (1953): Hydraulic Laboratory Practice, *Eng. Monogr.* no. 18.

U.S. WATERWAYS EXPERIMENT STATION (1935): Studies of River Bed Materials and Their Movement, with Special Reference to the Lower Mississippi River, *USWES, Vicksburg, Paper 17.*

VANONI, V. A. (1941): Some Experiments on the Transportation of Suspended Loads, *Trans. Am. Geophys. Union.*

——— (1946): Transportation of Suspended Sediment by Water, *Trans. Am. Soc. Civil Engrs.*, vol. 111.

——— (1953): Some Effects of Suspended Sediment on Flow Characteristics, *Proc. 5th Iowa Hydr. Conf.*

——— (1963): Sediment Transportation Mechanics: Suspension of Sediment (Progress Report of Task Committee), *Proc. Am. Soc. Civil Engrs.*, vol. 89, no. HY5.

——— (1964): Measurements of Critical Shear Stress, *Calif. Inst. Techn., Rep. no. KH-R-7.*

——— (1966): Sediment Transportation Mechanics: Initiation of Motion (Progress Report of Task Committee), *Proc. Am. Soc. Civil Engrs.*, vol. 92, no. HY2.

———, and N. H. Brooks (1957): Laboratory Studies of the Roughness and Suspended Load of Alluvial Streams, *Calif. Inst. Techn., MRD Sediment Ser. no. 11.*

———, and L. S. Hwang (1967): Relation between Bedforms and Friction in Streams, *Proc. Am. Soc. Civil Engrs.*, vol. 93, no. HY3.

———, and G. N. Nomicos (1960): Resistance Properties of Sediment-Laden Streams, *Trans. Am. Soc. Civil Engrs.*, vol. 125/I.

Vasiliev, O. F. (1969): Problems of Two-Phase Flow Theory, *Proc. 13th Congr. Intern. Assoc. Hydr. Res., Kyoto.*

Vennard, J. K. (1961): "Elementary Fluid Mechanics," Wiley, New York.

Vinckers, J. B., E. W. Bijker, and J. B. Schijf (1953): Bed-Load Transport Meter for Fine Sand "Sphinx," *Intern. Assoc. Hydr. Res., 5th Congr., Minneapolis.*

Vogel, H. D. (1935): Practical River Laboratory Hydraulics, *Trans. Am. Soc. Civil Engrs.*, vol. 100.

Ward, B. D. (1969): Relative Density Effects on Incipient Bed Movement, *Water Resources Res.*, vol. 5, no. 5.

Werner, P. W. (1951): On the Origin of River Meanders, *Trans. Am. Geophys. Union*, vol. 37.

White, C. M. (1940): The Equilibrium of Grains on the Bed of a Stream, *Proc. Roy. Soc. London*, vol. 174A.

Willis, J. C. (1969): An Error Function Description of the Vertical Suspended Sediment Distribution, *Water Resources Res.*, vol. 5, no. 6.

Wittmann, H. (1955): Flussbau in "Taschenbuch für Bauingenieure" (Schleicher, editor), Springer, Berlin.

Woodruff, N. P. (1965): Sediment Transportation Mechanics: Wind Erosion and Transportation (Progress Report of Task Committee), *Proc. Am. Soc. Civil Engrs.*, vol. 91, no. HY2.

Yalin, M. S. (1963): An Expression for Bed-Load Transportation, *Proc. Am. Soc. Civil Engrs.*, vol. 89, no. HY3.

——— (1964): Geometrical Properties of Sand Waves, *Proc. Am. Soc. Civil Engrs.*, vol. 90, no. HY5.

——— (1964): On the Average Velocity of Flow over a Movable Bed, *La Houille Blanche*, no. 1.

Yano, K., and A. Daido (1969): The Effect of Bed-Load Movement on the Velocity Distribution of Flow, *Proc. 13th Congr. Intern. Assoc. Hydr. Res., Kyoto.*

———, Y. Tsuchiya, and M. Michiue (1969): Tracer Studies on the Movement of Sand and Gravel, *Proc. 13th Congr. Intern. Assoc. Hydr. Res., Kyoto*, vol. 2.

YEN, C. L. (1970): Bed Topography Effect on Flow in a Meander, *Proc. Am. Soc. Civil Engrs.*, vol. 96, no. HY1.

ZELLER, J. (1963): Einführung in den Sedimenttransport offener Gerinne, *Schweiz. Bauzeitung*, Jgg. 81.

——— (1967): Meandering Channels in Switzerland, *Intern. Assoc. Sci. Hydr., Symp. on River Morphology, Bern.*

——— (1967*a*): Flussmorphologische Studie zum Mäanderproblem, *Geographica Helvetica*, Bd. XXII, no. 2.

——— (1969): Tonerosion, *Eidgen. Techn. Hochschule, EAFV, Interner Bericht*, T2, T3.

ZNAMENSKAYA, N. S. (1962): Calculations of Dimensions and Speed of Shifting of Channel Formations, *Soviet Hydrol.* (*Am. Geophys. Union*), no. 2.

——— (1965): The Use of the Laws of Sediment Dune Movement in Computing Channel Deformation, *Soviet Hydrol.* (*Am. Geophys. Union*), no. 5.

——— (1967): The Analyses and Estimation of Energy Loss, *Proc. Intern. Assoc. Hydr. Res., 12th Congr., Fort Collins.*

——— (1969): Morphological Principle of Modeling of River-Bed Processes, *Proc. 13th Congr. Intern. Assoc. Hydr. Res., Kyoto*, vol. 5.1.

part four

Sediment Transport in Closed Pipes

CONTENTS

INTRODUCTION
LIST OF SYMBOLS

15 FLOW OF SOLID-LIQUID MIXTURES IN PIPES
- 15.1 General Remarks
- 15.2 Determinations of the Head Loss
  - 15.2.1 Settling Mixtures in Horizontal Pipes—Pseudohomogeneous Flow
  - 15.2.2 Settling Mixtures in Horizontal Pipes—Heterogeneous Flow
    - 15.2.2.1 Durand-Condolios Relation
    - 15.2.2.2 Newitt et al. Relation
  - 15.2.3 Settling Mixtures in Horizontal Pipes—Bed Material Transport with a Bed (Deposit)
    - 15.2.3.1 Gibert-Condolios Relation
    - 15.2.3.2 Graf-Acaroglu Relation
    - 15.2.3.3 Further Studies
  - 15.2.4 Settling Mixtures in Inclined Pipes
    - 15.2.4.1 Vertical Pipes
    - 15.2.4.2 Inclined Pipes
  - 15.2.5 Nonsettling Mixtures
    - 15.2.5.1 Purely Viscous Fluids
    - 15.2.5.2 Time-dependent Fluids
    - 15.2.5.3 Viscoelastic Fluids
    - 15.2.5.4 Viscometry
    - 15.2.5.5 Concluding Remarks
- 15.3 Velocity and Concentration Distribution
  - 15.3.1 Settling Mixtures in Horizontal Pipes
  - 15.3.2 Settling Mixtures in Vertical Pipes
  - 15.3.3 Nonsettling Mixtures
- 15.4 Further Remarks
  - 15.4.1 Installations of Solid-Liquid-mixture Systems
  - 15.4.2 Operation of Solid-Liquid-mixture Systems
  - 15.4.3 Economics of Solid-Liquid-mixture Systems
  - 15.4.4 Pneumatic Conveying of Solid Material
  - 15.4.5 Hydraulic Transport in Containers
  - 15.4.6 Sample Calculation

16 MEASURING DEVICES FOR SOLID-LIQUID MIXTURES IN PIPES
- 16.1 Introductory Remarks

16.2 Methods and Equipment
16.2.1 Brine Injection Method
16.2.2 Trajectory Method
16.2.3 Whistle Meter
16.2.4 Contraction Device
16.2.5 Loop System
16.2.6 Modified Venturi Meter
16.2.7 Modified Elbow Meter
16.2.8 Contraction-Expansion Meter
16.2.9 Electrical Systems
16.2.10 Pneumatic Devices
16.3 Closing Remarks
REFERENCES FOR PART FOUR

INTRODUCTION

In recent years the interest in the hydraulic conveyance of solid material has increased. Its economic importance is recognized by different kinds of industries; major installations have been constructed all over the world. Mixtures of liquids, mainly water, and solids, such as sand, gravel, clay, coal, various ores, agricultural products, wood, plastics, etc.,[1] are transported over shorter and longer distances.

Both advantages and disadvantages of hydraulic conveyance systems have been listed by BREBNER (1962), FÜHRBÖTER (1961), CONDOLIOS et al. (1967), and others.

It is the task of Part Four to attempt to answer the following questions: (1) What is the head loss when solid-liquid mixtures are conveyed through pipelines? (2) What are the modes of sediment transport? (3) What are the limiting conditions for these modes? (4) How can one measure the flow of solid-liquid mixtures? The first three questions together with related topics will be discussed in Chap. 15. The last question will be discussed in Chap. 16.

At the end of Part Four a bibliography of the literature used in the preparation of Chaps. 15 and 16 is given. Further literature may be obtained from bibliographies by HASRAJANI (1962), WILLMOTT et al. (1963), and GRAF (1968).

LIST OF SYMBOLS

A area of flow cross section
C concentration
C_D drag coefficient
D pipe diameter
d particle diameter
f Darcy-Weisbach friction coefficient
g gravitational acceleration

[1] Although it is beyond the scope of this book, normal blood from living mammals may be considered as a solid-liquid mixture. It consists of red cells as well as of white cells and platelets which suspend in a plasma, a saline solution of proteins. An introduction to this interesting field of blood rheology was given by WHITMORE (1963).

g_{st}^{*} submerged weight of transported sediment particles for unit time and for unit wetted perimeter
Δh loss of energy or head loss
K coefficient or constant; consistency index of non-newtonian fluids
k coefficient or constant
k_s Nikuradse's equivalent sand roughness
L length
ΔL length, unit distance
n flow behavior index of non-newtonian fluids
N_R Reynolds number
N_R' generalized Reynolds number
P power
Δp pressure drop
Q flow rate
R_h hydraulic radius
S slope of energy grade line
s specific density
t time
u liquid velocity
$\bar{u}$ average liquid flow velocity
u_* shear velocity
V flow velocity
V_C critical velocity
V_H transition velocity, pseudohomogeneous heterogeneous flow
v_{ss} settling velocity
x horizontal distance
y vertical distance
γ unit weight
Δ change in concentration
θ angle of inclination of conduit
ν kinematic viscosity
ρ density
τ_0 shear stress
Φ_A transport parameter
φ dimensionless sediment transport
Ψ_A shear intensity parameter

Subscripts

g gas
l liquid
m mixture
s solid
cr critical conditions

15
Flow of Solid-Liquid Mixtures in Pipes

15.1 GENERAL REMARKS

Consider a horizontal pipe; the bottom of the pipe is covered with a plane, stationary bed of loose, cohesionless, and solid particles of uniform size. The remainder of the pipe cross section is filled with water. If liquid starts to flow, energy dissipation occurs which, in turn, appears in the form of a pressure drop. The loss of energy per unit distance $\Delta h/\Delta L$, often called the *head loss*, is proportional to the flow velocity V^n, or

$$\left(\frac{\Delta h}{\Delta L}\right) \propto V^n \qquad \text{where } n > 1 \tag{15.1}$$

and has been plotted for a specific case in Fig. 15.1. As soon as liquid flows, hydrodynamic forces are exerted on the solid particles of the bed. A further increase of the flow causes an increase in the magnitude of these forces. Hence, for a particular bed, a condition may be obtained where particles in the movable bed are unable to resist the hydrodynamic forces and, thus, get first dislodged and eventually start to move. The condition of the initial movement of some bed particles is determined by observation;

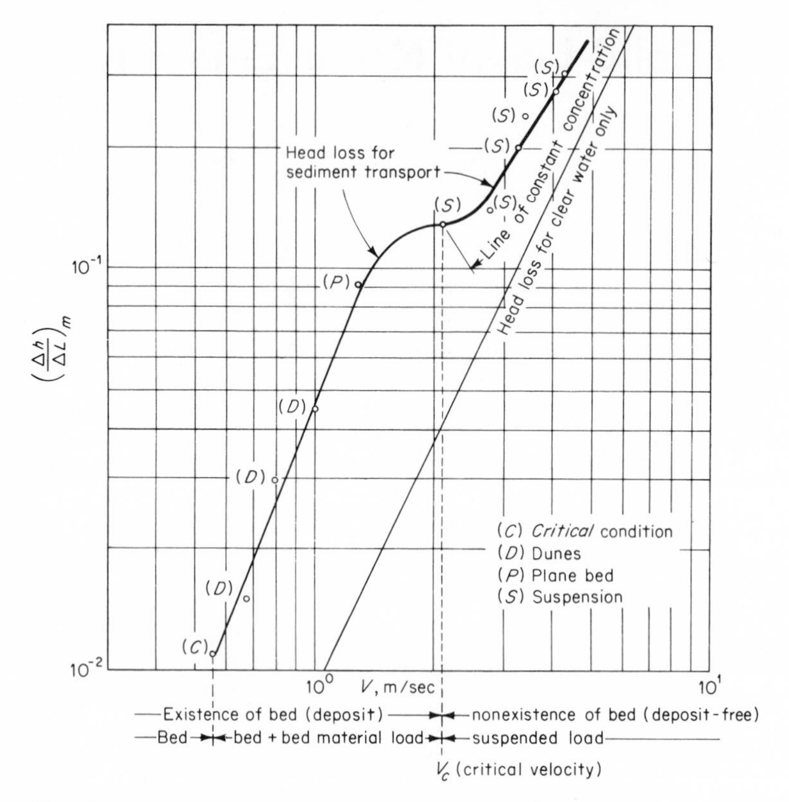

Fig. 15.1 Head loss vs. velocity relationship for closed-conduit flow, for sand with $d = 2.0$ mm. [*After* ACAROGLU *et al.* (*1969*).]

it is referred to as the *critical condition.* An extensive discussion on this and related topics is given in Chap. 6.

In Fig. 15.1 the data with the smallest head loss and velocity represent the critical condition for this particular experiment. The lower leg of the head loss vs. velocity relationship may be explained as follows. As the flow velocity is increased, the head loss increases proportionally, and so does the amount of moving solid particles, here referrred to as *concentration.* A bed (deposition) is always noticeable. THOMAS (1964) and ACAROGLU et al. (1968) observed that the plane bed deforms and forms dunes, but at high velocities these dunes are washed out. The movement of particles is at lower flow velocities restricted to a rather small zone close to the bed; at higher flow velocities the movement extends over the entire pipe cross section.

Between the lower and the upper leg of the relation there appears to be a discontinuity. Although the flow velocity is increased, no noticeable change in the head loss is encountered. It is in this flow region that the stationary bed starts to move, or, in other words, the stationary bed (the deposition) is scoured away.

Along the entire upper leg of the head loss vs. velocity relation, the transported concentration remains now constant. Increase of the flow velocity[1] results in a proportional increase of the head loss. All of the (former bed) particles are moving in suspension. At lower velocities the concentration distribution is such that most of the particles move in the lower half of the pipe cross section. At high velocities the distribution of particles may become rather uniform over the entire cross section.

Within the transition zone there is a velocity below which deposits will occur, but above which no deposit in the pipeline will be encountered. This velocity we shall call the critical velocity V_C; it is indicated in Fig. 15.1. For any design problem, information on this velocity is pertinent. As far as the sediment-transport problem is concerned, the critical velocity helps to define two overall zones. For flow velocities below the critical velocity, deposits occur; the sediment transport is due to an exchange mechanism between the stationary-bed and the moving-bed material. For flow velocities larger than the critical velocity, no deposit will occur; the sediment material is transported as suspended load. If it is uniformly distributed over the entire cross section, we refer to it as *pseudohomogeneous flow*; if nonuniformly distributed, we talk about *heterogeneous flow*.

The foregoing information is qualitatively indicated in Fig. 15.1. For sake of comparison, the head loss vs. velocity relation for water without sediment is also shown. Immediately we recognize what was to be expected, i.e., that the head loss for water and sediment is larger than the head loss for water without sediment. Head loss relations exhibiting a similar trend, i.e., two legs with a transition, are found in ACAROGLU et al. (1969), CONDOLIOS et al. (1963), DURAND (1953), and others. Finally, it is worthwhile to point to the fact that relations, as given in Fig. 15.1, are found in sediment transport through open channels as well as through conduits. A brief discussion on this was presented by ACAROGLU et al. (1969).

Figure 15.1 was obtained for an initial condition of a stationary bed of a given thickness. If the thickness of the stationary bed is changed and the experiment is performed again, another set of data becomes available. After a number of such experiments have been performed, points of equal (transport) concentration are connected; a family of equiconcentration lines is thus obtained. A rather typical plot is shown in Fig. 15.2.

From an observation of Fig. 15.2, it can be deduced that an additive relationship of[2]

$$\left(\frac{\Delta h}{\Delta L}\right)_m = \left(\frac{\Delta h}{\Delta L}\right)_l + \left(\frac{\Delta h}{\Delta L}\right)_s \tag{15.2}$$

[1] For small concentrations the flow velocity and the mixture velocity may be considered identical.

[2] $(\Delta h/\Delta L)_m$ and $(\Delta h/\Delta L)_l$ are the head loss of mixture and clear-liquid flow, respectively, expressed in terms of head of clear water.

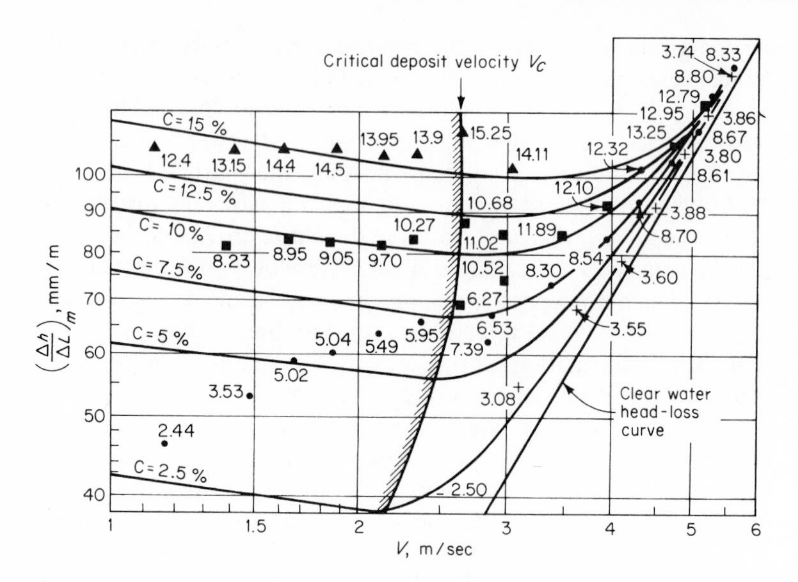

Fig. 15.2 Head loss vs. velocity relationship with equiconcentration lines, for sand graded to 0.44 mm. [*After* CONDOLIOS *et al.* (*1963*).]

represents a first approximation to the entire head loss due to mixture flow $(\Delta h/\Delta L)_m$. The head loss due to pure liquid is $(\Delta h/\Delta L)_l$, and the one due to presence of solid particles is $(\Delta h/\Delta L)_s$. The relationship has been found useful by WILSON (1942), DURAND (1953), NEWITT et al. (1955), and other investigators; it appears that it was suggested originally by BLATCH (1906) and DENT (1939). Although Eq. (15.2) is certainly convenient, it was remarked by NEWITT et al. (1955) that the contributions of liquid and solid particles to the head loss are not necessarily independent. Particles entrained in the flow will more or less alter the flow and have a pronounced effect on the velocity distribution.

Furthermore, both DURAND (1953) and NEWITT et al. (1955), in their pioneering studies on the hydraulic transport of solid particles through pipelines, found that representation of their data is facilitated if the head loss occurring with the mixture $(\Delta h/\Delta L)_m$ and with clear water $(\Delta h/\Delta L)_l$ was related to the concentration as

$$\frac{\left(\dfrac{\Delta h}{\Delta L}\right)_m - \left(\dfrac{\Delta h}{\Delta L}\right)_l}{C\left(\dfrac{\Delta h}{\Delta L}\right)_l} = \varphi \tag{15.3}$$

where φ is a dimensionless sediment-transport parameter to be determined experimentally, and C is the concentration of solids. It should be remarked that many of the data in sediment transport in pipes are represented according to Eq. (15.3).

15.2 DETERMINATIONS OF THE HEAD LOSS

To facilitate our discussion we may consider that solid particles are transported by liquids as either *settling* mixtures or as *nonsettling* mixtures. GOVIER et al. (1961) have classified nonsettling mixtures as those where the settling velocity of the solid particles is below 0.002 to 0.005 fps.[1] It should be remarked that the settling velocity depends not only on the usual parameters but also on the concentration. For a discussion on this topic, see Chap. 4. If the settling velocity is larger than the previously cited value, the mixtures are classed as settling mixtures.

In the following, the first part will be devoted to a discussion of the head loss encountered in transporting settling mixtures through horizontal pipes, and a subsequent part will deal with inclined pipes. The last part will focus on head loss determinations due to nonsettling mixtures.

15.2.1 SETTLING MIXTURES IN HORIZONTAL PIPES — PSEUDOHOMOGENEOUS FLOW

Settling mixtures are those mixtures having solid particles with a settling velocity above 0.002 to 0.005 fps.

However, due to strong turbulence at high flow velocities, small particles—which are classed to belong to a settling mixture—might virtually never get a chance to settle out. The solid particles become fully suspended in the liquid and are almost uniformly distributed over the entire pipe cross section. This kind of mixture flow may be considered to be a *pseudohomogeneous* one.

Invariably it has been suggested that pseudohomogeneous flow mixtures behave newtonian, and that the head loss can be obtained by using the density and viscosity of the mixture. O'BRIEN et al. (1937) were among the earliest investigators to suggest that the head loss of a pseudohomogeneous mixture flowing in a pipe of a diameter D may be given by

$$\left(\frac{\Delta h}{\Delta L}\right)_m \frac{\gamma}{\gamma_m} = f_m \frac{1}{D}\frac{V^2}{2g} \tag{15.4}$$

The left-hand side of the equation is the head loss of mixture flow in terms of column mixture. Equation (15.4) is but the usual Darcy-Weisbach head loss relation. The dimensionless friction factor f_m is the one for mixture flow. It may be obtained from the usual Moody-Diagram[2] by replacing

[1] At low concentrations such settling velocities are obtained with sand particles of $d \leq 20$ to 35 μ, a value also given by DURAND (1953) and NEWITT et al. (1955).

[2] For information on the Moody-Diagram, consult any book on fluid mechanics. The friction factor f is plotted against the Reynolds number VD/ν, with the relative roughness ϵ/D as a parameter.

the water viscosity ν with the mixture viscosity ν_{mixt}, which was discussed in Chap. 5. However, it has been shown with experiments of O'BRIEN et al. (1937), NEWITT et al. (1955), and GRAF et al. (1967) that friction factors of mixture flow and liquid flow may be assumed to be identical without serious error. Equation (15.4) may now be rewritten as

$$\left(\frac{\Delta h}{\Delta L}\right)_m = \left(\frac{\Delta h}{\Delta L}\right)_l [1 + C(s_s - 1)] \tag{15.5}$$

with s_m, the average density of the mixture, given as

$$s_m = \frac{\gamma_m}{\gamma} = [1 + C(s_s - 1)] \tag{15.6}$$

where C is the volume concentration and γ_s is the specific weight of solid particles. Comparing Eq. (15.5) with Eq. (15.3), we obtain

$$\frac{\left(\frac{\Delta h}{\Delta L}\right)_m - \left(\frac{\Delta h}{\Delta L}\right)_l}{C\left(\frac{\Delta h}{\Delta L}\right)_l} = (s_s - 1) = \varphi \tag{15.7}$$

Note that the dimensionless sediment-transport parameter φ is independent of the transport velocity and the particle diameter and dependent solely on the density of the solid particles. Experiments done by NEWITT et al. (1955) gave the following results. Pseudohomogeneous behavior is exhibited by sand (sand B) with $v_{ss} = 0.032$ fps, at all flow velocities, $3 < V < 12$ fps; at extreme flow velocities, certain deviations from Eq. (15.7) can be noticed. A larger sand (sand C) with $v_{ss} = 0.091$ fps exhibits pseudohomogeneous behavior at large velocities of $V \approx 11$ fps. However, a smaller sand (sand A) with $d = 0.0008$ in. was found to exhibit plastic properties and gave values in excess of the one predicted by Eq. (15.7). In NEWITT's et al. (1955) experimental study it was apparent that all the materials behave as pseudohomogeneous suspensions if the flow velocity can be sufficiently high. The latter was the case in some experiments conducted by O'BRIEN et al. (1937) and by BONNINGTON (1959).

As a result of an elaborate investigation with material of different densities, NEWITT et al. (1955) suggest a relation in terms of velocity between pseudohomogeneous and heterogeneous flow, or

$$V_H = \sqrt[3]{1{,}800 g v_{ss} D} \tag{15.8}$$

where V_H = transition velocity
v_{ss} = particle settling velocity
D = pipe diameter

As a practical rule, Eq. (15.8) implies that pseudohomogeneous flow is limited to particles of less than $d = 30\ \mu$.

In summary, we may state that Eq. (15.5) or (15.7) can be applied if the condition given with Eq. (15.8) is fulfilled. It is important to point to the fact that concentrations should be small so that the laws of newtonian flow are not too much violated, and that the friction factors of the liquid and the mixture are reasonably identical.

15.2.2 SETTLING MIXTURES IN HORIZONTAL PIPES — HETEROGENEOUS FLOW

Heterogeneous flow, where a suspension distribution over the cross section is recorded, has its upper limit at the transition velocity V_H, given by Eq. (15.8), and its lower limit at the critical velocity V_C, to be defined in the following. This region of flow was subject to many investigations; most prominent among these are the Durand-Condolios relation and the Newitt et al. relation.

15.2.2.1 Durand-Condolios relation. Upon examination of the experimental data obtained at SOGREAH (Société Grenobloise D'Etudes et D'Applications Hydrauliques) in Grenoble, France, DURAND (1953) suggested to express the "excess" head loss due to the presence of solid particles with Eq. (15.3), or

$$\frac{\left(\frac{\Delta h}{\Delta L}\right)_m - \left(\frac{\Delta h}{\Delta L}\right)_l}{C\left(\frac{\Delta h}{\Delta L}\right)_l} = \varphi \tag{15.3}$$

The dimensionless sediment-transport parameter φ could thus be given as a function of the flow velocity, the density, the size of the sediment particles, and the size of the conduit. After much replotting of the data, DURAND (1953) came up with the following dimensionless groups:

$$\varphi_D = K_D \cdot f_1(s_s - 1) \cdot f_2\left(\frac{V^2}{gD}\right) \cdot f_3\left(\frac{v_{ss}^2}{gd}\right) \tag{15.9}$$

where K_D and the functions f_1, f_2, and f_3 are to be determined from the data. This was done, and the resulting equation for the sediment-transport parameter of Durand-Condolios, φ_D, reads:

$$\varphi_D = K'_D\left[\frac{V^2}{gD(s_s - 1)}\sqrt{C_D}\right]^{-3/2} \tag{15.10}$$

where C_D is the drag coefficient of a particle. Since the drag coefficient C_D is given by[1]

$$C_D = \frac{4}{3}\frac{gd(s_s - 1)}{v_{ss}^2} \tag{15.11}$$

[1] Equation (15.11) is for a spherical particle; for nonspherical ones, consult Chap. 4.

Eq. (15.10) can be rewritten, or

$$\varphi_D = K_D\left[\frac{V^2}{gD(s_s - 1)}\sqrt{\frac{gd(s_s - 1)}{v_{ss}^2}}\right]^{-3/2} \tag{15.12}$$

Note that both Eqs. (15.12) and (15.10) are made up of dimensionless groups, given by Eq. (15.9). According to NEWITT et al. (1955) and BONNINGTON (1959), coefficients K_D and K'_D in previous equations are

$$K_D = 121 \qquad \text{or} \qquad K'_D = 150 \tag{15.13}$$

For sand with $s_s = 2.65$, Eq. (15.12) becomes

$$\varphi_D = K_{DS}\left[\frac{V^2}{gD}\sqrt{\frac{gd}{v_{ss}^2}}\right]^{-3/2} \tag{15.12a}$$

with the numerical value of $K_{DS} = 180$. The foregoing relations are empirical, and DURAND (1953) states that "nothing proves that such a formula is rigorously exact." However, these relations are backed by extensive tests performed at SOGREAH. Test points of 310 runs were obtained, where the pipe diameter D varied from 40 to 580 mm, the (uniform) grain size d varied from 0.2 to 25 mm, and the concentration C varied from 50 to 600 g/l; supplemented by a particle density variation from plastic ($s_s = 1.60$) to corundum ($s_s = 3.95$). Unfortunately, the original test data have not been published, which makes it difficult to compare other data with the SOGREAH data. A good (readily readable) plot of many of these data was given by GIBERT (1960) and is reproduced in Fig. 15.3. Note that this plot includes both heterogeneous flow data and data of the deposition regime for uniform sand and mixtures.

Many researchers have found satisfaction in comparing their experimental data with a Durand-Condolios relation. BONNINGTON (1959) conducted experiments at the British Hydromechanics Research Association. These data together with a variety of other data are plotted according to parameters of the Durand-Condolios relation, and deviation from the Durand-Condolios relation is not excessive. This may be seen in Fig. 15.4. An extension of both parameters of the Durand-Condolios relation was reported by SASSOLI (1963). Testing plastic and metallic pipes and concentrations up to $C = 32$ percent, the general function, as given by Durand-Condolios, was found to be in agreement. Also SMITH (1955) reports that his data, for concentrations up to 33 percent and sand with $0.089 < v_{ss} < 0.39$ fps, obtained with 2- and 3-in. pipelines, can be represented with a Durand-Condolios relation of Eq. (15.12) and for $K_D = 121$. WIEDENROTH (1967) compared his data, for concentrations up to 30 percent and sand with $0.0194 < v_{ss} < 0.340$ m/sec, obtained with a 125-mm conduit, with different available relations, and concluded that the Durand-Condolios

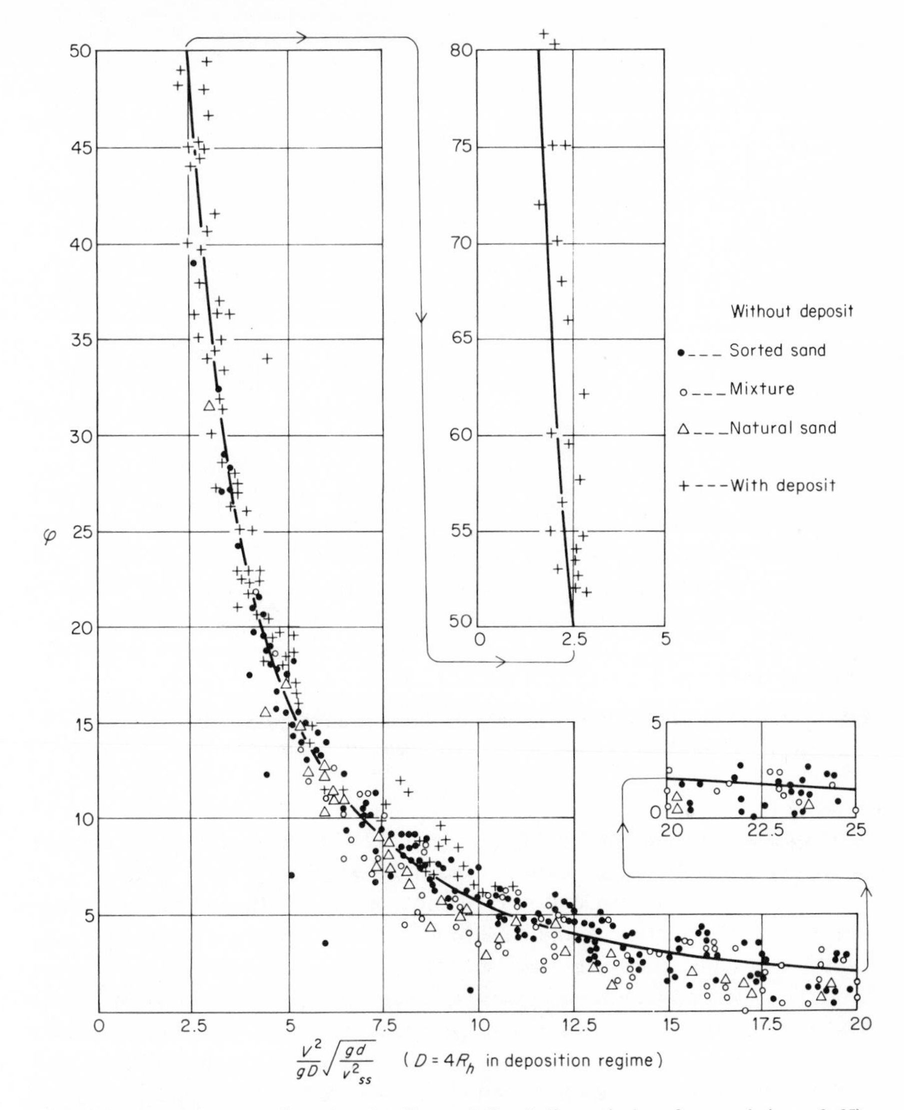

Fig. 15.3 Head loss as given by the Durand-Condolios relation for sand ($s_s = 2.65$); SOGREAH data are used. [*After* GIBERT (*1960*).]

relation explains the data quite well. TODA et al. (1969) find that the Durand-Condolios relations explain their data, but a certain deviation is noticeable.

A Durand-Condolios relation was used by COLORADO SCHOOL OF MINES (1963), WILSON (1965), and ZANDI et al. (1967); their coefficients are different from the coefficients of Eq. (15.13). WILSON (1965) conducted his

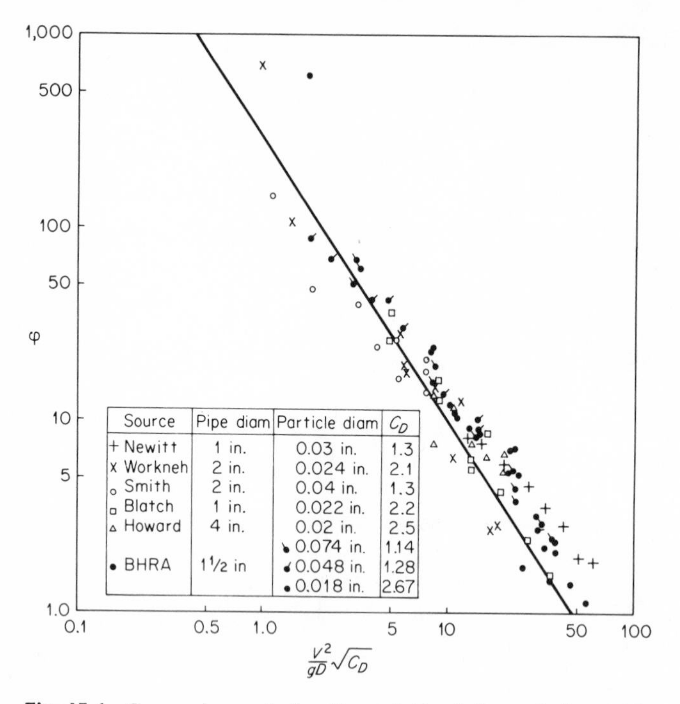

Fig. 15.4 Comparison of the Durand-Condolios relation with data from other sources; for sand $s_s = 2.65$. [*After* BONNINGTON (*1959*).]

experiments with sand and nylon ($s_s = 1.138$) in a square aluminum conduit of 3.691 in., a 2.067-in. aluminum pipe, and a 3.485-in. perspex pipe. It was shown that the Durand-Condolios relation of Eq. (15.10) explains the data, but the best-fit equation would have to have another coefficient and exponent. This is especially true for the nylon data.

The Durand-Condolios relation, given by Eq. (15.10) or (15.12), was first developed for uniform-sized material. However, CONDOLIOS et al. (1963) extended the relation to apply for nonuniform grain-size mixtures. It was suggested that the drag coefficient C_D in Eq. (15.10) be expressed as

$$\sqrt{C_D} = p_1\sqrt{C_{D_1}} + p_2\sqrt{C_{D_2}} + \cdots + p_n\sqrt{C_{D_n}} \tag{15.14}$$

where $p_1, p_2, \ldots, p_n$ are percentages by weight of material with corresponding drag coefficients $C_{D_1}, C_{D_2}, \ldots, C_{D_n}$. BONNINGTON (1959) proposed to define the drag coefficient as one which is equal to the weighted average of the drag coefficients of individual components, or

$$C_D = \frac{\sum C_{D_i} p_i}{\sum p_i} \tag{15.15}$$

Studying the transport of mixed-size sediments, BONNINGTON (1959) noted that the presence of a significant portion of fine material, which apparently exhibits pseudohomogeneous flow, is responsible for a noticeable decrease in the head loss of the mixture.

The Durand-Condolios relationship, given with Eq. (15.10) or (15.12), is applicable for heterogeneous flow. Its lower limit in terms of flow velocity was given by DURAND (1953) and DURAND et al. (1956) as

$$\frac{V_C}{\sqrt{2gD(s_s - 1)}} = F_L \tag{15.16}$$

where V_C is the critical velocity. This velocity separates the *deposit-free* regime from the *deposit* regime and is indicated in Fig. 15.1. Furthermore, it corresponds fairly accurately to the minimum head loss of a mixture flow $(\Delta h/\Delta L)_m$ of a given concentration and, thus, to favorable operating conditions. The variation of the parameter F_L is given in Fig. 15.5. Note that, for any sand with a size of $d > 1$ mm, the concentration but not the grain size has a pronounced effect upon this parameter. Experiments with a sand mixture of average diameter of $0.19 < d < 0.88$ mm in a 300-mm conduit by FÜHRBÖTER (1961) are in general agreement with Eq. (15.16). SPELLS (1955) applied dimensional analysis to data collected from the literature, and suggested an equation which may be compared with Eq. (15.16). An extension of Eq. (15.16) of DURAND was made by SINCLAIR (1962). The data were obtained for sand-water, iron-kerosine, and coal-water mixtures at concentrations up to 20 percent flowing in 0.5-, 0.75-, and 1.0-in. pipes, and the following relation was proposed

$$\frac{V_C}{\sqrt{gd_{85}(s_s - 1)^{0.8}}} = f_S\left(\frac{d_{85}}{D}\right) \tag{15.16a}$$

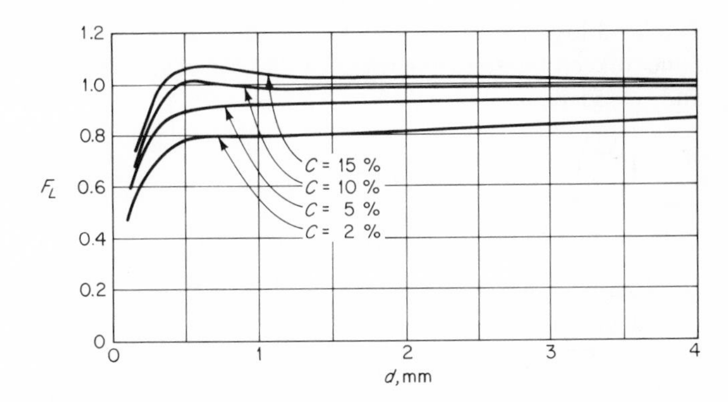

Fig. 15.5 F_L value vs. particle diameter, concentration as parameter. [*After* DURAND *et al.* (*1956*).]

A relation similar to Eq. (15.16*a*) was investigated by KRIEGEL et al. (1966). An extensive study on the critical velocity was recently carried out by GRAF et al. (1970). It was concluded that the critical velocity may safely be obtained for design purposes from the relation given with Eq. (15.16) and with Fig. 15.5.

Furthermore, a word of caution is perhaps in place. The critical velocity equation separates two regimes with a unique value. It must, however, be realized that transition between these regimes is gradual and depends very much on the entire pump-pipe system, and reproducibility is not always warranted. Therefore, BONNINGTON (1961) suggested to use a "working" limited velocity, which is greater than the one obtained from Eq. (15.16).

15.2.2.2 Newitt et al. relation. NEWITT et al. (1955) find it necessary to subdivide the heterogeneous flow. It is suggested that at higher velocities the solid particles travel with a velocity only slightly less than the liquid itself; the solid particles are fully suspended. At lower velocities, solid particles accumulate at the bottom section of the pipe. The particles saltate or form a bed which, as a whole, slides forward at velocities considerably lower than flow velocities. Thus NEWITT et al. (1955) consider a "suspension" flow and "flow with a moving bed." The head loss for either of these can be expressed by Eq. (15.2), or

$$\left(\frac{\Delta h}{\Delta L}\right)_m = \left(\frac{\Delta h}{\Delta L}\right)_l + \left(\frac{\Delta h}{\Delta L}\right)_s \tag{15.2}$$

with head loss for pure liquid given by the Darcy-Weisbach relation, or

$$\left(\frac{\Delta h}{\Delta L}\right)_l = f\frac{1}{D}\frac{V^2}{2g} \tag{15.17}$$

The head loss due to the presence of solid, $(\Delta h/\Delta L)_s$, is related to the work done to keep the solid particles, for the case of *suspension* flow, in suspension, and was given as

$$\left(\frac{\Delta h}{\Delta L}\right)_s = k_N C(s_s - 1)\frac{v_{ss}}{V} \tag{15.18}$$

In terms of Eq. (15.3), Eq. (15.18) becomes

$$\frac{\left(\frac{\Delta h}{\Delta L}\right)_m - \left(\frac{\Delta h}{\Delta L}\right)_l}{C\left(\frac{\Delta h}{\Delta L}\right)_l} = \varphi_N = K_N(s_s - 1)\frac{v_{ss}}{V}\frac{gD}{V^2} \tag{15.19}$$

where k_N and K_N are constants to be determined by experiments. This equation applies also to sand mixtures if an equivalent settling velocity for the mixture is calculated.

For the *flow with a moving bed* model the head loss $(\Delta h/\Delta L)_s$ is related to the work done in pushing the particles along the pipe, and was given as

$$\left(\frac{\Delta h}{\Delta L}\right)_s = k'_N \gamma C(s_s - 1) \tag{15.20}$$

In terms of Eq. (15.3) Eq. (15.20) becomes

$$\frac{\left(\frac{\Delta h}{\Delta L}\right)_m - \left(\frac{\Delta h}{\Delta L}\right)_l}{C\left(\frac{\Delta h}{\Delta L}\right)_l} = \varphi'_N = K'_N(s_s - 1)\frac{gD}{V^2} \tag{15.21}$$

where k'_N and K'_N are determined experimentally.

Experiments conducted by Newitt et al. (1955), used for evaluation of the constants in Eqs. (15.19) and (15.21), covered a wide variety of material, with $0.032 < v_{ss} < 0.83$ fps, with $1.18 < s_s < 4.60$ (perspex to zircon sand), and with $0 < C < 37$, but are limited to a single pipe size of $D = 1$ in. The following numerical values for the constants are thus obtained:

$$K_N = 1{,}100 \qquad K'_N = 66 \tag{15.22}$$

Of the two relations suggested by Newitt et al. (1955), Eq. (15.19) can be compared with the Durand-Condolios relation, given with Eq. (15.12). In either of the two equations, φ is proportional to the inverse of the flow velocity cubed. However, as far as the pipe diameter and the particle dimension are concerned, the two equations differ, but their differences are minor ones.

When equating Eqs. (15.19) and (15.21) with the appropriate coefficients given by Eq. (15.22), an expression in terms of flow velocity is obtained, or

$$V_B = 17v_{ss} \tag{15.23}$$

This velocity V_B represents the transition from *suspension* to *flow with a moving* bed. A similar relation was given by Thomas (1962). The transition velocity V_H between heterogeneous and pseudohomogeneous flow was given earlier as

$$V_H = \sqrt[3]{1{,}800gDv_{ss}} \tag{15.8}$$

Both Eqs. (15.23) and (15.8) were given by Newitt et al. (1955). However, the relation for the critical velocity V_C was adopted from Durand (1959), or

$$\frac{V_C}{\sqrt{2gD(s_s - 1)}} = F_L \tag{15.16}$$

It is not uncommon to find in the literature some confusion between V_C and V_B. This apparently results from the definition of *deposition*. Some investigators refer to deposition when sediment transport takes place as a moving bed, others refer to deposition—the author takes this stand—when a stationary bed exists already.

The three transition velocities V_C, V_B, and V_H are shown in Fig. 15.6, which illustrates the variation of the flow characteristics with particle size and flow velocity in a 1-in. pipe.

It is particularly NEWITT's et al. (1955) relation for the case of suspension, given with Eqs. (15.18) and (15.19), that has received attention by other researchers. Equations (15.2), (15.17), and (15.18) of the Newitt et al. relation have already been put forward by WILSON (1942). It was contested that a relationship exists between the head loss $(\Delta h/\Delta L)_s$ and the work done by the water on the particles to keep these in suspension. WILSON (1942) also presented an equation for the critical velocity V_C as follows:

$$V_C = k_W \sqrt[3]{f^{-1} C v_{ss} D g} \tag{15.24}$$

However, WILSON (1942) was unable to obtain consistent numerical values for the constant k_N in Eq. (15.18) and the constant k_W in Eq. (15.24).

The Newitt et al. relation, given with Eqs. (15.2), (15.17), and (15.18), was also used by FÜHRBÖTER (1961). The dimensionless sediment-transport coefficient was given as

$$\varphi_F = \frac{2s_k}{fV}\left(\frac{gD}{V^2}\right) \tag{15.25}$$

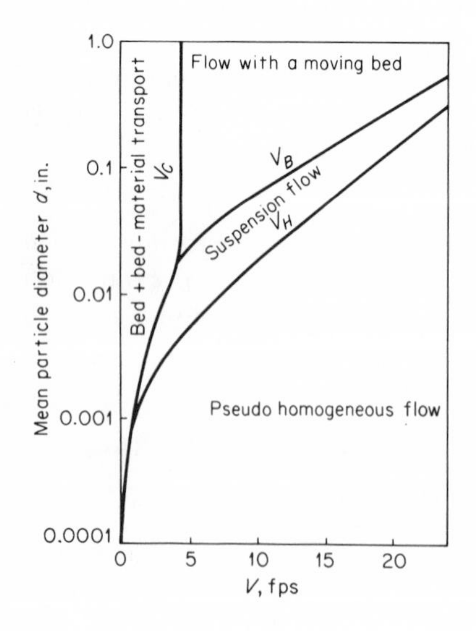

Fig. 15.6 Flow regimes and transition velocities, for $D = 1$ in. [*After* NEWITT *et al. (1955)*.]

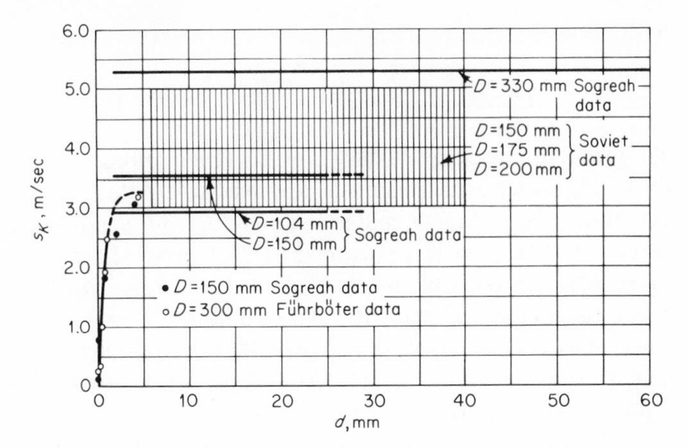

Fig. 15.7 Relationship of the sediment constant s_k. [*After* FÜHRBÖTER (*1961*).]

where s_k is a sediment constant given by

$$s_k = k_F(s_s - 1)v_{ss} \tag{15.26}$$

The sediment constant s_k is supposed to combine all properties of the sediment which are important for the transport through conduits, and is to be determined experimentally. FÜHRBÖTER (1961) attempted to show such a relation with his data and with data from the SOGREAH and Soviet experiments, which are reproduced in Fig. 15.7. Although a certain trend is noticeable, no definite relationship could be established. This is also the conclusion of a study by KRIEGEL et al. (1966) and WIEDENROTH (1967).

Another relation of the type developed by NEWITT et al. (1955) was proposed by BARTH (1960). A model for the transport of solid particles in liquids and gases is discussed, but KRIEGEL et al. (1966) have shown beyond doubt that BARTH's (1960) head loss equation has the form of the sum of Eqs. (15.18) and (15.17).

An improvement of the Newitt et al. relation was proposed by KRIEGEL et al. (1966), who implied that a relation given by Eq. (15.18) or (15.19) must remain unsatisfactory as long as the settling velocity v_{ss} in a quiescent medium is used. The difficulty in expressing particle settling velocities in a turbulent medium was discussed in Sec. 4.2.4.6. KRIEGEL et al. (1966) suggest a relation between the particle settling velocity in quiescent and turbulent media, which is based on rather strong assumptions. Using this relation, the head loss due to solid particles is given as

$$\left(\frac{\Delta h}{\Delta L}\right)_s = f_K \frac{1}{D} \frac{V^2}{2g} \tag{15.27}$$

with f_K as

$$f_K = 0.282C(s_s - 1)\left(\frac{v_{ss}^3}{g\nu}\right)^{1/3}\left(\frac{gD}{V^2}\right)^{4/3} \tag{15.28}$$

To check this relation KRIEGEL et al. (1966) produced an extensive set of data. In order to supplement the available data, it was felt to limit the experiment to sediment material different from sand and to sediment mixtures. The tests were performed in 26.2- and 53.5-mm pipes for material of $0.0375 < v_{ss} < 1.07$ m/sec and of $1.38 < s_s < 4.62$; the concentration was usually below 22 percent. These data together with those of FÜHRBÖTER (1961) and DURAND (1953) were used to develop the constant in Eq. (15.28). The data of NEWITT et al. (1955) and of WIEDENROTH (1967) do not check with the relation given by Eq. (15.28).

Throughout the literature of sediment transport in pipes, the Durand-Condolios and the Newitt et al. relations have been discussed, checked, and recommended for use by various independently working researchers. Although NEWITT et al. (1955) were rather successful in formulating a semi-theoretical approach, DURAND (1953) and CONDOLIOS et al. (1963) made use of dimensional analysis. However, the vast amount of data obtained at SOGREAH and at other laboratories justify the Durand-Condolios relation and give it, at the present, an advantage over the Newitt et al. relation, which is based only on limited experiments in a 1-in. pipe. WORSTER et al. (1955) have made pointed remarks on the limitations of the Newitt et al. relation when compared with the Durand-Condolios relation. Furthermore, the Durand-Condolios relation seems to fit the data for the entire range of heterogeneous flow.

15.2.3 SETTLING MIXTURES IN HORIZONTAL PIPES — BED MATERIAL TRANSPORT WITH A BED (DEPOSIT)

For flow velocities below the critical velocity V_C, deposition occurs on the bottom of the horizontal pipe. Sediment transport takes place as a result of an exchange mechanism between the bed (deposit) and the moving-bed material. As the flow velocity is reduced, deposition builds up and the exchange mechanism becomes less effective. Eventually this may lead, provided sufficient solid matter is in the system, to the clogging or plugging of the pipeline. In many applications of solid-liquid conveyance, a deposition of solid material is not desired. However, at certain times and at certain locations, a conveyance system may be such that some solid particles remain in deposition while others are transported through the conduit. The system may be in equilibrium, and the deposited material will not be scoured away nor will plugging of the conduit occur. A few investigations were concerned with the transport of a solid-liquid mixture over a bed (deposit).

15.2.3.1 Gibert–Condolios relation. The possibility of extending the Durand-Condolios relation, given with Eq. (15.10) or (15.12), to flow of solid-liquid mixtures with deposits is discussed by GIBERT (1960) and CONDOLIOS et al. (1960). The data used to establish this relation have not been reported; apparently, only data with sand were obtained. CONDOLIOS et al. (1963) report that the top of the deposit was approximately flat and the deposit height was measured. GIBERT (1960) has plotted the data, which are shown in Fig. 15.3. The data can be represented by a Durand-Condolios relation if the following arrangements are made. The flow velocity V is given by the ratio of flow discharge to the free-flow area over which the flow takes place. Increase in deposition reduces the free-flow area. Likewise, the hydraulic radius R_h, which replaces the pipe diameter D, is the hydraulic radius of the free-flow cross-sectional area. Note that, if the deposit height goes toward zero, the hydraulic radius becomes $R_h = D/4$. With this in mind, Eq. (15.10) reads

$$\varphi_D = K'_D\left[\frac{V^2}{g4R_h(s_s - 1)}\sqrt{C_D}\right]^{-3/2} \tag{15.29}$$

As is shown in Fig. 15.3, Eqs. (15.29) and (15.10) represent both data with and without deposit and have the same K'_D value, given with Eq. (15.13). The dimensionless sediment-transport parameter φ was given by Eq. (15.3), or

$$\frac{\left(\dfrac{\Delta h}{\Delta L}\right)_m - \left(\dfrac{\Delta h}{\Delta L}\right)_l}{C\left(\dfrac{\Delta h}{\Delta L}\right)_l} = \varphi \tag{15.3}$$

In the nondeposit regime the head loss due to water, $(\Delta h/\Delta L)_l$, could actually be measured. In the deposit regime this is not—at least easily—possible. GIBERT (1960) suggests calculation of this value with the Darcy-Weisbach formula, or

$$\left(\frac{\Delta h}{\Delta L}\right)_l = f\frac{1}{4R_h}\frac{V^2}{2g} \tag{15.17a}$$

where the friction factor f is given as

$$f = f\left(\frac{V4R_h}{\nu}, \frac{\epsilon}{R_h}\right) \tag{15.30}$$

Unfortunately, GIBERT (1960) and CONDOLIOS et al. (1963) refer to smooth-pipe flow only where the friction factor f is given by

$$f = f\left(\frac{V4R_h}{\nu}\right) \tag{15.30a}$$

For both rough-pipe flow and in the transition, the entire relation, as given with Eq. (15.30), ought to be considered. Its difficulty lies with the proper evaluation of the relative roughness value ϵ/D. Recently WILSON (1970) presented some interesting ideas along this line.

The Gibert-Condolios relation was found useful by WILSON (1965) to explain his set of data. WILSON (1965) finds it necessary to replace the hydraulic radius R_h, as it appears in Eq. (15.29), by the fraction of the hydraulic radius associated with the deposit R_{hb}. Furthermore, the best-fit line through his data shows a different exponent and coefficient in Eq. (15.29).

15.2.3.2 Graf-Acaroglu relation. The relation, proposed by GRAF et al. (1968), was developed for the prediction of the total sediment load in closed conduits as well as in open channels.

Consider a plane and stationary bed of loose and solid particles of uniform size and a fluid flowing over it; assume the cross-sectional area to be constant. If the fluid begins to flow, shearing stresses are created at the wetted perimeter of the conveyance system. An increase in the flow is responsible for an increase of the shearing stresses. Eventually a condition is reached at which shear stresses cannot be resisted by the movable particles of the bed and these particles start to move. This condition is referred to as *critical condition* or *initial movement of the bed*. A consequence of this is that further increase in the flow will cause more particles to be removed from the bed, and conceivably all the particles in the bed can become entrained into the fluid body if the flow is adequately increased.

The foregoing is a brief account of the sediment-transport phenomenon which can be readily observed under natural and laboratory conditions. To analyze the problem, the forces acting on the particle must be considered. These forces are the drag force and the submerged weight. The condition of incipient motion can be given, according to Eq. (6.2), as

$$W \tan \varphi_r = F_D \tag{15.31}$$

where φ_r is the angle of repose. The submerged weight of the particle is given as

$$W = k_3(\rho_s - \rho)gd^3 \tag{15.32}$$

where k_3 is a shape factor. The drag force is expressed as

$$F_D = C_D k_1 d^2 \frac{\rho u_b^{\,2}}{2} \tag{15.33}$$

where the drag coefficient C_D is a function of the particle Reynolds number and of a shape factor k_2, or

$$C_D = f_1\left(\frac{du_b}{\nu}, k_2\right) \tag{15.34}$$

and k_1 is another shape factor. The bed velocity u_b for turbulent flow with solid-liquid mixtures can be expressed as[1]

$$u_b = u_* f_2\left(\frac{k_s u_*}{\nu}, \frac{y}{k_s}, c\right) \tag{15.35}$$

where c = concentration of solid particles
k_s = Nikuradse's equivalent sand roughness
u_* = shear or friction velocity

The shear or friction velocity is given as

$$u_* = \sqrt{\frac{\tau_0}{\rho}} = \sqrt{gR_h S} \tag{15.36}$$

where τ_0 = shearing stress
S = slope of the energy grade line
R_h = hydraulic radius defined by the ratio of the cross-sectional area in which the flow takes place to the wetted perimeter

The ratio y/k_s can be expressed as

$$\frac{y}{k_s} = \frac{k_4 d}{d} = k_4 \tag{15.37}$$

if the following assumptions are made. Roughness dimensions and particle diameter are equal, and the bottom velocity u_b occurs at a distance of $y = k_4 d$.

Equation (15.31) can now be rewritten as

$$k_3(\rho_s - \rho)gd^3 \tan \varphi_r$$
$$= f_1\left(\frac{du_b}{\nu}, k_2\right) k_1 d^2 \frac{\rho}{2}\left[gR_h S\, f_3\left(\frac{du_*}{\nu}, k_4, c\right)\right]$$

or, rearranging,

$$\frac{(s_s - 1)d}{SR_h} = \frac{1}{2}\frac{k_1}{k_5} f_1\left(\frac{du_b}{\nu}, k_2\right) f_3\left(\frac{du_*}{\nu}, k_4, c\right) \tag{15.38}$$

where $k_5 = k_3 \tan \varphi_r$. It is safe to assume a relationship between the shear velocity u_* and the bottom velocity u_b. Furthermore, after applying Eq. (15.38) to a particular sediment particle, we may write

$$\frac{(s_s - 1)d}{SR_h} = f_4\left(\frac{du_*}{\nu}, c\right) \tag{15.39}$$

The argument on the left-hand side of Eq. (15.39) was called the *shear intensity parameter* Ψ_A, or

$$\Psi_A = \frac{(s_s - 1)d}{SR_h} \tag{15.39a}$$

[1] Experimental data in channels by EINSTEIN et al. (1955), and in conduits by NEWITT et al. (1955), and ACAROGLU (1968), satisfy this relation.

It is similar to the intensity of shear on particle Ψ', introduced by EINSTEIN (1942, 1950) for the free surface flow. However, EINSTEIN (1950) used only the part of the hydraulic radius that refers to the bed, whereas GRAF et al. (1968) use the entire hydraulic radius. This becomes necessary because it is very difficult, both experimentally and theoretically, to subdivide the hydraulic radius in a pipe with a deposited sediment bed. Equation (15.39*a*) is identical with the inverse of the dimensionless shear given by SHIELDS (1936), which was discussed in detail in Sec. 6.3.2.2.

Equations (15.39) and (15.39*a*) give information on the intensity of shear for a variety of flow conditions. This is illustrated in Fig. 15.8 where the shear intensity parameter is plotted against the particle Reynolds number, with concentration c as a parameter. Two regions can be distinguished. If $\Psi'_A > \Psi'_{A_{cr}}$ or $\tau_0 < (\tau_0)_{cr}$, no movement of particles takes place; if $\Psi'_A < \Psi'_{A_{cr}}$ or $\tau_0 > (\tau_0)_{cr}$, particle movement will take place. The zone of separation between the two regions is the *critical condition* given by $\Psi'_A = \Psi'_{A_{cr}}$ or $\tau_0 = (\tau_0)_{cr}$.

From the previous discussion it may be expected that the shear intensity parameter Ψ'_A can be used as a sediment-transport criterion. This shall be undertaken next.

The available power P is given by

$$P = \gamma Q H \tag{15.40}$$

where H is the total head available at any particular cross section. The power P^* per unit length and the per unit wetted perimeter p may be given by

$$P^* = \gamma \bar{u} A S \frac{1}{p} \tag{15.41}$$

where, owing to continuity, $Q = \bar{u}A$, and S is the energy grade-line slope.

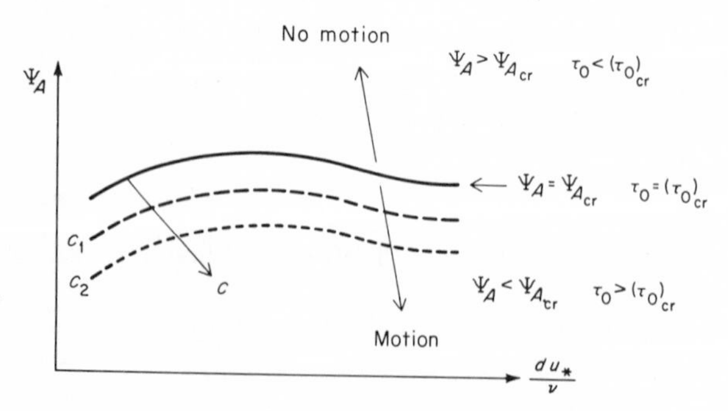

Fig. 15.8 Representation of Eq. (15.39). [*After* GRAF *et al.* (*1968*).]

Substituting the hydraulic radius R_h for A/p, and replacing τ_0 for $\gamma R_h S$, Eq. (15.41) becomes

$$P^* = \tau_0 \bar{u} \tag{15.42}$$

Only part of this power (work rate) P^* is used to transport solid particles if the available shear stress τ_0 exceeds the critical shear stress $(\tau_0)_{cr}$. Hence the work rate used for the sediment transport is given as

$$P_s^* = P^* b \tag{15.43}$$

where b is a factor expressing sediment-transport ability within a given flow. Determination of this factor b by a rational approach is presently impossible. However, a functional relationship with the shear intensity parameter Ψ_A may very well have some merit; thus we write

$$b = f_5(\Psi_A) \tag{15.44}$$

Further, the rate of work done to transport the sediment must be related to the transport rate of the sediment. The submerged weight of the transported particles per unit time and per unit wetted perimeter is given by

$$g_{st}^* = (\rho_s - \rho) g C V A \frac{1}{p} \tag{15.45}$$

BAGNOLD (1966) considered this equation to

> . . . have dimensions and quality of work rates, being the products of weight force per unit bed area times velocity. As they stand, however, they are not in fact work rates, for the stress is not in the same direction as the velocity of its action.

BAGNOLD (1966) suggested that the "transport rates become actual work rates when multiplied by notional conversion factors." GRAF et al. (1968) proposed to use as the conversion factor the ratio of the average liquid velocity to the particle settling velocity. Thus the sediment-transport (work) rate is given by

$$g_{st}^* \frac{\bar{u}}{v_{ss}} \tag{15.46}$$

In an equilibrium the work rate used for the sediment transport P_s^*, given with Eq. (15.43), can be related to the sediment-transport rate, or

$$P^* b = g_{st}^* \frac{\bar{u}}{v_{ss}} \tag{15.47}$$

Substituting Eqs. (15.42) and (15.44) into Eq. (15.47), we get

$$g_{st}^* = \tau_0 v_{ss} f_5(\Psi'_A) \tag{15.48}$$

Provided the value of $b = f_5(\Psi'_A)$ can be evaluated, Eq. (15.48) gives the total transport rate of a sediment material described by its settling velocity v_{ss} at a given shear stress τ_0 in a particular conveyance system.

Equation (15.48) is made dimensionless by dividing both sides by $(\tau_0)_{cr}\sqrt{(\tau_0)_{cr}/\rho}$, such that

$$\frac{g_{st}^*}{(\tau_0)_{cr}\sqrt{(\tau_0)_{cr}/\rho}} = \frac{\tau_0}{(\tau_0)_{cr}} \frac{v_{ss}}{\sqrt{(\tau_0)_{cr}/\rho}} f_5(\Psi'_A) \tag{15.49}$$

The following substitution can be made: g_{st}^* can be expressed with Eq. (15.45); the ratio $\tau_0/(\tau_0)_{cr}$ is a function of Ψ'_A; the critical shear value $(\tau_0)_{cr}$ can be expressed, for example, with the value due to SHIELDS (1936), or $(\tau_0)_{cr} = k_6(\rho_s - \rho)gd$, and the settling velocity is

$$v_{ss} = \sqrt{\frac{4(\rho_s - \rho)gd}{3\rho C_D}}$$

Equation (15.49) becomes

$$\frac{(\rho_s - \rho)gCVR_h}{[k_6(\rho_s - \rho)gd]\sqrt{k_6(\rho_s - \rho)gd(1/\rho)}} = f_6(\Psi'_A)\frac{\sqrt{[4(\rho_s - \rho)gd]/(3\rho C_D)}}{\sqrt{k_6(\rho_s - \rho)gd(1/\rho)}} f_5(\Psi'_A) \tag{15.50}$$

Rearrangement of Eq. (15.50) yields

$$\frac{CVR_h}{\sqrt{(s_s - 1)gd^3}} = f_5(\Psi'_A)f_6(\Psi'_A)\sqrt{\frac{4}{3C_D}}\, k_6 \tag{15.51}$$

Since the shear intensity parameter Ψ'_A includes the various particle coefficients such as C_D and k_6 to some extent, we may safely write

$$\frac{CVR_h}{\sqrt{(s_s - 1)gd^3}} = f_7(\Psi'_A) \tag{15.52}$$

The left-hand side of this equation may be considered a dimensionless *transport parameter* given the symbol Φ_A, so that

$$\Phi_A = f_7(\Psi'_A) \tag{15.53}$$

This functional relationship between the transport parameter Φ_A and the shear parameter Ψ'_A suggests a physical model for sediment transport in conveyance systems. The exact form of the function has to be determined by using the available data obtained from model and prototype conveyance

Table 15.1 Data used to establish Eq. (15.54) [*after* GRAF *et al.* (*1968*)]

Source	*Conveyance system*	*No. points*	*d, mm*	*Material*	s_s
Ismail (1952)	Square steel conduit (0.27 × 0.076 m)	32	0.091	Sand	2.650
		28	0.147	Sand	2.650
Wilson (1965)	Square aluminium conduit (0.0937 × 0.0937 m)	67	0.710	Sand	2.650
Acaroglu (1968)	Aluminium pipe ($D = 0.076$ m)	111	2.000	Sand	2.670
		112	2.780	Sand	2.670
Gilbert (1914)	Laboratory flumes (various sizes)	59	0.305	Sand	2.690
		197	0.375	Sand	2.690
		158	0.505	Sand	2.690
		115	0.787	Sand	2.690
		48	1.710	Sand	2.690
Guy, Simons, Richardson (1966)	Laboratory flumes (2.41 × 0.61 m and 0.61 × 0.762 m)	39	0.450	Sand	2.650
		35	0.280	Sand	2.650
		33	0.190	Sand	2.650
		19	0.270	Sand	2.650
		36	0.930	Sand	2.650
		29	0.320	Sand	2.650
		13	0.330	Sand	2.650
		15	0.330	Sand	2.650
Ansley (1963)	Laboratory flume (0.153 × 0.153 m)	26	0.223	Sand	2.650
Einstein (1944)	(Natural) river	81	0.900	Sand	2.670

systems. The applicability of Eq. (15.53) depends on the quality and the range of the available data. GRAF et al. (1968) used total-load data obtained in open channels, in rivers, and in closed conduits. These data are summarized in Table 15.1. All of these data were used in a regression analysis and gave the following result:

$$\Phi_A = 10.39(\Psi'_A)^{-2.52} \tag{15.54}$$

Equation (15.54) and its significance to open-channel flow were briefly discussed in Sec. 9.3.3. There it was shown how the open-channel data agree with Eq. (15.54); this can be seen in Fig. 9.11. In Fig. 15.9 the relationship of Eq. (15.54) is compared with closed-conduit data.[1] In spite of the scatter it can be said that the function given with Eq. (15.54) is a representative one; it explains 59 percent of the data. Realizing the difficulties involved in measuring and defining the quantities, this is a good result. A further discussion of the relationship and the data was given by GRAF et al. (1968).

[1] Note that all closed-conduit data used for GRAF's et al. (1968) study are data where deposition occurs. Only such data can be compared with open-channel data.

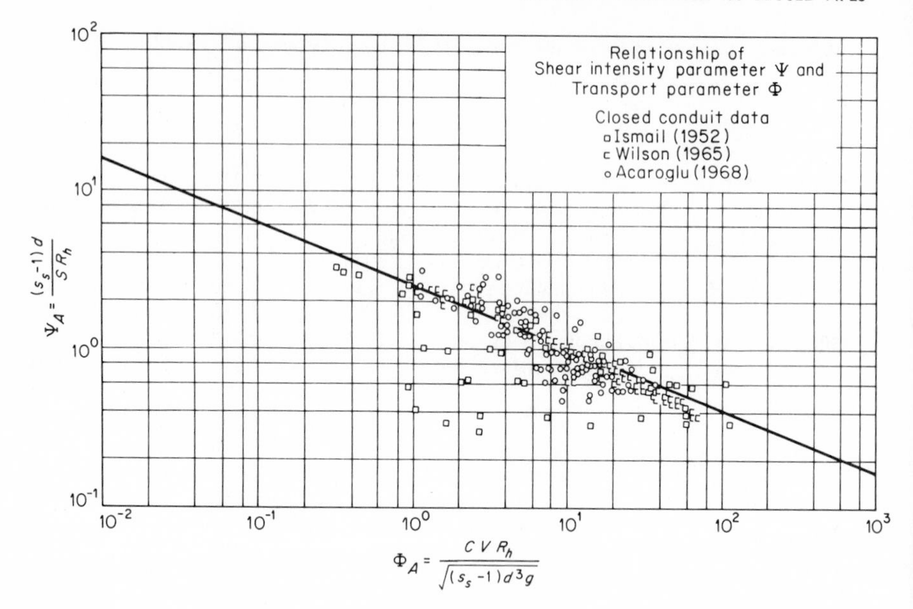

Fig. 15.9 Φ_A vs. Ψ_A relation, with closed-conduit data only. [*After* GRAF *et al.* (*1968*).]

A relationship similar to Eq. (15.54) was suggested for the determination of bedload (see Sec. 7.4) and for the determination of the total load (see Sec. 9.2.1) by EINSTEIN (1950). These relations are, however, for flow in open channels. SAKTHIVADIVEL (1967) has shown that Einstein's relation can be used, with some success, in closed-conduit flow; but only a limited amount of data were used. WILSON (1966) used the Φ vs. Ψ' relation as proposed by EINSTEIN (1950) and obtained the best-fit line for data with sand. In the latter investigation only this part of the hydraulic radius is used that is associated with the deposit. Both SAKTHIVADIVEL (1967) and WILSON (1966) used the data reported by WILSON (1965), some of which are briefly described in Table 15.1.

15.2.3.3 Further studies. GIBERT (1960, p. 342) has postulated that in the deposition regime—for $V < V_C$—a relationship does exist, such as

$$\frac{V}{\sqrt{4gR_h}} = \frac{V_C}{\sqrt{gD}} \tag{15.55}$$

where R_h is the hydraulic radius of the free-flow area (this is the entire pipe cross section reduced by the deposition area). For a given flow velocity V, pipe size D, and critical velocity V_C, the corresponding hydraulic radius R_h is determined. Knowledge of the hydraulic radius R_h and of the pipe size

D permits calculation of the depth of deposition in the pipe.[1] The relation given with Eq. (15.55) presents a necessary information for the head loss calculations in the deposition regimes if the procedures of Gibert-Condolios or Graf-Acaroglu are followed.

Besides the *Gibert-Condolios* and the *Graf-Acaroglu* relations for the solid-liquid transport with a deposition, there are a few other investigations to be found in the literature. However, none of these relations has an extensive experimental backing.

CRAVEN (1952) studied the problem by application of dimensional procedures and by determination of the functional relationship. It was proposed that the following relation is sufficient to describe the sediment-transport problem, or

$$\left(\frac{\Delta h}{\Delta L}\right)_m = f\left(\frac{Q_{st}}{D^{2.5}\sqrt{g}}, \frac{\gamma_s - \gamma}{\gamma}, \frac{y_s}{D}, \frac{k_s}{D}, N_R\right)$$

where Q_{st} is the volume rate of sand flow and y_s is the depth of sand in the conduit. However, only a relatively small amount of data was used to establish the function and, thus, this study remains of limited use. The work by CRAVEN (1952) was extended to free surface flow in conduits by AMBROSE (1952). Dimensional considerations to correlate data were also discussed by CHAMBERLAIN et al. (1960).

Another investigation by SHOOK et al. (1965) attempts to represent the data with a theory developed by BAGNOLD (1956) for dilatant suspensions. A simplified version of the concept of BAGNOLD (1956) was given more recently by BAGNOLD (1966), and was discussed earlier in Sec. 9.2.3. SHOOK et al. (1965) developed an experimental setup and produced their own data. With these data ranging from sand to lead with $0.054 < v_{ss} < 2.52$ fps, it was suggested that the empirical constant in BAGNOLD's (1956) relation could be explained.

Recently CARSTENS (1969) has advanced a theory for heterogeneous flow. Some of the theoretical indications are, however, in conflict with accepted concepts and, furthermore, lack any kind of experimental evidence. For example, it was assumed that heterogeneous flow is always accompanied with a bed (deposit); the head loss due to solids, $(\Delta h/\Delta L)_s$, is largely extended as a result of the bed itself. Furthermore, it was argued that at the occurrence of the critical velocity the bed deposit attains more than one-third of the pipe diameter.

15.2.4 SETTLING MIXTURES IN INCLINED PIPES

It facilitates discussion to consider first the case of vertical pipes and, subsequently, the general case of inclined pipes.

[1] It should be mentioned that GIBERT (1960) warns that Eq. (15.55) does not make sense(!) if the deposition becomes larger than the pipe radius.

15.2.4.1 Vertical pipes. According to studies by DURAND (1953*a*), WORSTER et al. (1955), and CONDOLIOS et al. (1963), it was suggested that the pressure drop of a solid-liquid mixture is the same as for water flowing at the same average flow velocity if allowance is made for the static head of the solids in the pipe. The head loss of the solid-liquid mixture may be expressed as follows, recalling Eq. (15.2), or

$$\left(\frac{\Delta h}{\Delta L}\right)_m = \left(\frac{\Delta h}{\Delta L}\right)_l + \left(\frac{\Delta h}{\Delta L}\right)_s \tag{15.2}$$

where the first term on the right-hand side is

$$\left(\frac{\Delta h}{\Delta L}\right)_l = f\frac{1}{D}\frac{V^2}{2g} \tag{15.17}$$

and the second term is

$$\left(\frac{\Delta h}{\Delta L}\right)_s = (s_m - 1) = C(s_s - 1) \tag{15.56}$$

after considering Eq. (15.6), with $s_m = [1 + C(s_s - 1)]$. Equation (15.17) is the Darcy-Weisbach relation; Eq. (15.56) represents the static head of the solids. Substitution of Eqs. (15.17) and (15.56) into Eq. (15.2) gives

$$\left(\frac{\Delta h}{\Delta L}\right)_m = f\frac{1}{D}\frac{V^2}{2g} + C(s_s - 1) \tag{15.57}$$

Note that Eq. (15.57) defines the head loss. The pressure drop between two elevations, which are a distance ΔL apart, can be obtained as

$$\frac{\Delta p}{\gamma} = (\Delta h)_m + \Delta L\frac{\gamma_m}{\gamma} \tag{15.58}$$

It is important to point out that Eq. (15.57) is only correct if the settling velocity v_{ss} is considerably smaller than the average flow velocity V. For the more general case, NEWITT et al. (1961) suggested to modify Eq. (15.56) to

$$\left(\frac{\Delta h}{\Delta L}\right)_s = C(s_s - 1)\frac{V}{u} \tag{15.59}$$

where V is the average mixture velocity and u is the velocity of the liquid. At this point it is appropriate to consider the influence of the particle settling velocity on flow in a vertical pipe. This shall be done by considering the upward flow in Fig. 15.10 where the different velocities are represented. The liquid velocity u is larger than the velocity of the solids u_s, which slip backward by the amount of the settling velocity v_{ss}. The average flow velocity of the mixture V is larger than the solids velocity, but smaller than the liquid

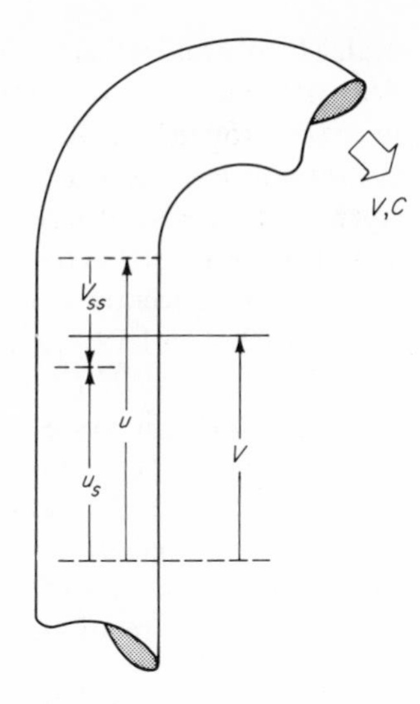

Fig. 15.10 Definition sketch of vertical flow.

velocity. In upward flow the concentration in the pipe C'_u is larger than the "delivered" concentration C; in downward flow the opposite is true, or

$$\begin{aligned} &\text{Upward flow:} \quad C'_u > C \\ &\text{Downward flow:} \quad C'_d < C \end{aligned} \tag{15.60}$$

The ratio of C'/C is in functional relationship to v_{ss}/V and was given by WORSTER et al. (1955); it is reproduced in Fig. 15.11. It is obvious that, for small settling velocities, the ratio of V/u approaches unity, and Eq. (15.59) is sufficiently well approximated with Eq. (15.56). This approximated relation of Eq. (15.56) was suggested for use by DURAND (1953*a*), WORSTER

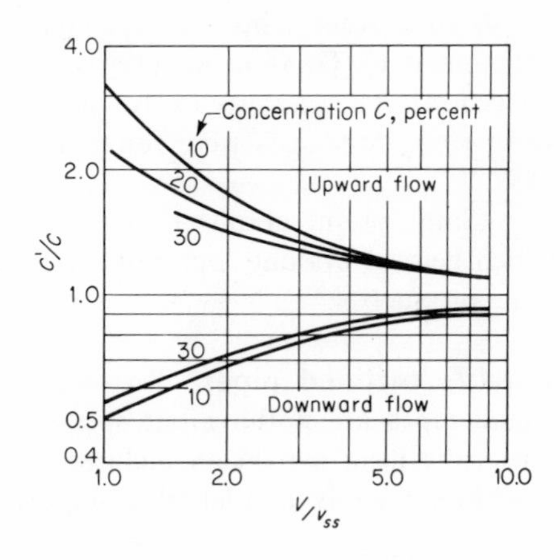

Fig. 15.11 Effect of particle slip in vertical flow. [*After* WORSTER *et al.* (*1955*).]

et al. (1955), and others. The following experimental results are in agreement with Eq. (15.57). DURAND's (1953*a*) experiments were done in a conduit, $D = 150$ mm, with sand of a median diameter of $0.18 < d < 4.36$ mm and for concentrations to more than 7 percent. BONNINGTON (1959) did an experiment with sand having a settling velocity $v_{ss} = 0.5$ fps, and for concentrations up to 20 percent in a 1.5-in. pipe. NEWITT et al. (1961) experimented with concentrations up to 35 percent in 1- and 2-in. pipes, and with solid particles of $1.19 < s_s < 4.20$ and of $0.13 < v_{ss} < 1.09$ fps. Furthermore, NEWITT et al. (1961) concluded that solid-liquid mixtures, whose particle settling velocity is in the transition or turbulent region, are found to be in an overall agreement with Eq. (15.57).

However, in the same study by NEWITT et al. (1961), solid-liquid mixtures composed of small particles $0.033 < v_{ss} < 0.082$ fps are in disagreement with Eq. (15.57). The results are found to be well described with an empirical relation, or

$$\varphi''_N = 0.037 \frac{\sqrt{gD}}{V} \frac{D}{d} s_s^2 \tag{15.61}$$

with the φ value as defined in Eq. (15.3). It must be recognized that Eq. (15.61) was established with limited data.

EINSTEIN et al. (1966) suggested that the pseudohomogeneous head loss relation, given earlier with Eq. (15.5), may be applied if allowance is made for the static head of the solids in the conduit. This relation was given as

$$\left(\frac{\Delta h}{\Delta L}\right)_m = \left(\frac{\Delta h}{\Delta L}\right)_l [1 + C(s_s - 1)] + C(s_s - 1) \tag{15.62}$$

and should be compared with Eq. (15.57). EINSTEIN et al. (1966) used two uniform sands with $d = 1.15$ mm and $d = 1.70$ mm, the concentrations were up to 18 percent, and the pipe diameter was $D = 0.076$ m. In another experiment by GRAF et al. (1967), a sand of $d = 2.85$ mm was used, the concentrations were up to 10 percent, and the pipe was $D = 0.076$ m. Both laboratory tests were found to be in reasonable agreement with Eq. (15.62).

Some useful graphs have been presented by BREBNER et al. (1964), which help to evaluate optimum design conditions for vertical pumping of mine products.

15.2.4.2 Inclined pipes. In one sense, the vertical pipes and the horizontal pipes are nothing but special cases of inclined pipes. In another sense, the head loss in an inclined pipe may be considered as the sum of head loss in equivalent length of horizontal and vertical pipes. This can be written as

$$\left(\frac{\Delta h}{\Delta L}\right)_m = \left(\frac{\Delta h}{\Delta L}\right)_l + \left[\frac{\Delta h}{\Delta L}\right]_s^h \cos\theta + \left[\frac{\Delta h}{\Delta L}\right]_s^v \sin\theta \tag{15.63}$$

with θ being the angle of inclination. Note that the last term on the right-hand side vanishes for a horizontal pipe, and the second term for a vertical pipe.

A relation similar to Eq. (15.63) was proposed and investigated by WORSTER et al. (1955). The third term of Eq. (15.63) was expressed with Eq. (15.56), and the second term with a type of Durand-Condolios relation. Limited tests were made with a 1.5-in. pipe, with various inclinations. The experimental results were in agreement with the relation.

GIBERT (1960) has suggested the following relation:

$$\varphi_D = K_{DS}\left(\frac{V^2}{gD}\sqrt{\frac{gd}{v_{ss}^2}}\right)^{-3/2}(\cos\theta)^{3/2} \tag{15.64}$$

This relation together with Eq. (15.3) is sufficient to determine the second term of Eq. (15.63). When $\theta = 0$, the horizontal-pipe equation, given earlier with Eq. (15.12*a*), is obtained; for $\theta = 90°$, the vertical-pipe equation, given earlier with Eq. (15.57), is obtained. Neither GIBERT (1960) nor CONDOLIOS et al. (1963) report the numerical value of the coefficient K_{DS}. Experiments with a 150-mm pipe on horizontal, descending, and ascending pipe slopes for 0.85-mm sand were found to substantiate Eq. (15.64) rather well.

GRAF et al. (1967) have modified Eq. (15.62) to apply for inclined conduits, or

$$\left(\frac{\Delta h}{\Delta L}\right)_m = \left(\frac{\Delta h}{\Delta L}\right)_l [1 + C(s_s - 1)] + C(s_s - 1)\sin\theta \tag{15.65}$$

However, for horizontal pipes, Eq. (15.65) becomes Eq. (15.7), which was shown earlier to give good results only for pseudohomogeneous flow. In the tests a 2.85-mm sand was used with a 0.076-m pipe of different slopes. For the vertical pipes good agreement was obtained; for inclined pipes the measured head loss was usually larger than the one predicted with Eq. (15.65).

Furthermore, it is worthwhile to note that SHIH (1964) and BARR et al. (1968) have outlined a procedure by choosing dimensionless parameters important to the problem.

15.2.5 NONSETTLING MIXTURES

Solid-liquid mixtures with solid particles having a settling velocity[1] of $v_{ss} < 0.002$ to 0.005 fps are considered as *nonsettling* mixtures; we may

[1] The settling velocity depends in this case on the concentration as well as on the usual parameter.

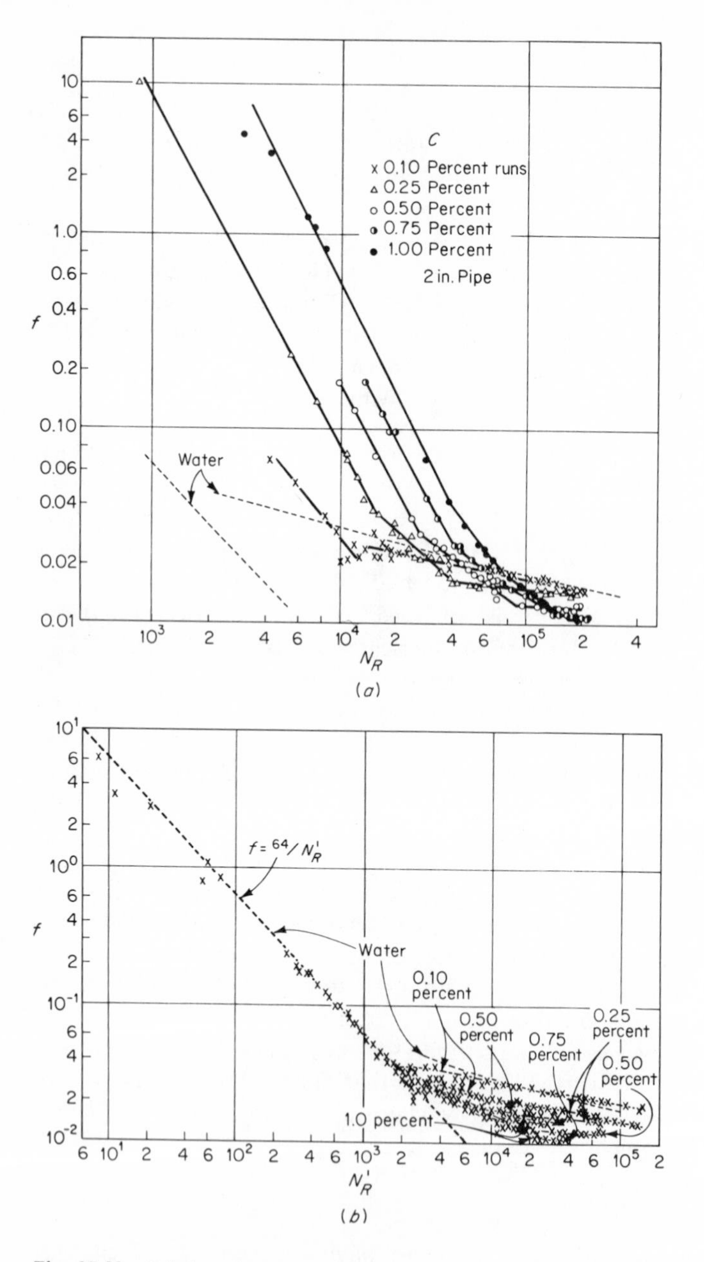

Fig. 15.12 Friction factor vs. (*a*) Reynolds number vs. (*b*) generalized Reynolds number; both for "Kraft softwood." [*After* BUGLIARELLO *et al.* (*1961*).]

safely ignore the settling tendency. Such mixtures frequently fail to obey Newton's law of viscosity and behave as non-newtonian fluids. The non-newtonian fluids exhibit homogeneous flow behavior in most cases.

METZNER (1961) distinguished six different categories of non-newtonian fluids, the first four of which are: (1) purely viscous fluids; (2) time-dependent fluids; (3) viscoelastic systems; and (4) systems which combine the complexities of categories 1, 2, and 3. It was also remarked that most fluids fall into categories 3 and 4, both of which are least understood. This presents some dilemma and, thus, it is not surprising that the different mathematical models developed for non-newtonian fluids hold true only under very restrictive conditions.

Non-newtonian fluids have been treated in numerous books on fluid mechanics; the ones by METZNER (1956, 1961), BIRD et al. (1960), and FREDRICKSON (1964) are of major importance.

With a superficial look at head loss relations for non-newtonian fluid, the following is observed. In the laminar-flow regime deviations between the head loss of newtonian and non-newtonian fluids are considerable; in general, the head loss of a newtonian fluid is smaller than that of a non-newtonian fluid for any given Reynolds number. In the turbulent regime deviations exist but are small; it is not uncommon to find that a non-newtonian fluid has a smaller head loss than a newtonian one. Experimental data by DAILY et al. (1961) and THOMAS (1962) show this trend. A typical relation exhibiting the previous statement is given in Fig. 15.12*a*, where the friction factor is plotted against the Reynolds number. The same data are plotted (see Fig. 15.12*b*) in terms of a generalized Reynolds number, and a better correlation is obtained.

15.2.5.1 Purely viscous fluids. Among the many models which were developed for this category of fluids, there are two prominent ones, i.e., the Bingham plastics and the pseudoplastic and dilatant fluids with power-law behavior. A schematic sketch of the shear stress vs. shear rate relation of these models is given in Fig. 15.13.

Bingham plastics are characterized by a shear stress vs. shear rate relation such as

$$\tau = \tau_y + \eta\left(\frac{du}{dy}\right) \tag{15.66}$$

where τ_y is the yield stress and η is the coefficient of rigidity. Flow takes place if the shear stress τ exceeds the yield stress τ_y. BINGHAM (1922) suggested the Eq. (15.66), which became quite popular. METZNER (1961) and CLEGG et al. (1966) recognized its historical importance, but submitted that no convincing evidence exists that a fluid behaves according to Eq. (15.66). Head loss relations have been established theoretically and/or

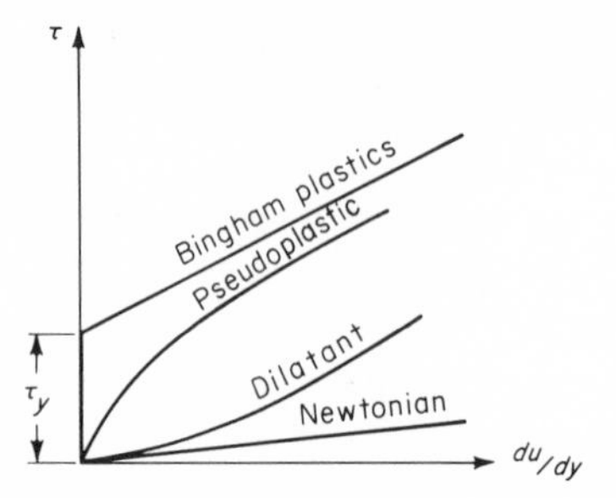

Fig. 15.13 Type of rheological behavior.

experimentally for laminar and turbulent flow, and have been summarized by METZNER (1956) and BEHN (1960). According to BABITT et al. (1939), mixtures of clay and water, sewage sludges, and similar aqueous suspensions of fine particles obeyed the Bingham-plastics model. BEHN (1960*a*) remarked that digested sewage sludges have Bingham-plastics behavior, but a possibility of pseudoplastic and/or time-dependent behavior exists. Raw sewage sludges are usually assumed to have Bingham-plastics behavior, although there exists but meager evidence. METZNER (1956) lists the following fluids that approximate Bingham-plastics behavior: drilling mud, sewage sludge, cement rock slurry, grain suspensions, and chalk suspensions. THOMAS (1961) found that flocculated-suspension data could be satisfactorily explained with the Bingham-plastics model.

Pseudoplastic and dilatant fluids have been represented by a variety of mathematical models, some of which are listed in BIRD et al. (1960) and THOMAS (1962). However, the *power-law* model is the one that has received considerable attention because of its usefulness and versatility. METZNER (1956) remarked that such a model could facilitate understanding of the behavior of all fluids, both newtonian and non-newtonian, the latter including the Bingham plastics as well. The power-law equation is given as

$$\tau = K\left(\frac{du}{dy}\right)^n \tag{15.67}$$

where K is the consistency index and n is the flow-behavior index. The flow-behavior index n is a dimensionless quantity, and is $n < 1$ for pseudoplastic fluids or $n > 1$ for dilatant fluids. Invariably it was observed that the n value changes with a change in the shear rate. [Examples are given in METZNER (1961).] For example, a certain fluid may exhibit newtonian behavior at low and high shear rates, while at intermediate shear rates, it might exhibit non-newtonian behavior, being pseudoplastic or dilatant, or both, depending on the shear rates.

Head losses due to non-newtonian fluids can be obtained with the Darcy-Weisbach equation, given by Eq. (15.17). The friction factor f vs.

the Reynolds-number relation was developed by DODGE et al. (1959). However, it becomes necessary to express the Reynolds number as

$$N'_R = \frac{D^n V^{2-n} \rho}{K^*} \tag{15.68}$$

with

$$K^* = K8^{n-1}\left(\frac{3n+1}{4n}\right)^n$$

This relation became known as the generalized Reynolds number; for the special case of a newtonian fluid with $n = 1$ and $K = \mu$, Eq. (15.68) becomes

$$N_R = \frac{DV\rho}{\mu} \tag{15.68a}$$

The friction factor for laminar flow is given as

$$f = \frac{64}{N'_R} \tag{15.69}$$

and for fully turbulent flow in smooth pipes as[1]

$$f^{-1/2} = A' \log\,(N'_R f^{(1-0.5n)}) + C' \tag{15.70}$$

Note again that both Eqs. (15.69) and (15.70) give, for $n = 1$, the well-known friction factor relations of newtonian fluids. METZNER (1956) reported that the utility of the previous relations was checked with actual data; this can also be seen in Fig. 15.12*b*. A useful design chart for power-law fluids was given and is reproduced in Fig. 15.14. The applicability of such a graph has been verified according to METZNER et al. (1964) in a number of independent studies. It is evident from this graph that the transition from laminar to turbulent flow is dependent on the flow-behavior index. Also note that the larger the n value, the larger the friction factor and head loss. Simultaneously to the research of DODGE et al. (1959), SHAVER et al. (1959) also conducted an experimental study.

As far as turbulent flow through rough conduits is concerned, METZNER (1961) suggested to use the usual newtonian friction factor. Experimental evidence is not sufficient to justify the use of a generalized relation.

It has been stated that most of the non-newtonian fluids have or can be approximated by pseudoplastic behavior. METZNER (1956) listed the following fluids: "polymeric solutions or melts, as rubbers, cellulose and

[1] METZNER (1961) remarked that by making $n = 1$ in Eq. (15.70), a good approximation is still obtained. This is in agreement with the plot of Fig. 15.12*a* and *b*. In Eq. (15.70), A' and C' are constants.

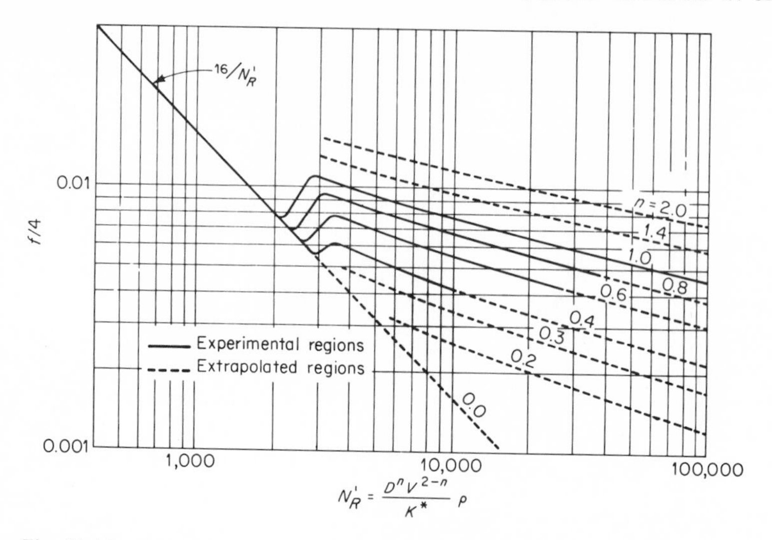

Fig. 15.14 Friction factor vs. generalized Reynolds number. [*After* DODGE *et al.* (*1959*).]

Napalm; suspensions such as paints, paper pulp, mayonnaise and detergent slurries; dilute suspensions of inert, unsolvated solids." BEHN (1962) reported that digested sewage sludge fitted a pseudoplastic (power-law) relationship.

According to METZNER (1961), dilatancy is "common to extremely concentrated suspensions." Earlier METZNER (1956) listed starch, potassium silicate, and gum arabic in water. BEHN (1960*a*) has stated that dilatant fluids apparently do not appear in the waste-treatment process.

15.2.5.2 Time-dependent fluids. A non-newtonian fluid for which the shear rate depends on the magnitude and duration of the shear stress is referred to as a *time-dependent fluid.* A sketch of such a behavior is shown in Fig. 15.15. We can distinguish between the following two cases of time-dependent fluids. If the shear stress increases with time and everything else

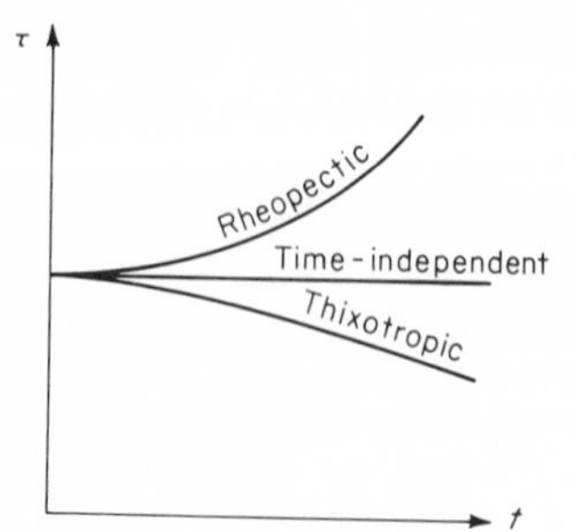

Fig. 15.15 Types of time-dependent fluids.

is constant, the fluid is termed *rheopectic.* A kind of stiffening action takes place in the fluid. If the shear stress decreases with time and everything else remains constant, the fluid is referred to as being *thixotropic.* This is commonly attributed to a relaxing action within the fluid. It has been remarked by METZNER (1961) that thixotropic-fluid systems are frequently encountered in the food and paint industries. The experimental work by HERUM et al. (1964), with two different feed materials for livestock and with a meat mixture, showed time-dependency effects. BEHN (1960*a*) has stated that some thixotropic behavior also exists in most sewage sludges. Rheopectic fluids are relatively uncommon, but, according to METZNER (1956), they are observed in certain sols and in bentonite clay suspensions.

There appears to be no model in existence for time-dependent fluids; time-dependency effects are as yet very unpredictable. The literature on the flow of such fluid through pipes is meager for laminar and turbulent flow. However, as far as the design of processing equipment is concerned, METZNER (1961) remarked that "... equipment must usually be designed to accommodate extremes of physical properties, which ... are encountered after very short times of shear." From this METZNER (1961) concluded that "time dependency is of no importance since the design becomes the same as for purely viscous fluid."

15.2.5.3 Viscoelastic fluids. If the shear rate depends on the shear stress and the extent of deformation, the fluid is referred to as a *viscoelastic* one. In other words, viscoelastic fluids exhibit normal stress effects. Various mathematical models have been proposed, but none of these seems to offer immediate engineering results. Some of the models are briefly reviewed by METZNER (1961).

Head loss studies have been reported and showed the following results. For laminar flow, the relation of the generalized Reynolds number vs. the friction factor f, given with Eq. (15.69), explains the data. For turbulent flow, the data follow an extension of the laminar line up to very high Reynolds numbers and only then start to diverge. Both laminar- and "turbulent" flow data of a viscoelastic fluid are given in Fig. 15.16. From these data[1] METZNER et al. (1964) suggest that the flow transition from laminar to turbulent flow is increased considerably. Furthermore, friction factors in the turbulent range are considerably lower than those of either newtonian or pseudoplastic data. Much more data will be essential for a generalization of these findings. However, for the time being, METZNER et al. (1964) offer some explanation for their results. It is proposed that the presence of small particles "promote(s) stability of the laminar flow field or dampen(s) the

[1] According to METZNER et al. (1964), "a test has been made of the hypothesis that the turbulent-drag-reduction characteristics of polymeric additives are a result of their viscoelastic properties."

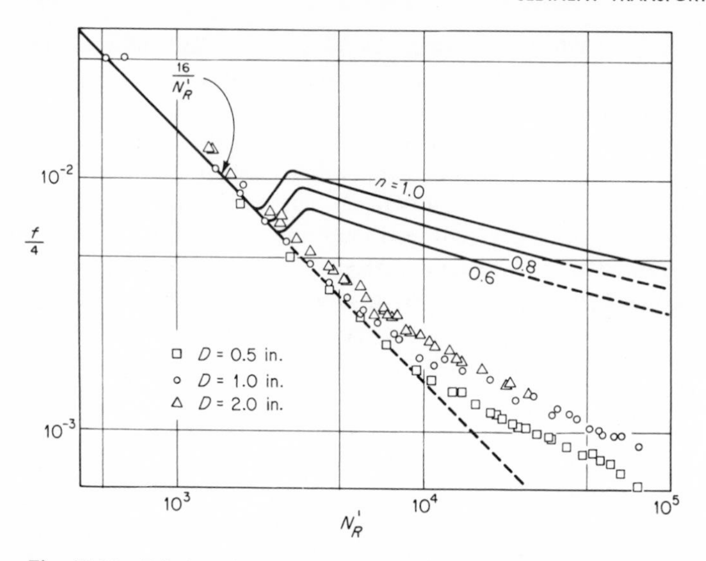

Fig. 15.16 Friction factor vs. generalized Reynolds number; data from a viscoelastic fluid. [*After* METZNER *et al.* (*1964*).]

turbulence, or both." It is of interest to note that METZNER et al. (1964) stressed that only certain polymeric additives give the effect shown in Fig. 15.16, which is caused by the addition of very small quantities of the polymer. O'BRIEN et al. (1937) have observed a similar phenomenon in studying sand-liquid transport through conduits. This was also noticed by VANONI (1946) for open-channel flow. Another explanation, offered by METZNER et al. (1964), is that the separational mechanism may lead to an annular layer of low-viscosity solvent, thus developing a *slip* mechanism at the pipe wall. Yet a third explanation is given that is concerned with the viscoelastic properties of the fluid. Recently SEYER et al. (1967), PATTERSON et al. (1968), and SEYER et al. (1969) have advanced some analytical thoughts.

Rather similar experimental results are reported by BUGLIARELLO et al. (1961). Three wood pulp and two synthetic fibers with a wide range of physical characteristics and concentrations up to 1 percent have been used. It was remarked that suspensions of fibers develop networks with elastic strength and, therefore, the suspensions show a viscoelastic behavior. When the results were plotted with the generalized parameter, some agreement with METZNER's et al. (1964) data is obvious; some data are reproduced in Fig. 15.12.

BOBKOWICZ et al. (1965) made observations similar to the ones given in Fig. 15.16. Marked deviations are reported for large, solid concentrations and high ratios of fiber length to pipe diameter. The fibers used were nylon, cut to length from 0.5 to 1.25 mm; 10 different fiber sizes were used.

Recently POREH et al. (1970) reported that head losses of solid-liquid mixtures of moderate concentration can be reduced by about 70 percent by addition of small amounts of complex soap.

Common to all studies with viscoelastic fluid is the noticeable drag reduction in the turbulent-flow regime. Its potential value to the processing industry is recognized, and was recently stressed by PATTERSON et al. (1969). In a review paper the authors point out that not only small amounts of soluble polymers but also soap solutions cause this drag reduction. In the same paper many experiments are referenced, and friction factor relations similar to Eq. (15.70) are reviewed. A similar drag reduction was shown by HERSHEY et al. (1967) for a flow of a polymer solution.

15.2.5.4 Viscometry. The rheological behavior of non-newtonian fluids is obtained by testing with viscometers. There are two parameters that have to be obtained. For a Bingham-plastics fluid, they are the yield stress τ_y and the coefficient of rigidity η, and for a power-law fluid, they are the consistency index K and the flow-behavior index n.

It has been reported that rotational and capillary-tube viscometers are the most useful and popular ones. There are different designs of rotational viscometers in existence. In principle, the fluid to be tested experiences shear from the inner or outer rotating surface of a container. A bob-and-cup rotational viscometer is shown in Fig. 15.17. Another viscometer is the cone-and-plate rotational viscometer. In the capillary-tube viscometer the laminar flow through a long, smooth, cylindrical pipe is tested. The theory of viscometers was reviewed by METZNER (1961) and BEHN (1960). These investigations also cover problems encountered in measurement and interpretation of the data. METZNER (1961) remarked that the capillary-tube viscometer is the most versatile for either viscous or viscoelastic fluids. BEHN (1960) implied that both types are recognized as yielding equivalent data. BEHN (1962) investigated procedures and techniques in obtaining the two parameters for sewage sludge with both types of viscometers.

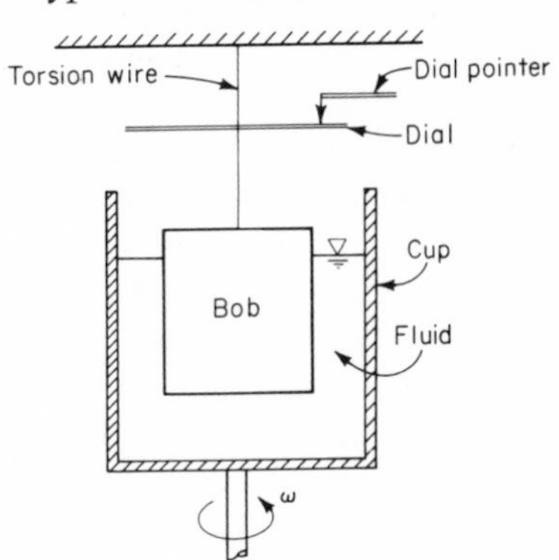

Fig. 15.17 Schematic representation of a bob-and-cup rotational viscometer.

15.2.5.5 Concluding remarks. A non-newtonian fluid is one which fails to behave according to Newton's law of viscosity. Such behavior is not uncommon, especially with solutions of polymeric materials and with highly concentrated solid-liquid suspensions. The non-newtonian fluids have been classified, but such a classification must remain, at least to some extent, arbitrary. It appears that for engineering purposes most of the non-newtonian fluids can be approximated with a power-law relation, given with Eq. (15.67); the head loss may be obtained with generalized Reynolds number and generalized friction factor relations, given with Eqs. (15.68), (15.69), and (15.70). The rheological behavior of a fluid is measured with viscometers. For further topics, such as pumping and mixing of non-newtonian fluids, high-viscosity suspensions, and economic aspects, the review articles by THOMAS et al. (1964) are recommended for reading.

15.3 VELOCITY AND CONCENTRATION DISTRIBUTION

15.3.1 SETTLING MIXTURES IN HORIZONTAL PIPES

Visual observations together with photographic evidence and hydraulic considerations allow a schematic representation of both concentrations and velocity distributions, as shown in Fig. 15.18. Note that we distinguish three kinds of flow, namely, pseudohomogeneous flow, heterogeneous flow, and bed material transport with a bed (deposit). Within the heterogeneous flow, two extremes are sketched. For $V > V_C$, the suspended load will be rather uniformly distributed, but for $V \approx V_C$, the suspended load will move close to the bottom of the horizontal pipe. Distributions of the local concentration and of the local liquid velocity are drawn for each kind of flow. Decreasing the flow velocity—which means going from graph *A* to *D* in Fig. 15.18—results in less uniform concentration distributions. A similar observation can be made with the velocity distribution. If the flow velocity V is below the critical velocity V_C, as shown in Fig. 15.18*D*, deposition occurs.

Photographic evidence for Fig. 15.18 has been given by KRIEGEL et al. (1966) and ACAROGLU et al. (1968). Hydraulic measurements which document the shape of Fig. 15.18 have been reported and will be discussed subsequently.

In an investigation by HOWARD (1940), both the velocity and concentration distributions in a pipe transporting sand and gravel are reported. The results obtained are similar to Fig. 15.18*C*.

A very detailed study considering mainly the hydraulics of solid-liquid transport was done by ISMAIL (1952). The experiments were performed with sand of $d_{50} = 0.10$ mm and $d_{50} = 0.16$ mm in a rectangular (10.5 × 3 in.) channel which was 40 ft long. A pitot tube was employed to measure velocity; samplers in the shape of a pitot tube were used to siphon the sample from the flow. Under consideration were flow conditions comparable to

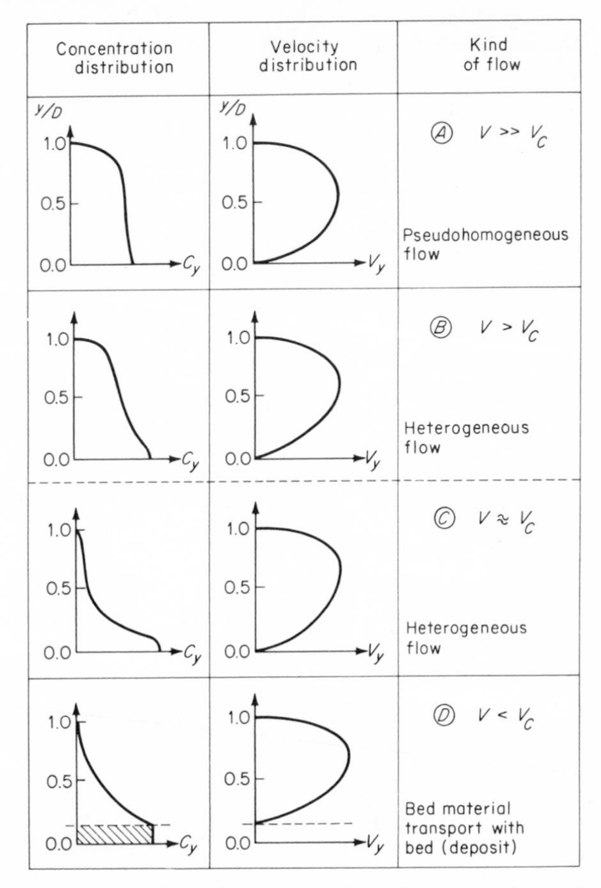

Fig. 15.18 Schematic representation of concentration and velocity distributions.

graphs B, C, and D of Fig. 15.18. ISMAIL (1952) applied to the problem the theory of suspended load, which was discussed in detail in Chap. 8. The governing equation for the vertical concentration distribution at steady-state condition was given as

$$0 = v_{ss}C + \epsilon_s \frac{dC}{dy} \tag{8.20a}$$

where v_{ss} is the settling velocity and ϵ_s is the diffusivity of solid particles. A relation between the diffusivity of solid particles, ϵ_s, and the diffusivity of liquid particles, ϵ, can be expressed by

$$\epsilon_s = \beta\epsilon \tag{8.19}$$

Invariably it has been assumed that $\beta = 1$, and Eq. (8.20a) can be solved, as was done in Sec. 8.3.2.3.1. The concentration distribution is thus given as

$$\frac{C}{C_a} = \left(\frac{D - y}{y} \frac{a}{D - a}\right)^z \tag{8.35}$$

where the quantity z is defined as

$$z = \frac{v_{ss}}{ku_*} \tag{8.33}$$

where k is Karman's constant and u_* is the friction velocity. It was suggested by ISMAIL (1952) that Eq. (8.35) explains the data fairly well, but the β value of Eq. (8.19) should be taken as $\beta = 1.5$ for the 0.10-mm sand and $\beta = 1.3$ for the 0.16-mm sand. As far as the velocity distribution is concerned, the following was concluded. If the logarithmic-velocity-distribution law given by

$$\frac{V_y - V_{max}}{u_*} = \frac{1}{k} \ln \frac{y}{D} \tag{15.71}$$

is to explain the data, it becomes necessary for the Karman constant k to decrease with an increase of the concentration. The decrease of k was explained as an indication of damping of turbulence. A similar trend in the variation of the k value was reported by PAINTAL et al. (1966). The k value was correlated with a function similar to the one plotted in Fig. 8.6. A conduit with a rectangular cross section was also investigated by SHOOK et al. (1968).

Local concentration and velocity measurements are reported by DURAND (1953). The velocity measurements were obtained as usual with a pitot tube, and the concentration measurements with an electric gage recording conductiveness of the mixture. The experimental results were similar to graphs B and C in Fig. 15.18. An attempt was made to explain suspended-load transport with Eq. (8.35).

An extensive program of measurements of concentration and velocity profiles was reported by NEWITT et al. (1962). Investigated were five different sediments moving through a 1-in. pipe. For gravel with $v_{ss} =$ 0.695 fps, the concentration and velocity distributions were similar to graph C; for fine sand with $v_{ss} = 0.064$ fps, the concentration distribution was similar to graph B of Fig. 15.18. This is reproduced in Fig. 15.19. Also investigated was the particle slip. It was concluded that slip velocities are small for fine particles, but are appreciable for coarse particles. TODA et al. (1969) made measurements of the velocity distribution of a water-glass-spheres mixture. The results, given in Fig. 15.20, are comparable to graphs

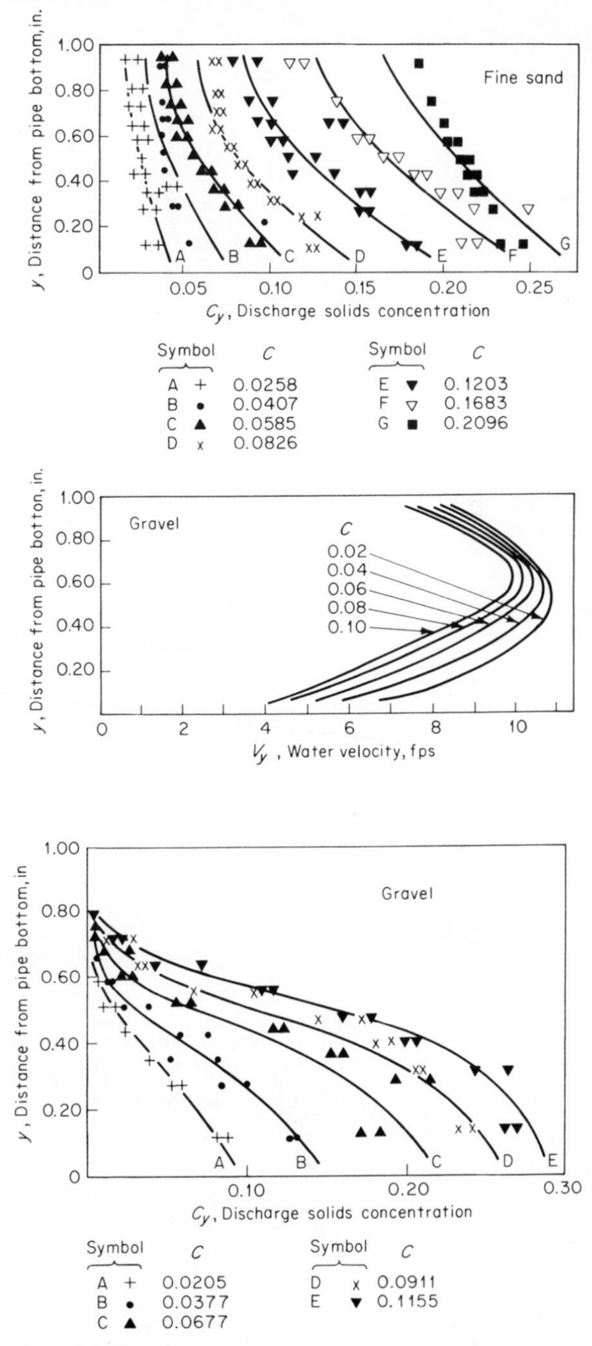

Fig. 15.19 Concentration and velocity profiles. [*After* NEWITT *et al.* (*1962*).]

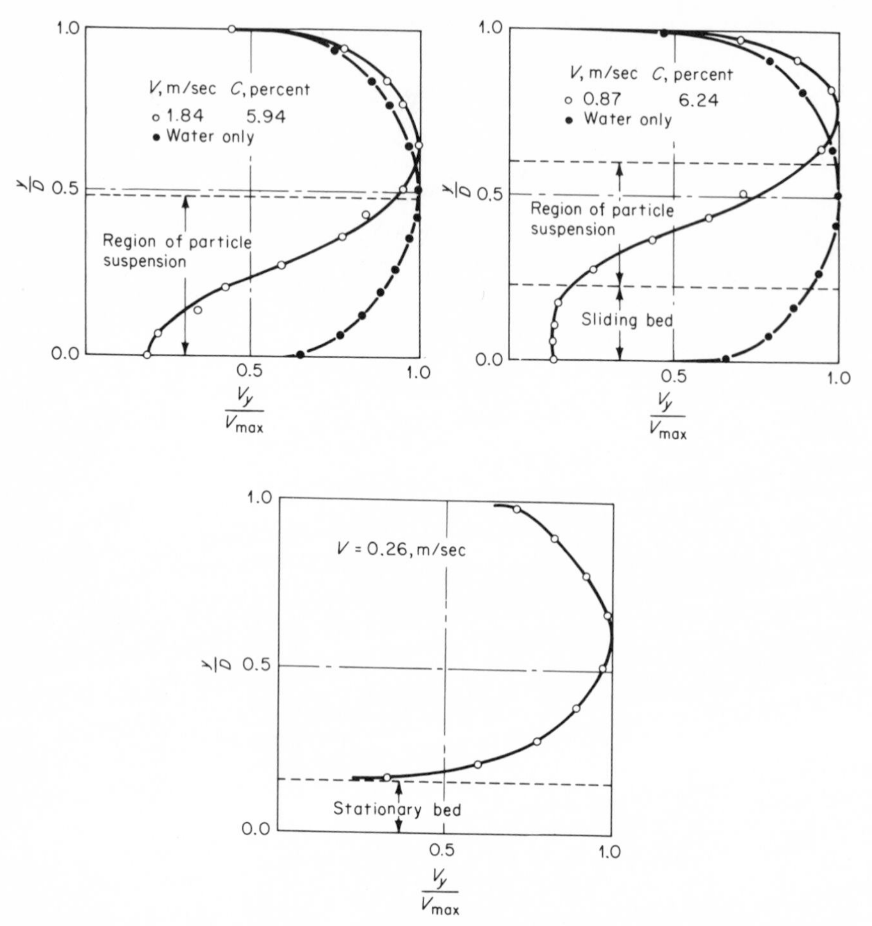

Fig. 15.20 Velocity profiles for three different flow conditions; glass spheres, $d = 2.07$ mm. [*After* TODA *et al.* (*1969*).]

B, *C*, and *D* of Fig. 15.18. Similar measurements of velocity and concentration distributions were reported by SILIN et al. (1969). This study was also concerned with the distribution of turbulence characteristics.

An investigation with neutrally buoyant spherical particles in a 2-in. pipe was reported by DAILY et al. (1966). The resulting information must be considered as surprising as well as preliminary. For coarse, suspended particles the velocity profiles are fuller, whereas for fine ones the profiles are sharper. A mechanism was suggested which helps explain this paradox. The velocity of spherical particles transported through a horizontal pipe was studied experimentally by BINNIE et al. (1966). It was concluded that the mean velocity of nylon spheres ($s_s = 1.20$) was less than that of neutrally buoyant wax spheres.

15.3.2 SETTLING MIXTURES IN VERTICAL PIPES

Both the concentration and the velocity profile are axisymmetric and sensibly uniform over the entire cross section. Experimental investigations by DURAND (1953*a*) document this statement. NEWITT's et al. (1961) experiment showed the following additional results. When the effect of the particles on the velocity distribution was studied, it was found that an addition of 5 percent of perspex or sand particles did not materially alter the velocity profile. A further addition of these particles, now to 15 percent, caused the velocity profile to become flatter. Similar observations have been made by TODA et al. (1969) and are reproduced in Fig. 15.21. The latter investigation shows also that the magnitude of flow velocity decides as to when the concentration has a sizable effect. NEWITT et al. (1961) and TODA et al. (1969) supply photographic evidence that with lower flow velocities the particles are evenly distributed over the cross section; at high velocities the particle movement takes place more in the core of the pipe.

15.3.3 NONSETTLING MIXTURES

Laminar flow. For fluids which can be explained with the power-law of Eq. (15.67), METZNER (1961) gave the following relation:

$$\frac{u_r}{V} = \frac{1 + 3n}{1 + n}\left[1 - \left(\frac{r}{D/2}\right)^{(n+1)/n}\right] \tag{15.72}$$

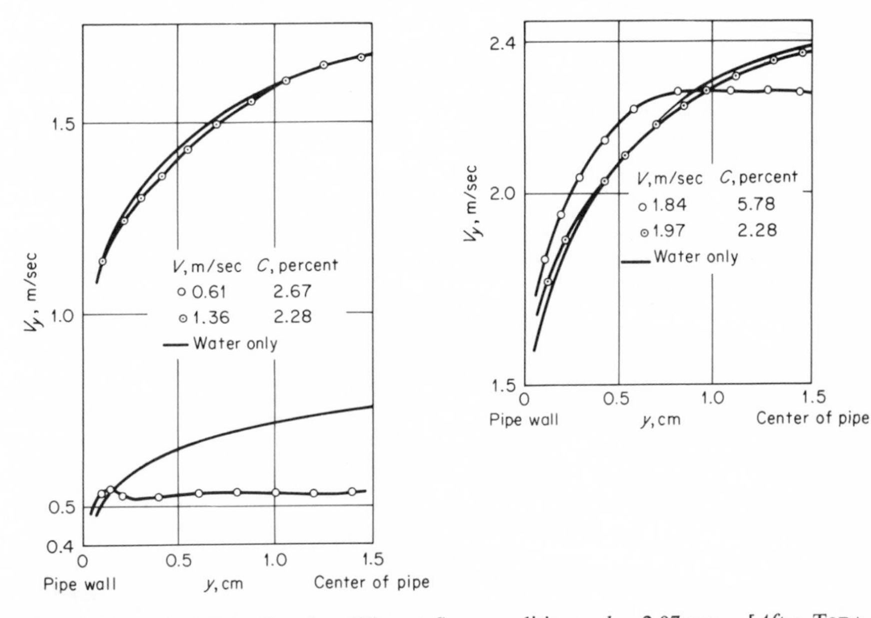

Fig. 15.21 Velocity profiles for different flow conditions, $d = 2.07$ mm. [*After* TODA *et al.* (*1969*).]

Note that for $n = 1$, the special case of newtonian fluid flow, the well-known parabolic distribution is obtained. With a precise technique for measuring velocity distributions, CARVER et al. (1965) obtained the following results. Aqueous solutions of non-newtonian additives showed no deviation from the velocity profiles of distilled water except near the boundary (a distance of up to 50 μ) where the velocities were measurably higher.

Turbulent flow. METZNER (1956, p. 107) remarked that "there is little reason to believe that non-Newtonians have velocity profiles which differ appreciably from those of Newtonian fluids." DODGE et al. (1959) suggested a method to predict the velocity profiles. It became evident that the turbulent-velocity profiles for non-newtonian fluids are similar to newtonian ones, the latter being slightly flatter. An attempt to predict the velocity distribution analytically was also made by BRODKEY et al. (1961) and ZANDI et al. (1965).

15.4 FURTHER REMARKS

In the present section such topics will be briefly discussed which are of importance, but for which a detailed treatment is considered beyond the present scope.

15.4.1 INSTALLATIONS OF SOLID-LIQUID-MIXTURE SYSTEMS

Types of systems. Over the years numerous different systems have been proposed, designed, and built. It is impossible to discuss all of their features. However, CONDOLIOS et al. (1963) have given typical diagrams of different concepts of such installations; these diagrams are reproduced in Fig. 15.22. CONDOLIOS et al. (1963) describe their installation as follows:

Type *A* are gravity installations.
Type *B* are installations equipped with pumping stations, the solid material passing through the pumps.
Type *C* are installations equipped with pumping stations, but the solid material does not pass through the pumps.
Type *D* are installations that allow underwater pickup, with passage of the material through the pumps.
Type *E* are installations similar to type *D*, but the material does not pass through the pumps.
Type *F* installations allow pickup from great depths underwater.

Since a type *A* installation works without a pump, it has the lowest operating costs. The "classical installations," as CONDOLIOS et al. (1963) call the type *B* and type *D* installations, are those using solid-liquid-mixture

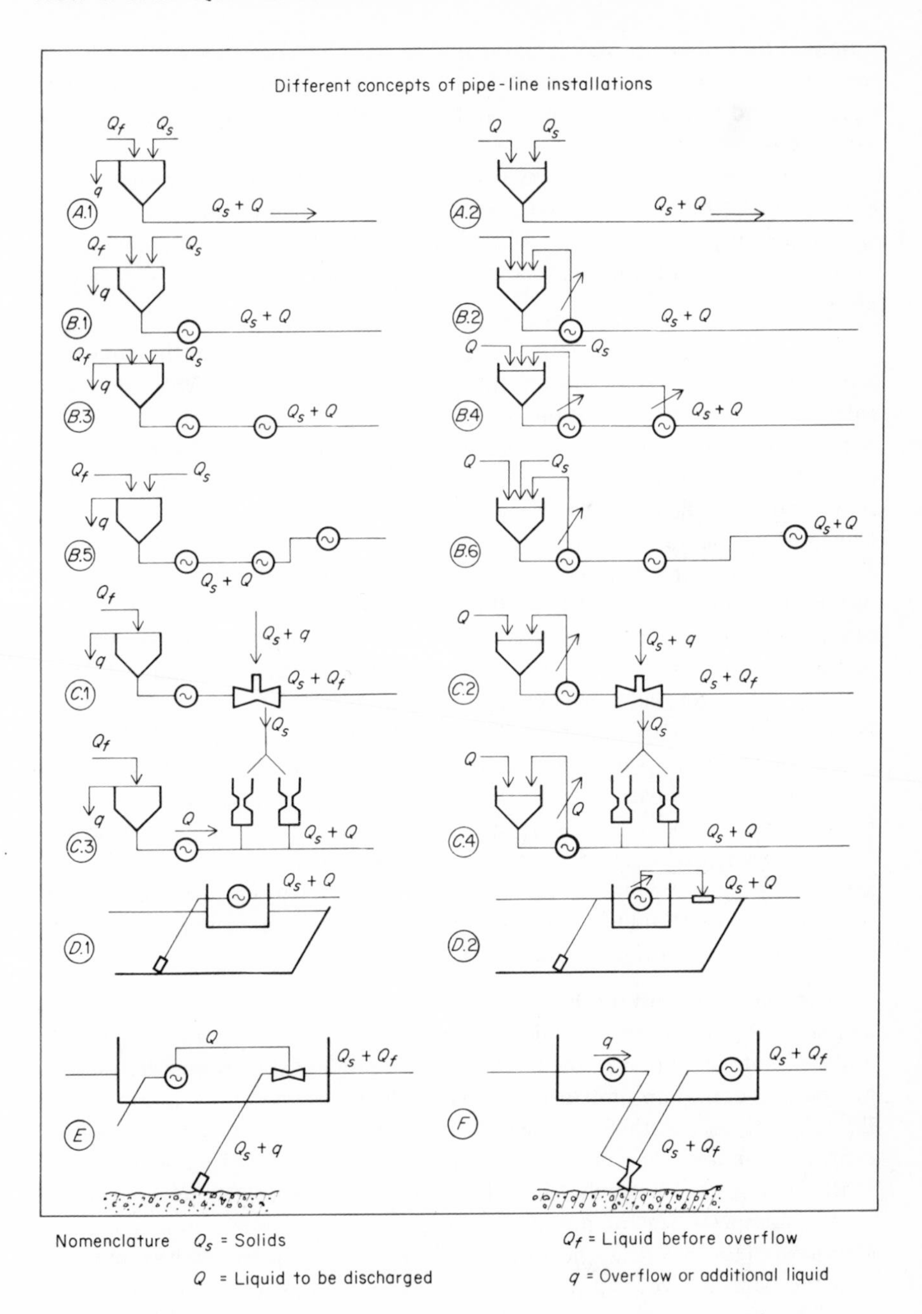

Fig. 15.22 Typical diagrams of different concepts of solid-liquid-mixture installations. [*After* Condolios *et al.* (*1963*).]

pumps. Sometimes it may be desirable to have a pump of standard design for clear liquid and type *C* and type *E* installations become necessary. For dredging and filling operations, installations of type *D*, type *E*, and type *F* are chosen.

Technical data on designed systems can be found in the literature. The COLORADO SCHOOL OF MINES (1963) discussed 52 plants mainly concerned with the transport of minerals; CONDOLIOS (1967) listed 17 different operations. Additional listing may be obtained in BARTH (1960) and CONDOLIOS et al. (1967).

Pipes. The success of any system will depend mainly on the pipes and the pump(s). The diameter of the pipe can be obtained from hydraulic calculations; structural considerations will determine the pipe thickness. It is recommended to make the pipe diameter at least three times larger than the size of the material to be transported. CHAMBERLAIN et al. (1960) remarked that rifled or helical corrugated conduits are rather efficient when transporting medium- to large-size fractions. FÜHRBÖTER (1961) suggested to use noncircular conduits. The material of which the pipe is made will depend, among other things, on the solid matter to be transported. More often than not it is found that wear of the pipe material is minimal and a slow process. Proper selection of the material and its coating will minimize wear considerably. Should wear occur, it is recommended to rotate the pipeline or to allow for solids material to accumulate on the bottom. Laboratory studies on the wear of pipelines are reported by COLORADO SCHOOL OF MINES (1963) and by GARSTKA (1961). Field experience is given by WORSTER et al. (1955), CONDOLIOS et al. (1963), and SCHMIDT et al. (1965). Wear is considerably increased at pipe fittings and at any other obstruction within the smooth, straight pipeline system. Excessive wear is noticed in elbows. If possible, elbows should be replaced by T-sections, because the mixture flow will form an appropriate elbow.

Pumps. According to CONDOLIOS et al. (1963), three types of pumps can be used to pump solid-liquid mixtures: the reciprocating, the diaphragm, and the centrifugal pump, the latter one being the most common design. In any case, these pump units usually have to be redesigned for pumping solid-liquid mixtures. Although research is being done on the proper design of centrifugal pumps, findings are frequently contradictory. In an early study, O'BRIEN et al. (1937) concluded that, at a given discharge, the head developed by the pump (in feet of mixture) decreases as the concentration increases. FAIRBANK (1942) found the same tendency; additionally, it was reported that the particle size, expressed with the settling velocity, is of importance. Similar conclusions are reported in experimental studies by WIEDENROTH (1967). A dependency of the efficiency on the concentration and grain size is reported by FAIRBANK (1942) and WIEDENROTH (1967). Pump characteristics,

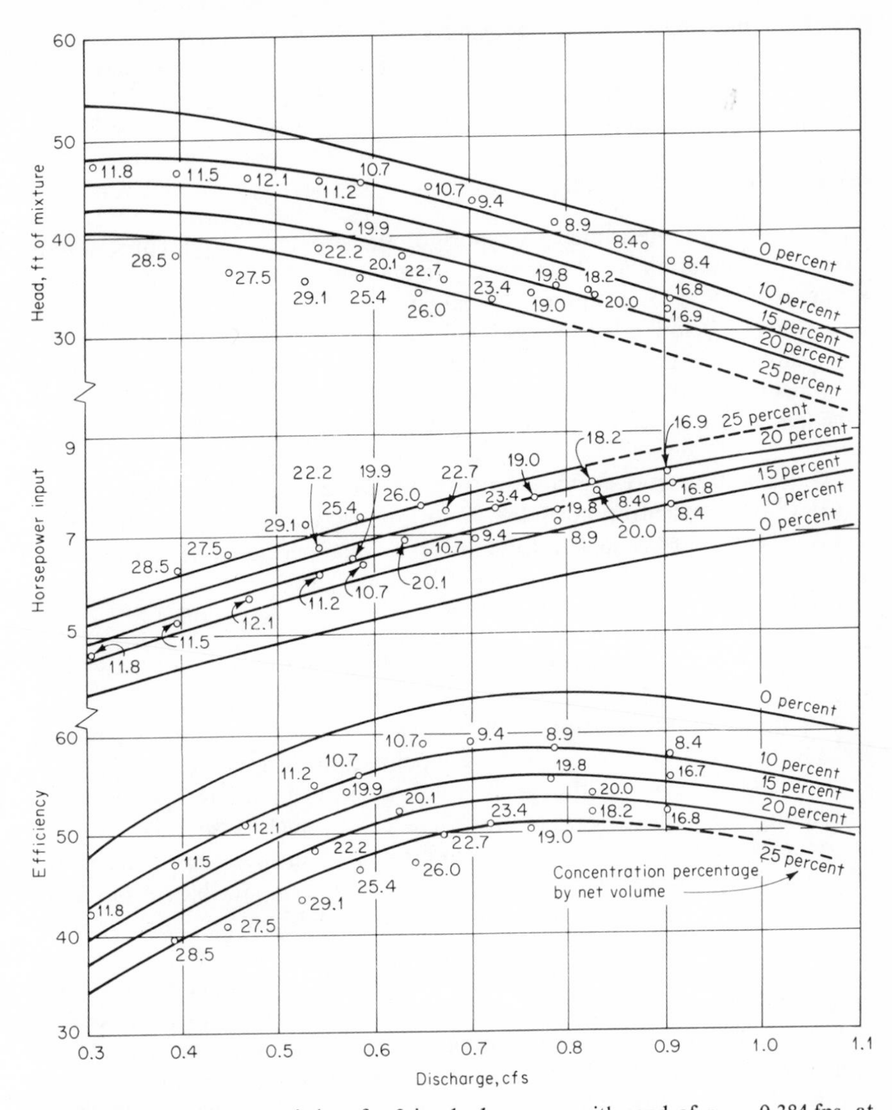

Fig. 15.23 Pump characteristics of a 3-in. dredge pump with sand of $v_{ss} = 0.384$ fps, at 1,000 rpm. [*After* FAIRBANK (*1942*).]

as reported by FAIRBANK (1942), are reproduced in Fig. 15.23. Furthermore, WIEDENROTH (1967) has given experimental evidence for a dependency of the head and the specific pump speed. Unusually heavy wear can be noticed at the impeller and the casing of the centrifugal pump, as was reported by WIEDENROTH (1967), SCHMIDT et al. (1965), and others. Wear resistance may be achieved by using special steels or rubber for coating exposed parts. Also the proper design of the impeller, as suggested by HERBICH (1964), will minimize wear. For a further discussion of pumps and pumping of mixtures, the reader is referred to STEPANOFF (1965).

15.4.2 OPERATION OF SOLID-LIQUID-MIXTURE SYSTEMS

To ensure an uninterrupted transport of the mixture, certain precautions have to be undertaken. Under any circumstance, the actual flow velocity V has to be equal to, but preferably larger than, the critical velocity V_C, the very velocity below which a stationary bed begins to accumulate. If the flow velocity V falls below the critical velocity V_C, blockage (plugging) of the pipeline occurs. SCHMIDT et al. (1965), judging from field experience, have listed four reasons for blockage:

(1) If the transport velocity drop below the critical flow velocity, . . .
(2) If the concentration of the mixture is raised over the maximum.
(3) If the pipeline is left full . . . for a long period after a sudden stop, . . .
(4) If leakage occurs . . .

The first two points have been discussed in detail by CONDOLIOS et al. (1963) and involve the stability of the system. The stability of the operation can be determined by comparison of the head vs. flow-rate curves for the pump(s) and the head loss vs. flow rate for the pipeline; this is done in Fig. 15.24 for a horizontal conduit. Although it is desirable to operate the pump at the minimum of the head loss vs. discharge curve, such a situation usually presents strong instabilities. The slightest change in the head of the pump and/or a change in the concentration may readily be responsible for blockage of the conduit. Suppose a plant is to transport solid-liquid mixtures ranging from concentrations $C_1 < C < C_2$ with a centrifugal pump (I), or a piston pump (II), or by means of a gravity system (III). Consider the stage vs. discharge curve for a centrifugal pump with constant speed (I). The system will be a stable one if the lower concentration C_1 is transported (I_1). However, it will be barely stable at the upper concentration C_2, since the slightest change of head or concentration could cause complete blockage (I_2). Consider the stage vs. discharge relation for a variable-speed centrifugal pump or piston pump (II). At both concentrations stability of operation is safely ensured (II_1, II_2). Consider the stage vs. discharge relation of a gravity installation (III). The system will be fairly stable at a lower concentration (III_1). At the upper concentration it becomes eventually blocked.

Invariably it has been pointed out that the pipe diameter should be about three times as large as the coarsest sediment, which, in turn, should be half the size of the pump passage. It was recommended to install screens to prevent the limiting solid fraction from entering the system.

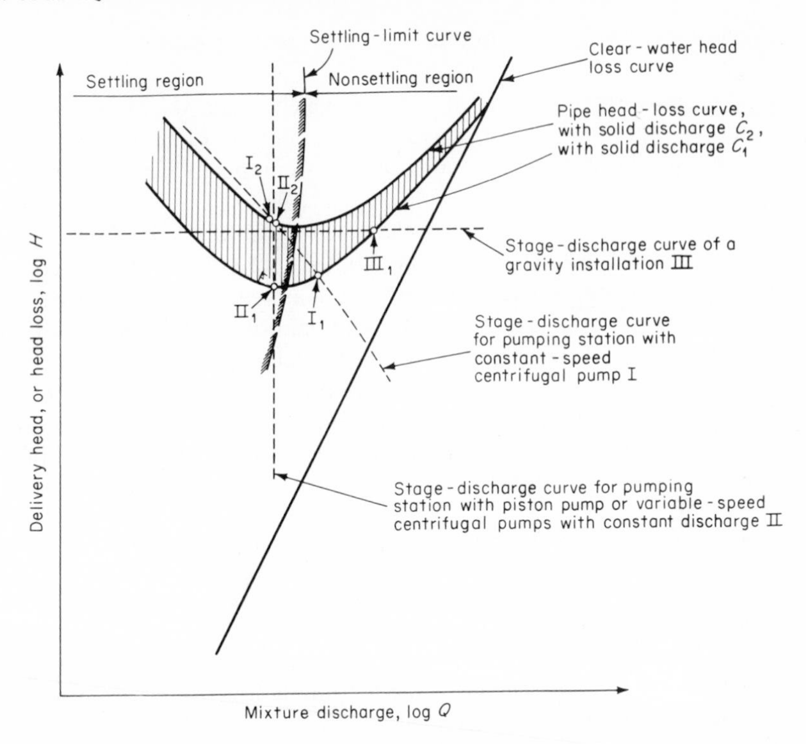

Fig. 15.24 Operating stability of a solid-liquid-mixture system; head of pump and head loss of conduit flow vs. flow rate. [*After* CONDOLIOS *et al.* (*1963*).]

Furthermore, starting and stopping the system may cause considerable trouble. This should be done when the conduit flows with clear water or with very small concentrations. In case of inclined pipes, starting and stopping, and especially the latter, must be done with clear water; otherwise blockage is likely to occur. Sudden closure of valves can be responsible for the water-hammer effect. Safety devices should be installed in the system to minimize this undesirable side effect. Whenever a conduit gets plugged, it must be unplugged before operation continues. The COLORADO SCHOOL OF MINES (1963) discusses this cleaning operation in some length. An effective practice is to flush the pipeline with clean water.

15.4.3 ECONOMICS OF SOLID-LIQUID-MIXTURE SYSTEMS[1]

Although it is beyond the scope of this book, a few remarks on the economics are in place. For further discussion the reader is referred to

[1] Practical information on economics as well as on installations and operations of solid-liquid-mixture systems is reported in the *Proceedings of WODCON*, World Dredging Conference.

CONDOLIOS (1967). An optimization study was recently done by HUNT et al. (1968).

There are economic advantages which make the transport of solid-liquid mixtures rather attractive. Both operating and maintenance costs can usually be kept at a minimum when compared with a transportation system.

The capital cost of a pipeline will definitely vary from country to country. Cost estimates for the United States have been reported by CONDOLIOS (1967). The operating and maintenance costs can be kept low for at least two reasons, namely, automation and dependability of the system. Most of the currently installed pipelines are operated automatically by remote control devices; thus they are rather independent of an inflationary labor cost. Furthermore, past experience has shown that the systems are usually dependable. There are also certain advantages in obtaining a continuous transport and in having compact plant dimensions.

15.4.4 PNEUMATIC CONVEYING OF SOLID MATERIAL

Pneumatic conveying of solid particles takes place in a gaseous environment. Hydraulic and pneumatic conveying is similar in some respects. However, due to the large differences between the specific density ratios of gas-solid mixtures and liquid-solid mixtures, extrapolation from one into the other mixture is not easily accomplished.

For determination of the head loss, relations similar to Eq. (15.3) have been used. A typical plot is shown with Fig. 15.25. GASTERSTÄDT (1924) proposed a relation such as

$$\frac{(\Delta h/\Delta L)_m}{(\Delta h/\Delta L)_g} = 1 + \mu \tan \alpha \tag{15.73}$$

where μ is the loading ratio given as

$$C = \frac{\mu}{\gamma_s/\gamma_g}$$

and with α as a function of the gas-flow rate. Equation (15.73) can readily be transformed into the form of Eq. (15.3), or

$$\left(\frac{\Delta h}{\Delta L}\right)_m = \left(\frac{\Delta h}{\Delta L}\right)_g + \left(\frac{\Delta h}{\Delta L}\right)_s \tag{15.74}$$

Numerous researchers have more or less accepted that head loss in gaseous-solid flow can be represented by Eq. (15.73) or (15.74). Such investigations include the works of VOGT et al. (1948), FARBAR (1949), ZENZ (1949), BELDEN et al. (1949), ALBRIGHT et al. (1951), PINKUS (1952), ROSE et al. (1957), MEHTA et al. (1957), RICHARDSON et al. (1960), SPENCER et al. (1966),

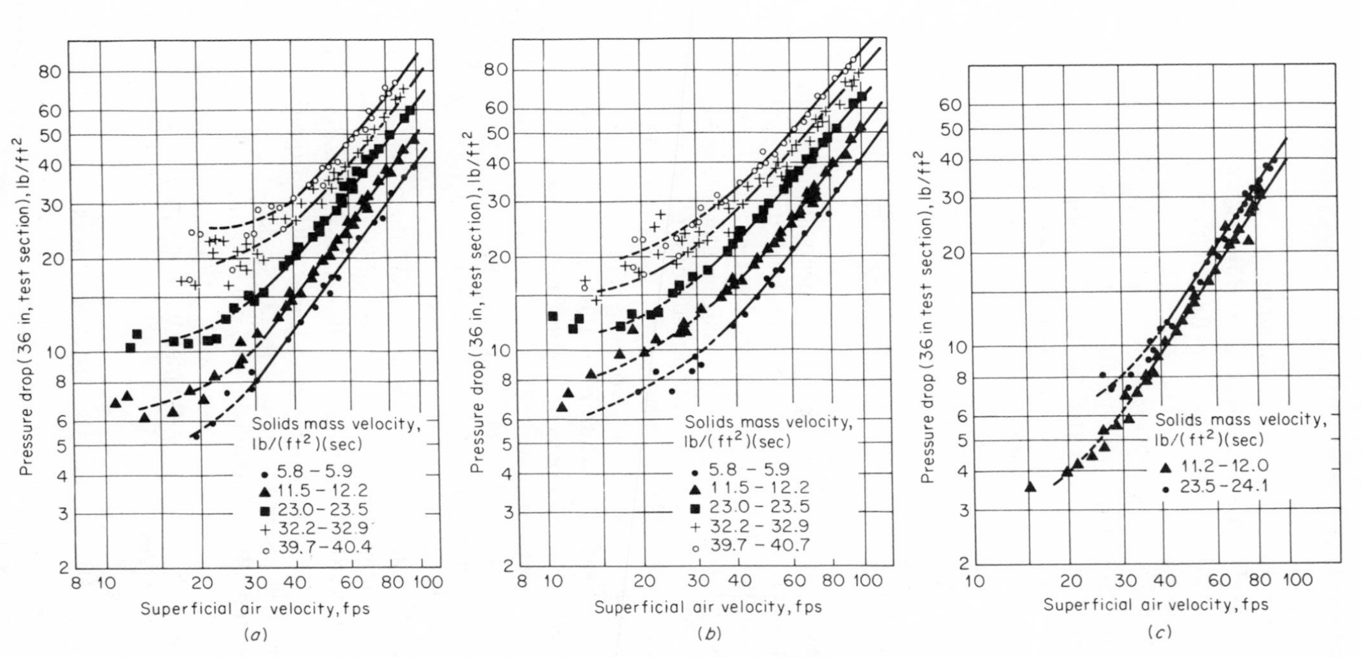

Fig. 15.25 Pressure drop in air-solid flow systems. (*a*) Horizontal pipe, $d = 97\ \mu$; (*b*) vertical pipe, $d = 97\ \mu$; (*c*) horizontal pipe, $d = 36\ \mu$. [*After* Mehta *et al.* (*1957*).]

and others. Many contributions have been recently reviewed by OWEN (1969).

These studies have common findings. For example, tan α, given with Eq. (15.73), is not a simple function of the gas-flow rate but involves other variables, such as the pipe-flow Reynolds number, particle density and particle size, and others. The influence of the particle size is shown with Fig. 15.25. The parameters influencing the head loss have been developed with dimensional analyses by ROSE et al. (1969) and BOOTHROYD (1969). Interesting ideas are reported by BARTH (1960), BOHNET (1965), and WEBER et al. (1967). The latter investigators extended BARTH's (1960) relation to the case of highly concentrated solids flow. STEPANOFF (1965) has suggested to use the Durand-Condolios parameter, discussed in Sec. 15.2.2. It was found that a relation of

$$\varphi = 81 \frac{gD}{V^2} \frac{s_s}{\sqrt{C_D}} \tag{15.75}$$

can be obtained; data of many experiments have been used.

Studies at vertical pipe sections are reported by HARIU et al. (1949), MEHTA et al. (1957), BARTH (1960), ROSE et al. (1969), and others. It was pointed out that an important contribution to the head loss, due to the presence of solids, $(\Delta h/\Delta L)_s$, is a result of the particle acceleration.

STEPANOFF (1965) and ROSE et al. (1967) discussed briefly the significance of the critical velocity V_C. However, no relation is proposed.

Worthwhile information on the concentration and mass-flow distribution of gas-solid suspension is contained in SOO (1967, p. 162). Pumping solids-air suspensions with blowers is introduced by STEPANOFF (1965). Pneumatic transport systems and their advantages and disadvantages have been discussed by the COLORADO SCHOOL OF MINES (1963) and by WEBER et al. (1967).

Both solid-liquid and solid-gaseous flows represent two-phase flow systems. Yet another system is the liquid-gaseous flow. The knowledge gained by studying liquid-gaseous flow is usually only of limited value for solid-liquid flow investigations. Important contributions to the liquid-gaseous flow problems are due to LOCKHART et al. (1949), GOVIER et al. (1957), and others. Among books which treat this topic are those by KUTATELADZE et al. (1958), TONG (1965), and WALLIS (1969). These books as well as a review chapter by TEK (1961) should be consulted for further references.

15.4.5 HYDRAULIC TRANSPORT IN CONTAINERS

This type of transport may be defined as the movement of a succession of individual particles, which approach the size of the conduit, in a fluid carrier.

Solid particles are packed into containers or capsules which, in turn, are fed into a pipeline and transported through the pipeline by a moving fluid. No limitation exists on the geometry and dimension of the container. However, there should be a layer of fluid of reasonable thickness between the container and conduit wall. Experimental investigations of the transport of water of single cylindrical and spherical capsules with densities equal to and greater than water have been reported by ELLIS (1964). It was found that the average water-capsule ratio depends on the flow Reynolds number, on the capsule length-diameter ratio, on the capsule-pipe diameter ratio, on the density ratio, and on the relative roughness ratio. Graphical relationships among these ratios are presented. None of the results are at the present being generalized since they seem to be restricted to the limited test parameters. From experimental data reported by ELLIS et al. (1963), it can be deduced that pressure gradients increase with an increase in concentration or with an increase in capsule velocity, but decrease rapidly with an increase in fluid velocity. A theoretical analysis of capsule flow has been initiated by CHARLES (1963) and was extended by KENNEDY (1966). It was concluded that the capsule velocity exceeds the average velocity by 22 percent for laminar and by 3 percent for turbulent conditions, provided the capsule-pipe diameter ratio is 0.8. The pressure gradient may increase or decrease, depending on the type of flow. The hydraulic transport of solids in containers is usually superior to the hydraulic conveyance as a mixture, if the solids or the fluid get contaminated or grinded or if the solids are reactive or abrasive. Another theoretical study was reported by KRUYER et al. (1967) where the relevance of the analysis is indicated with experimental values. The economic aspect of capsule transport was discussed by ELLIS et al. (1963) and CONDOLIOS et al. (1967). The latter paper includes information on designed installations.

15.4.6 SAMPLE CALCULATION

Example 15.1 A sand-water mixture flows through a horizontal galvanized iron pipe with a 4-in. diameter. The granulometric curve of the sand analysis indicates that it is a fairly uniform material with a size of $d_{50} = 0.88$ mm. Compute the head loss of the settling mixture flow for an 11.8-ft-long pipe for volumetric transport concentrations up to $C = 10$ percent and for mixture velocities up to $V = 20$ fps.

Preliminary calculations

$d_{50} = 0.88 \text{ mm} = 0.0346 \text{ in.} = 0.0029 \text{ ft}$

$v_{ss} = 0.0915 \text{ m/sec} = 0.3 \text{ fps}$

$D = 0.1017 \text{ m} = 4 \text{ in.} = 0.333 \text{ ft}$

$\epsilon = 0.00011 \text{ ft}$

$\epsilon/D = 0.00033$

$\nu = 1.06 \times 10^{-5} \text{ ft}^2\text{/sec}$ (assume $\nu_{mixt} = \nu$)

Head loss calculations Head losses are to be calculated for $0 < C < 10$ percent and 1 fps $< V <$ 20 fps. It can be anticipated that different flow regimes are encountered. Therefore the critical velocity V_C, which separates heterogeneous flow from flow with deposit, and the transition velocity V_H, which separates heterogeneous flow from pseudohomogeneous flow, have to be calculated first. The equations used for these computations are listed in Table 15.2. The resulting numerical values are tabulated in Table 15.3 and subsequently plotted in Fig. 15.26.

For each flow regime a different head loss relation must be used. For pseudohomogeneous flow Eq. (15.7) is used. For heterogeneous flow, the Durand-Condolios relation, given with Eqs. (15.3) and (15.12*a*), and the Newitt et al. relation, given with Eqs. (15.3) and (15.19), are investigated. For a flow with deposit, the Gibert-Condolios relation, given with Eqs. (15.3) and (15.29), and the

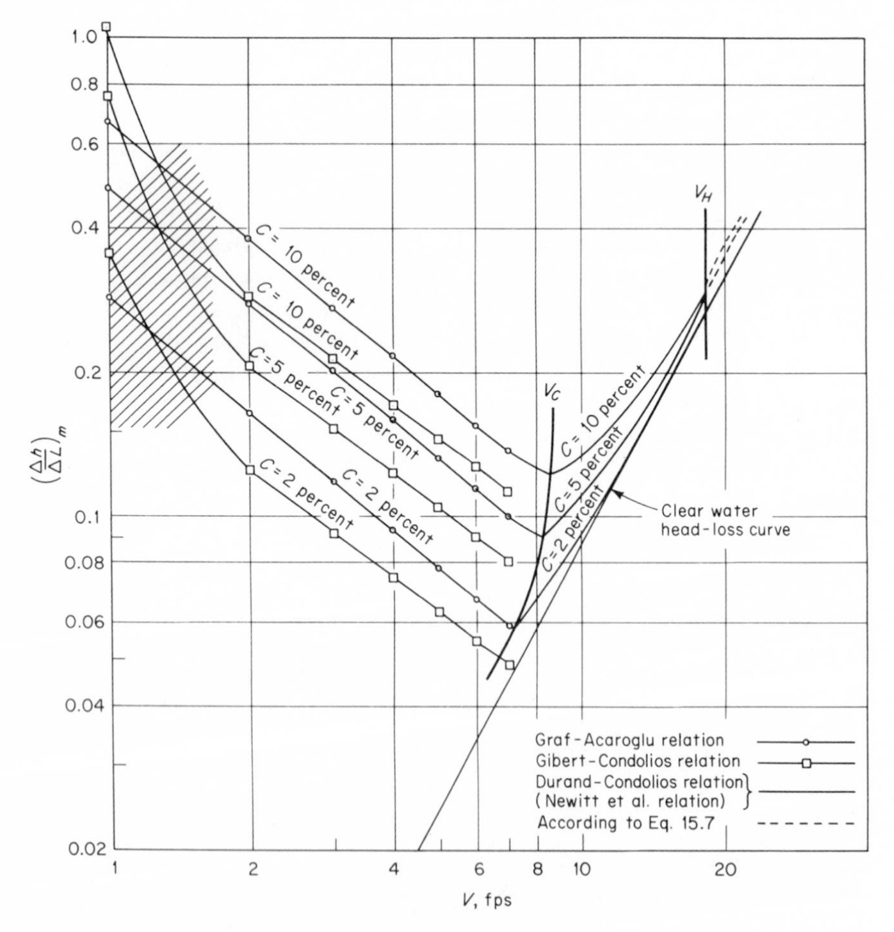

Fig. 15.26 Head loss vs. velocity relationship with equiconcentration lines for the sample calculation of $D = 4$ and $d_{50} = 0.88$ mm.

Table 15.2 Summary of the relations for different regimes

Regime	*Head loss and boundary velocity equations*
Flow with deposit, $V_C > V$	$\dfrac{\left(\frac{\Delta h}{\Delta L}\right)_m - \left(\frac{\Delta h}{\Delta L}\right)_l}{C\left(\frac{\Delta h}{\Delta L}\right)_l} = \varphi$ (15.3) $\quad \dfrac{V}{\sqrt{4R_h g}} = \dfrac{V_C}{\sqrt{Dg}}$ (15.55) ↓ $\varphi_D = 150\left[\dfrac{V_2}{g4R_h(s_s - 1)}\sqrt{C_D}\right]^{-3/2}$ (15.29) $\dfrac{CVR_h}{\sqrt{(s_s - 1)gd^3}} = 10.39\left[\dfrac{(s_s - 1)d}{SR_h}\right]^{-2.52}$ (15.54) $\dfrac{V}{\sqrt{4R_h g}} = \dfrac{V_C}{\sqrt{Dg}}$ (15.55)
V_C	$\dfrac{V_C}{\sqrt{2gD(s_s - 1)}} = F_L$ (15.16) with $F_L = f(d,C)$ from Fig. 15.5
Heterogeneous flow, $V_C < V < V_H$	$\dfrac{\left(\frac{\Delta h}{\Delta L}\right)_m - \left(\frac{\Delta h}{\Delta L}\right)_l}{C\left(\frac{\Delta h}{\Delta L}\right)_l} = \varphi$ (15.3) ↓ $\varphi_D = 180\left[\dfrac{V^2}{gD}\sqrt{\dfrac{gd}{v_{ss}^2}}\right]^{-3/2}$ (15.12*a*) or ↓ $\varphi_N = 1100(s_s - 1)\dfrac{v_{ss}}{V}\dfrac{gD}{V^2}$ (15.19)*
V_H	$V_H = \sqrt[3]{1800 g v_{ss} D}$ (15.8)
Pseudohomogeneous flow, $V > V_H$	$\dfrac{\left(\frac{\Delta h}{\Delta L}\right)_m - \left(\frac{\Delta h}{\Delta L}\right)_l}{C\left(\frac{\Delta h}{\Delta L}\right)_l} = \varphi$ (15.3) ↓ $\varphi = (s_s - 1)$ (15.7)

* It has been previousl checked that Eq. (15.19) covers the entire regime between $V_C < V < V_H$.

Table 15.3 Head loss relation for different velocities and concentrations for sample calculation

V, fps \ *C, %*	*0*	*2*	*5*	*10*	*Regime*
1	0.0072	0.353* 0.287	0.751* 0.488	1.042* 0.665	$V_C > V$ Gibert-Condolios relation Graf-Acaroglu relation
2	0.0078	0.125 0.164	0.206 0.278	0.289 0.379	
3	0.0129	0.092 0.118	0.152 0.201	0.213 0.273	
4	0.0185	0.075 0.094	0.123 0.159	0.272 0.216	
5	0.0245	0.063 0.078	0.104 0.133	0.145 0.181	
6	0.0307	0.055 0.067	0.091 0.114	0.127 0.156	
7	0.0372	0.049 0.059	0.081 0.101	0.113 0.138	
$V_C = 7.25$	—	0.060	—	—	V_C
$V_C = 8.35$	—	—	0.091	—	
$V_C = 8.60$	—	—	—	0.126	
9	0.0677	0.080 0.079	0.098 0.096	0.128 0.124	$V_C < V < V_H$ Durand-Condolios relation Newitt et al. relation
11	0.0996	0.109 0.109	0.124 0.122	0.149 0.145	
13	0.137	0.142 0.145	0.158 0.160	0.178 0.175	
15	0.181	0.185 0.188	0.199 0.201	0.216 0.214	
17	0.231	0.234 0.237	0.247 0.248	0.262 0.260	
$V_H = 18.2$	—	0.269	0.278	0.292	V_H
19	0.287	0.296	0.311	0.334	$V > V_H$
20	0.317	0.327	0.343	0.369	

* Laminar flow data.

Graf-Acaroglu relation, given with Eq. (15.54), are used; either of these relations needs an additional equation given with Eq. (15.55). All the equations for the head loss calculations are tabulated in Table 15.2. Numerical values obtained with these equations are listed in Table 15.3 and plotted on Fig. 15.26.

Discussion of the results In the heterogeneous and deposit flow regime two relations have been used to predict the head loss. In either regime agreement between these two relations is rather good. There are two transition regions evident. At very low flow velocities the Reynolds numbers indicate laminar flow; any sand-water mixture concept breaks down. Furthermore, at flow velocities slightly below critical velocity, there exist discontinuities which are presently not fully understood. [Part of the dilemma might rest with the difficulties in properly evaluating Eq. (15.17*a*) together with Eqs. (15.30) and (15.30*a*).]

16

Measuring Devices for Solid-Liquid Mixtures in Pipes[1]

16.1 INTRODUCTORY REMARKS

In recent years, interest in and knowledge pertaining to the measurement of two-phase flow in pipes have increased rapidly. Dredging systems, industrial and mining operations, and laboratory investigators have a need for devices with which to measure two of the basic quantities associated with this mixture flow, namely, the flow rate and solids concentration of the solid-liquid mixture.

The simplest and most accurate method to measure the flow rate and the concentration of a mixture would be to use the calibrated volumetric or weighing tank. Indeed, this is often done, especially in the laboratory. However, these methods, which frequently require interruption of the continuous mixture flow, are seldom practicable. In the following the discussion will focus on such measuring devices which continuously (without interrupting the flow) record the solid-liquid-mixture flow.

[1] This chapter represents an extended version of the paper by R. N. Weisman and W. H. Graf (1968). (*By permission of the American Society of Civil Engineers.*)

16.2 METHODS AND EQUIPMENT

16.2.1 BRINE INJECTION METHOD

HOWARD (1940) discussed the "brine injection method" for measuring the flow rate of the mixture for use on dredging operations. By injecting a salt solution into the discharge line, i.e., downstream from the pump, and then marking the time that the solution takes to pass two electrodes which are a certain known distance apart, the velocity of the mixture can be determined. It was mentioned that, if there is a sliding bed, i.e., a large concentration gradient with most of the solids traveling close to the bottom of the pipe, the velocity profile will be distorted. Therefore the velocity that is measured would tend to be the water velocity rather than the mixture velocity, and the measurement would be higher than for the mixture velocity. Another problem with brine injection is that complete mixing of the brine must be ensured before reaching the first electrode. If mixing is not accomplished, the measurement might be that of the velocity of a few streamlines and not one representative of the average mixture velocity. HOWARD (1940) also mentioned that the brine injection method cannot be used in a dredge operating in the ocean or any other salt-water environment. The brine injection method was successfully used by FÜHRBÖTER (1961) and WIEDENROTH (1967).

16.2.2 TRAJECTORY METHOD

The trajectory method described by HOWARD (1940) is a simple application of kinematics. With a measuring stick called the *velocity stick*, as seen in Fig. 16.1, the horizontal and vertical distances to a point in the stream of mixture emitted from the end of a discharge line are measured. The stream shooting from the discharge line follows a parabolic arc and falls because of the acceleration of gravity g. The horizontal displacement of an element of

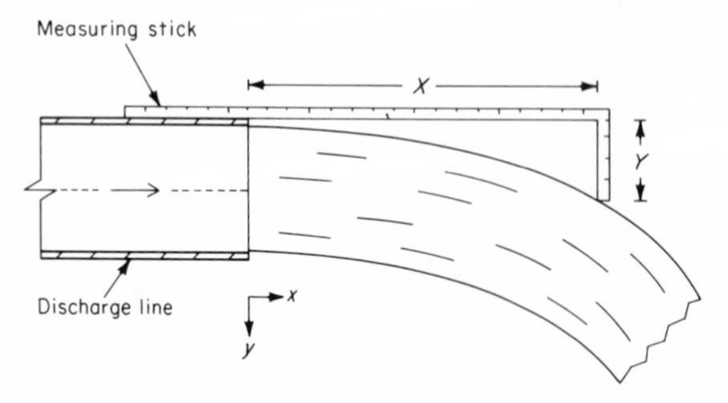

Fig. 16.1 Sketch of trajectory method, using a measuring stick.

the stream after a time t is

$$x = vt \tag{16.1}$$

where v is the velocity of the stream element in the discharge line. The y displacement is

$$y = \tfrac{1}{2}gt^2 \tag{16.2}$$

Eliminating time t from Eqs. (16.1) and (16.2) and solving for the velocity gives

$$v = \sqrt{\frac{gX^2}{2Y}} \tag{16.3}$$

where X and Y are the distances read on the measuring stick. As can be seen from the position of the measuring stick in Fig. 16.1, only the velocity of the top streamline is actually being determined. Also, after leaving the pipe, the stream becomes aerated; thus the placing of the measuring stick becomes subjective. Therefore this method of velocity measurement should be considered inaccurate.

16.2.3 WHISTLE METER

The LABORATOIRE DAUPHINOIS D'HYDRAULIQUE (1953) at Grenoble, France, has developed a whistle meter, as seen in Fig. 16.2, to measure the flow rate of mixture. This device is placed at the end of the discharge line in dredging operations. The meter consists of a short length of pipe with the top of the pipe flattening and tapering down as the outlet is approached.

Neglecting energy losses and using the energy equation, the pressure difference corresponding to the difference in kinetic heads at the outlet and the measurement section just upstream of the meter can be expressed as

$$\left(\frac{\Delta p}{\gamma}\right)_m = \frac{V_1^2}{2g}\left(\frac{A_1^2}{A_2^2} - 1\right) \tag{16.4}$$

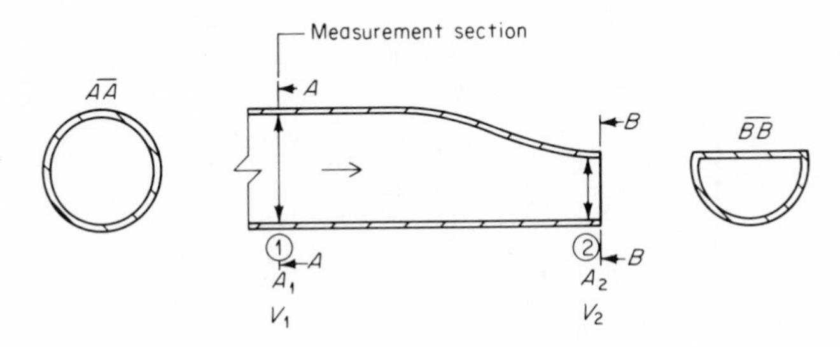

Fig. 16.2 Sketch of a whistle meter.

where $(\Delta p/\gamma)_m$ = pressure drop in length of clear water
A_1 = area of discharge line
A_2 = area of outlet
V_1 = average flow velocity of the mixture in discharge line

Whistle meters of various sizes were tested. It is reported that cast-iron chips, soot, and three types of sands were used as transported material in the tests. It was found that it is unnecessary to convert the pressure drop Δh from length of clear water to length of mixture; this implies that the pressure drop is apparently independent of solids concentration in the range tested, i.e., $C < 20$ percent. A whistle meter was used by WILSON (1965).

16.2.4 CONTRACTION METER

The LABORATOIRE DAUPHINOIS D'HYDRAULIQUE (1953*a*) used the concept that, for a solid-liquid mixture flowing upward in a vertical pipe, the head loss is the linear sum of the loss due to friction with clear water and the loss due to the influence of the specific gravity of the solids. With reference to Fig. 16.3, it was stated that the head loss for clear water between sections 1 and 2 a distance ΔL apart is

$$(\Delta h)_l = \frac{p_1 - p_2}{\gamma} - \Delta L \tag{16.5}$$

while the loss due to the presence of solids is

$$(\Delta h)_s = (s_m - 1)\,\Delta L \tag{16.6}$$

where s_m is the specific gravity of the mixture. The total loss for the upward-moving flow, therefore, is

$$(\Delta h)_m = (\Delta h)_l + (\Delta h)_s \tag{16.7}$$

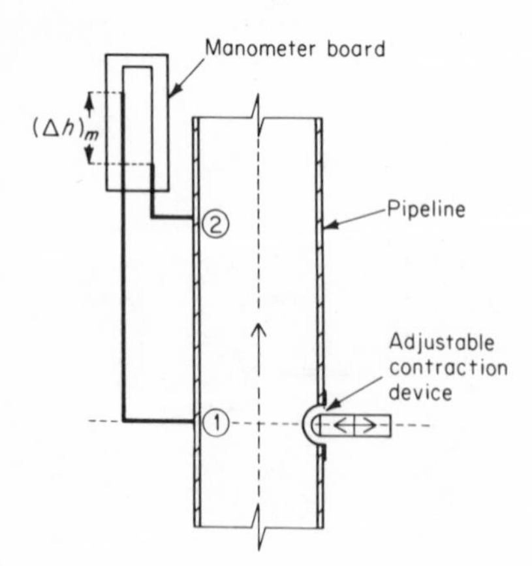

Fig. 16.3 Sketch of a contraction meter.

Previous Eqs. (16.5), (16.6), and (16.7) reflect the concept discussed earlier in Sec. 15.2.4.1.

If the flow rate were known from the whistle meter, the head loss due to clear water flow could be calculated. The total loss given by Eq. (16.7) would reduce to the head loss due to the solids only and could be solved for the concentration C if the specific weight of the solid material, γ_s, were known. The LABORATOIRE DAUPHINOIS D'HYDRAULIQUE (1953*a*) developed a contraction device, as seen in Fig. 16.3, to eliminate the necessity of having to know the flow rate in order to determine the concentration. The contraction device, which consists of a diaphragm and a plunger to stretch or release it, is placed at cross section 1. When the diaphragm is pushed in, a local increase in velocity occurs. Thus, for clear-water flow, the local velocity rise can be made large enough so that no head loss may be observed between sections 1 and 2. Now, with mixture flow, the loss between these sections is solely due to the solids, or $(\Delta h)_s = (s_m - 1)\,\Delta L$, and can be read off directly. The specific gravity of the mixture, s_m, as well as the concentration can be readily determined.

16.2.5 LOOP SYSTEM

GRAF (1962) and EINSTEIN et al. (1966) have developed a method for the simultaneous determination of both the mixture flow rate Q_m and the solids concentration C with a loop system, as seen in Fig. 16.4. The loop system consists essentially of two identical vertical pipe sections with opposite flow direction. Head differences are obtained over the vertical pipe sections. The two principles involved in the development of measuring equations are based on the continuity of the solid and liquid phases, and the concept that the head loss for flow in a vertical pipe is due to two terms, namely, the loss due to clear-water flow and the loss due to the presence of solids, as expressed by Eq. (15.2).

Assuming that the total-flow friction in running length of mixture equals the total-flow friction in running length of water, manometer equations

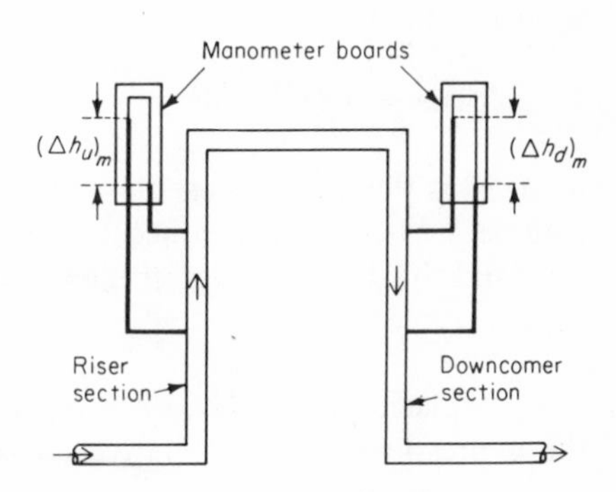

Fig. 16.4 Sketch of the loop system.

can be written for the head differences for flow through two identical vertical pipe sections with opposite flow directions. For flow in the riser,

$$(\Delta h_u)_m = LC_u(s_s - 1) + f\frac{L}{D}\frac{Q_m{}^2}{A^2}\frac{1}{2g}[1 + C_u(s_s - 1)] \tag{16.8}$$

and in the downcomer,

$$(\Delta h_d)_m = -LC_d(s_s - 1) + f\frac{L}{D}\frac{Q_m{}^2}{A^2}\frac{1}{2g}[1 + C_d(s_s - 1)] \tag{16.9}$$

where s_s = specific gravity of the solids
L = length of measurement section
f = Darcy-Weisbach friction factor
D = pipe diameter
A = pipe area
subscript u = upward flow
subscript d = downward flow

Equations (16.8) and (16.9) can be compared with Eq. (15.62).

Taking the sum of Eqs. (16.8) and (16.9) gives

$$(\Delta h_u)_m + (\Delta h_d)_m = L\{(s_s - 1)(C_u - C_d) + \Theta[2 + (s_s - 1)(C_u + C_d)]\} \tag{16.10}$$

and their difference gives

$$(\Delta h_u)_m - (\Delta h_d)_m = L\{(s_s - 1)(C_u + C_d) + \Theta[(s_s - 1)(C_u - C_d)]\} \tag{16.11}$$

where

$$\Theta = \frac{f}{D}\left(\frac{Q_m}{A}\right)^2\frac{1}{2g} \tag{16.12}$$

Applying the concept of continuity for both the solid and liquid phases, Einstein et al. (1966) showed that the change in concentration Δ in the riser or downcomer section due to the settling velocity v_{ss} of the solids can be given as

$$\Delta = \frac{v_{ss}A}{Q}C(1 - C)^2 \tag{16.13}$$

where Q is the liquid-phase flow rate and C is the average concentration by volume. Also, relative concentration equations can now be written for the riser and the downcomer respectively as

$$C_u = C + \Delta \qquad C_d = C - \Delta \tag{16.14}$$

Replacing in the head loss equations, given by Eqs. (16.8) and (16.9), the expressions for the relative concentrations C_u and C_d, given by Eq.

(16.14), and then replacing the concentration difference Δ by Eq. (16.13), Eqs. (16.10) and (16.11) become

$$\frac{(\Delta h_u)_m + (\Delta h_d)_m}{2L} = (s_s - 1)\Delta + \Theta[1 + (s_s - 1)C] \tag{16.15}$$

and

$$\frac{(\Delta h_u)_m - (\Delta h_d)_m}{2L} = (s_s - 1)(C + \Delta\Theta) \tag{16.16}$$

Equations (16.15) and (16.16) can be solved simultaneously for concentration C, and the flow rate of the mixture Q_m can be solved by a trial-and-error method.

As a first approximation, however, assume that the change in concentration within the loop system is insignificant, or $\Delta \ll 1$. Equations (16.15) and (16.16) reduce to

$$\frac{(\Delta h_u)_m + (\Delta h_d)_m}{2L} = \Theta[1 + (s_s - 1)C] \tag{16.17}$$

and

$$\frac{(\Delta h_u)_m - (\Delta h_d)_m}{2L} = (s_s - 1)C \tag{16.18}$$

The foregoing two equations can be solved directly for concentration C and flow rate Q_m. Beginning with these values, better ones may be obtained from Eqs. (16.15) and (16.16) by step calculation. Equation (16.15) or (16.17) shows that an addition of the differential manometer readings in head of mixture provides information on the frictional head loss across the vertical sections, which, in turn, gives information on the mixture flow rate. The subtraction of the manometric differences between two elevations, given with Eq. (16.16) or (16.18), is equal to the immersed weight of the solid material held in suspension between these two elevations.

The loop system was checked in an experimental apparatus developed by EINSTEIN et al. (1963). Experiments were conducted using two sands with mean diameters of 1.15 and 1.70 mm, and were studied in a loop system having 3-in.-diameter pipes. From the results of the experiments, it was concluded that the approximate Eq. (16.18) gives very good results if solved for the concentration C. The maximum error between calculated and experimental concentration was given as 5 percent for the range tested, i.e., $C < 20$ percent. However, the approximate Eq. (16.17) for finding the flow rate Q_m did not give such good results. Therefore the authors suggest the use of Eq. (16.15) for determining the flow rate of mixture.

It should be mentioned here that DURAND (1953*a*) and BROOK (1962) developed a device similar to the one developed by GRAF (1962) and EINSTEIN et al. (1966), which was referred to as a *counterflow meter*. BROOK's (1962) development of the manometer equations was similar to the one by EINSTEIN et al. (1966), but no account was taken of the change of concentration in the riser and downcomer pipe sections in the derivation of the equations. Because of this fact, BROOK (1962) stated quite correctly that his equations are limited, and are useful only when the mean flow velocity is much greater than the settling velocity of the solids. This statement leads to the same result as saying that relative concentrations do not exist in the loop. BROOK's (1962) equations are essentially of the type given with Eq. (15.57), and have been originally proposed by DURAND (1953*a*). Furthermore, it should be pointed out that two systems based on principles similar to the loop system have been reviewed by HUFF et al. (1961).

The loop system was successfully used by ACAROGLU (1968) and WEISMAN (1968) for determination of mixture flow rate and concentration. WILSON (1965) used the counterflow meter for determining the concentration; for all but small concentrations it was found to work satisfactorily.

16.2.6 MODIFIED VENTURI METER

GRAF (1962, 1967) tested the practicability of a venturi meter to measure both mixture velocity and the concentration in a solid-liquid mixture. The research was limited to the study of a single venturi meter, as shown in Fig. 16.5. The pressure drop between sections 1 and 2 and the energy loss across the meter between sections 3 and 4, were observed.

Experimentally it was found that the pressure drop in the column of mixture between sections 1 and 2 is a function of the flow rate. It is shown that the laws explaining the venturi effect with clear-water flow also explain the flow of mixture through the meter. Using the equations of continuity and of energy, the flow rate of mixture Q_m was given as

$$Q_m = c_v \frac{A_2}{\sqrt{1 - \left(\frac{A_2}{A_1}\right)^2}} \sqrt{\frac{2g \, \Delta p}{\gamma_m}} \tag{16.19}$$

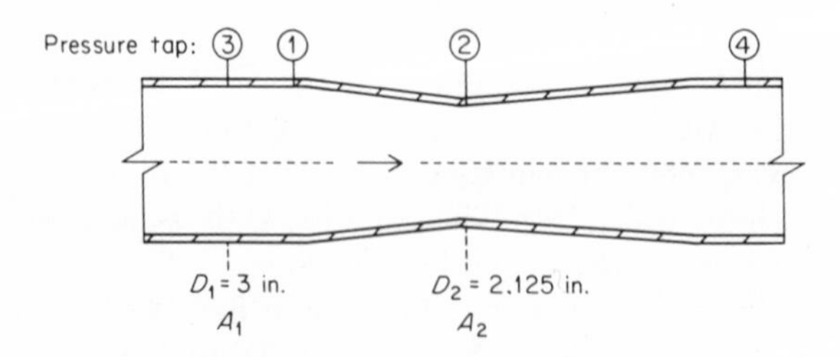

Fig. 16.5 Modified venturi meter. [*After* GRAF (*1962*).]

where c_v = venturi discharge coefficient
A_1 = the pipe area
A_2 = throat area
Δp = pressure drop
γ_m = specific weight of mixture

Furthermore, it was found that the total head loss through the venturi is a function of concentration and gives a linear relationship for a constant flow rate. Since this head loss is the linear sum of a clear-water loss and a loss due to solids, GRAF (1967) separated the two effects and plotted head loss solely due to solids against concentration of solids. This relation is a single, straight line covering all flow rates. In an attempt to understand these experimental findings, it is necessary to describe the resistance of grains in water for an accelerated and decelerated motion.

Two relations now exist: a pressure drop as a function of mixture flow rate in terms of concentration, and a head loss as a function of both flow rate and concentration. These two equations can be solved simultaneously or a nomogram may be constructed, as seen in Fig. 16.6. The limited experimental results reported by GRAF (1962, 1967) were obtained with a single venturi meter having a pipe diameter of 3 in. and a throat diameter of 2.125 in. Two sands were used, one having a density of $s_s = 2.607$ and a mean diameter of $d_{50} = 1.16$ mm, and the other sand with $s_s = 2.726$ and $d_{50} = 1.70$ mm.

BROOK (1962) also tested venturi meters, both in horizontal and vertical positions, but only for the use of measuring the mixture flow rate. He recorded the pressure drop from the pipe to the throat or between sections 1 to 2 in Fig. 16.5. It was found that the pressure drop vs. flow rate relation for mixture flow obeyed the same law as for clear-water flow. Such an instrument was also mentioned by WORSTER et al. (1955).

16.2.7 MODIFIED ELBOW METER

The possibility of using a pipe bend to measure the mixture flow rate was suggested by BROOK (1962). In a further investigation by GRAF et al. (1969) a 90° bend, an elbow meter, was checked out to measure flow rate and concentration of a mixture. The elbow meter is shown in Fig. 16.7; pressure drop across the elbow, taps I and II, and head loss over the pipe bend, taps *A* and *B*, were recorded. It was found that the pressure drop vs. flow rate relation was the same for clear-water flow and mixture flow. The total head loss over the pipe bend for mixture flow was found to be the sum of a head loss due to clear water and a head loss due to the presence of solids; this is similar to Eq. (15.2). The head loss due to solids was shown to be a linear function of the concentration. Thus the elbow meter can be calibrated

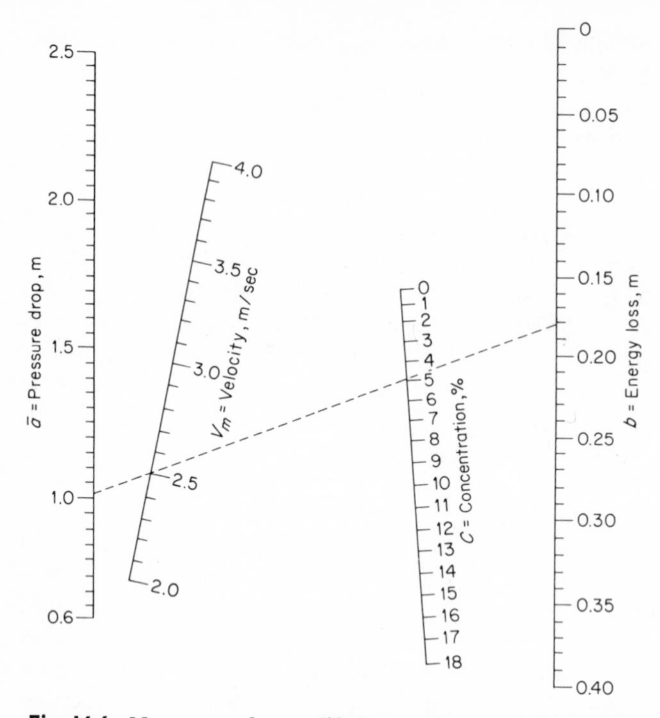

Fig. 16.6 Nomogram for modified venturi meter [*after* LEVENS (*1962*)]. Relationship between pressure drop, energy loss, flow velocity, and concentration of mixture in a modified venturi meter. (*Example:* Given are readings, $\bar{a} = 1.01$ m and $b = 0.182$ m. Find: $V_m = 2.5$ m/sec and $C = 5$ percent.)

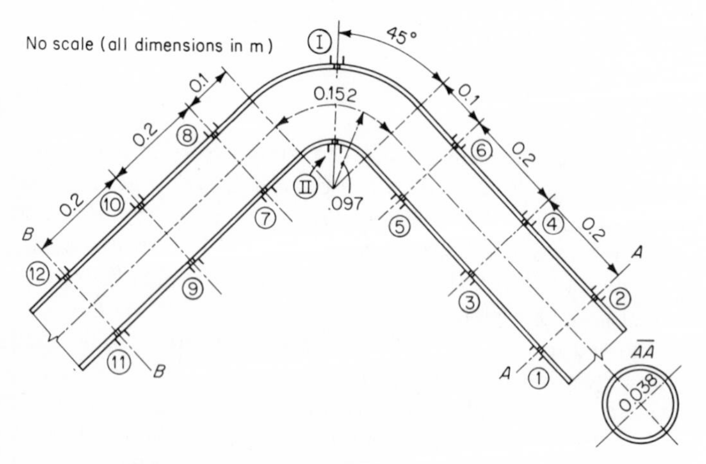

Fig. 16.7 Elbow meter. [*After* GRAF *et al.* (*1969*).]

and used as a solid-liquid-mixture measuring device. The pressure drop leads to the flow rate, and the knowledge of flow rate and total head loss permits calculation of the concentration. Simplicity of construction, installation, and use make the modified elbow meter an efficient and inexpensive device.

16.2.8 CONTRACTION-EXPANSION METER

ACKERMAN et al. (1964) proposed a meter, as shown in Fig. 16.8. The meter consists of a tapered section connected to a sudden expansion section. ACKERMAN et al. (1964) developed equations for the meter, which is placed in a vertical manner with flow moving downward.

In the development of equations, ACKERMAN et al. (1964) first consider continuity of both the solid and liquid phases. Then an energy relation between sections 1 and 2 is considered by equating the flux in kinetic energy across sections 1 and 2 with the rate of work done on the mixture. The resultant elaborate equation can be expressed as a functional relationship, or

$$\Delta h_{\overline{12}} = f_a\left(\frac{Q_s}{Q}, Q\right) \tag{16.20}$$

where $\Delta h_{\overline{12}}$ = head difference between sections 1 and 2
Q = liquid flow rate
Q_s = solids flow rate

Next, by considering the flow between sections 3 and 4, a momentum equation is written. This results in another large and complex equation, which is expressed functionally as

$$\Delta h_{\overline{34}} = f_b\left(\frac{Q_s}{Q}, Q\right) \tag{16.21}$$

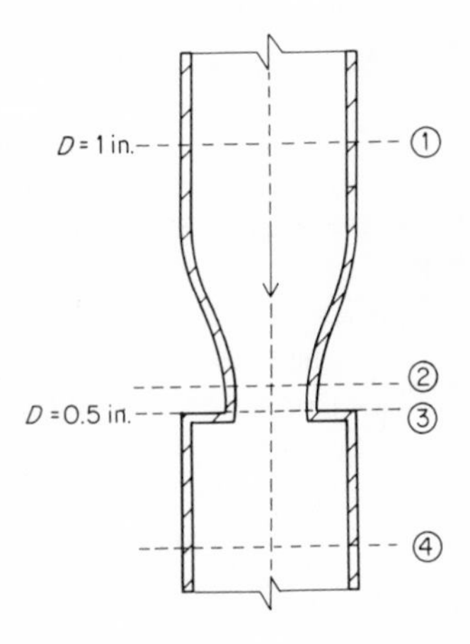

Fig. 16.8 Sketch of contraction-expansion meter.

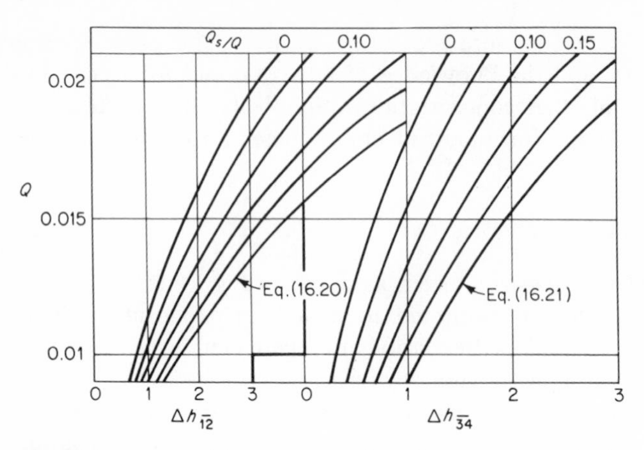

Fig. 16.9 Relationship between $\Delta h_{\overline{12}}$, $\Delta h_{\overline{34}}$, Q, and Q_s/Q. [*After* ACKERMAN *et al.* (*1964*).]

where $\Delta h_{\overline{34}}$ is the head difference between sections 3 and 4. Equations (16.20) and (16.21) may be solved simultaneously for the flow rate ratio Q_s/Q and the liquid flow rate Q. Graphs were drawn, as is shown in Fig. 16.9, which facilitate the simultaneous solution of the equations. To use the graphs, enter the abscissa with the values $\Delta h_{\overline{12}}$ and $\Delta h_{\overline{34}}$ and find the values of flow rate ratio Q_s/Q and liquid flow rate Q that satisfy both readings. ACKERMAN et al. (1964) experimentally verified their equations and graphs by testing a single meter with pipe diameter $D = 1$ in. and a throat diameter of $D = 0.5$ in. The sand used was nonuniform, with diameters in a size range of 0.0232 to 0.0331 in. and a median settling velocity of $v_{ss} = 0.285$ fps. The specific gravity of the sand was $s_s = 2.63$. It was reported that the simultaneous solution of Eqs. (16.20) and (16.21) predict the mixture discharge within ± 2 percent accuracy and the concentration to within $\frac{1}{2}$ percent of the measured values.

16.2.9 ELECTRICAL SYSTEMS

An electrical system, drawn in Fig. 16.10, is a combination of a magnetic flow meter to measure the flow rate of mixture and a radioactive source and receiving cell to measure its density.

The magnetic flow meter uses Faraday's law of induction as its principle. The mixture acts as a conductor moving through an electrically propagated magnetic field and, thereby, induces a voltage. This induced voltage is proportional to the mixture velocity. The flow rate can then be obtained. The electromagnetic flow meter was either suggested for use or used by HUFF et al. (1961), WILSON (1965), and FORTINO (1966).

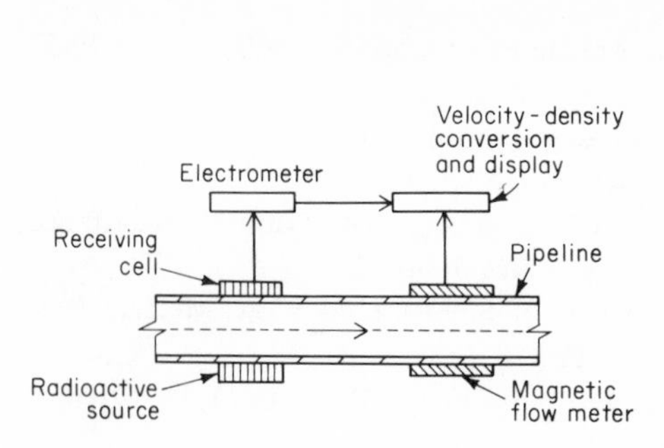

Fig. 16.10 Sketch of an electrical system.

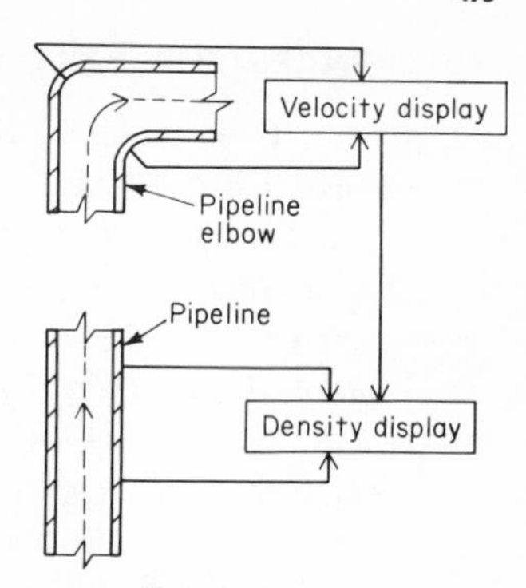

Fig. 16.11 Sketch of a pneumatic system.

For the density measurements, a radioactive source is placed in a shielding box at the bottom of a pipe and a receiving cell is placed right opposite the source at the top. With proper calibration the density of the mixture can be correlated with the amount of radiation getting through the mixture to the receiving cell.

An electrical system has been discussed by Fortino (1966), who reported that it is operated on several dredges in the United States. A similar system was reviewed by Brook (1962).

16.2.10 PNEUMATIC DEVICES

Fortino (1966) also suggested the following device to measure concentration and flow rate. It was suggested to use an elbow meter to determine the flow rate of mixture. With reference to Fig. 16.11, two pressure taps are placed in a vertical length of pipe and pressure gages or differential pressure instruments were used. To avoid clogging, a purge system was installed. The concentration can be determined in a manner similar to that used for the contraction device developed by the Laboratoire Dauphinois d'Hydraulique (1953).

16.3 CLOSING REMARKS

Various measuring devices have been developed and the uses of some have been reported. Some of the studies have been rather narrow in scope, i.e., the testing of some devices has not been done under a wide range of conditions. But, generally speaking, some of the measuring devices reported herein can be used confidently, especially in operations where high degrees of accuracy are not needed, such as dredging installations. Other devices need to be tried and tested in the future.

REFERENCES FOR PART FOUR

ACAROGLU, E. R. (1968): "Sediment Transport in Conveyance Systems," Ph.D. Thesis, Cornell Univ.

———, and W. H. GRAF (1968): Sediment Transport in Conveyance Systems, Part 2, *Bull. Intern. Assoc. Sci. Hydr.* vol. XIII, no. 3.

———, and ——— (1969): The Effect of Bedforms on the Hydraulic Resistance, *Intern. Assoc. Hydr. Res., 13th Congr., Kyoto*, vol. 2.

ACKERMAN, N. et al. (1964): Development of Solid-Liquid Flow Meters, *Proc. Am. Soc. Civil Engrs.*, vol. 90, no. HY2.

ALBRIGHT, C. W. et al. (1951): Pressure Drop in Flow of Dense Coal-Air Mixture, *Ind. Eng. Chem.*, vol. 43, no. 8.

AMBROSE, H. H. (1952): The Transportation of Sand in Pipes; Free Surface Flow, *Proc. 5th Hydr. Conf., Univ. of Iowa.*

BABITT, H. E., and D. H. CALDWELL (1939): Laminar Flow of Sludges in Pipes with Special Reference to Sewage Sludge, *Bull. V. XXXVII/12, Univ. Illinois.*

BAGNOLD, R. A. (1956): Flow of Cohesionless Grains in Fluids, *Phil. Trans. Roy. Soc. London*, ser. A, vol. 249, no. 964.

——— (1966): An Approach to the Sediment Transport Problem from General Physics, *U.S. Geol. Survey, Prof. Paper* 422-J.

BARR, D., and J. RIDELL (1968): Homogeneous Suspensions in Circular Conduits: A Discussion, *Proc. Am. Soc. Civil Engrs.*, vol. 94, no. PL1.

BARTH, W. (1960): Physikalische und wirtschaftliche Probleme des Transportes von Festteilchen in Flüssigkeiten und Gasen, *Chem. Ing. Techn.*, Jgg. 32, no. 3.

BEHN, V. C. (1960): Flow Equations for Sewage Sludges, *J. Water Poll. Control Fed.*, July.

——— (1960*a*): Derivation of Flow Equations for Sewage Sludges, *Proc. Am. Soc. Civil Engrs.*, vol. 86, no. SA6.

——— (1962): Experimental Determination of Sludge-Flow Parameters, *Proc. Am. Soc. Civil Engrs.* vol., 88, no. SA3.

BELDEN, D. H., and L. S. KASSEL (1949): Pressure Drops, *Ind. Eng. Chem.*, vol. 41, no. 6.

BINGHAM, E. C. (1922): "Fluidity and Plasticity," McGraw-Hill, New York.

BINNIE, A. M., and T. C. KU (1966): The Mean Velocity of Nylon Spheres Transported in a Horizontal Water Pipe, *Brit. J. Appl. Phys.*, vol. 17, no. 7.

BIRD, R. B., W. E. STEWART, and E. N. LIGHTFOOT (1960): "Transport Phenomena," Wiley, New York.

BLATCH, N. S. (1904): Works for the Purification of the Water Supply of Washington: A Discussion, *Trans. Am. Soc. Civil Engrs.*, vol. 57.

BOBKOWICZ, A. J., and W. N. GAUVIN (1965): The Turbulent Flow Characteristics of Model Fibre Suspension, *Can. J. Chem. Eng.*, vol. 43.

BOHNET, M. (1965): Experimentelle und theoretische Untersuchungen . . . , *VDI Forschungsh.*, no. 507.

BONNINGTON, S. T. (1959): Experiments on the Hydraulic Transport of Mixed-Sized Solids, *British Hydrom. Res. Assoc.*, RR637.

BOOTHROYD, R. G. (1969): Similarity in Gas-Borne Flowing Particulate Suspensions, *Trans. Am. Soc. Mech. Engrs., J. Eng. Ind.*, vol. 91/B, no. 2.

BREBNER, A. (1962): An Introduction to Aqueous Hydraulic Conveyance of Solids in Pipelines, *Civil Eng. Dept., Queen's Univ., Kingston (Ontario), Rept. no. 21.*

———, and K. WILSON (1964): Theoretical Considerations of Vertical Pumping of Mine Products, *Civil Eng. Dept., Queen's Univ., Kingston (Ontario), Rept. no. 33.*

BRODKEY, R. S., J. LEE, and R. C. CHASE (1961): A Generalized Velocity Distribution for Non-Newtonian Fluids, *Am. Inst. Chem. Eng.*, vol. 7/3.

BROOK, N. (1962): Flow Measurements of Solid-Liquid Mixtures Using Venturi and Other Meters, *Proc. Inst. Mech. Engrs.*, vol. 176, no. 6.

BUGLIARELLO, G., and J. W. DAILY (1961): Rheological Models and Laminar Shear Flow of Fiber Suspensions, *J. Techn. Assoc. Pulp Paper Ind.*, vol. 44/12.

CARSTENS, M. R. (1969): A Theory of Heterogeneous Flow of Solids in Pipes, *Proc. Am. Soc. Civil Engrs.*, vol. 95, no. HY1.

CARVER, C. E., and R. H. NAVOLINK (1965): Measurement of Laminar Velocity Profiles with Non-Newtonian Additives Using Photomicroscopy, *Eng. Res. Inst., Univ. of Mass., Rept. no. 1.*

CHAMBERLAIN, A. R. et al. (1960): Transport of Material by Fluids in Pipes, *Civil Eng. Sec., Colo. State Univ., Fort Collins, CER60 ARC43.*

CHARLES, M. E. (1963): The Pipeline Flow of Capsules; Part 2, *Can. J. Chem. Eng.*, vol. 41/2.

CLEGG, D. B., and R. L. WHITMORE (1966): Boundary Layers in Bingham Plastics, *Rheologica Acta*, Bd. 5/2.

COLORADO SCHOOL OF MINES (1963): The Transportation of Solids in Steel Pipelines, *Am. Iron Steel Inst., Steel.*

CONDOLIOS, E. (1967): Transport of Materials in Bulk or in Container by Pipelines, *United Nations Publ. no. 66-VIII-1.*

———, and E. E. CHAPUS (1963): Transporting Solid Materials in Pipelines, *Chem. Eng.*, June-July.

———, ———, and J. CONSTANS (1967): New Trends in Solids Pipelines, *Chem. Eng.*, May.

CRAVEN, J. P. (1952): The Transportation of Sand in Pipes; Full Pipe Flow, *Proc. 5th Hydr. Conf., Univ. of Iowa.*

DAILY, J. W., and G. BUGLIARELLO (1961): Basic Data for Dilute Fiber Suspensions in Uniform Flow with Shear, *J. Techn. Assoc. Pulp Paper Ind.*, vol. 44/7.

———, and P. R. ROBERTS (1966): Solid Particle Suspensions in Turbulent Shear Flow, *J. Techn. Assoc. Pulp. Paper Ind.*, vol. 49/3.

DENT, E. J. (1939): Transportation of Sand and Gravel in a Four-Inch Pipe: A Discussion, *Trans. Am. Soc. Civil Engrs.*, vol. 104.

DODGE, D. W., and A. B. METZNER (1959): Turbulent Flow of Non-Newtonian Systems, *J. Am. Inst. Chem. Engrs.*, vol. 5/2.

DURAND, R. (1953): Basic Relationships of the Transportation of Solids in Pipes—Experimental Research, *Intern. Assoc. Hydr. Res., 5th Congr. Minneapolis.*

——— (1953*a*): Ecoulements de Mixture en Conduites Verticales, *La Houille Blanche*, no. Spec. A.

———, and E. CONDOLIOS (1956): Données Techniques sur le Refoulement hydraulique des Matériaux solides en Conduite, *Rev. l'Industrie Minerals*, Special Number 1F, June.

EINSTEIN, H. A. (1942): Formulas for the Transportation of Bed-Load, *Trans. Am. Soc. Civil Engrs.*, vol. 107.

——— (1950): The Bed-Load Function for Sediment Transportation in Open Channel Flow, *U.S. Dept. Agric., Tech. Bull. no. 1026.*

———, and N. CHIEN (1955): Effects of Heavy Sediment Concentration Near the Bed on Velocity and Sediment Distribution, *Univ. of Calif., Berkeley, MRD Sed. Ser. no. 8.*

———, and W. H. GRAF (1963): Experimental Apparatus Studies Sediment Transport in Closed Conduits, *Civil Eng.* (*Am. Soc. Civil Engrs.*), October.

———, and ——— (1966): Loop System for Measuring Sand-Water Mixtures, *Proc. Am. Soc. Civil Engrs.*, vol. 92, HY1.

ELLIS, H. S. (1964): The Pipeline Flow of Capsules; Part 3, Part 4, and Part 5, *Can. J. Chem. Eng.*, vol. 42, nos. 1, 2, 4.

———, P. J. REDBERGER, and L. BOLT (1963): Transporting Solids by Pipelines, *Chem. Eng.*, vol. 55, no. 9.

FAIRBANK, L. C. (1942): Effect on the Characteristics of Centrifugal Pumps, *Trans. Am. Soc. Civil Engrs.*, vol. 107.

FARBAR, L. (1949): Flow Characteristics of Solids-Gas Mixtures, *Ind. Eng. Chem.*, vol. 41, no. 6.

FORTINO, E. (1966): Flow Measurement Technique for Hydraulic Dredges, *Proc. Am. Soc. Civil Engrs.*, vol. 92, WW1.

FÜHRBÖTER, A. (1961): Über die Förderung von Sand-Wasser-Gemischen in Rohrleitungen, *Mitteil. Franzius-Inst., Techn. Hochschule Hannover*, Heft 19.

FREDRICKSON, A. G. (1964): "Principles and Applications of Rheology" Prentice-Hall, Englewood Cliffs, N.J.

GARSTKA, W. U. (1961): Erosion Studies of Pipe Lining Materials, *U.S. Dept. of Int., Bureau of Recl., Chem. Eng. Lab. Rept. no. P-79.*

GASTERSTÄDT, J. (1924): Die experimentelle Untersuchung des pneumatischen Fördervorganges, *Forsch. Arb. Ing. Wesen, no. 265*, VDI-Verlag, Berlin.

GIBERT, R. (1960): Transport Hydraulique et Refoulement des Mixtures en Conduit, *Ann. Pontes Chaussees*, 130e année, nos. 12, 17.

GOVIER, G. W., and M. E. CHARLES (1961): The Hydraulics of the Pipeline Flow of Solid-Liquid Mixtures, *Eng. J.* (*Can.*), August.

———, B. A. RADFORD, and J. DUNN (1957): The Upward Vertical Flow of Air-Water Mixtures, *Can. J. Chem. Eng.*, August.

GRAF, W. H. (1962): Investigation on a Two-Phase Problem in Closed Pipes, *Univ. of Calif., Inst. Eng. Res., HEL 2-2*, December.

——— (1967): A Modified Venturi Meter for Measuring Two-Phase Flow, *J. Hydr. Res.*, vol. 5, no. 3.

——— (1968): List of Sources of Information of Pipeline Design: A Discussion, *Proc. Am. Soc. Civil Engrs.*, vol. 94, no. PL1.

———, and E. R. ACAROGLU (1967): Homogeneous Suspensions in Circular Conduits, *Proc. Am. Soc. Civil Engrs.*, vol. 93, no. PL2.

———, and ——— (1968): Sediment Transport in Conveyance Systems; Part 1, *Bull. Intern. Assoc. Sci. Hydr.*, XIIIe année, no. 2.

———, and R. N. WEISMAN (1969): Continuous Measurement in Water-Sand Mixtures, *22d Intern. Nav. Congr.*, Sec. II/2.

———, M. ROBINSON, and Ö. YÜCEL (1970): The Critical Deposit Velocity for Solid-Liquid Mixtures, *Proc. Intern. Conf. Hydr. Transport of Solids in Pipes (Brit. Hydrom. Res. Assoc.), Coventry.*

HARIU, O. H., and M. C. MOLSTAD (1949): Pressure Drop in Vertical Tubes in Transport of Solids by Gases, *Ind. Eng. Chem.*, vol. 41.

HASRAJANI, S. U. (1962): A Bibliography on the Flow of Liquid-Solid Mixtures . . . , *British Hydrom. Res. Assoc., Harlow, Essex, BIB 8.*

HERBICH, J. B. (1963): Effects of Impeller Design Changes on Characteristics of a Model Dredge Pump, *Am. Soc. Mech. Engrs., Paper no. 63-AHGT-33.*

HERSHEY, H. C., and J. L. ZAKIN (1967): Existence of Two Types of Drag Reduction, *I and EC Fundamentals*, vol. 6, no. 3.

HERUM, F. L., G. W. ISAACS, and R. M. PEART (1964): Flow Properties of Highly Viscous Organic Wastes, *Am. Soc. Agric. Engrs., Annual Meeting.*

HOWARD, G. (1940): Measuring Velocities in Dredge Pipes, *Mech. Eng. (Am. Soc. Mech. Engrs.)*, April.

HUFF, W. R., and L. F. WILLMOTT (1961): Development and Operation of a Pilot Plant . . . , *U.S. Dept. of Int., Bureau of Mines, RI 5719.*

HUNT, W. A., and J. C. HOFFMANN (1968): Optimization of Pipelines Transporting Solids, *Proc. Am. Soc. Civil Engrs.*, vol. 94, no. PL1.

ISMAIL, H. M. (1952): Turbulent Transfer Mechanism and Suspended Sediment in Closed Channels, *Trans. Am. Soc. Civil Engrs.*, vol. 117.

KENNEDY, R. J. (1966): Towards an Analysis of Plug Flow through Pipes, *Can. J. Chem. Engrs.*, vol. 44/6.

KRIEGEL, E., and H. BRAUER (1966): Hydraulischer Transport körniger Feststoffe durch waagerechte Rohrleitungen, *VDI-Forschungsh.*, no. 515.

KRUYER, J., P. J. REDBERGER, and H. S. ELLIS (1967): The Pipeline Flow of Capsules: Part 9, *J. Fluid Mech.*, vol. 30/3.

KUTATELADZE, S. S., and M. A. STYRIKOVICH (1958): "Gidravlika Gaso-Zhidkostiykh Sistem," Gosudar. Energet. Izdat., Moscow. [Translation (1960) Tech. Inf. Center, Wright-Patterson Air Force Base, Ohio.]

LABORATOIRE DAUPHINOIS D'HYDRAULIQUE (1953): Contraction Flow Meter for Mixtures of Water and Materials, *La Houille Blanche*, January-February.

LABORATOIRE DAUPHINOIS D'HYDRAULIQUE (1953*a*): Concentration Measuring Instrument for Hydraulic Transport Installations, *La Houille Blanche*, May.

LEVENS, A. S. (1962): "Graphical Methods of Research," Wiley, New York.

LOCKHART, R. W., and R. C. MARTINELLI (1949): Proposed Correlation of Data for Isothermal Two-Phase, Two Component Flow, *Chem. Eng. Progr.*, vol. 45, no. 1.

MEHTA, N. C., J. M. SMITH, and E. M. COMINGS (1957): Pressure Drop in Air-Solid Flow Systems, *Ind. Eng. Chem.*, vol. 49.

METZNER, A. B. (1956): Non-Newtonian Technology, in "Advances in Chemical Engineering," vol. 1, Academic, New York.

——— (1961): Flow of Non-Newtonian Fluids, in "Handbook of Fluid Dynamics" (Streeter, editor), McGraw-Hill, New York.

———, and M. G. PARK (1964): Turbulent Flow Characteristics of Viscoelastic Fluids, *J. Fluid Mech.*, vol. 20/2.

NEWITT, D. M., et al. (1955): Hydraulic Conveying of Solids in Horizontal Pipes, *Trans. Inst. Chem. Engrs.*, vol. 33/2.

——— et al. (1961): Hydraulic Conveying of Solids and Vertical Pipes, *Trans. Inst. Chem. Engrs.*, vol. 39/2.

———, J. RICHARDSON, and C. SHOOK (1962): Hydraulic Conveying of Solids in Horizontal Pipes; Part 2, *Proc. Symp. Interaction between Fluids and Particles, Inst. Chem. Engrs.*

O'BRIEN, M. P., and R. G. FOLSOM (1937): The Transportation of Sand in Pipelines, *Univ. of Calif., Eng. Publ.*, vol. 3/7.

OWEN, P. R. (1969): Pneumatic Transport, *J. Fluid Mech.*, vol. 39/2.

PAINTAL, A. S., and R. J. GARDE (1966,67): Effect of Fine Sediment on the Characteristics of Flow in Smooth Pipes, *Univ. of Roorkee, Res. J.*, vol. IX, no. 3.

PATTERSON, G. K., and J. K. ZAKIN (1968): Prediction of Drag Reduction with a Viscoelastic Model, *J. Am. Inst. Chem. Eng.*, vol. 14, no. 3.

———, ———, and J. M. RODRIGUEZ (1969): Drag Reduction, *Ind. Eng. Chem.*, vol. 61, no. 1.

PINKUS, O. (1952): Pressure Drop in the Pneumatic Conveyance of Solids, *Trans. Am. Soc. Mech. Engrs.*, vol. 74; *J. Appl. Mech.*, p. 425.

POREH, M. et al. (1970): Drag Reduction in Hydraulic Transport of Solids, *Proc. Am. Soc. Civil Engrs.*, vol. 96, no. HY4.

RICHARDSON, J. F., and M. MCLEMAN (1960): Pneumatic Conveying—Part 2, *Trans. Inst. Chem. Engrs.*, vol. 38, no. 151.

ROSE, H. E., and H. E. BARNACLE (1957): Flow of Suspensions of Non-Cohesive Spherical Particles in Pipes, *The Engr. (London)*, vol. 203, nos. 5290 and 5291.

———, and R. A. DUCKWORTH (1969): Transport of Solid Particles in Liquids and Gases, *The Engr. (London)*, vol. 227, nos. 5903, 5904, 5905.

SAKTHIVADIVEL, R. (1967): Bed-Load Transport at High Shear Stress: A Discussion, *Proc. Am. Soc. Civil Engrs.*, vol. 93, HY4.

SASSOLI, F. (1963): Transporto solido nella correnti in pressione, anche con concentrazioni notevoli, *L'Acqua*, vol. XLI/6.

SCHMIDT, J. J., and G. LIMEBEER (1965): Preliminary Experiences in Operating a 120-Ton-Per-Hour Hydraulic Coal Transport System, *The South African Mech. Engr.*, February.

SEYER, F. A., and A. B. METZNER (1967): Turbulent Flow Properties of Viscoelastic Fluids, *Can. J. Chem. Eng.*, vol. 45/3.

———, and ——— (1969): Turbulence Phenomena in Drag Reducing Systems, *Am. Inst. Chem. Eng.*, vol. 15, no. 3.

SHAVER, R. G., and E. W. MERILL (1959): Turbulent Flow of Pseudoplastic Polymer Solutions in Straight Cylindrical Tubes, *Am. Inst. Chem. Engrs.*, vol. 5/2, January.

SHIELDS, A. (1936): Anwendung der Ähnlichkeitsmechanik und Turbulenzforschung auf die Geschiebebewegung, *Mitteil.*, *PVWES*, *Berlin*, no. 26.

SHIH, C. C. (1964): Hydraulic Transport of Solids in a Sloped Pipe, *Proc. Am. Soc. Civil Engrs.*, vol. 90, no. PL2.

SHOOK, C. A., and S. M. DANIEL (1965): Flow of Suspensions of Solids in Pipelines; Part 1, *Can. J. Chem. Engrs.*, vol. 43.

———, et al. (1968): Flow of Suspensions in Pipelines, *Can. J. Chem. Eng.*, vol. 46.

SILIN, N. A. et al. (1969): Research of the Solid-Liquid Flows with High Consistence, *13th Congr. Intern. Assoc. Hydr. Res.*, *Kyoto*, vol. 2.

SINCLAIR, C. G. (1962): The Limit Deposit-Velocity of Heterogeneous Suspensions, *Proc. Symp. on the Interaction between Fluids and Particles*, *Inst. Chem. Engrs.*

SMITH, R. A. (1955): Experiments on the Flow of Sand-Water Slurries in Horizontal Pipes, *Trans. Inst. Chem. Engrs.*, vol. 33, no. 2.

SPELLS, K. E. (1955): Correlations for Use in Transport of Aqueous Suspensions of Fine Solids through Pipes, *Trans. Inst. Chem. Engrs.*, vol. 33.

SPENCER, J. D., et al. (1966): Pneumatic Transportation of Solids, *U.S. Dept. Int.*, *Bur. Mines*, *IC 8314*.

SOO, S. L. (1967): "Fluid Dynamics of Multiphase Systems," Blaisdell, Waltham, Mass.

STEPANOFF, A. J. (1965): "Pumps and Blowers; Two-Phase Flow," Wiley, New York.

TEK, M. R. (1961): Two-Phase Flow, in "Handbook of Fluid Dynamics" (Streeter, editor), Sec. 17, McGraw-Hill, New York.

THOMAS, D. G. (1961): Laminar-Flow Properties of Flocculated Suspensions, *J. Am. Inst. Chem. Engrs.*, vol. 7/3.

——— (1962): Transport Characteristics of Suspensions; Part 6, *Am. Inst. Chem. Engrs.*, vol. 8, no. 3.

——— (1962*a*): Significant Aspects of Non-Newtonian Technology and Transport Characteristics of Suspensions, in "Progress in International Research on Thermodynamic and Transport Properties," *Am. Soc. Mech. Engrs.*, New York.

THOMAS, D. G. (1964): Periodic Phenomena Observed with Spherical Particles in Horizontal Pipes, *Am. Assoc. Adv. Sci., Science*, vol. 144, May.

———, et al. (1964): Research in Non-Newtonian Fluid Mechanics, *Chem. Eng. Progr.*, vol. 60, no. 8.

TODA, M., et al. (1969): Hydraulic Conveying of Solids through Horizontal and Vertical Pipes, *Kagaku Kogaku* (*Chem. Eng., Japan*), vol. 33/1.

TONG, L. S. (1965): "Boiling Heat Transfer and Two-Phase Flow," Wiley, New York, p. 75.

VANONI, V. (1946): Transportation of Suspended Sediment by Water, *Trans. Am. Soc. Civil Engrs.*, vol. 111.

VOGT, E. G. and R. R. WHITE (1948): Friction in Flow of Suspensions, *Ind. Eng. Chem.*, vol. 40, no. 9.

WALLIS, G. B. (1969): "One-Dimensional Two-Phase Flow," McGraw-Hill, New York.

WEBER, M., and N. SCHAUKI (1967): Pneumatische und hydraulische Förderung, *Aufbereitungs-Techn.*, 37. Jgg., no. 10.

WEISMAN, R. N. (1968): "An Elbow Meter in a Solid-Liquid Mixture Flow," M.S. Thesis, Cornell Univ.

———, and W. H. GRAF (1968): Measuring Solid-Liquid Mixture in Closed Conduits, *Proc. Am. Soc. Civil Engrs.*, vol. 94, no. WW4.

WHITMORE, R. L. (1963): Hemorheology and Hemodynamics, *Biorheology*, vol. 1.

WIEDENROTH, W. (1967): "Untersuchungen über die Förderung von Sand-Wasser-Gemischen durch Rohrleitungen und Kreiselpumpen," Dr. Diss., Techn. Hochschule, Hannover.

WILLMOTT, L. F., W. R. HUFF, and W. E. CROCKETT (1963): Aqueous Slurries of Coal and Granular Materials: A Bibliography, *U.S. Dept. of Int., Bureau of Mines, IC 8165.*

WILSON, K. C. (1965): Derivation of the Regime Equations from Relationships for Pressurized Flow . . . , *Civil Eng. Dept., Queen's Univ., Kingston* (*Ontario*), *Rept. no. 51.*

——— (1966): Bed-Load Transport at High Shear Stress, *Proc. Am. Soc. Civil Engrs.*, vol. 92, no. HY6.

——— (1970): Slip Point of Beds in Solid-Liquid Pipeline Flow, *Proc. Am. Soc. Civil Engrs.*, vol. 96, no. HY1.

WILSON, W. E. (1942): Mechanics of Flow, with Noncolloidal Inert Solids, *Trans. Am. Soc. Civil Engrs.*, vol. 107.

WORSTER, R. C., and D. F. DENNY (1955): The Hydraulic Transport of Solid Material in Pipes, *Proc. Inst. Mech. Engrs.*, vol. 169/32.

ZANDI, I., and G. GOVATOS (1967): Heterogeneous Flow of Solids in Pipelines, *Proc. Am. Soc. Civil Engrs.*, vol. 93, no. HY3.

———, and R. H. RUST (1965): Turbulent Non-Newtonian Velocity Profiles in Pipes, *Proc. Am. Soc. Civil Engrs.*, vol. 91, no. HY6.

ZENZ, F. A. (1949): Two-Phase Fluid Solid Flow, *Ind. Eng. Chem.*, vol. 41, no. 12.

Name Index

Abdel-Rahman, N. M., 340
Acaroglu, E. R., 281, 426, 427, 442–448, 462, 490
Ackerman, N., 493–494
Ackers, P., 248
Adachi, S., 269
Albertson, M., 52
Albright, C. W., 474
Allen, J., 385, 390, 393
Allen, J. R. L., 274, 276, 279, 285, 341
Ambrose, H. H., 449
Ananian, A. K., 188, 269
Anderson, A. G., 175, 267, 294–295
Ansley, R. W., 219
Apmann, R. P., 172
Arnborg, L., 368

Babitt, H. E., 456
Bagnold, R. A., 150–151, 153–155, 208–211, 274, 286, 445–449
Bajorunas, L., 315
Barekyan, A. S., 89, 133
Barnes, H. H., 305
Barr, D. I., 387, 453
Barth, W., 439, 470, 476
Baschin, O., 274
Basset, A., 31, 34
Bates, R. E., 261
Bazin, H. E., 21
Behn, V. C., 456, 458, 459, 461
Belden, D. H., 474
Bell, H. S., 353
Benedict, P. C., 268, 380
Beverage, J. P., 352
Bhattacharya, P. K., 379
Bingham, E. C., 453
Binnie, A. M., 266
Bird, R. B., 455
Bisal, F., 154
Bishop, A. A., 215–218
Biswas, A. K., 3
Blasius, H., 274
Blatch, N. S., 428
Blench, T., 243–245, 248–251, 253–255, 259–260, 272, 393–394
Bobkowicz, A. J., 460
Bodman, G. B., 330
Bogardi, J. L., 98, 215, 281
Bohnet, M., 476
Bondurant, D. C., 214
Bonnington, S. T., 430, 432, 434–436, 452
Boothroyd, R. G., 476
Boussinesq, J., 31
Brahms, A., 14
Brandeau, G., 367
Brebner, A., 422, 452
Brenner, H., 53, 57, 62–63
Bretting, A. E., 121
Brodkey, R. S., 468
Brook, N., 489–491, 495
Brooks, N. H., 114, 194–195, 317
Brown, C. B., 128, 145
Brunings, C., 19
Brush, L. M., 34, 177, 179, 272
Bugliarello, G., 199, 454, 460
Bürja, A., 19
Byrne, J. V., 383

Callander, R. A., 268
Camp, T. R., 171
Carey, W. C., 321
Carlson, E. J., 337
Carstens, M. R., 114, 177, 449
Carver, C. E., 468
Casey, H. J., 135
Castelli, B., 13

Chamberlain, A. R., 449, 470
Chang, F. M., 211–212
Chang, Y., 93, 127
Charles, M. E., 477
Chepil, W. S., 105, 154
Cherry, J., 383
Chezy, A., 14
Chien, N., 99, 100, 101, 149–150, 177–178, 180, 248
Chitale, S. V., 260, 264
Chiu, C. L., 182
Chow, V. T., 121, 305, 343
Christiansen, J. E., 175
Clamen, A., 61
Clegg, D. B., 455
Colby, B. R., 175, 204–208, 221, 233–234, 238, 240, 241
Coleman, N. L., 105
Colorado School of Mines, 433, 470, 473, 476
Condolios, E., 422, 427, 431–436, 440–442, 450, 468–470, 472–474, 477
Corrsin, S., 32
Courtois, G., 381, 382
Craven, J. P., 449
Crickmore, M. J., 152–279, 286, 381, 383, 384
Ctesias, 9
Culbertson, J. K., 281, 321
Cummings, R. S., 383
Cunha, L. V., 312

Daily, J. W., 455, 466
Dallavalle, J., 42
Darcy, H. P., 21, 304
D'Aubuisson De Voisins, J. F., 20
DaVinci, L., 11–12
Davis, C. V., 343
Dawdy, D. R., 321
Debski, K., 364, 367
Dent, E. J., 428
DeProny, R., 19
DeVries, M., 159, 384
Dobbins, W. E., 169–171
Dobson, G. C., 365
Dodge, D. W., 457, 468
Donat, J., 125, 126, 131, 132
DuBoys, M. P., 22, 124–126, 130–131, 396
DuBuat, P., 13–20, 22, 86, 90, 274, 338
Dunn, I. S., 334, 341
DuPlessis, M. P., 57
Dupuit, H. P., 21, 162
Durand, R., 427, 428, 431–436, 440, 450–452, 464, 467, 489–490

Egiazaroff, J. V., 98, 212
Ehrenberger, R., 159, 360
Einstein, A., 65–66, 267
Einstein, H. A., 105, 139–150, 154, 174, 178–180, 182–186, 190–194, 204, 206–208, 216, 219, 224, 226–227, 232, 237–238, 240, 269, 304, 305, 311–312, 315, 317, 321, 347, 349–351, 359, 365, 368, 380, 394–398, 444, 448, 452, 487–490
Eisner, F., 91
Elata, C., 179
Elder, J. W., 197, 200–201
Ellis, H. S., 477
Engels, H., 23, 91, 265
Engelund, F., 198, 299–303, 313
Enger, P. F., 113, 339, 343
Ertel, H., 303
Etter, R. J., 351
Eveson, G. F., 56, 68, 69
Exner, F. M., 152, 267–269, 287–291
Eytelwein, J., 15, 19

Fairbank, L. C., 470–472
Farbar, L., 474
Fargue, L., 23
Fidleris, V., 55, 57
Fischer, H. B., 196–200
Flachsbart, O., 3, 11
Flaxman, E. M., 341–342
Fleming, H. W., 3, 11

Forchheimer, P., 2, 11, 21, 23, 86, 90, 114, 121, 130, 162, 270, 274, 275
Fortier, S., 86–88, 342–343
Fortino, E., 494, 495
Franco, J. J., 239, 241
Frank, P., 164
Franzius, L., 23, 86
Fredrickson, A. G., 455
Freeman, J. R., 11, 23, 385, 393
Friedkin, J. F., 266, 385
Fromm, J. E., 41
Frontinus, S. J., 10
Führböter, A., 303, 422, 435, 438–439, 470, 484

Ganguillet, E. O., 21
Garbrecht, G., 343
Garde, R. J., 215, 281, 314
Garstka, W. U., 470
Gasterstädt, J., 474
Gessler, J., 99, 101–103
Gessner, J. B., 182
Ghetti, A., 121
Gibbs, H. J., 345
Gibert, R., 432–433, 441–442, 448–449, 453
Gilbert, G. K., 24, 89, 131–133, 153, 163, 275–276
Glover, R. E., 119
Goda, T., 172
Goldstein, S., 39
Gosh, S., 113
Govier, G. W., 429, 476
Graf, W. H., 40, 42, 45, 50, 53, 62, 95, 218–220, 224, 230, 305, 422, 430, 436, 442–448, 452, 453, 483, 487–492
Grass, A. J., 99, 100
Grim, R. E., 326–329, 346
Grissinger, E. H., 346
Grover, N. C., 355
Guglielmini, D., 13
Guy, H. P., 219, 232, 234, 276, 358

Hack, J. T., 272
Hagen, G., 16, 20, 21
Halbronn, G., 185
Hanson, E., 152
Happel, J., 46, 51–54, 56, 63, 66, 68, 69
Hariu, O. H., 476
Harleman, D. R., 352
Harms, J. C., 274
Harrison, A. S., 175, 311
Hasrajani, S. U., 422
Hayashi, T., 299
Helley, E. J., 89
Helmholtz, H., 274
Henderson, F. M., 252
Herbertson, J. G., 387
Herbich, J. B., 472
Hershey, H. C., 461
Herum, F. L., 459
Heyndrickx, G. A., 89, 131
Heywood, H., 51, 53
Higginbotham, G. H., 68
Hill, H. M., 303
Hino, M., 179–180
Hinze, J. O., 32, 34, 62, 165, 353
Hjelmfelt, A. T., 34
Hjulström, F., 88–89
Hoerner, S. F., 52, 57, 58
Householder, M. K., 177
Howard, G., 462, 484
Hubbell, D. W., 358–359, 364, 365, 367, 368, 377, 381–384
Huff, W. R., 490, 494
Hughes, R. R., 34
Humphreys, A. A., 21
Hunt, J. N., 185
Hunt, W. A., 474
Hurst, H. E., 168
Hwang, L., 111

Ingersoll, A. C., 341
Inglis, C. C., 244, 250, 251, 261–262, 266
Interagency Committee on Water Resources, 366–367, 370, 371, 373–376, 379

Ippen, A. T., 98, 111
Ismail, H. M., 462–464
Iversen, H. W., 33
Ivicsics, L., 367

Jakuschoff, P., 163, 172, 370–373, 379
Jarocki, W., 89, 133, 186, 237, 360, 367, 371, 377
Jefferson, M., 261
Jeffreys, H., 104
Jensen, V. G., 41
Jobson, H. E., 167
Johnson, J. W., 155, 371, 373
Jorissen, A., 135

Kabelac, O., 267
Kalinske, A. A., 99, 100, 105, 129–130, 167, 169, 396
Karaki, S. S., 368
Karasev, I. F., 337–338
Karolyi, Z., 361–363
Kennedy, J. F., 296–299, 316, 317
Kennedy, R. G., 24, 163, 235, 245–246
Kennedy, R. J., 477
Kidson, C., 382
Klingeman, P. C., 383
Ko, S. C., 62
Kolupaila, S., 3
Kondrat'ev, N., 187, 259, 276, 282, 286, 296, 303
Kotschin, N. J., 35
Kramer, H., 91, 99, 100, 276
Kresser, W., 204
Krey, H., 23, 89, 91, 163
Kriegel, E., 62, 436, 439, 462
Krone, R. B., 340, 341, 349–351, 382, 383
Krumbein, W. C., 274, 383
Kruyer, J., 477
Kruyt, H. R., 347–349
Kuenen, P. H., 352
Kutateladze, S. S., 476
Kutter, W. R., 21

Laboratoire Dauphinois D'Hydraulique, 485–487, 495
Lacey, G., 246–248, 253
Lamb, H., 46, 47, 274
Lane, E. W., 86, 90, 94–95, 104, 107, 108, 114–116, 119, 189–190, 238, 247, 250, 343
Langbein, W. B., 259, 268, 272
Laursen, E. M., 213–214, 224, 228, 229
Leighly, J. B., 107, 172
Leliavsky, S., 94, 107, 182, 202, 246, 264–266
Leopold, L. B., 236–239, 241–242, 253–262, 265, 268, 272, 279
Leutheusser, H. J., 111, 113
Levens, A. S., 492
Liggett, J. A., 113
Lindley, E. S., 246
Liu, H. K., 281
Lockhart, R. W., 476
Lovera, F., 218
Lundgren, H., 107
Lyle, W. M., 338

MacDougall, C. H., 134
McDowell, L. L., 382–384
McNown, J. S., 44, 47–50, 52, 54–56
Maddock, T., 234
Majumdar, H., 177
Makkaveev, V. M., 172
Mamak, W., 46, 47
Manning, R., 21, 304
Marchillon, E. K., 50
Masch, F. D., 335–336
Matsunashi, J., 303
Matthes, G. H., 264, 272
Matyukhin, V. J., 177
Maude, A. D., 56
Mavis, F. T., 89
Mehta, N. C., 474–476
Mei, C. C., 172
Metzner, A. B., 455–461, 467–468
Meyer, L. D., 203
Meyer-Peter, E., 136–139, 150, 396

Michael, P., 50
Miller, R. L., 86
Milne-Thomson, L. M., 292
Moore, W. L., 334
Moreland, C., 63, 69
Morris, H. M., 120
Mühlhofer, L., 159, 360
Müller, R., 139
Muramoto, Y., 269
Murphree, C. E., 379

Nash, J. E., 259
Neill, C. R., 89
Nelson, M. E., 375
Nemenyi, P., 91, 182
Nesper, F., 159, 360
Newitt, D. M., 428, 430, 432, 436–440, 450, 452, 464, 467
Newton, I., 36–37
Nixon, M., 259
Nordin, C. F., 151, 175, 179, 237, 283, 286, 311, 317, 321
Norgaard, R. B., 237
Novak, P., 360, 362–364

O'Brien, M. P., 91, 127, 133, 172, 352, 429, 430, 460, 470
Oliver, D. R., 68
Olson, R., 42
Orlob, G. T., 198
Oseen, C., 31, 39, 46
Owen, P. R., 153, 476

Paintal, A. S., 464
Pang, Y. H., 98
Partheniades, E., 98, 340, 341, 346, 350
Patterson, G. K., 460, 461
Pettis, C. R., 247
Pettyjohn, E. S., 51
Pezzoli, G., 98
Piest, R. F., 205
Pinkus, O., 474
Pollio, V., 9
Poreh, M., 461
Prandtl, L., 36, 37, 41, 47, 57, 59, 60, 164, 182
Preston, J. H., 110
Proudman, I., 39
Putzinger, J., 270

Raichlen, F., 287
Raudkivi, A. J., 221, 286, 299, 303, 314, 315, 394
Rehbock, T., 23
Reitz, W., 104
Rektorik, R. J., 335
Replogle, J. A., 107, 111–113
Reynolds, A. J., 299
Reynolds, O., 23
Richardson, E. G., 175
Richardson, E. V., 317–319
Richardson, J. F., 474
Rose, H. E., 474, 476
Rouse, H., 3, 11, 38, 168, 173
Rozovskii, I. L., 264
Rubey, W. W., 42, 53, 90, 97, 131
Rzhanitsyn, N. A., 264, 385, 392, 398

Sakthivadivel, R., 448
Sassoli, F., 432
Sayre, W. W., 165, 196–197, 201
Schaak, E. M., 361
Schaffernak, F., 89, 91, 132, 360
Scheidegger, A., 121, 186, 254, 274, 355
Schiller, L., 42–44, 58, 59, 63
Schlichting, H., 41, 58, 59, 186
Schlichting J., 23
Schmidt, J. J., 470, 472
Schmidt, W., 164, 172
Schmitt, H., 151
Schoklitsch, A., 23, 90–91, 93, 116, 125–126, 131, 133–136, 270–271, 396
Schroeder, K. B., 208
Schumm, S. A., 264

Seyer, F. A., 460
Shanks, D., 39
Shaver, R. G., 457
Shen, H. W., 260, 269, 312–313
Shields, A., 95–99, 127–128, 276, 281, 394, 396
Shih, C. C., 453
Shinohara, K., 282
Shook, C. A., 449, 464
Shukry, A., 264
Shulits, S., 131, 135, 270
Silberman, E., 305, 309–310, 318
Silin, N. A., 466
Sinclair, C. G., 435
Sineltshikov, V., 188
Simons, D. B., 152–153, 251–252, 276–281, 311, 316, 317, 321, 345
Smerdon, E. T., 241, 338
Smith, R. A., 432
Smith, W. O., 354
Smoltczyk, H. U., 367
Smythe, W. R., 63
Soo, S. L., 32, 62, 476
Sooky, A. A., 199
Spells, K. E., 435
Spencer, J. D., 474
Stall, J. B., 221
Stanton, T. E., 110
Stebbings, J., 248
Stepanoff, A. J., 472, 476
Sternberg, H., 86, 270
Stevens, J. C., 385, 391–393
Stewartson, K., 39
Stokes, G. G., 38–39
Straub, L. G., 126–132, 175, 235
Strauss, V., 109
Strickler, A., 21
Sutherland, A. J., 106

Taylor, G., 196–197
Tchen, C., 31, 33–34, 62, 176
Tek, M. R., 476
Teleki, P. G., 382
Terzaghi, K., 330–333
Thackston, E. L., 197
Thomas, C. W., 336, 343
Thomas, D. G., 426, 437, 455, 462
Tiffany, J. B., 91–92
Tison, L. J., 98
Toda, M., 464, 467
Todorovic, P., 381
Toffaleti, F. B., 208
Tong, L. S., 476
Torobin, L. B., 42, 58, 60, 61
Tracy, H. J., 182
Trask, P. D., 350
Truesdell, C., 36
Tsubaki, T., 281
Tsuchiya, A., 299
Tsuchiya, Y., 153
Tulla, J. G., 23

U.S. Bureau of Reclamation, 391
U.S. Waterways Experiment Station, 127, 145

Valentik, L., 57
Van Dyke, M., 39
Vanoni, V. A., 98, 100, 105, 174–175, 179, 181–182, 311, 317, 318, 377, 460
Vasiliev, O. F., 188
Velikanov (Welikanoff), M., 89, 172, 187, 291
Vennard, J. K., 108
Vernon-Hartcourt, L., 23
Vinckers, J. B., 361
Vogel, H. D., 385
Vogt, E. G., 474

Wadell, H., 51
Wallis, G. B., 476
Ward, B. D., 98
Ward, S. G., 66–69
Weber, M., 476
Weisbach, J., 21, 304
Weisman, R. N., 483, 490
Werner, P. W., 267
White, C. M., 98

Whitmore, R. L., 69, 422
Wiedenroth, W., 432, 470–471, 484
Willis, J. C., 174
Willmott, L. F., 422
Wilson, K. C., 433, 442, 448, 486, 490, 494
Wilson, W. E., 428, 438
Wittmann, H., 343
Woltman, R., 19
Woodruff, N. P., 155
Worster, R. C., 440, 450–451, 453, 470, 491

Yalin, M. S., 105, 151, 153, 283–285, 312
Yano, K., 183, 380
Yen, C. L., 269

Zandi, I., 433, 468
Zeller, J., 97, 126, 139, 175, 262–263, 341
Zenz, F. A., 474
Znamenskaya, N. S., 285, 286, 315

Subject Index

Abrasion of particle, 271
Antidunes, 277–279
  (*See also* Bedforms, types of)
Armoring of bed, 101–103

Bagnold's model:
  bedload, 150–151
  total load, 208–211
Bars, 277–279
  (*See also* Bedforms, types of)
Bed material load, 203–205
  (*See also* Total load)
Bedforms:
  creation of, 274, 276, 286–287
  and flow resistance, 303, 320–321
    (*See also* Table of Contents, 77, 78)
  geometry of, 274, 277, 281–285, 287
  movement of, 275, 281, 287
  theoretical studies of: Exner's models, 287–291
    potential flow models, 291–299
    real fluid-sediment models, 299–303
  types of, 275–279
  (*See also* Table of Contents, 77)
Bedload:
  DuBoys-type equations, 124–130
  Einstein's bedload equations, 139–151
  equation considering bedform motion, 151–153
  measuring devices for, 358–368
  model laws for, 396–398
  Schoklitsch-type equations, 130–139
  (*See also* Table of Contents, 75)
Braided channels, 260

Chezy's C-value, 304, 317–318
Clay minerals, 326–328
Cohesive-material channels:
  cohesive materials, 325–332
  scour criteria, 332–347
  sedimentation problem, 347–351
  transportation problem, 351–355
  (*See also* Table of Contents, 78)
Concentration distribution in pipes, 462–468
Container transport in pipes, 476–477

Density currents, 351–355
Drag coefficient, 36–38
  (*See also* Settling velocity)
Dunes, 277–279
  (*See also* Bedforms, types of)
Durand-Condolios relations for pipes, 431–436

Einstein's bedload function, 139–151, 206–207
Equation of particle motion:
  ideal fluid, 34–36
  linear resistance, 31–32, 34
  nonlinear resistance, 32–34
  real fluid, 34, 36–38
  (*See also* Table of Contents, 27)
Erosion (*see* Scour criteria)
Estuary, flocculation in, 349–351

Flocculation, 347–351
Flow resistance:
  with bedforms, 310–320
  without bedforms, 304–310
Friction factor f:
  with bedforms, 316–320

Friction factor f:
without bedforms, 304–305, 309–310
Friction velocity, 95
in presence of bedforms, 310–316

Gibert-Condolios relation for pipes, 441–442
Graf-Acaroglu relation, 218–220, 442–448
Grassed channels, 347

Head-loss of solid-liquid mixtures in pipes:
horizontal pipes, 429–449
inclined pipes, 449–453
nonsettling mixtures, 453–462
sample calculation, 477–481
settling mixtures, 429–453
(*See also* Table of Contents, 421)
History of sediment transport:
China, 6
DuBuat and contemporaries, 13–20
Egypt, 5
Greece, 7–9
India, 6–7
Italian School, 12–13
Leonardo da Vinci, 11–12
Mesopotamia, 4
Middle Ages, 10
Nineteenth Century, 20–24
Persia, 6–7
Renaissance, 11–12
Roman Empire, 9–10
(*See also* Table of Contents, 1)

Loop system, 487–490

Manning's n-value, 304–305, 309–310, 315
tabulation of, 306–309
Meandering:
analytical contributions, 267–269
definition of, 260
geometry of, 261–265
origin of, 265–267
Measuring devices:
for bedload, 358–368
for solid-liquid mixtures in pipes, 483–495
for suspended load, 368–379
for total load, 379–384
(*See also* Table of Contents, 78, 421)
Meyer-Peter et al. bedload equations, 136–139
Model laws:
fixed-bed models, 387–392
movable-bed models, 392–398
similitude and dimensionless numbers, 385–387
(*See also* Table of Contents, 79)

Newitt et al. relation for pipes, 436–440
Non-newtonian fluid flow in pipes, 453–462

Plane bed, 277, 278–279
(*See also* Bedforms, types of)
Pneumatic conveying of solids in pipes, 474–476

Regime canals:
Blench's contributions, 248–251
Kennedy's study, 245
Lacey's contributions, 246–248
Lindley's study, 246
Simon's et al. contributions, 251–252
Regime concept:
general remarks, 243–244
regime canals, 245–253
regime rivers, 253–272
(*See also* Table of Contents, 77)

Regime rivers:
longitudinal profiles, 270–272
regime equations, 254–260
river versus canal, 253
Ripples, 277, 278–279
(*See also* Bedforms, types of)
River profiles longitudinal, 270–272

Saltation, 153–155
Scour criteria:
bank scour, 113–116
cohesive soils, 332–347
critical shear, 90–101
critical velocity, 84–90
lift force mechanism, 101–106
model laws for, 394–396
stable channel design, 116–122
(*See also* Table of Contents, 76)
Secondary flow, 181–182, 268–269
Sediment-rating curves, 234–238
Settling velocity:
effects of: boundaries, 53–56
particle shape, 46–53
particles, 56–57
rotation, 57–58
roughness, 57
turbulence, 58–62
various effects, 44, 46, 62–63
empirical equations, 42–43
terminal, 43–45
theoretical considerations, 38–41
(*See also* Table of Contents, 27)
Shear stress:
critical shear stress, 90–101
distribution of, 106–108, 111–113
measurement of, 108–111
Shields' diagram, 94–98
Sinuous channels, 260
Solid-liquid (pipe) systems:
economics of, 473–474
installations of, 468–472
operation of, 472–473
Stable channel, design of, 116–122
(*See also* Scour criteria)
Stage-discharge relation, 321

Suspended load:
cohesive matter, 347–351
longitudinal distribution of, 195–201
neutrally buoyant dispersant, 196–200
sediment particles, 200–201
measuring device for, 368–379
model laws for, 396–398
vertical distribution of, 166–195
calculation of, 189–195
diffusion-dispersion model of, 164–166
with nonuniform turbulence distribution, 172–195
with uniform turbulence distribution, 167–172
(*See also* Table of Contents, 76)

Total load:
Bagnold's approach, 208–211
Bishop's approach, 215–218
Chang's approach, 211–212
Einstein's bedload functions, 206–207
Graf et al. approach, 218–219
hydrological effects on, 232–242
Laursen's approach, 213–215
measuring devices for, 379–384
model laws for, 396–398
modified Einstein's approach, 207–208
sample calculation, 219–232
(*See also* Table of Contents, 76–77)
Tracing methods, 380–384

Velocity distribution in pipes, 182–184, 462–468
Viscosity of particle suspensions:
experimental, 67–69
theoretical, 64–67
(*See also* Table of Contents, 27)

Washload, 204–205, 208, 232–234